W9-BKT-685

# Agroecology

## The Ecology of Sustainable Food Systems

Second Edition

# Agroecology

## The Ecology of Sustainable Food Systems

Second Edition

Stephen R. Gliessman
University of California
Santa Cruz

CRC Press
Taylor & Francis Group
Boca Raton London New York

CRC Press is an imprint of the
Taylor & Francis Group, an informa business

CRC Press
Taylor & Francis Group
6000 Broken Sound Parkway NW, Suite 300
Boca Raton, FL 33487-2742


Printed in the United States of America on acid-free paper
10 9 8 7 6 5 4 3

International Standard Book Number-10: 0-8493-2845-4 (Hardcover)
International Standard Book Number-13: 978-0-8493-2845-9 (Hardcover)



**Library of Congress Cataloging-in-Publication Data**

Gliessman, Stephen R.
Agroecology : the ecology of sustainable food systems / Steven R. Gliessman ; editor & technical illustrator, Eric Engles ; contributing writer, Robin Krieger ; editorial researcher, Ernesto Mendez. -- 2nd ed.
p. cm.
Rev. ed. of: Agroecology : ecological processes in sustainable agriculture. 1998.
Includes bibliographical references and index.
ISBN 0-8493-2845-4 (alk. paper)
1. Agricultural ecology. 2. Sustainable agriculture. 3. Agricultural systems. I. Engles, Eric. II. Krieger, Robin. III. Title.

S589.7.G58 2006
630.2'77--dc22 2006042923

**Visit the Taylor & Francis Web site at**
**http://www.taylorandfrancis.com**

**and the CRC Press Web site at**
**http://www.crcpress.com**

# Dedication

*This book is dedicated to Alf and Ruth Heller,*
*for believing in agroecology, encouraging community,*
*and fostering positive change*
*at every opportunity*

# Foreword

Farming and food systems are changing rapidly. Modern technologies have kept food production apace with population growth, but large challenges of inequity in food distribution still plague many families, countries, and regions. Growing awareness of the finite nature of critical nonrenewable resources, the undesirable impacts of current conventional agricultural systems, and the costs and other shortcomings of a globalized food system are causing us to rethink our basic assumptions about how and where to grow food. At the cutting edge of this critical awareness is agroecology. As a field of study and action, it presents both a critique of the present system and alternatives.

Steve Gliessman, an eminent teacher and scholar in this important arena, is uniquely qualified to convey what agroecology is all about. He has conducted research and worked extensively with farmers in the irrigated agriculture of California's coastal valleys, as well as in the rainfed lowland regions of Mexico and Central America. This background, coupled with long years of close collaboration with Latin American colleagues and students, has provided the foundation for much of what is evolving into the integrative field we now call agroecology.

*Agroecology: Ecological Processes in Sustainable Agriculture* was first published in 1998 and has become the most widely used text on the subject in the U.S., with popular editions also published in Spanish and Portuguese. While many of the examples in the first edition came from California, Mexico, and Central America, the principles presented and the comprehensive discussions in the book are sound and widely applicable. *Agroecology* has been the textbook of choice for many of us across the U.S. who are teaching this course. The "Food for Thought" and "Recommended Reading" sections at the end of each chapter have provided additional stimulation for students to delve more deeply into the topics. The reference list at the end of the book has served as a basic reading resource for students new to the subject, and the comprehensive glossary is one that I have used in several courses both in the U.S. and in the Nordic region.

As in any rapidly moving field, however, some information quickly becomes outdated, new perspectives emerge, and more current research is published. The second edition of this textbook provides the necessary updating and more. Several important substantive additions will be welcomed by teachers as well as those looking for an introductory overview of agroecology. There is a new chapter on the role of animals in sustainable agroecosystems, and animals have been given increased attention throughout. The introductory chapters boast several new sections, including one on how agroecology can be interpreted broadly as the ecology of food systems. There is increased attention to the growing debate about transgenic crops, and discussion about how genetically modified foods (GMOs) do not fit into many people's concept of sustainability. The importance of genetic diversity in both crops and animals is given more space in response to our growing awareness of the rapid move toward monoculture and loss of diversity across the agricultural landscape. In a major section on the transition to more sustainable systems, one former chapter has been split into two, with one providing specific information about the development of indicators and the measurement of sustainability and the other discussing the process of converting to more sustainable practices. Finally, a new chapter near the end of the book gives considerable coverage to the role of communities and how agroecology can contribute to local food systems and a new agrarianism.

It is perhaps this last dimension that brings even more vibrant life and spirit to the second edition. The tone of the revised text indicates a level of impatience on the part of Dr. Gliessman to implement change, to apply the principles of agroecology to current agriculture in order to make change happen, in short, to create a more sustainable agriculture and food system. The author's commitment to equity of benefits from the food system for those who labor in the field is reflected in the work he and his students are doing with coffee growers in Chiapas and Central America to get them a fair price for their product. Such dedication to a worthwhile contemporary cause is admirable in itself. It is especially exciting when there is a clear link in the text between learning and action and Gliessman's call for change through the development of a new agrarianism in the rural sector. Some would consider this a move away from academic integrity and the requisite impartiality we have all grown up with. Yet, some of us in education today recognize that there is often a much larger gap between knowledge and action than between ignorance and knowledge, and we should urge our students and others who read our writings to put their education to work.

In an age when science is viewed by many of its practitioners as a value-free and even sterile professional activity, it is refreshing to welcome a new textbook that makes transparent the values of the author.

We need to clearly recognize that none of what we do in science or in other pursuits is entirely free of the influence of our choices, interpretations, prior experiences, ethics, and worldviews. It is delightful to find a text with a call for action that grows from thoughtful analysis, careful scholarship, and an appreciation of the complex and multiple production, economic, environmental, and social dimensions of agriculture and the food system. I applaud Steve Gliessman's courage and dedication both to sharing his considerable wisdom and perspective on agroecology and his willingness to urge us to implement needed changes in agriculture. This is a valuable lesson for the current generation of students and provides a model for others in academia who feel passionate about the implications of their research and teaching. We can hope that this text will be even more successful than the first edition.

**Charles A. Francis**
*Professor of Agronomy and Horticulture*
*University of Nebraska — Lincoln*

# Preface to the Second Edition

As I look back on the events and activities in the field of agroecology since the publication of the first edition of *Agroecology*, I am pleased and encouraged by all that has happened. Some of these developments have occurred in the allied field of ecology. It was a big step forward when the president of the Ecological Society of America (ESA), in her address to the annual meeting of the Society this past summer in Montreal, called for linking the science of ecology with the social sciences, so that ecologists could become more active in helping solve issues and problems facing society today and in the future. This was preceded in 2002 by the establishment of a specific section in the ESA for agroecology and the formalization of a series of symposia, presented papers, and posters on ecological research in agroecology at the annual meetings leading up to the Montreal gathering. The late Gene Odum, who wrote the foreword to the first edition of this book, was often in attendance at those sessions, and his strong support of agroecology is evident in the emphasis given to agroecosystem design and management that runs throughout the recent fifth edition of his classic text *Fundamentals of Ecology*.

Over on the "agro" side of agroecology, parallel events have been occurring. Symposia, presented paper sessions, and posters focused on agroecology have become commonplace at the annual trisociety meetings of the American Society of Agronomy (ASA), Soil Science Society of America, and the Crop Science Society. This movement culminated in 2004 with the publication of Diane Rickerl and Chuck Francis' book *Agroecosystem Analysis* in the ASA Monograph Series, which makes a challenging call for the agricultural and soil sciences to more directly engage the difficult interplay between ecology and society, which an agroecological approach to agriculture requires. At the same time, David Clements and Anil Shretha's edited volume, *New Dimensions in Agroecology*, was published in the *Journal of Crop Production,* further advancing agroecological thinking for researchers in agronomy.

At the juncture of agronomy and ecology, Taylor & Francis/CRC Press has played a vital role in promoting and advancing the field of agroecology by publishing its *Advances in Agroecology* series under the leader-ship of Clive Edwards and John Sulzycki, as well as many other books related to agroecology. Journals dedicated to agroecology and sustainable agriculture have appeared in many languages, including English, Spanish, and Portuguese. The availability of agroecologically based information grows daily.

In 2002, the National Organic Standards were put into effect in the U.S., with much fanfare and controversy, highlighting issues addressed by the agroecological analysis put forward in this new edition. The critique of conventional agriculture implicit in the "organic movement" has raised consumer awareness, while at the same time stimulating discussions about the need to go beyond "organic farming" in our understanding of sustainability.

During the past eight years, many different programs in agroecology — or at least with a strong agroecological component — have appeared at colleges and universities in the U.S. and around the world. My friend and colleague Miguel Altieri has inspired multiple programs throughout Latin America and begun the development of the Society for Latin American Agroecology (Sociedad Latinoamericana de Agroecología, or SOCLA). Another friend and colleague, Eduardo Sevilla Guzmán, continues to promote the vital link between agroecology and rural development with his excellent graduate programs in agroecology and sustainable rural development at the University of Cordoba, Andalucia, Spain. An incredible group of past students, friends, and colleagues continue to weave a remarkable agroecological network in Mexico, with degree programs in agroecology at Chapingo University, the Colegio de Postgraduados, Universidad Autónoma de Yucatán, ITESO in Guadalajara, and many more. Spearheaded by a group of agroecologists from the southern state of Rio Grande do Sul, there is now a Brazilian Society for Agroecology. One can imagine that this growth reflects a global agroecology movement.

A key step forward for agroecology was the 2003 publication in the *Journal of Sustainable Agriculture* of "Agroecology: The Ecology of Food Systems" by Chuck Francis. He coordinated a large group of us in the process of reevaluating and expanding the definition of agroecology to include the social and economic sides of agriculture, something that was latent in the first edition of this textbook. I have tried to bring this larger focus forward in the new edition. I am proud to have Chuck write the foreword to this new edition.

Despite all of this progress for agroecology, the threats to the sustainability of agriculture are more serious than ever. Increased competition for scarcer water supplies, the continued loss of small and family farms, migration from rural areas to the cities, poor labor conditions, degradation of natural resources, consolidation of the food industry, outsourcing, international trade policy, threats to the safety of the food supply, continued hunger in many parts of the world, the rise of genetic engineering, and many of the other issues discussed in Chapter 1 continue to impact our food systems. We need agroecology even more than when the first edition of this text appeared in 1997.

For this text to continue to play a leading role in expanding knowledge and use of agroecological principles, it must reflect the many changes that have occurred in the field. The need for a new, revised edition has been apparent for many years, and I am pleased that it is now a reality. This second edition is updated throughout, with new data, new readings, new issues and case studies, and new options. Two completely new chapters have been added, one on the role of livestock animals in agroecosystems, and one about the cultural and community aspects of sustainable food systems. Another chapter has been expanded into two, and other chapters have been completely reworked. The first edition was oriented toward changing the way we grow food, but readers of this edition will find an even greater focus on change and action, especially in the final chapter.

The process of developing this second edition really began immediately after the publication of the first. Throughout, I have been aided by a huge network of organizations, colleagues, friends, and loved ones.

Much inspiration and practical knowledge have come from a new undergraduate program that colleagues Jenny Anderson, Karen Nordstrom, Sara Rabkin, Don Rothman, and I formed at University of California Santa Cruz (UCSC) — the Program in Community and Agroecology (PICA). With support from Alfred Heller, the Clarence E. Heller Charitable Foundation, and the Halliday Foundation, we established a residential living/learning program where students from diverse academic majors find common ground in growing, harvesting, cooking, and eating food as a community. On-campus community gardens, seminars, and projects focused on sustainable living provide a context for learning how to be a positive part of our food systems. I am constantly encouraged by the students taking part in this program, as well as by the swarm of undergraduates who have used the first edition of this text in my agroecology courses at UCSC.

I have truly been blessed to be able to work with the team of agroecologists who have received their training through the Ph.D. program in the Environmental Studies Department at UCSC. Grounded in strong ecological training, but capable of linking to the economic, social, political, and cultural elements of food systems, they have brought richness and true interdisciplinarity to the program. I owe so much to the constant exchange and enrichment that come from working with such a dynamic and committed group of graduate students and postdoctoral colleagues. Our team efforts in teaching the International Short Courses in Agroecology, beginning in the summer of 1999 in Santa Cruz, and now reaching out to Mexico, Costa Rica, Nicaragua, Brazil, and Spain, have been incredibly enriching as we help provide others with an agroecological foundation for their own work. This amazing team includes Nicholas Babin, Chris Bacon, Rose Cohen, Wes Colvin, Ariane de Bremond, Carlos Guadarrama, Julie Jedlicka, Hilary Melcarek, Ernesto Mendez, Joji Muramoto, and Laura Trujillo.

I also owe a great deal to our nonprofit Community Agroecology Network (CAN). During the early part of 2001, while a Fulbright Fellow at the Universidad Autónoma de Yucatán, I was reconnected to the incredible efforts that Maya communities around the Yucatan Peninsula have made to maintain and build their culture of food production in the face of immense challenges. Our friends Jimmy and Judi Sandler suggested that a nonprofit organization might be a helpful addition to these efforts. This led to the development of a network that spreads from Mexico to Costa Rica, pulling in similar community-based efforts being carried out by other agroecologists in our team. We are now engaged in a grassroots effort to use agroecology to link research, education, and market innovation and reconnect the most widely separated components of the food system — the growers and the eaters. Our CAN researchers are Nicolas Babin, Chris Bacon, Rebecca Cole, Carlos Guadarrama, Karen Holl, Julie Jedlicka, Juan Jose Jimenez-Osornio, Ernesto Mendez, and Laura Trujillo. Sharing this effort with the cofounder of CAN, Robbie Jaffe, is immensely fulfilling.

I am extremely grateful to my editor and writing adviser Eric Engles, one of the best writers I know. His encouragement, guidance, and cajoling kept me on task, making sure that this edition actually came about. He managed the project from the beginning, contributed much new material, and worked as an equal partner in the development of the new chapters on animals and community and culture. Ernesto Mendez provided invaluable assistance as research editor, navigating information sources with ease. John Sulzycki, at CRC Press and the Taylor & Francis Group, has been that essential element in making sure that this book and many others in agroecology reach those who need them the most. He has a true agroecological vision.

And finally, it is almost impossible to find the words to show my appreciation for the support I have received from my wife Robbie Jaffe in all aspects of developing this second edition. Daily, she shows me the deep values of relationship, commitment, and sharing. Be it at CAN, Life Lab, or in talks over tea at night, she has been a true co-conspirator in this process. As our ideas and experience in agroecology evolve and mature, so has our own agroecological experiment at our farm Condor's Hope. Plants that our son Erin helped us plant as the first edition went to press are now producing wine and olive oil, reflecting the community spirit that runs throughout this book.

Agroecology truly has extended beyond farming practices and the boundaries of the individual farm. It now reaches the entire food system. In some ways, the most significant change in this new edition is an explicit focus on a topic that was as yet undeveloped in the first edition — the values, beliefs, and ethics of sustainable food systems. This topic forms the foundation of a culture of sustainability, a crucial part of the process of transitioning to a sustainable society.

**Steve Gliessman**
*Cuyama Valley, California*

# Preface to the First Edition

If the process leading to the publication of this text had a beginning, it occurred on a Costa Rican hillside in the early 1970s. I had just completed two months of doctoral research in the area aroun d the small town of Santa Maria de Dota and was silently saying my own personal good-bye to the place and its people, thinking about how they had inspired me to look at ecology in a different way.

It was early morning, and a blanket of smoke from cooking fires hung low over the small valley, partially shrouding the town square. The hotel where I had been living was at one corner of the square. I could see the windows of my two rooms, upstairs from the small restaurant. I had converted one of those rooms into a makeshift laboratory for studying plant–plant interactions and smile even today as I recall the odd looks I received from the family that ran the hotel when I made one of my many trips downstairs through the kitchen to the one faucet with running water in the patio behind the building, where I could wash my beakers, petri dishes, and pipettes.

My thesis project was focused on understanding the ecological mechanisms of dominance of bracken (*Pteridium aquilinum*), an especially noxious weed that invades lands following the clearing and burning of forests in many parts of the world. This very aggressive plant seemed to be able to take great advantage of the disturbance created by humans in their quest for agricultural or forest products. Over several years, I had determined that the plant's ability to produce compounds that inhibited the growth of other plants, combined with several other factors, allowed it to be the successful invader it was in so many places, including this upland region of the tropics.

But while in Santa Maria de Dota, I had begun to realize that although the ecological knowledge I had gained about bracken's abilities might provide new information for ecologists, it did not necessarily help the local farmers, who usually had little choice but to move to new land and clear more forests once bracken took over. I became concerned that if my ecological knowledge did not become useful to the people who had the most impact on the land, then it was doing little good aside from producing more academic knowledge. So, up there on the hillside, I decided that I would study ecology not just to learn about how plants and animals interact with the environment, but to provide useful tools to help farmers better manage their farms.

When I brought this idea back to my thesis advisor, C. H. Muller, at the University of California, Santa Barbara, he was somewhat skeptical about my trying to bridge the gap between basic and applied research in ecology. Nevertheless, I realized that it was in part his influence that had pushed me in this direction. During eight challenging years, I had pursued both undergraduate and graduate studies with Dr. Muller and received added guidance from such professors as Bob Haller, Maynard Moseley, Dale Smith, and Wally Muller. An encouraging and challenging group of fellow graduate students shared in this experience, including Jim McPherson, Roger del Moral, Bob Tinnin, Dave Bell, C. H. Chou, Himayet Naqvi, Nancy Vivrette, Norm Christensen, and Jim Hull. From all these colleagues I gained a whole-systems view of plants in the environment and a concern for the world in which we live.

I completed my graduate studies at about the same time as the first Earth Day. Ecology was being faced with the challenge of helping to heal the earth from the impacts of humans and to restore the balance we had learned to respect in the natural world. But rather than choose to pursue such goals from an academic setting, I accepted a different kind of challenge. I joined Darryl Cole and his family at Finca Loma Linda in the uplands of southern Costa Rica, where I entered the field of applied ecology by becoming a farmer of vegetables and coffee. For two and a half years, I worked alongside Darryl, dealing with the range of problems that face farmers everywhere — pests, disease, fertility management, erosion, unpredictable weather, difficult markets. Paying the bills had to be balanced with ecological practice, and we had lots of time to discuss the potential for combining ecology and agriculture as we returned from market runs to the tropical lowlands over several hours of difficult roads. I gained great respect for Darryl's wealth of knowledge and vision and am grateful for what he shared with me.

During this time, Loma Linda also became a testing ground for some early research in the ecology of agriculture. Students and faculty from the Organization of Tropical Studies — including Ron Carroll, John Vandemeer, Chuck Schnell, Barbara Bentley, Steve Risch, and Leslie Real, to name just a few — used Loma Linda as a field study site, allowing me to share my accumulating knowledge with ecologists and to benefit in turn from their perspective. In retrospect, we were laying down the foundations of agroecology as we tested different mulching techniques, used different organic soil amendments, mapped the distribution of insects away from the forest margin, and did comparative plantings of cabbage plants in the forest and nearby farmland. In this way, on-farm research became an important early component of the ecology of agricultural systems, and a framework began to form upon which this book would be built.

My experience with managed ecosystems expanded when I moved to Guadalajara, Mexico, and became the general manager of a large commercial nursery business that produced a great variety of fruit trees and ornamental plants for the local and regional horticultural trade. Propagation, plant maintenance, landscape design, and business management all formed part of my new training in yet another area of applied ecology. My teachers were the dedicated staff of the business — Martin Muñoz, Jose Ruiz, Joaquin Guzman, Augustin Muñoz, and others; their love of plants formed the core of both a way of life and a way of making a living. For almost three years, we worked together in a partnership dedicated to providing customers with plants that would add both beauty and utility to the environments in which they lived.

Perhaps the key step in my development as an agroecologist occurred when I left the nursery and joined the faculty of the Colegio Superior de Agricultura Tropical (CSAT) in Cárdenas, Tabasco, México. CSAT was designed to train agronomists from the tropics in the tropics, and the vision of the director, Angel Ramos Sanchez, was evident in the diversity of programs that were developed to support this goal. Ricardo Almeida Martinez, an agronomist with a vision of how the systems approach of ecology could provide long-term solutions to the problems facing tropical agriculture, had established the Department of Ecology at the school. Another key person was Roberto Garcia Espinosa, a plant pathologist who understood that instead of continuing to focus so much attention on getting rid of diseases once they had become problems, we needed to see disease as a problem inherent in the design of agricultural systems. His respect for and knowledge of the small, traditional farming systems belonging to farmers of Mayan descent surrounding the highly modified conventional farming systems at the Colegio's experimental fields encouraged us to think locally.

One day, Roberto and I were driving the 21 km up the highway from Cárdenas to the Colegio. He pointed out a corn planting in an area that a few months earlier had been a swamp inundated with at least a meter of water and vegetated with plants typical of the wetlands of the region. The corn looked extremely healthy and productive, so we decided to stop. We talked to the farmers who were tending the field, and to our amazement, a story of a sustainable agroecosystem based on local knowledge began to unfold. This system is described later in this book, but the most remarkable part of the story is that agronomists of the Colegio had been driving past the planting for years, never once stopping to investigate why farmers planted in such an area in the first place, nor discovering that farmers were able to obtain — year after year on the same ground — five to ten times the conventional average in corn yields, with no other cultural inputs besides local seed, machetes, and their own labor. Sitting with Roberto and listening to a man who was more than 100 years old describe the intricacies of managing the system, how he had learned it as a child, and his role as the "keeper of the seed" for the system had dramatic impact on my thinking about agroecology.

For five years, I worked with an incredible group of new agroecologists at CSAT, working as a team to build what we called *agroecología*. A master's program in tropical agroecology was established. Partnerships with students, researchers, and faculty formed, and most importantly, local farmers joined in sharing knowledge about how to merge ecology and agriculture in the tropics. Farmers who, for generations, had sustained their farms without the use of mechanization, hybrid seed, or synthetic chemical fertilizers or pesticides became our teachers. Research projects aimed at testing ecological principles in an agricultural setting began to take place, and the analysis of farming systems as ecosystems, or agroecosystems, was initiated. The beginnings of the concept of sustainability were laid down. Key colleagues during this exciting time were Moises Amador Alarcon, Angel Martinez Becerra, Radamez Bermudez, Juan Carlos Chacon, Rosalinda del Valle, Judith Espinosa, Fausto Inzunza, David Jimenez, Silas Romero, Francisco Rosado-May, and Octavio Ruiz Rosado. Much of what is in this text began as discussions with these colleagues.

Outside of CSAT, ecologist Dan Janzen's writings about tropical agroecosystems and Maestro Efraím Hernández Xolocotzi's insistence on the value of traditional agricultural knowledge and the importance of using an interdisciplinary approach for studying these agroecosystems provided added impetus for expanding our research and training programs at CSAT into the ecological realm. Tropical ecologist Arturo Gomez-Pompa began carrying out important tests of models for sustainable agriculture, and anthropologist and cultural ecologist Alba Gonzalez Jacome provided essential guidance in how to include the cultural component in our analysis. Other Mexican colleagues who played important roles during this time include Miguel Angel Martinez Alfaro, Rodolfo Dirzo, Epifanio Jimenez, Ana Luisa Anaya, and Silvia Del Amo. They all made significant contributions in framing the broader implications of the work presented throughout the text.

In 1980, I came to the University of California, Santa Cruz (UCSC), where I joined the faculty of the interdisciplinary Board of Studies in Environmental Studies. I found a large number of students eager to embrace agroecology and apply the concepts and principles of ecology to the design and management of sustainable agroecosystems.

I also found a remarkable field facility in the UCSC Farm and Garden, established many years before by the eccentric horticulturist and dramatist Alan Chadwick. Its fields and gardens served as training grounds for apprentices who learned the techniques and spirit of organic agriculture and horticulture. I endeavored to link the farm and garden facilities with the academic programs on campus by making teaching and research in agroecology and sustainable agriculture part of our interdisciplinary curriculum.

The enthusiasm, interest, and excitement of undergraduate students at UCSC became a large part of the stimulus to complete this textbook, and I deeply appreciate the challenge and motivation they have given me. The graduate students who came to UCSC before we had an official graduate program in environmental studies (and who carried out a range of masters and doctoral projects through the boards of studies in biology and education) also greatly influenced the evolution of this text by contributing to the research base of agroecology. This group includes Jan Allison, Rich Berger, Marc Buchanan, Judith Espinosa, Roberta Jaffe, Juan Jose Jimenez-Osornio, Rob Kluson, Leslie Linn, Jim Paulus, Francisco Rosado-May, Martha Rosemeyer, Octavio Ruiz Rosado, and Hollis Waldon.

Linking the Farm and Garden to the academic programs at UCSC proved to be both a key foundation of the book's development and a constant challenge. Without the help of a great number of people, it would not have happened. My primary mentor in this ongoing process was Ken Norris, colleague and teacher of natural history, who always understood the value of the connections between humans and nature. Kay Thornley, friend and fellow farmer, spearheaded the organizational and grant writing efforts that helped establish the Agroecology Program at UCSC. Many other people and organizations played an important role in supporting the Agroecology Program, and thus the writing of this text; they include Janie Scardina Davis, Kima Muiretta Fenn, Sharon Ornellas, Louise Cain, Marc Buchanan, Sean Swezey, Matt Werner, the staff of the Agroecology Program, the faculty and staff of Environmental Studies, Huey Johnson, the Colombia Foundation, the Goldman Fund, the Heller Charitable and Educational Fund, the Clarence E. Heller Charitable Trust, the Educational Foundation of America, the W. K. Kellogg Foundation, and John Halliday.

Since 1982, Alf Heller's support for the development of the Agroecology Program at UCSC has been an essential part of our success. His encouragement, questioning, and financial backing allowed us to move beyond normal institutional limits and ensured that agroecology could grow and develop on a firm foundation. When we were scheduled to meet for the first time, at the Stock Exchange Club in San Francisco, to discuss funding some of our early activities, the maitre d' would not let me in because I was not wearing a tie. Thanks to some spares kept in a drawer for such occasions, however, we were able to meet and initiate what has become a remarkable partnership for the development of agroecology.

Throughout my career, and especially during the past several years of writing, members of my family have been a crucial source of support and encouragement, and I extend my thanks and appreciation to all of them. My first wife Nannette gave her support and companionship in Costa Rica and Mexico; my sons Erin and Alex shared with me the roller coaster ride of growing up; my mother Mary and sister Leslie gave their understanding and love; my brother Eric encouraged me to put my beliefs on the line and join him in our current farming venture; my late father Lester set an example of how to make a difference in the world; and most importantly, my wife Robbie gave her special love and encouragement, especially during the times of growth. My support group, Superglue, has continually provided the trust and encouragement I have needed to keep my commitments.

Many other people deserve thanks and appreciation. Foremost among them is Eugene Odum, who has always exerted a strong influence on me, beginning with the textbook I read in my first undergraduate ecology course, and continuing with his most recent textbook of ecology, which I use to teach a general ecology course bridging science and society. It was Eugene who suggested many years ago that I write this book, and who gave me the impetus to actually begin writing it. I also owe a great deal to my friend and colleague Miguel Altieri, whose prolific writing in the area of agroecology has served as an excellent motivator for many of us working in this field.

The writing of this book could not have happened without the assistance of Michael Arenson, who gave me computer literacy and helped get an early draft of the book on line for use in my teaching. Over the past year, I have benefited greatly from expert peer reviews of all or part of the book by Alba Gonzalez Jacome, David Dumaresq, Diane Gifford-Gonzalez, and Ken Norris. I am deeply indebted to my graduate students and postdoctoral researchers, who have painstakingly reviewed draft after draft of each chapter in our weekly lab seminar meetings, contributing quite significantly to the accuracy and readability of the book. This group includes Erle Ellis, Phillip Fujiyoshi, Carlos Guadarrama, Eric Holt-Jimenez, Robin Krieger, Gabriel Labbate, Marc Los Huertos, Joji Muramoto, Ricardo Santos, Claudia Schmitt, and Laura Trujillo.

I could not have completed the manuscript without the assistance of my skillful editor, Eric Engles, who patiently helped organize my thoughts, ideas, and experiences, and transformed my prose into a workable form. Skip DeWall, Jr. at Ann Arbor Press, and Lynne

Sterling at Sleeping Bear Press, did a remarkable job of turning our manuscript into a textbook in record time.

The agroecological network continues to expand. My hope is that this textbook will accelerate this expansion, and in the process, help halt and reverse the loss of the agricultural communities and the small and family farms that form the basis for sustainable agriculture worldwide. If we can use agroecology to firmly establishing the ecological foundation for sustainability, the rest of the components will fall into place.

**Steve Gliessman**
*Santa Cruz, California*

*June, 1997*

# Author

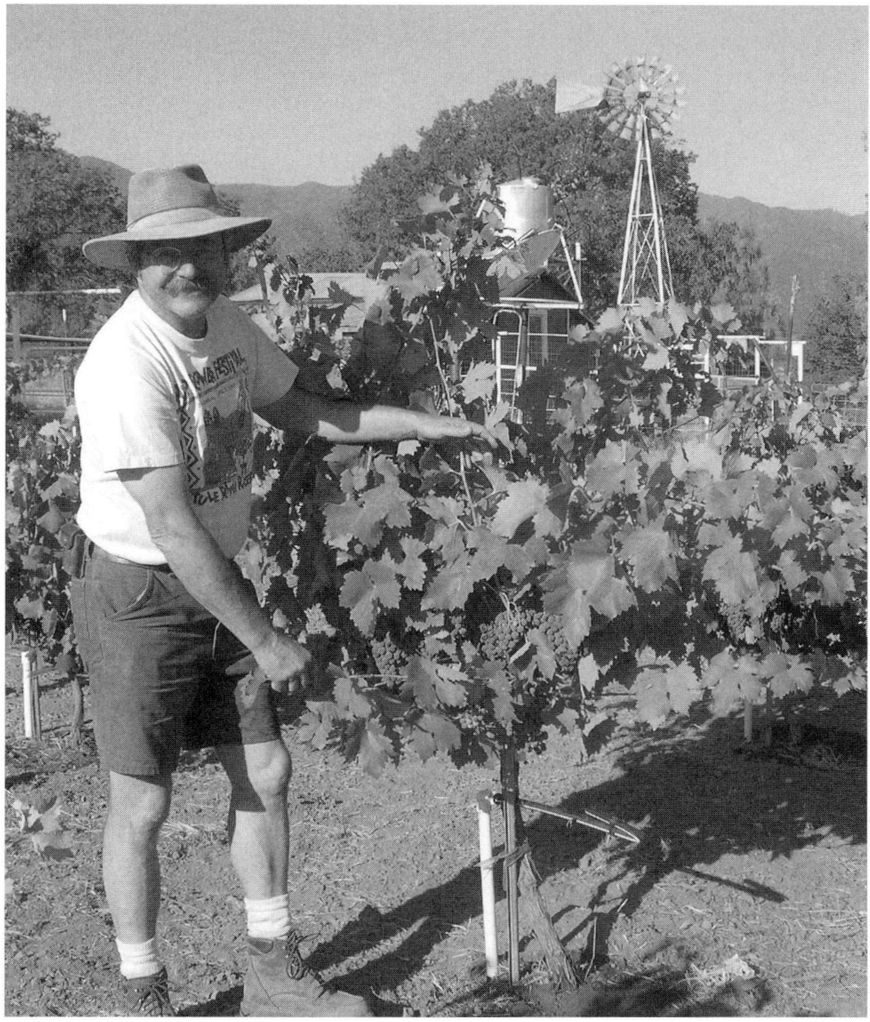

With graduate degrees in botany, biology, and plant ecology from the University of California, Santa Barbara, Stephen R. Gliessman has accumulated more than 33 years of teaching, research, and production experience in the field of agroecology. His international experiences in tropical and temperate agriculture, small-farm and large-farm systems, traditional and conventional farm management, hands-on and academic activities, nonprofit and for-profit work experience, and organic and synthetic chemical farming approaches have provided a unique combination of materials and perspectives to incorporate into this text. He was the founding director of the University of California, Santa Cruz, Agroecology Program, one of the first formal agroecology programs in the world, and is the Alfred and Ruth Heller Professor of Agroecology in the Department of Environmental Studies at University of California Santa Cruz (UCSC). He is the cofounder of the nonprofit Community Agroecology Network (CAN), and director of the undergraduate residential learning Program in Community and Agroecology (PICA). He also dry farms organic wine grapes and olives with his wife Robbie and son Erin in northern Santa Barbara County, California.

# Recommendations for Using This Textbook

Reflecting agroecology's origins in both the pure-science field of ecology and the applied field of agronomy, this text has a dual identity: in one sense, it is designed to teach ecology in the context of agriculture; in another sense, it teaches about agriculture from an ecological perspective.

Despite its attention to the practice of growing food, however, this is not a book on how to farm. Farming is an activity that must be adapted to the particular conditions of each region of the world, and this text's mission is to create an understanding of concepts that are of universal applicability.

The text has been written to accommodate a range of experience and knowledge levels in both ecology and agriculture. Sections I and II assume only a basic knowledge of ecology and biology, and even those students with minimal college-level science training should have little difficulty comprehending the material, if they are diligent. Intensive study of Chapter 1 through Chapter 12 will prepare any student for the more complex chapters of Sections III and IV.

Readers with extensive background in ecology will benefit most from the two latter sections. They may want to skim Chapter 2 for review, and then read Chapter 3 through Chapter 12 selectively before turning their attention to Sections III and IV. Readers with advanced training in both ecology and agriculture, including advanced undergraduates, may want to pursue this strategy as well, supplementing the text with additional materials that provide more extensive literature review and reports on research findings.

The text can be used in either a one-quarter or one-semester course, but the rate at which material is covered will depend greatly on the instructor, the students, and the curriculum. Ideally, a laboratory section will complement the lecture section of any course using this textbook, allowing the testing of ecological concepts in agriculture, and the demonstration of how the tools of ecology can be applied to the study of agroecosystems. The accompanying lab manual, *Field and Laboratory Investigations in Agroecology*, is designed to fill this role. Its investigations are keyed to the chapters in this text, and the two work together to create an integrated course.

Suggested readings and a list of Internet resources at the end of each chapter provide further materials for the curious reader. The questions following each chapter are open ended, designed to encourage the reader to consider the ideas and concepts presented in the broader context of sustainability.

The concepts and principles in this text can be applied to agroecosystems anywhere in the world. Just as a farmer must adjust to local and changing conditions, readers of this book are challenged to make the necessary adaptations to apply its contents to their own situations — finding appropriate examples and case studies in the research literature and working with local farmers to connect principles to actual practices.

# Table of Contents

## Section IV: The Transition to Sustainability

# Section I

## Introduction to Agroecology

Agriculture is in crisis. Although the world's agricultural lands continue to produce at least as much food as they have in the past, there are abundant signs that the foundations of their productivity are in danger.

The first chapter of this section describes the many problems confronting agriculture today and explains their roots in modern agricultural practices. It concludes with an explanation of how applying ecological concepts and principles to the design and management of systems of food production — the essence of agroecology — can help us produce food more sustainably. Chapter 2 outlines the basic conceptual and theoretical framework of agroecology, which will be used to study and analyze food production systems (agroecosystems) throughout the remainder of the text.

**An intensive vegetable-based agroecosystem on the urban fringe of Shanghai, China.** *In systems such as this, food is produced for local markets without much of the fertilizer, pesticides, and machinery characteristic of large-scale, single-crop agroecosystems.*

# 1 The Need for Sustainable Food Production Systems

On a global scale, agriculture was very successful in meeting a growing demand for food during the latter half of the 20th century. Yields per hectare of staple crops such as wheat and rice increased dramatically, food prices declined, the rate of increase in food production generally exceeded the rate of population growth, and chronic hunger diminished. This boost in food production was due mainly to scientific advances and technological innovations, including the development of new plant varieties, the use of fertilizers and pesticides, and the growth of extensive infrastructure for irrigation.

Now, in the first decade of the 21st century, our system of global food production must grapple with a sobering fact as it attempts to feed a world population that continues to grow: the techniques, innovations, practices, and policies that have allowed increases in productivity have also undermined the basis for that productivity. They have overdrawn and degraded the natural resources upon which agriculture depends — soil, water resources, and natural genetic diversity. They have also created a dependence on nonrenewable fossil fuels and helped forge a system that increasingly takes the responsibility for growing food out of the hands of farmers and farm workers, who are in the best position to be stewards of agricultural land. In short, our system of agricultural production is unsustainable — it cannot continue to produce enough food for the global population over the long term because it deteriorates the conditions that make agriculture possible.

At the same time, our global food system faces threats not entirely of its own making, most notably the emergence of new agricultural diseases (such as mad cow and Nipah virus) and climate change. These threats underline the importance of moving towards more sustainable agricultural practices.

## PRACTICES OF CONVENTIONAL AGRICULTURE

Conventional agriculture is built around two related goals: the maximization of production and the maximization of profit. In pursuit of these goals, a host of practices have been developed without regard for their unintended, long-term consequences and without consideration of the ecological dynamics of agroecosystems. Seven basic practices — intensive tillage, monoculture, irrigation, application of inorganic fertilizer, chemical pest control, genetic manipulation of domesticated plants and animals, and "factory farming" of animals — form the backbone of modern industrial agriculture. Each is used for its individual contribution to productivity, but as a whole, the practices form a system in which each depends on the others and reinforces the necessity of using all in concert.

These practices are also integrated into a framework with its own particular logic. Food production is treated like an industrial process in which plants and animals assume the role of miniature factories: their output is maximized by supplying the appropriate inputs, their productive efficiency is increased by manipulation of their genes, and the environments in which they exist are as rigidly controlled as possible.

### Intensive Tillage

Conventional agriculture has long been based on the practice of cultivating the soil completely, deeply, and regularly. The purpose of this intensive cultivation is to loosen the soil structure to allow better drainage, faster root growth, aeration, and easier sowing of seed. Cultivation is also used to control weeds and to turn under crop residues. Under typical practices — that is, when intensive tillage is combined with short rotations — fields are plowed or cultivated several times during the year, and in many cases this leaves the soil free of any cover for extended periods. It also means that heavy machinery makes regular and frequent passes over the field.

Ironically, intensive cultivation tends to degrade soil quality in a variety of ways. Soil organic matter is reduced as a result of accelerated decomposition and the lack of cover, and the soil is compacted by the recurring traffic of machinery. The loss of organic matter reduces soil fertility and degrades soil structure, increasing the likelihood of further compaction and making cultivation and its temporary improvements even more necessary. Intensive cultivation also greatly increases rates of soil erosion by water and wind.

### Monoculture

Over the last century, agriculture all over the world has moved relentlessly toward specialization. Farming once meant growing a diversity of crops and raising livestock, but now farmers are far more likely to specialize, growing corn for livestock feed, for example, or raising hogs. In crop agriculture, specialization means monoculture — growing only one crop in a field, often on a very extensive scale. Monoculture allows more efficient use of farm machinery for cultivation, sowing, weed control, and harvest, and can create economies of scale with regard to purchase of seeds, fertilizer, and pesticides. Monoculture is a natural outgrowth of an industrial approach to agriculture, where labor inputs are minimized and technology-based inputs are maximized in order to increase productive efficiency. Monoculture techniques mesh well with the other practices of modern agriculture: monoculture tends to favor intensive cultivation, application of inorganic fertilizer, irrigation, chemical control of pests, and specialized plant varieties. The link with chemical pesticides is particularly strong; vast fields of the same plant are more susceptible to devastating attack by specific pests and diseases and require protection by pesticides.

### Application of Synthetic Fertilizer

The spectacular increases in yields in the second half of the 20th century were due in large part to the widespread and intensive use of synthetic chemical fertilizers. In the U.S., the amount of fertilizer applied to fields each year increased rapidly after World War II, from 9 million tons in 1940 to more than 47 million tons in 1980. Worldwide, the use of fertilizer increased tenfold between 1950 and 1992; since then, the increase has moderated, but in 2002, the total world consumption of fertilizers was estimated to be 141.6 million metric tons (FAOSTAT, 2005).

Produced in large quantities at relatively low cost using fossil fuels and mined mineral deposits, fertilizers can be applied easily and uniformly to crops to supply them with ample amounts of the most essential plant nutrients. Because they meet plants' nutrient needs for the short term, fertilizers have allowed farmers to ignore long-term soil fertility and the processes by which it is maintained.

The mineral components of synthetic fertilizers, however, are easily leached out of the soil. In irrigated systems, the leaching problem may be particularly acute; a large amount of the fertilizer applied to fields actually ends up in streams, lakes, and rivers, where it causes *eutrophication* (excessive growth of oxygen-depleting plant and algal life). Fertilizer can also be leached into groundwater used for drinking, where it poses a significant health hazard. Furthermore, the cost of fertilizer is a variable over which farmers have no control since it rises with increases in the cost of petroleum.

### Irrigation

An adequate supply of water is the limiting factor for food production in many parts of the world. Thus supplying water to fields from underground aquifers, reservoirs, and diverted rivers has been key to increasing overall yield and the amount of land that can be farmed. Although only 18% of the world's crop land is irrigated (FAOSTAT, 2005), this land produces 40% of the world's food (Serageldin, 1995; FAO, 2002). Currently, there are more than 44 ha of irrigated land per 1000 people in the world (FAOSTAT, 2005).

All sectors of society have placed rapidly increasing demands on fresh water supplies over the past half-century, but agricultural purposes account for the lion's share of the demand — about 70% of water use worldwide (Postel and Vickers, 2004). Unfortunately, agriculture is such a prodigious user of water that in many areas where land is irrigated for farming, irrigation has a significant effect on regional hydrology. One problem is that groundwater is often pumped faster than it is renewed by rainfall. This overdraft can cause land subsidence, and near the coast it can lead to saltwater intrusion. In addition, overdrafting groundwater is essentially borrowing water from the future. Where water for irrigation is drawn from rivers, agriculture is often competing for water with water-dependent wildlife and urban areas. Where dams have been built to hold water supplies, there are usually dramatic effects downstream on the ecology of rivers. Irrigation has another type of impact as well: it increases the likelihood that fertilizers will be leached from fields and into local streams and rivers, and it can greatly increase the rate of soil erosion.

### Chemical Pest and Weed Control

After World War II, chemical pesticides were widely touted as the new, scientific weapon in humankind's war against plant pests and pathogens. These chemical agents had the appeal of offering farmers a way to rid their fields once and for all of organisms that continually threatened their crops and literally ate up their profits. But this promise has proven to be false. Pesticides can dramatically lower pest populations in the short term, but because they also kill pests' natural predators, pest populations can often quickly rebound and reach even greater numbers than before. The farmer is then forced to use even more of the chemical agents. The dependence on pesticide use that results has been called the "pesticide treadmill." Augmenting the dependence problem is the phenomenon of increased resistance: pest populations continually exposed to pesticides are subjected to intense natural selection for pesticide resistance. When resistance among the pests increases, farmers are forced to apply larger amounts of pesticide or to use different pesticides, further contributing to the conditions that promote even greater resistance.

Although the problem of pesticide dependence is widely recognized, many farmers — especially those in developing nations — do not use other options. Even in the U.S., the amount of pesticides applied to major field crops, fruits, and vegetables each year remains at twice

**FIGURE 1.1 Furrow irrigation with gated pipe in coastal central California.** Overdraft of the underground aquifers from which the irrigation water is pumped has caused salt water intrusion, threatening the sustainability of agriculture in the region.

the level it was in 1962, when Rachel Carson published *Silent Spring* (Kimbrell, 2002). Ironically, total crop losses to pests have stayed fairly constant despite increasing pesticide use (Pimentel et al., 1991; Pimentel, 2005).

Besides costing farmers a great deal of money, pesticides — including herbicides and fungicides — can have a profound effect on the environment and often on human health. Pesticides applied to fields are easily washed and leached into surface water and groundwater, where they enter the food chain, affecting animal populations at every level and often persisting for decades.

### Manipulation of Plant and Animal Genomes

Humans have selected for specific characteristics among crop plants and domesticated animals for thousands of years; indeed, human management of wild species was one of the foundations of the beginning of agriculture. In recent decades, however, technological advances have brought about a revolution in the manipulation of genes. First, advances in breeding techniques allowed for the production of hybrid seeds, which combine the characters of two or more plant strains. Hybrid plant varieties can be much more productive than similar nonhybrid varieties and have thus been one of the primary factors behind the yield increases achieved during the so-called "green revolution." The hybrid varieties, however, often require optimal conditions — including intensive application of inorganic fertilizer — in order to realize their productive potential, and many require pesticide application to protect them from extensive pest damage because they lack the pest resistance

**FIGURE 1.2 Broadcast spraying to control codling moth in an apple orchard in the Pajaro Valley, California.**

of their nonhybrid cousins. In addition, hybrid plants cannot produce seeds with the same genome as their parents, making farmers dependent on commercial seed producers.

More recently, breakthroughs in genetic engineering have allowed the customized production of plant and animal varieties through the ability to splice genes from a variety of organisms into the target genome. The resulting organisms are referred to as *transgenic*, *genetically modified* (GM), or *genetically engineered* (GE).

Only a few animal species used for food have been genetically engineered as yet — these include pigs with spinach genes that produce lower-fat bacon and cows that produce milk with higher casein levels — but transgenic crop plants are now widespread and important in agricultural production. Between 1996 and 2003, the area planted to genetically engineered crops worldwide increased almost 40-fold, from 1.7 million ha to 67.7 million ha (James, 2003). The U.S., Argentina, Canada, Brazil, China, South Africa, Australia, and India all planted at least 100,000 ha to transgenic crops in 2003. Of the world's soybean crop, 55% was transgenic in 2003, as was 21% of the world's cotton crop (James, 2003).

Although genetically engineered organisms hold many promises — reducing the use of pesticides and irrigation, allowing agriculture on soils too saline for normal crops, and increasing the nutritional value of some crops — there are many concerns about the spread of this and related biotechnologies. The main source of concern is the potential for the migration of modified genes into other populations, both wild and domestic. This could result, for example, in more aggressive weeds or the introduction of toxins into crop plants. Increased use of transgenic crops may also diminish biodiversity, as traditional cultivars are abandoned, and increase the dependence of farmers on the transnational corporations owning the patents on the new organisms.

### Factory Farming of Animals

If you live in a developed country, a large portion of the meat, eggs, and milk that you eat probably comes from large-scale, industrialized operations driven by the goal of bringing these food products to market at the lowest possible unit cost. The animals in these "confined animal feeding operations" (CAFOs) are typically crowded so tightly they can barely move, given antibiotics to prevent the spread of disease, and fed highly processed feed supplemented with hormones and vitamins. Even though they are completely dependent on crop agriculture for the production of feed, CAFOs are disconnected — spatially and functionally — from the fields in which the feed grains are grown.

Factory-farm livestock production is another manifestation of the specialization trend in agriculture. In many ways, factory farming is for pigs, cattle, and poultry what monoculture is for corn, wheat, and tomatoes. The livestock in CAFOs are more susceptible to disease, just as monocropped corn plants are to pest damage, and both require chemical inputs (pharmaceuticals for livestock and pesticides for crops) to compensate. Both factory farming and monoculture encourage the use of organisms bred or engineered for productive efficiency and dependent on the artificial conditions of the industrial process.

Factory farming is criticized by animal rights groups as cruel and inhumane. Laying hens and broiler chickens are routinely de-beaked to keep them from pecking each other; hogs are often kept in pens so small they cannot turn around; beef cattle commonly suffer slow and painful deaths at the slaughterhouse.

There are many other reasons to be critical of the industrial approach to raising livestock. CAFOs, for example, have serious impacts on the environment. Disposal of the massive amounts of manure and urine generated by the confined animals is a huge problem, usually dealt with by treating the wastes in large anaerobic lagoons that leak nitrates into surface streams and groundwater and allow ammonia to escape into the atmosphere. This problem arises because CAFOs by their very nature cannot recycle nitrogen within the system, as is the case on smaller traditional farms where animals and crop plants are raised together. Thus nitrogen becomes a problematic waste product instead of a valuable plant nutrient.

The rise in factory farming is coupled with a worldwide trend toward diets higher in meat and animal products. As demand for meat increases, industrialized methods of animal food production become more profitable and more widespread, replacing more sustainable pastoral and mixed crop–livestock systems.

**FIGURE 1.3 A confined animal feeding operation in California's Central Valley.**

## WHY CONVENTIONAL AGRICULTURE IS NOT SUSTAINABLE

The practices of conventional agriculture all tend to compromise future productivity in favor of high productivity in the present. Therefore, signs that the conditions necessary to sustain production are being eroded should be increasingly apparent over time. Today, there is in fact a growing body of evidence that this erosion is underway. In the last 15 years, for example, all countries in which Green Revolution practices were adopted at a large scale have experienced declines in the annual growth rate of the agricultural sector. Further, in many areas where modern practices were instituted for growing grain in the 1960s (improved seeds, monoculture, and fertilizer application), yields have begun to level off and have even decreased following the initial spectacular improvements in yield. Mexico, for example, has seen little change in wheat yields since 1980, after climbing from about 0.9 tons/ha in 1950 to 4.4 tons in 1982 (Brown, 2001). For the world as a whole, the rise in land productivity has slowed markedly since about 1990. In the 40 years before 1990, world grain yield per hectare rose an average of 2.1% a year, but between 1900 and 2000, the annual gain was only 1.1 percent (Brown, 2001). From 2000 to 2003, global grain reserves shrank alarmingly every year, from 635 million tons (a 121-d supply), to 382 million tons (a 71-d supply).

Figure 1.4 shows the world's annual per capita grain production for each year from 1961 to 2004, as calculated by the Food and Agriculture Organization (FAO) of the United Nations. These data indicate that after trending upward for many years, per capita production of cereal grains has trended downward since reaching a peak in 1984. This situation is the result of reduced annual yield increases combined with continued logarithmic population growth.

The ways in which conventional agriculture puts future productivity at risk are many. Agricultural resources such as soil, water, and genetic diversity are overdrawn and degraded, global ecological processes on which agriculture ultimately depends are altered, human health suffers, and the social conditions conducive to resource conservation are weakened and dismantled. In economic terms, these adverse impacts are called *externalized costs*. They are real and serious, but because their consequences can be temporarily ignored or absorbed by society in general, they are excluded from the cost–benefit calculus that allows conventional agricultural operations to continue to make economic "sense."

### Soil Degradation

Every year, according to the Food and Agriculture Organization of the United Nations, between 5 and 7 million ha of valuable agricultural land are lost to soil degradation. Other estimates run as high as 10 million ha per year (e.g., World Congress on Conservation Agriculture, 2001). Degradation of soil can involve salting, waterlogging, compaction, contamination by pesticides, decline in the quality of soil structure, loss of fertility, and erosion by wind and water. Although all these forms of soil degradation are severe problems, erosion is the most widespread. Worldwide, 25,000 million tons of topsoil are washed away annually (Loftas et al., 1995). Soil is lost to wind and water erosion at the rate of 5 to 10 tons/ha

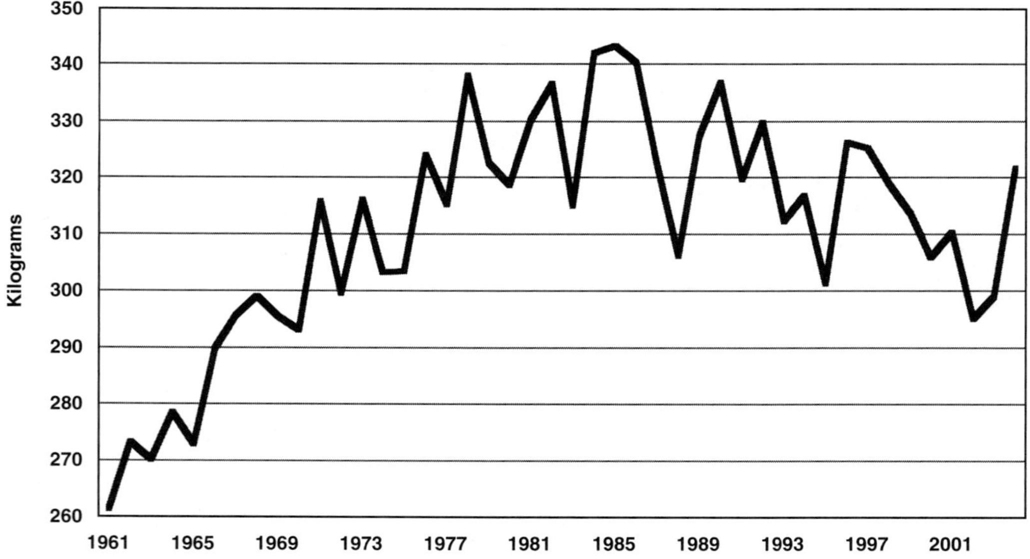


**FIGURE 1.4 Worldwide grain production per capita, 1961 to 2004.** *Data source*: Food and Agricultural Organization, FAOSTAT database; Worldwatch Institute.

per year in Africa, South America, and North America, and almost 30 tons/ha annually in Asia. In comparison, soil is created at the rate of about 1 ton/ha per year, which means that in just a short period, humans have wasted soil resources that took thousands of years to be built up.

The cause–effect relationship between conventional agriculture and soil erosion is direct and unambiguous. Intensive tillage, combined with monoculture and short rotations, leaves the soil exposed to the erosive effects of wind and rain. The soil lost through this process is rich in organic matter, the most valuable soil component. Similarly, irrigation is a direct cause of much water erosion of agricultural soil.

Combined, soil erosion and the other forms of soil degradation render much of the agricultural soil of the world increasingly less fertile. Some land — severely eroded or too salty from evaporated irrigation water — is lost from production altogether. The land that can still produce is kept productive by the artificial means of adding synthetic fertilizers. Although fertilizers can temporarily replace lost nutrients, they cannot rebuild soil fertility and restore soil health; moreover, their use has a number of negative consequences, as discussed above.

Since the supply of agricultural soil is finite, and because natural processes cannot come close to renewing or restoring soil as fast as it is degraded, agriculture cannot be sustainable until it can reverse the process of soil degradation. Current agricultural practices must undergo a vast change if the precious soil resources we have remaining are to be conserved for the future.

**FIGURE 1.5 Severe soil erosion on a sloping hillside following intense winter rains.** In this strawberry growing region in the Elkhorn Slough watershed of central California, soil losses exceed 150 tons/acre in some years.

## Overuse of Water and Damage to Hydrological Systems

Fresh water is becoming increasingly scarce in many parts of the world as industry, expanding cities, and agriculture compete for limited supplies. Some countries have too little water for any additional agricultural or industrial development to occur. To meet demands for water in many other places, water is being drawn from underground aquifers much faster than it can be replenished by rainfall, and rivers are being drained of their water to the detriment of aquatic and riparian ecosystems and their dependent wildlife. Many of the world's major rivers — including the Colorado, Ganges, and Yellow — now run dry for part of the year as a result.

Agriculture accounts for more than two thirds of global water use. For every person on the planet, there are more than 0.04 ha of irrigated land. Agriculture uses so much water in part because it uses water wastefully. More than half of the water applied to crops is never taken up by the plants it is intended for (Van Tuijl, 1993). Instead, this water either evaporates or drains out of fields. Some wastage of water is inevitable, but a great deal of waste could be eliminated if agricultural practices were oriented toward conservation of water rather than maximization of production. For example, crop plants could be watered with drip irrigation systems, and production of water-intensive crops such as rice could be shifted away from regions with limited water supplies.

The increasing importance of meat in human diets worldwide is another factor in agriculture's rising demand for water, as is the trend toward concentrated grain feeding of livestock. Animal factories use prodigious amounts of water for cooling the animals and flushing their wastes, and many animals drink large amounts of water. Hogs, for example, can consume up to 8 gallons per animal per day (Marks and Knuffke, 1998). And these are just the direct uses of water for raising livestock. Factoring in the water needed to grow the biomass fed to animals, animal-derived food requires at least twice as much water to produce as plant-derived food, and usually much more.

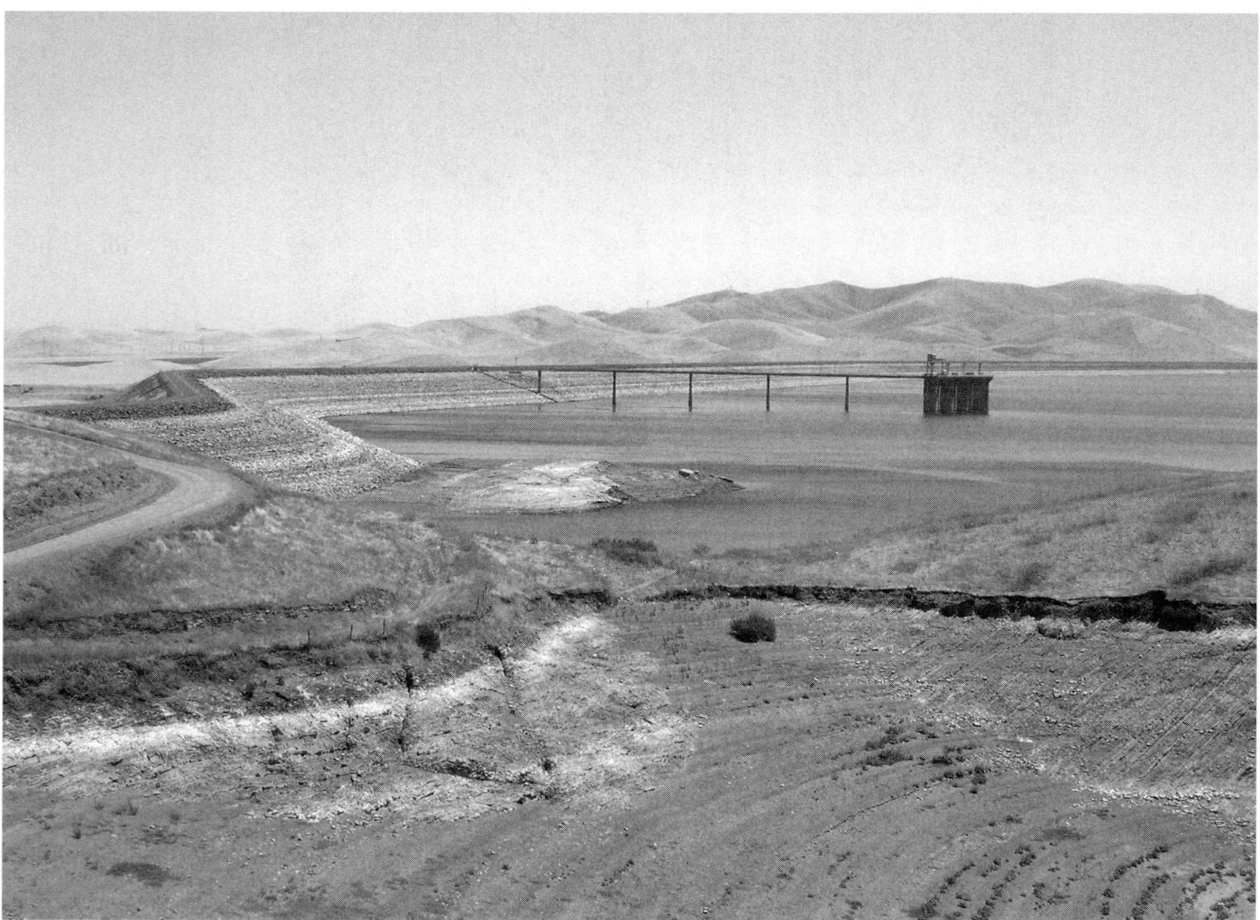

**FIGURE 1.6 The San Luis Dam in California.** Built in part to hold irrigation water for farms on the west side of the San Joaquin Valley, it is one of an estimated 800,000 dams in the world that trap life-giving silt, destroy riverine and riparian ecosystems, and completely alter natural hydrological functioning.

The difference between the amount of water needed to grow calorie-equivalent amounts of plant food and animal food can be extreme. For example, it takes only 89 liters of water to grow 500 calories of potatoes, but an astonishing *55 times* more, or 4902 liters, to raise 500 calories of grain-fed beef (Postel and Vickers, 2004). If we look at protein alone, the ratio is even more skewed: on average, producing 1 kg of animal protein requires about 100 times as much water as producing 1 kg of grain protein (Pimentel and Pimentel, 2003).

In addition to using a large share of the world's fresh water, conventional agriculture has an impact on regional and global hydrological patterns and the aquatic, riparian, and marine ecosystems dependent on them. First, by drawing such large quantities of water from natural reservoirs on land, agriculture has caused a massive transfer of water from the continents to the oceans. A 1994 study concluded that this transfer of water involves about 190 billion $m^3$ of water annually and has raised sea level by an estimated 1.1 cm (Sahagian et al., 1994). Moreover, the amount of water that agriculture causes to be moved from the land to the oceans is only increasing; by one estimate the net flow will increase by as much as 30% over present rates (Sahagian, 2000). Second, where irrigation is practiced on a large scale, agriculture brings about changes in hydrology and microclimate. Water is transferred from natural watercourses to fields and the soil below them, and increased evaporation changes humidity levels and may affect rainfall patterns. These changes in turn significantly impact natural ecosystems and wildlife. Third, the dams, aqueducts, and other infrastructure created to make irrigation possible have dramatically altered many of the world's rivers, causing enormous ecological damage. Rivers that once provided valuable "ecosystem services" to human society cannot do so anymore — their wetland, aquatic, and floodplain ecosystems can no longer absorb and filter out pollutants or provide habitat for fish and waterfowl, and they can no longer deposit the rich sediment so important for restoring the fertility of agricultural soils in floodplain areas (Postel and Richter, 2003).

If conventional agriculture continues to use water in the same ways, our rivers will become increasingly crippled and regional water crises will become increasingly common, either shortchanging the environment, marginalized peoples, and future generations, or limiting irrigation-dependent food production.

### Pollution of the Environment

More water pollution comes from agriculture than from any other single source. Agricultural pollutants include pesticides, herbicides, other agrochemicals, fertilizer, animal wastes, and salts.

Pesticides and herbicides — applied in large quantities on a regular basis, often from aircraft — are easily spread beyond their targets, killing beneficial insects and wildlife directly and poisoning farmers and farmworkers. The pesticides that make their way into streams, rivers, and lakes — and eventually the ocean — can have serious deleterious effects on aquatic ecosystems. They can also affect other ecosystems indirectly. Fish-eating raptors, for example, may eat pesticide-laden fish, reducing their reproductive capacity and thereby impacting terrestrial ecosystems. Although persistent organochloride pesticides such as DDT — known for their ability to remain in ecosystems for many decades — are being used less in many parts of the world, their less-persistent replacements are often much more acutely toxic.

Pesticides also pose a significant human health hazard. They spread throughout the environment by hydrological, meteorological, and biological means, and so it is impossible for humans to avoid exposure. In its 2003 edition of *Human Exposure to Environmental Chemicals*, the Centers for Disease Control reported that all of the 9282 people they tested had pesticides and their breakdown products in their bodies, and the average person had detectable amounts of 13 different pesticides (Schafer et al., 2004). Pesticides enter our bodies through our food and our drinking water. Pesticide contamination of groundwater has occurred in at least 26 states, and an EPA study in 1995 found that of 29 cities tested in the Midwest, 28 had herbicides present in their tap water. If all the drinking water sources in the U.S. at risk for pesticide contamination were properly monitored for the presence of harmful agents, the cost would be well over U.S.$15 billion (Pimentel, 2005).

Fertilizer leached from fields is less directly toxic than pesticides, but its effects can be equally damaging ecologically. In aquatic and marine ecosystems it promotes the overgrowth of algae, causing eutrophication and the death of many types of organisms. Nitrates from fertilizers are also a major contaminant of drinking water in many areas. Rounding out the list of pollutants from crop lands are salts and sediments, which in many locales have degraded streams, helped destroy fisheries, and rendered wetlands unfit for bird life.

Where factory farming has become the dominant form of meat, milk, and egg production, animal waste has become a huge pollution problem. Farm animals in the U.S. produce far more waste than do humans (Marks and Knuffke, 1998). The large size of feedlot and other factory farming operations poses challenges for the treatment of these wastes. As noted above, the wastes are typically treated in large anaerobic lagoons not well suited to protection of the environment. Some of the nitrogen from the wastes leaks out of the lagoons and into underlying aquifers, adding large quantities of nitrates to the groundwater and eventually to rivers. Even more nitrogen from the

## THE GULF OF MEXICO'S HYPOXIC "DEAD ZONE"

Every summer, a large area of the Gulf of Mexico near the mouth of the Mississippi River loses most of its dissolved oxygen and thus its ability to support nearly all species of marine life. It has been appropriately named the "dead zone." The size of the dead zone varies, but in recent years it has been alarmingly large; in 2002 it encompassed about 8500 square miles, nearly the size of New Jersey. The dead zone has many direct negative effects on human society, most notably threatening the important commercial fisheries of the Gulf coast region by killing fish and shrimp directly, compromising the ability of many species to reproduce, and altering migration patterns.

The dead zone is a direct result of massive amounts of nitrogen and phosphorus leaching out of the agricultural lands of the Mississippi River basin and causing excessive growth ("blooms") of phytoplankton in the Gulf. When the phytoplankton die, their decomposition by bacteria uses up much of the oxygen dissolved in the water. The relatively calm summer weather prevents mixing of the water column, resulting in the sustained hypoxic (low oxygen) conditions that kill fish and bottom-dwelling organisms.

The dead zone phenomenon shows the multifaceted and interrelated ways in which conventional agriculture impacts the environment. Irrigation, intensive tillage, monoculture, over-application of inorganic fertilizer, and factory farming of animals all play a role in causing unnaturally large amounts of nitrogen and phosphorus to flow into the Gulf of Mexico.

A little more than half of the excess nitrogen (an estimated 56%) comes from the inorganic fertilizer applied to fields in Kansas, Missouri, the Dakotas, Arkansas, and the other agricultural states in the Mississippi's vast watershed. Much of this nitrogen leaches into the region's rivers because much more nitrogen is applied than can be taken up by plants or chemically bound in the soil; excess fertilizer is applied because monocropped high-yield varieties require it for maximum production. And even more nitrogen ends up in the rivers because of irrigation and the erosion caused by intensive tillage.

About 25% of the excess nitrogen, and an even greater proportion of the excess phosphorus, comes from the animal waste produced by hog, poultry, and cattle CAFOs. These nutrients find their way into the rivers from manure spills, leaching of manure-treatment lagoons, and leaching from the excess treated manure applied to fields.

Ironically, if the Mississippi River and its tributaries were not so thoroughly engineered for human purposes — dammed for flood control and irrigation, channelized and locked for shipping — its healthy aquatic and wetland ecosystems and functioning floodplains would be able to remove much of the excess nitrogen and phosphorus from the rivers before these nutrients reached the Gulf of Mexico. Since much of the altering of the rivers in the Mississippi's watershed was done for the sake of agriculture — irrigation and transport of agricultural commodities — this is just one more way in which conventional agriculture is implicated in a continuing environmental disaster with huge impacts on human society.

**FIGURE 1.7 Satellite image of the "dead zone" in the Gulf of Mexico.** The darker areas indicate highly turbid waters with high concentrations of phytoplankton fed largely by agricultural runoff from the huge Mississippi River basin. The phytoplankton in the blooms will die and sink to the bottom, causing bacterial decay that removes oxygen from the surrounding water. *Source*: NASA.

wastes converts to ammonia and enters the atmosphere, where it combines with water droplets to form ammonium ions. As a result, the rainwater downwind of livestock feeding operations often has extremely high concentrations of ammonium ions. Although most treated animal waste is ultimately applied to fields as fertilizer, the phosphorus and nitrogen it contains is beyond useful levels for most crops. Furthermore, factory farms often have so much waste to get rid of that they apply more treated waste to fields than the soil can accommodate, and do so year-round, even at times in the crop cycle when fields and crops are unable to absorb it. The excess nitrogen and phosphorus finds its way into streams, rivers, and the local drinking water supply.

Through all these various avenues, tons of nitrogen and phosphorus from animal waste and inorganic fertilizer make their way into lakes and rivers and then into the oceans, creating large "dead zones" near river mouths. More than 50 of these dead zones exist seasonally around the world, with some of the largest — in the Chesapeake Bay, Puget Sound, and Gulf of Mexico — off the coast of the U.S.

### Dependence on External Inputs

Conventional agriculture has achieved its high yields mainly by increasing agricultural inputs. These inputs comprise material substances such as irrigation water, fertilizer, pesticides, and processed feed and antibiotics; the energy used to manufacture these substances, to run farm machinery and irrigation pumps, and to climate-control animal factories; and technology in the form of hybrid and transgenic seeds, new farm machinery, and new agrochemicals. These inputs all come from outside the agroecosystem itself; their extensive use has consequences for farmers' profits, use of nonrenewable resources, and the locus of control of agricultural production.

The longer conventional practices are used on farmland, the more the system becomes dependent on external inputs. As intensive tillage and monoculture degrade the soil, continued fertility depends more and more on the input of fossil-fuel-derived nitrogen fertilizer and other nutrients.

Agriculture cannot be sustained as long as this dependence on inputs remains. First, the natural resources from which many of the inputs are derived are nonrenewable and their supplies finite. Second, dependence on external inputs leaves farmers, regions, and whole countries vulnerable to supply shortages, market fluctuations, and price increases. In addition, excessive use of inputs has multiple negative off-farm and downstream impacts, as noted above.

### Loss of Genetic Diversity

Throughout most of the history of agriculture, humans have increased the genetic diversity of crop plants and livestock worldwide. We have been able to do this both by selecting for a variety of specific and often locally adapted traits through selective breeding, and by continually recruiting wild species and their genes into the pool of domesticated organisms. In the last 100 years or so, however, the overall genetic diversity of domesticated plants and animals has declined. Many varieties of plants and breeds of animals have become extinct, and a great many others are heading in that direction. About 75% of the genetic diversity that existed in crop plants in 1900 has been lost (Nierenberg and Halweil, 2004). The United Nations Food and Agriculture Organization estimates that as many as two domesticated animal breeds are being lost each week worldwide (FAO, 1998).

In the meantime, the genetic bases of most major crops and livestock species have become increasingly uniform. Only six varieties of corn, for example, account for more than 70% of the world's corn crop, and 99% of the turkeys raised in the U.S. belong to a single breed (FAO, 1998).

The loss of genetic diversity has occurred mainly because of conventional agriculture's emphasis on short-term productivity gains. When highly productive varieties and breeds are developed, they tend to be adopted in favor of others, even when the varieties they displace have many desirable and potentially desirable traits. Genetic homogeneity among crops and livestock is also consistent with the maximization of productive efficiency because it allows standardization of management practices.

For crop plants, a major problem with increasing genetic uniformity is that it leaves each crop as a whole more vulnerable to attack by pests and pathogens that acquire resistance to pesticides and to the plants' own defensive compounds; it also makes crops more vulnerable to changes in climate and other environmental factors. These are not insignificant or hypothetical threats. Every year, crop pests and pathogens destroy an estimated 30 to 40% of potential yield. Plant pathogens can evolve rapidly to overcome a crop's defenses, and global commerce and genetically uniform farm fields allow these new virulent strains to spread rapidly from field to field and continent to continent. In a report on crop diversity and disease threats released in 2005, researchers identified four diseases with the potential to devastate the U.S. corn crop, five that could threaten potatoes, and three with the potential to harm U.S.-grown wheat (Qualset and Shands, 2005). In late 2004, for example, a new soybean rust (a type of fungus) appeared in the southern U.S. and began to attack the soybean crop. None of the commercial soybean varieties planted in the U.S. are resistant to it, and scientists are concerned about the potential impact on the U.S.$18 billion soybean harvest as the rust spreads north.

Throughout the history of agriculture, farmers — and more recently, plant scientists — have responded to outbreaks of disease by finding and planting resistant varieties of the affected crop. But as the size of each crop's genetic reservoir declines, there are fewer and fewer varieties from which to draw resistant or adaptive genes. The importance of having a large genetic reservoir can be illustrated by example. In 1968, greenbugs attacked the U.S. sorghum crop, causing an estimated $100 million in damage. The next year, insecticides were used to control the greenbugs at a cost of about $50 million. Soon thereafter, however, researchers discovered a sorghum variety that carried resistance to the greenbugs. No one had known of the greenbug resistance, but it was there nonetheless. This variety was used to create a hybrid that was grown extensively and not eaten by greenbugs, making the use of pesticides unnecessary. Such pest resistance is common in domesticated plants, "hiding" in the genome but waiting to be used by plant breeders. As varieties are lost, however, the valuable genetic reservoir of traits is reduced in size, and certain traits potentially invaluable for future breeding are lost forever. There may very well be a soybean variety somewhere in the world resistant to the new soybean rust, but will plant scientists locate it before it goes extinct?

Increasing vulnerability to disease is also a serious concern for domesticated animal species as they lose their genetic diversity, but perhaps more serious is increased dependence on methods of industrial food production. Livestock breeds that are not adapted to local conditions require climate-controlled environments, doses of antibiotics, and large amounts of high-protein feed.

### Loss of Local Control Over Agricultural Production

Accompanying the concentration of agriculture into large-scale monocultural systems and factory farms has been a dramatic decline in the number of farms and farmers, especially in developed countries where mechanization and high levels of external inputs are the norm. From 1920 to the present, the number of farms in the U.S. has dropped from more that 6.5 million to just over 2 million, and the percentage of the population that lives and works on farms has dropped below 2%. Data from the 2000 U.S. census show that only 0.4% of the employed civilians in the U.S. listed their occupation as "farmer or rancher" (U.S. Census Bureau, 2005). In developing countries as well, rural people who work primarily in agriculture continue to abandon the land to move to urban and industrial areas, which will hold an estimated 60% of the world's population by 2030. As shown in Figure 1.8, there are now far more people in the world whose livelihoods are nonagricultural than there are people who grow food, and this gap continues to widen over time.

Besides encouraging an exodus from rural areas, large-scale commodity-oriented farming tends to wrest control of food production from rural communities. This trend is disturbing because local control and place-based knowledge and connection are crucial to the kind of management required for sustainable production. Food production carried out according to the dictates of the global market, and through technologies developed elsewhere, inevitably severs the connection to ecological principles. Experience-based management skill is replaced by purchased inputs requiring more capital, energy, and use of nonrenewable resources. Farmers become mere instruments of technology application, rather than independent decision-makers and managers.

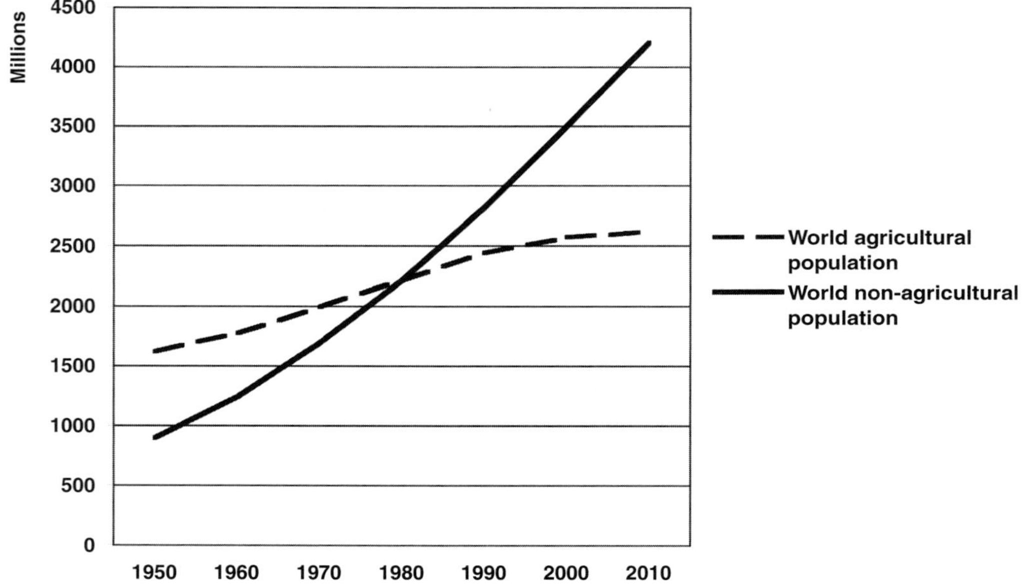


**FIGURE 1.8 Number of people worldwide involved in agriculture and not involved in agriculture.** *Source:* Data from FAOSTAT (2005). Figures for 2010 are projections.

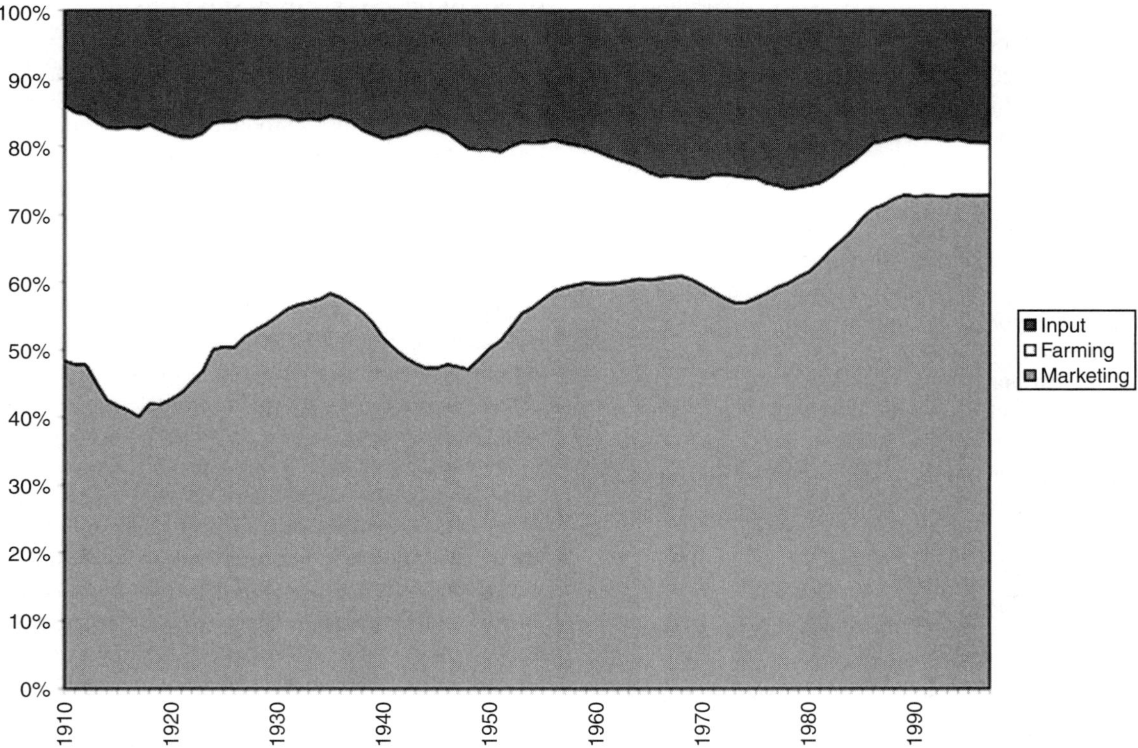


**FIGURE 1.9 U.S. farmers' declining share of the consumer food dollar, 1910 to 1997.** Marketing represents all services performed after food leaves the farm gate. The farmers' share includes payments to local governments and hired labor. Inputs include all purchased, nonfarm inputs. *Source:* Data from Stewart Smith, University of Maine, 2005.

Smaller-scale farmers seem to have little power against the advancement of industrial agriculture. Smaller farms cannot afford the cost of upgrading their farm equipment and technologies in order to compete successfully with the large farm operations. Moreover, the increase in the share of the food dollar going to distributors and marketers, coupled with cheap food policies that have kept farm prices relatively stable, has left many farmers in a tightening squeeze between production costs and marketing costs. Their share of the consumer food dollar, as shown in Figure 1.9, has dropped from almost 38% to less than 8% (Smith, pers. comm.).

Faced with such economic uncertainty, there is less incentive for farmers to stay on the land. One trend is for larger farmers to buy out their smaller neighbors. But when agricultural land is adjacent to rapidly expanding urban centers, such as in California, the incentive instead is to sell farmland at the inflated value it has as urban land. Because of this dynamic, the agriculturally rich Great Central Valley of California has seen the loss of hundreds of thousands of hectares of farmland to development since 1950, and the rate of loss of agricultural land in the state as a whole averaged 49,700 acres annually from 1988 to 1998 (Kuminoff et al., 2001).

In less developed countries, the growth of large-scale export agriculture has an even more ominous effect. As rural people — who were once able to feed themselves adequately *and* sell surplus food to city-dwellers — are pushed off the land, they migrate to cities, where they become dependent on others for their food. Since more of the food produced in the countryside is destined for export, increasing amounts of food for the expanding urban areas must be imported. Because of this dynamic, exports of food to developing countries from developed countries increased fivefold between 1970 and 1990, and during the 1990s, developing countries increased their food imports at the rate of 5.6% per year (FAO, 2003). In the period between 1980 and 2000, the quantity of coarse grains exported from developed nations to developing nations more than tripled (FAOSTAT, 2005). This imbalance threatens the food security of less-developed countries and makes them even more dependent on developed countries.

## Global Inequality

Despite increases in productivity and yields, hunger persists all over the globe. In some countries, such as India and much of Africa, the percentage of chronically hungry people has actually increased in recent years (FAO, 2004). There are also huge disparities in calorie intake and food security between people in developed nations and those in developing nations. At the beginning of the 21st century, the world reached a dubious milestone: the number of overweight people (about 1.1 billion) grew roughly equal to the number of underweight people (Gardner and Halweil, 2000). This statistic indicates that the unequal distribution of food — which is both a cause and a consequence of global inequality — is at least as serious a problem as the threats to global food production.

Developing nations too often grow food mainly for export to developed nations, using external inputs purchased from the developed nations. While the profits from the sale of the export crops enrich small numbers of elite landowners, many people in the developing nations go hungry — an estimated 815 million in 2002 (FAO, 2004). In addition, those with any land are often displaced as the privileged seek more land on which to grow export crops.

Besides causing unnecessary human suffering, relationships of inequality tend to promote agricultural policies and farmer practices that are driven more by economic considerations than by ecological wisdom and long-term thinking. For example, subsistence farmers in developing nations, displaced by large landowners increasing production for export, are often forced to farm marginal lands. The results are deforestation, severe erosion, and serious social and ecological harm.

Although inequality has always existed between countries and between groups within countries, the modernization of agriculture has tended to accentuate this inequality because its benefits are not evenly distributed. Those with more land and resources have had better access to the new technologies. Therefore, as long as conventional agriculture is based on First World technology and external inputs accessible to so few, the practice of agriculture will perpetuate inequality, and inequality will remain a barrier to sustainability.

## RUNNING OUT OF SOLUTIONS

During the 20th century, food production was increased in two ways: by bringing more land under production and by increasing the land's productivity — the amount of food produced per unit of land. As detailed above, many of the techniques that have been used to increase productivity have a great many negative consequences that in the long term work to undermine the productivity of agricultural land, and in many cases these techniques have approached their physical and practical limits. Conventional means of increasing productivity, therefore, cannot be relied on to help meet the increasing food needs of an expanding global population — a population that surpassed 6 billion in 2004, according to U.N. estimates.

However, increasing food production by cultivating more land is also problematic. Most of the land on the Earth's surface that is amenable to agriculture has already been converted to human use, and of this chunk of land, the proportion that can be farmed is actually shrinking due to urban expansion, soil degradation, and desertification. In the coming years, the growth of cities and industrialization will continue to claim more agricultural land — and often the best land, too. In addition, climate change threatens to take large areas of agricultural land out of production, especially in the tropics, where warming and drying may accelerate desertification in some areas and rising sea levels will inundate low-lying land.

Figure 1.10 shows the problem graphically. In the mid-1980s, the regular annual increases in the area of arable land worldwide observed since the 1970s (and earlier) ceased, and shrinkages have been observed in the periods 1988 to 1992, 1994 to 1995, 1997 to 1999, and 2001 to 2003.

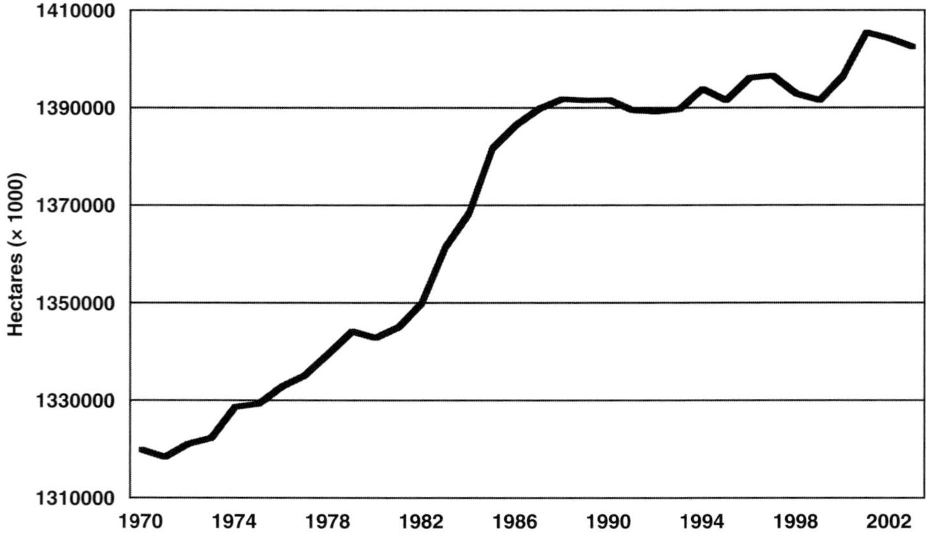


**FIGURE 1.10 Worldwide arable land area, 1970 to 2003.** As the total amount of arable land remains about the same each year, population growth continues its upward trend. *Source:* Data from Food and Agriculture Organization, FAOSTAT database, 2006.

Neither is it possible to bring much more land under cultivation through irrigation. In most drier regions, water is already scarce and there is no surplus available for increased agricultural use. Developing new supplies of water, moreover, has increasingly severe environmental consequences. In some areas that rely on groundwater for irrigation, such as Saudi Arabia and parts of the U.S., the amount of water available for irrigation will actually decrease in the future because of overdrafting and increasing nonagricultural demands.

There remain small but significant areas of land that could be farmed but are now covered by natural vegetation. Some of this land is in the process of being converted to agricultural use, but this way of increasing the amount of cultivated land also has its limits. First, much of this land is tropical rain forest, the soil of which cannot support continual agricultural production. Second, this land is increasingly being recognized for its value to global biological diversity, the carbon dioxide balance of the atmosphere, and maintenance of the Earth's climatic patterns. Because of this recognition and the efforts of environmental groups, a large proportion of the planet's remaining wild lands will be off-limits to agricultural conversion.

Exacerbating the problem of limited arable land is a trend toward meat-intensive diets worldwide. In the 30 years between 1973 and 2003, the world's population increased 61%, while at the same time worldwide meat production increased by more than 133% (FAOSTAT, 2005). The amount of meat produced per person, which has risen steadily since data on meat production began being collected in 1961, surpassed the 40 kg/person level in 2004.

Because the conversion of plant biomass into animal protein is highly inefficient, a large amount of plant biomass is needed to produce meat. For example, about 43 kg of plant biomass go into creating 1 kg of beef flesh (Pimentel and Pimentel, 2003). This means that a diet rich in meat requires much more land (and much larger expenditures of fossil fuel energy) to support than a vegetarian diet. Already, more corn and soybeans go to fattening livestock worldwide than to feeding human beings (in the U.S., seven times more grain is fed to livestock than is consumed by humans). As people increase both the total number of calories they consume, and the proportion of these calories that come from meat, they place increasing demands on the Earth's limited supply of arable land.

## THE PATH TOWARD SUSTAINABILITY

The only option we are left with is preserving the long-term productivity of the world's agricultural land while changing consumption and land use patterns to more equitably benefit everyone, from farmers to consumers. The first part of this challenge for the future defines the subject of most of this book; the latter part, touched on in the final two chapters, will rely to a large extent on the reconceptualizations of agriculture offered herein.

Preserving the productivity of agricultural land over the long term requires sustainable food production. Sustainability is achieved through alternative agricultural practices informed by in-depth knowledge of the ecological processes occurring in farm fields and the larger contexts of which they are a part. From this foundation we can move towards the social and economic changes that promote the sustainability of all sectors of the food system.

### What Is Sustainability?

Sustainability means different things to different people, but there is general agreement that it has an ecological basis. In the most general sense, sustainability is a version of the concept of sustained yield — the condition of being able to harvest biomass from a system in perpetuity because the ability of the system to renew itself or be renewed is not compromised.

Because "perpetuity" can never be demonstrated in the present, the proof of sustainability always remains in the future, out of reach. Thus it is impossible to know for sure if a particular practice is in fact sustainable or if a particular set of practices constitutes sustainability. However, it is possible to demonstrate that a practice is moving away from sustainability.

Based on our present knowledge, we can suggest that a sustainable agriculture would, at the very least:

- have minimal negative effects on the environment and release insignificant amounts of toxic or damaging substances into the atmosphere, surface water, or groundwater
- preserve and rebuild soil fertility, prevent soil erosion, and maintain the soil's ecological health
- use water in a way that allows aquifers to be recharged and the water needs of the environment and people to be met
- rely mainly on resources within the agroecosystem, including nearby communities, by replacing external inputs with nutrient cycling, better conservation, and an expanded base of ecological knowledge
- work to value and conserve biological diversity, both in the wild and in domesticated landscapes
- guarantee equality of access to appropriate agricultural practices, knowledge, and technologies and enable local control of agricultural resources

## The Role of Agroecology

The agriculture of the future must be both sustainable *and* highly productive if it is to feed the growing human population. This twin challenge means that we cannot simply abandon conventional practices wholesale and return to traditional or indigenous practices. Although traditional agriculture can provide models and practices valuable in developing sustainable agriculture, it cannot produce the amount of food required to supply distant urban centers and global markets because of its focus on meeting local and small-scale needs.

What is called for, then, is a new approach to agriculture and agricultural development that builds on the resource-conserving aspects of traditional, local, and small-scale agriculture while at the same time drawing on modern ecological knowledge and methods. This approach is embodied in the science of *agroecology*, which is defined as "the application of ecological concepts and principles to the design and management of sustainable food systems."

Agroecology provides the knowledge and methodology necessary for developing an agriculture that is on the one hand environmentally sound and on the other hand highly productive and economically viable. It opens the door to the development of new paradigms for agriculture, in part because it undercuts the distinction between the production of knowledge and its application. It values the local, empirical knowledge of farmers, the sharing of this knowledge, and its application to the common goal of sustainability.

Ecological methods and principles form the foundation of agroecology. They are essential for determining (1) if a particular agricultural practice, input, or management decision is sustainable, and (2) the ecological basis for the functioning of the chosen management strategy over the long term. Once these are known, practices can be developed that reduce purchased external inputs, lessen the impacts of such inputs when they are used, and establish a basis for designing systems that help farmers sustain their farms and their farming communities.

Even though an agroecological approach begins by focusing on particular components of a cropping system and the ecology of alternative management strategies, it establishes in the process the basis for much more. Applied more broadly, it can help us examine the historical development of agricultural activities in a region and determine the ecological basis for selecting more sustainable practices adapted to that region. It can also trace the causes of problems that have arisen as a result of unsustainable practices. Even more broadly, an agroecological approach helps us explore the theoretical basis for developing models that can facilitate the design, testing, and evaluation of sustainable agroecosystems. Ultimately, ecological knowledge of agroecosystem sustainability must reshape humanity's approach to growing and raising food in order for sustainable food systems to be achieved worldwide.

### THE HISTORY OF AGROECOLOGY

The two sciences from which agroecology is derived — ecology and agronomy — had an uneasy relationship during the 20th century. Ecology had been concerned primarily with the study of natural systems, whereas agronomy dealt with applying the methods of scientific investigation to the practice of agriculture. The boundary between pure science and nature on the one hand, and applied science and human endeavor on the other, has kept the two disciplines relatively separate, with agriculture ceded to the domain of agronomy. With a few important exceptions, little attention was devoted to the ecological analysis of agriculture until the mid-1990s.

An early instance of cross-fertilization between ecology and agronomy occurred in the late 1920s with the development of the field of crop ecology. Crop ecologists were concerned with where crops were grown and the ecological conditions under which they grew best. In the 1930s, crop ecologists actually proposed the term *agroecology* as the applied ecology of agriculture. However, since ecology was becoming more of an experimental science of natural systems, ecologists left the *applied ecology* of agriculture to agronomists, and the term agroecology seems to have been forgotten.

Following World War II, while ecology moved in the pure science direction, agronomy became increasingly results-oriented, in part because of the growing mechanization of agriculture and the greater use of agricultural chemicals. Researchers in each field became less likely to see any commonalties between the disciplines and the gulf between them widened.

In the late 1950s, the maturing of the ecosystem concept prompted some renewed interest in crop ecology and some work in what was termed agricultural ecology. The ecosystem concept provided, for the first time, an overall framework for examining agriculture from an ecological perspective, although few researchers actually used it in this way.

Through the 1960s and 1970s, interest in applying ecology to agriculture gradually gained momentum with intensification of community and population ecology research, the growing influence of systems-level approaches, and the increase in environmental awareness among members of the public. An important sign of this interest at the international level occurred in 1974 at the first International Congress of Ecology, when a working group developed a report entitled "Analysis of Agroecosystems."

As more ecologists in the 1970s began to see agricultural systems as legitimate areas of study, and more agronomists saw the value of the ecological perspective, the foundations of agroecology grew more rapidly. By the beginning of the 1980s, agroecology had emerged as a distinct methodology and conceptual framework for the study of agroecosystems. An important influence during this period came from traditional farming systems in developing countries, which began to be recognized by many researchers as important examples of ecologically based agroecosystem management (e.g., Gliessman, 1978a; Gliessman et al., 1981).

As its influence grew, agroecology helped contribute to the development of the concept of sustainability in agriculture. While sustainability provided a goal for focusing agroecological research, agroecology's whole-systems approach and knowledge of dynamic equilibrium provided a sound theoretical and conceptual basis for sustainability. In 1984, a variety of authors laid out the ecological basis of sustainability in the proceedings of a symposium (Douglass, 1984); this publication played a major role in solidifying the connection between agroecological research and the promotion of sustainable agriculture.

During the 1990s, agroecology matured into a well-recognized approach for the conversion to sustainable food systems. Agroecological research approaches emerged (Gliessman, 1990), several textbooks were published (Altieri, 1995; Pretty, 1995; Gliessman, 1998), websites were developed (www.agroecology.org), and academic research and education programs were put into motion. The establishment of an Agroecology Section for the Ecological Society of America in 1998 signaled a major change in how ecologists thought about agriculture, and the regular presentation of symposia, oral papers, and posters on agroecology at annual meetings of the American Society of Agronomy showed the embracing of the ecological approach.

Today, agroecology continues to straddle established boundaries. On the one hand, agroecology is the study of ecological processes in agroecosystems. On the other, it is a change agent for the complex social and ecological shifts that may need to occur in the future to move agriculture to a truly sustainable basis. Together, these complementary thrusts forge the way toward achieving sustainable food systems.

**Important Works in the History of Agroecology**

| Year | Author(s) | Title |
|---|---|---|
| 1928 | K. Klages | "Crop ecology and ecological crop geography in the agronomic curriculum" |
| 1938 | J. Papadakis | *Compendium of Crop Ecology* |
| 1939 | H. Hanson | *"Ecology in agriculture"* |
| 1942 | K. Klages | *Ecological Crop Geography* |
| 1956 | G. Azzi | *Agricultural Ecology* |
| 1962 | C.P. Wilsie | *Crop Adaptation and Distribution* |
| 1965 | W. Tischler | *Agrarökologie* |
| 1973 | D.H. Janzen | "Tropical agroecosystems" |
| 1974 | J. Harper | "The need for a focus on agro-ecosystems" |
| 1976 | INTECOL | *Report on an International Programme for Analysis of Agro-Ecosystems* |
| 1977 | O.L. Loucks | "Emergence of research on agro-ecosystems" |
| 1978b | S. Gliessman | *Memorias del Seminario Regional sobre la Agricultura Agricola Tradicional* |
| 1979 | R.D. Hart | *Agroecosistemas: Conceptos Basicos* |
| 1979 | G. Cox and M. Atkins | *Agricultural Ecology: An Analysis of World Food Production Systems* |
| 1981 | S. Gliessman, R. Garcia-Espinosa, and M. Amador | "The ecological basis for the application of traditional agricultural technology in the management of tropical agroecosystems" |
| 1983 | M. Altieri | *Agroecology* |
| 1984 | R. Lowrance, B. Stinner, and G. House | *Agricultural Ecosystems: Unifying Concepts* |
| 1984 | G. Douglass (ed.) | *Agricultural Sustainability in a Changing World Order* |
| 1990 | S. Gliessman (ed.) | *Agroecology: Researching the Ecological Basis for Sustainable Agriculture* |
| 1995 | M. Altieri | *Agroecology: The Science of Sustainable Agriculture (3rd edition)* |
| 1995 | J. Pretty | *Regenerating Agriculture: Policies and Practice for Sustainability and Self-Reliance* |
| 1998 | S. Gliessman | *Agroecology: Ecological Processes in Sustainable Agriculture* |
| 2004 | D. Rickerl and C. Francis (eds.) | *Agroecosystem Analysis* |
| 2004 | D. Clements and A. Shrestha (eds.) | *New Dimensions in Agroecology* |

## FOOD FOR THOUGHT

1. How does the holistic approach of agroecology allow for the integration of the three most important components of sustainability: ecological soundness, economic viability, and social equity?
2. Why has it been so difficult for humans to see that much of the environmental degradation caused by conventional agriculture is a consequence of the lack of an ecological approach to agriculture?
3. What common ground is there between agronomy and ecology with respect to sustainable agriculture?
4. What are the issues of greatest importance that threaten the sustainability of agriculture in the town or region in which you live?

## INTERNET RESOURCES

Agroecology
www.agroecology.org
A primary site for information, concepts, and case studies in the field of agroecology.

Earth Policy Institute
www.earth-policy.org
Led by the well-known eco-economist Lester Brown, this organization is dedicated to providing a vision of an eco-economy and a roadmap on how to get there. The website provides information on major milestones and setbacks in building a sustainable society.

Food and Agriculture Organization of the United Nations
www.fao.org

Food First: Institute for Food and Development Policy
www.foodfirst.org
Food First is a nonprofit think-tank and "education-for-action center" focused on revealing and changing the root causes of hunger and poverty around the world.

Pesticide Action Network International
www.pan-international.org
Pesticide Action Network (PAN) is a network of over 600 participating nongovernmental organizations, institutions and individuals in over 90 countries working to replace the use of hazardous pesticides with ecologically sound alternatives.

Sustainable Table
www.sustainabletable.org
Sustainable Table is a consumer campaign developed by the Global Resource Action Center for the Environment.

Worldwatch Institute
www.worldwatch.org
A nonprofit public policy research organization dedicated to informing policy makers and the public about emerging global problems and trends, and the complex links between the world economy and its environmental support systems. Food and farming are key support systems they monitor.

## RECOMMENDED READING


Altieri, M.A. 1995. *Agroecology: The Science of Sustainable Agriculture*. 3rd ed. Boulder, CO: Westview Press. An important pioneering work on the need for sustainability and a review of the kinds of agroecosystems that will help lead us toward it.

Brown, L. 2001. Feeding everyone well. *Eco-Economy: Building an Economy for the Earth.* New York and London: W.W. Norton & Co, 145–168. An in-depth analysis of the crises facing food production systems and the kinds of strategies needed to eradicate hunger and achieve food security.

Clements, D., and A. Shrestha (eds.) 2004. *New Dimensions in Agroecology.* New York: Food Products Press. An important collection of contributions from prominent agroecologists that covers the state of the art in agroecological research, showing the progress that has been made over the last decade in scientific thinking and research in agroecology.

Douglass, G.K., (ed.) 1984. *Agricultural Sustainability in a Changing World Order.* Boulder, Colorado: Westview Press. Proceedings of a landmark symposium that helped define the trajectory for future work on the interdisciplinary nature of agricultural sustainability.

Freyfogle, Eric T., (ed.) 2001. *The New Agrarianism: Land, Culture, and the Community of Life.* Washington, D.C.: Island Press. An exciting collection of essays and writing that paint a hopeful vision for reestablishing a new relationship between humans, their food, and the communities in which they live.

Gliessman, S.R., (ed.) 1990. *Agroecology: Researching the Ecological Basis for Sustainable Agriculture.* Ecological Studies Series #78. New York: Springer-Verlag. An excellent overview of what research is needed to identify the ecological basis for sustainable agroecosystems.

Halweil, B. 2004. Eat Here: Reclaiming Homegrown Pleasures in a Global Supermarket. Washington, D.C.: Worldwatch Institute. An engaging analysis of the current crisis in farm and food systems, accompanied by a convincing argument for reconnecting what we eat with how and where food is grown.

Jackson, W., W. Berry, and B. Colman (eds.) 1986. *Meeting the Expectation of the Land*. Berkeley CA: Northpoint Press. A collection of contributions from a diverse set of experts, designed to inform the general public of the people- and culture-based elements that are needed to make the transition to a sustainable agriculture.

Kimbrell, A., (ed.) 2002. *The Fatal Harvest Reader: The Tragedy of Industrial Agriculture*. Washington, D.C.: Island Press. An important collection of essays that vividly portray the devastating impacts of the current industrial agricultural system on the environment, human health, and farm communities, and present a compelling vision for a healthy, humane, and sustainable agriculture for the future.

Miller, G.T., Jr. 2004. *Living in the Environment: Principles, Connections, and Solutions*. 14th ed. Belmont, CA: Brooks/Cole, One of the most up-to-date textbooks in the field of environmental science, with a focus on problem solving.

Postel, S. and B. Richter, 2003. *Rivers for Life: Managing Water for People and Nature*. Washington, D.C.: Island Press. Explains the ecological and economic value of healthy riverine systems and how human alteration of rivers — in part to provide water for irrigation — has completely altered the ecology and hydrology of rivers, imperiling both their dependent wildlife and the human societies that depend on healthy rivers for their "ecosystem services."

Pretty, Jules N. 1995. *Regenerating Agriculture: Policies and Practice for Sustainability and Self-Reliance*. Washington, D.C.: Joseph Henry Press. An extensive review of the need for redirection agricultural policy and practice, and the steps that are taking place to create the change.

Rickerl, D. and C. Francis (eds.) 2004. *Agroecosystem Analysis*, Monograph #43 in the *Agronomy Series*. Madison, Wisconsin: American Society of Agronomy. A valuable review of agroecology as a field of inquiry that seeks to provide an ecologically based assessment of the structure, function, multidimensionality, and spatial scale of sustainable food systems.

# 2 The Agroecosystem Concept

An *agroecosystem* is a site or integrated region of agricultural production — a farm, for example — understood as an ecosystem. The agroecosystem concept provides a framework with which to analyze food production systems as wholes, including their complex sets of inputs and outputs and the interconnections of their component parts.

Because the concept of the agroecosystem is based on ecological principles and our understanding of natural ecosystems, the first topic of discussion in this chapter is the ecosystem. We examine the structural aspects of ecosystems — their parts and the relationships among the parts — and then turn to their functional aspects — how ecosystems work. Agroecosystems are then described in terms of how they compare, structurally and functionally, with natural ecosystems.

The principles and terms presented in this chapter will be applicable to our discussion of agroecosystems throughout this book.

## STRUCTURE OF NATURAL ECOSYSTEMS

An *ecosystem* can be defined as a functional system of complementary relations between living organisms and their environment, delimited by arbitrarily chosen boundaries, which in space and time appear to maintain a steady yet dynamic equilibrium. An ecosystem thus has physical parts with particular relationships — the *structure* of the system — that together take part in dynamic processes — the *function* of the system.

The most basic structural components of ecosystems are *biotic factors*, living organisms that interact in the environment, and *abiotic factors*, nonliving physical and chemical components of the environment such as soil, light, moisture, and temperature.

### LEVELS OF ORGANIZATION

Ecosystems can be examined in terms of a hierarchy of organization of their component parts, just as the human body can be examined at the level of molecules, cells, tissues, organs, or organ systems. At the simplest level is the individual organism. Study of this level of organization is called autecology or physiological ecology. It is concerned with how a single individual of a species performs in response to the factors of the environment and how the organism's particular degree of tolerance to stresses in the environment determine where it will live. The adaptations of the banana plant, for example, restrict it to humid, tropical environments with a particular set of conditions, whereas a strawberry plant is adapted to a much more temperate environment.

At the next level of organization is groups of individuals of the same species. Such a group is known as a *population*. The study of populations is called population ecology. An understanding of population ecology becomes important in determining the factors that control population size and growth, especially in relation to the capacity of the environment to support a particular population over time. Agronomists have applied the principles of population ecology in the experimentation that has led to the highest-yielding density and arrangement of individual crop species.

Populations of different species always occur together in mixtures, creating the next level of organization, the *community*. A community is an assemblage of various species living together in a particular place and interacting with each other. An important aspect of this level is how the interactions of organisms affect the distribution and abundance of the different species that make up a particular community. Competition between plants in a cropping system or the predation of aphids by lady beetles are examples of interactions at this level in an agroecosystem. The study of the community level of organization is known as community ecology.

The most inclusive level of organization of an ecosystem is the ecosystem itself, which includes all of the abiotic factors of the environment in addition to the communities of organisms that occur in a specific area. An intricate web of interactions goes on within the structure of the ecosystem.

These four levels can be directly applied to agroecosystems, as shown in Figure 2.1. Throughout this text, reference will be made to these levels: individual crop plants (the organism level), populations of crop species or other organisms, farm field communities, and whole agroecosystems.

An important characteristic of ecosystems is that at each level of organization properties emerge that were not present at the level below. These emergent properties are the result of the interaction of the component "parts" of that level of ecosystem organization. A population, for example, is much more than a collection of individuals of the same species, and has characteristics that cannot be

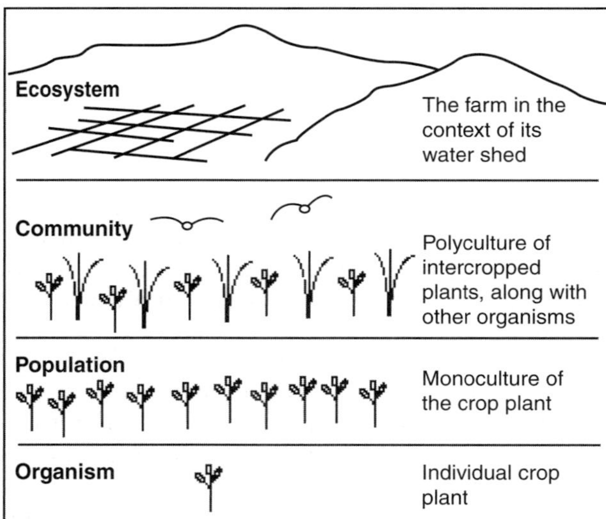


**FIGURE 2.1 Levels of ecosystem organization applied to an agroecosystem.** The diagram could be extended in the upward direction to include regional, national, and global levels of organization, which would involve such things as markets, farm policy, even global climate change. In the downward direction, the diagram could include the cellular, chemical, and atomic levels of organization.

understood in terms of individual organisms alone. In an agroecosystem context, this principle means in essence that the farm is greater than the sum of its individual crop plants. Sustainability can be considered the ultimate emergent quality of an ecosystem approach to agriculture.

## STRUCTURAL PROPERTIES OF COMMUNITIES

A community comes about on the one hand as a result of the adaptations of its component species to the gradients of abiotic factors that occur in the environment, and, on the other hand, as a result of interactions between populations of these species. Since the structure of the community plays such an important role in determining the dynamics and stability of the ecosystem, it is valuable to examine in more detail several properties of communities that arise as a result of interactions at this level.

### Species Diversity

Understood in its simplest sense, species *diversity* is the number of species that occur in a community. Some communities, such as that of a freshwater pond, are exceedingly diverse; others are made up of very few species.

### Dominance and Relative Abundance

In any community, some species may be relatively abundant and others less abundant. The species with the greatest impact on both the biotic and abiotic components of the community is referred to as the *dominant species.* Dominance can be a result of an organism's relative abundance, its size, its ecological role, or any of these factors in combination. For example, since a few large trees in a garden can dramatically alter the light environment for all the other species in the garden, the tree species is dominant in the garden community even though it may not be the most abundant species. Natural ecosystems are often named for their dominant species. The redwood forest community of coastal California is a good example.

### Vegetative Structure

Terrestrial communities are often characterized by the structure of their vegetation. This is determined mostly by the form of the dominant plant species, but also by the form and abundance of other plant species and their spacing. Thus vegetative structure has a vertical component (a profile with different layers) and a horizontal component (groupings or patterns of association), and we learn to recognize how different species occupy different places in this structure. When the species that make up vegetative structure take on similar growth forms, more general names are given to these assemblages (e.g., grassland, forest, shrubland).

### Trophic Structure

Every species in a community has nutritive needs. How these needs are met in relation to other species determines a structure of feeding relationships. This structure is called the community's *trophic structure.* Plants are the foundation of every terrestrial community's trophic structure because of their ability to capture solar energy and convert it, through photosynthesis, into stored chemical energy in the form of *biomass*, which can then serve as food for other species. Because of this trophic role, plants are known as *producers*. Physiologically, plants are classified as *autotrophs* because they satisfy their energy needs without preying upon other organisms.

The biomass produced by plants becomes available for use by the *consumers* of the community. Consumers include *herbivores,* which convert plant biomass into animal biomass, *predators*, which consume herbivores and other predators, *parasites*, which consume blood or tissues of a host but usually do not kill it, and *parasitoids*, which are insects whose larvae live within and consume their host, which is usually another insect. All consumers are classified as *heterotrophs* because their nutritive needs are met by consuming other organisms.

Each level of consumption is considered to be a different trophic level. The trophic relationships among a community's species can be described as a food chain or a food web, depending on their complexity. As we will see later, trophic relationships can become quite complex and are of considerable importance in agroecosystem processes such as pest and disease management (Table 2.1).

**TABLE 2.1**
**Trophic Levels and Roles in a Community**

| Type of Organism | Trophic Role | Trophic Level | Physiological Classification |
|---|---|---|---|
| Plants | Producers | First | Autotrophic |
| Herbivores | First-level consumers | Second | Heterotrophic |
| Predators and parasites | Second-level (and higher) consumers | Third and higher | Heterotrophic |

### STABILITY

Over time, the species diversity, dominance structure, vegetative structure, and trophic structure of a community usually remain fairly stable, even though individual organisms die and leave the area, and the relative sizes of populations shift. In other words, if you were to visit and observe a natural community and then visit it again 20 years later, it would probably appear relatively unchanged in its basic aspects. Even if some kind of *disturbance* — such as fire or flooding — killed off many members of many species in the community, the community would eventually recover, or return to something close to the original condition and species composition.

Because of this ability of communities to resist change and to be resilient in response to disturbance, communities — and the ecosystems of which they are a part — are sometimes said to possess the property of stability. The relative stability of a community depends greatly on the type of community and the nature of the disturbances to which it is subjected. Ecologists disagree about whether or not stability should be considered an inherent characteristic of communities or ecosystems.

## FUNCTIONING OF NATURAL ECOSYSTEMS

Ecosystem function refers to the dynamic processes occurring within ecosystems: the movement of matter and energy and the interactions and relationships of the organisms and materials in the system. It is important to understand these processes in order to address the concepts of ecosystem dynamics, efficiency, productivity, and development, especially in agroecosystems where function can determine the difference between the success and failure of a particular crop or management practice.

The two most fundamental processes in any ecosystem are the flow of energy among its parts and the cycling of nutrients.

### ENERGY FLOW

Each individual organism in an ecosystem is constantly using energy to carry out its physiological processes, and its sources of energy must be regularly replenished. Thus energy in an ecosystem is like electricity in a home: it is constantly flowing into the system from outside sources, fueling its basic functioning. The energy flow in an ecosystem is directly related to its trophic structure. By examining energy flow, however, we are focusing on the sources of the energy and its movement within the structure, rather than on the structure itself.

Energy flows into an ecosystem as a result of the capture of solar energy by plants, the producers of the system. This energy is stored in the chemical bonds of the biomass that plants produce. Ecosystems vary in their ability to convert solar energy to biomass. We can measure the total amount of energy that plants have brought into the system at a point in time by determining the *standing crop* or biomass of the plants in the system. We can also measure the rate of the conversion of solar energy to biomass: this is called *gross primary productivity*, which is usually expressed in terms of kilocalories per square meter per year. When the energy plants use to maintain themselves is subtracted from gross primary productivity, a measure of the ecosystem's *net primary productivity* is attained.

Herbivores (primary consumers) consume plant biomass and convert it into animal biomass, and predators and parasites (secondary and higher-level consumers) who prey on herbivores or other consumers continue the biomass conversion process between trophic levels. Only a small percentage of the biomass at one trophic level, however, is converted into biomass at the next trophic level. This is because a large amount of energy is expended in maintaining the organisms at each level (as much as 90% of the consumed energy). In addition, a large amount of biomass at each level is never consumed (and some of what is consumed is not fully digested); this biomass (in the form of dead organisms and fecal matter) is eventually broken down by *detritivores* and *decomposers*. The decomposition process releases (in the form of heat) much of the energy that went into creating the biomass, and the remaining biomass is returned to the soil as organic matter.

In natural ecosystems, the energy that leaves the system is mostly in the form of heat, generated in part by the respiration of the organisms at the various trophic levels and in part by the decomposition of biomass. Other forms of energy output are quite small. The total energy output (or energy loss) of an ecosystem is usually balanced by the energy input that comes from plants capturing solar energy (Figure 2.2).

### NUTRIENT CYCLING

In addition to energy, organisms require inputs of matter to maintain their life functions. This matter — in the form of nutrients containing a variety of crucial elements and compounds — is used to build cells and tissues and the complex organic molecules required for cell and body functioning.

The cycling of nutrients in ecosystems is obviously linked to the flow of energy: the biomass transferred

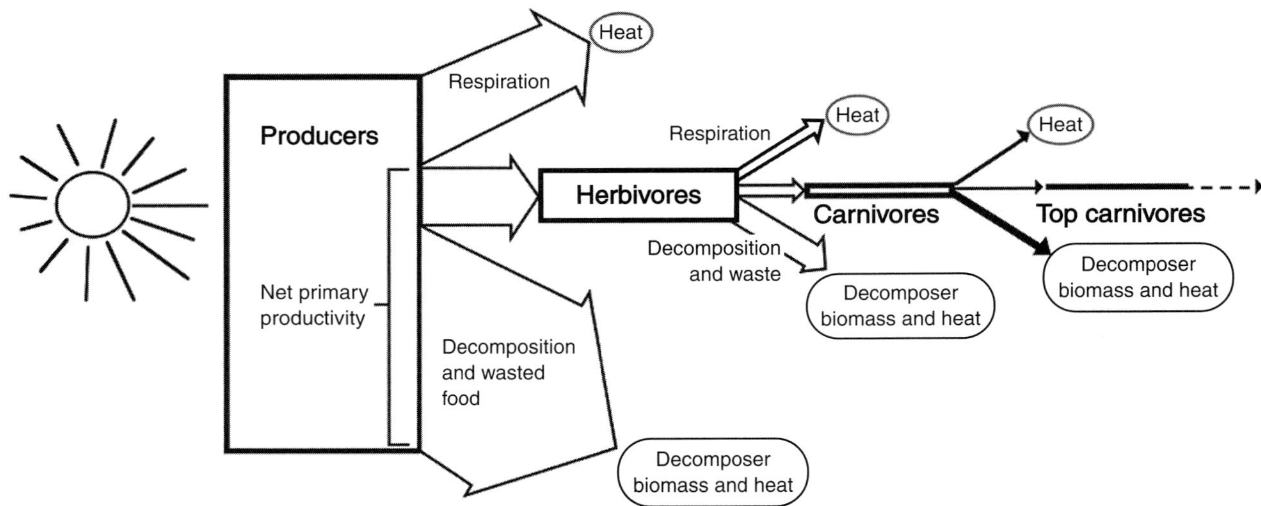


**FIGURE 2.2 Ecosystem energy flow. The size of each box represents the relative amount of energy flowing through that trophic level.** In the average ecosystem, only about 10% of the energy in a trophic level is transferred to the next trophic level. Nearly all the energy that enters an ecosystem is eventually dissipated as heat.

between trophic levels contains both energy in chemical bonds and matter serving as nutrients. Energy, however, flows in one direction only through ecosystems — from the sun to producers to consumers to the environment. Nutrients, in contrast, move in cycles — through the biotic components of an ecosystem to the abiotic components, and back again to the biotic. Since both abiotic and biotic components of the ecosystem are involved in these cycles, they are referred to as *biogeochemical cycles*. As a whole, biogeochemical cycles are complex and interconnected; in addition, many occur at a global level that transcends individual ecosystems.

Many nutrients are cycled through ecosystems. The most important are carbon (C), nitrogen (N), oxygen (O), phosphorus (P), sulfur (S), and water. With the exception of water, each of these is known as a *macronutrient*. Each nutrient has a specific route through the ecosystem depending on the type of element and the trophic structure of the ecosystem, but two main types of biogeochemical cycles are generally recognized. For carbon, oxygen, and nitrogen, the atmosphere functions as the primary abiotic reservoir, so we can visualize cycles that take on a global character. As an example, a molecule of carbon dioxide respired into the air by an organism in one location can be taken up by a plant halfway around the planet. Elements that are less mobile, such as phosphorus, sulfur, potassium, calcium, and most of the trace elements, cycle more locally, and the soil is their main abiotic reservoir. These nutrients are taken up by plant roots, stored for a period of time in biomass, and eventually returned to the soil within the same ecosystem by decomposers.

Some nutrients can exist in forms that are readily available to organisms. Carbon is a good example of such a material, easily moving between its abiotic form in the atmospheric reservoir to a biotic form in plant or animal matter as it cycles between the atmosphere as carbon dioxide and biomass as complex carbohydrates. Carbon spends varying lengths of time in living or dead organic matter, or even humus in the soil, but it returns to the atmospheric reservoir as carbon dioxide before it is recycled again. Figure 2.3 is a simplified depiction of the carbon cycle, focusing on terrestrial systems and leaving out the reservoir of carbon found in carbonate rocks.

Nutrients in the atmospheric reservoir can exist in forms much less readily available and must be converted to some other form before they can be used. A good example is atmospheric nitrogen ($N_2$). The conversion of molecular nitrogen ($N_2$) to ammonia ($NH_3$) through biological fixation by microorganisms begins the process that makes nitrogen available to plants. Once incorporated into plant biomass, this "fixed" nitrogen can then become part of the soil reservoir and eventually be taken up again by plant roots as nitrate ($NO_3$). As long as this soil-cycled nitrogen is not reconverted back to gaseous $N_2$ or lost as volatile ammonia or gaseous oxides of nitrogen, it can be actively cycled within the ecosystem (Figure 2.4). The agroecological significance of the biotic interactions involved in this cycle are discussed in more detail in Chapter 16.

Phosphorus, on the other hand, has no significant gaseous form. It is slowly added to the soil by the weathering of rock, and once there, can be taken up by plants as phosphate and then form part of the standing crop, or be returned to the soil by excretion or decomposition. This cycling between organisms and soil tends to be very localized in ecosystems, with two major exceptions: (1) phosphates may leach out of ecosystems in ground water if they

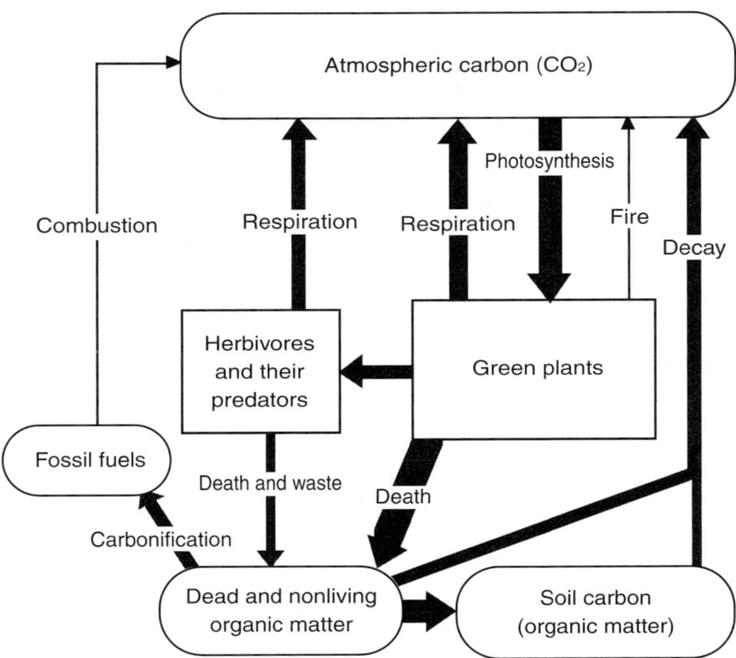


**FIGURE 2.3 The carbon cycle.**

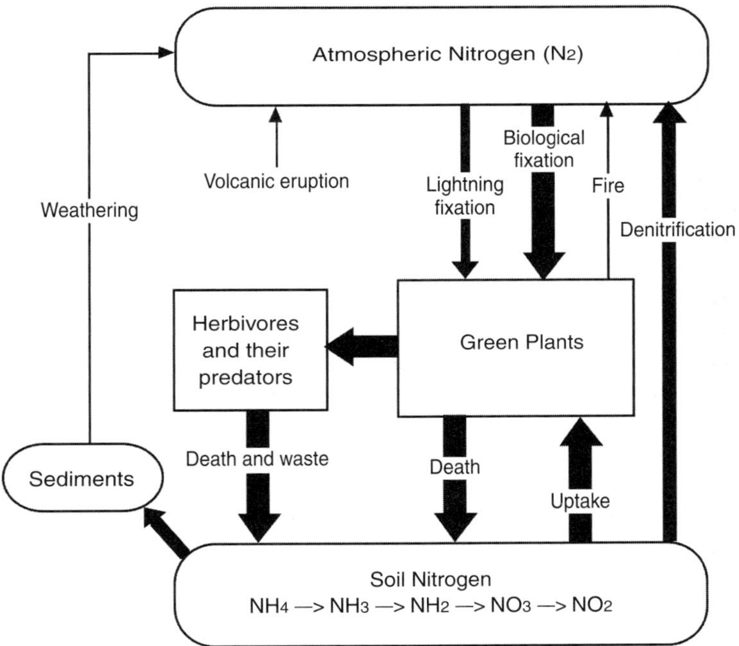


**FIGURE 2.4 The nitrogen cycle.**

are not absorbed or bound and (2) phosphates adhering to soil particles may be removed by erosion. In both of these cases, the phosphates leave the ecosystem and end up in the oceans. Once phosphorus is deposited into the sea, the time frame required for it to cycle back into terrestrial systems enters the geological realm, hence the importance of the localized cycles that keep phosphorus in the ecosystem (Figure 2.5).

In addition to the macronutrients, a number of other chemical elements must be present and available in the ecosystem for plants to grow. Even though they are needed in very small quantities, they are still of great importance for living organisms. They include iron (Fe), magnesium (Mg), manganese (Mn), cobalt (Co), boron (B), zinc (Zn), and molybdenum (Mo). Each of these elements is known as a *micronutrient*.

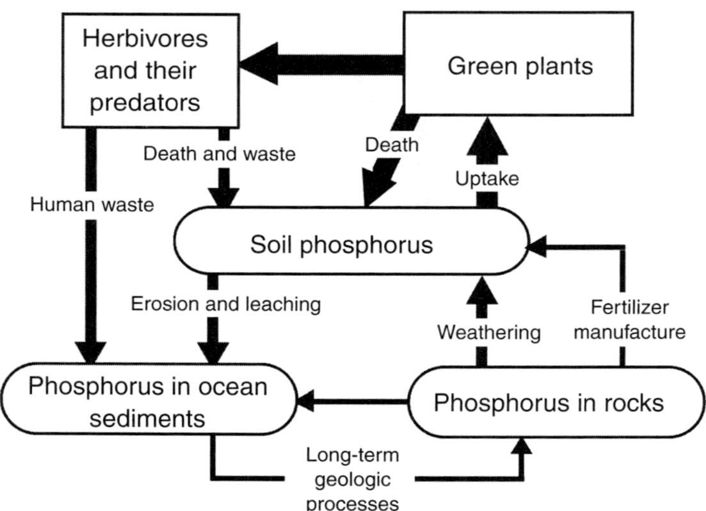


**FIGURE 2.5 The phosphorus cycle.**

Both types of nutrients are taken up by organisms and are stored in living or dead biomass or organic matter. If too much of a nutrient is lost or removed from a particular system, it can become limiting for further growth and development. Biological components of each system are very important in determining how efficiently nutrients move, ensuring that the minimum amount is lost and the maximum amount recycled. Productivity can become very closely linked to the rates at which nutrients are able to be recycled.

## REGULATION OF POPULATIONS

Populations are dynamic: their size and the individual organisms that make them up change over time. The demographics of each population are a function of that species' birth and death rates, rate of population increase or decrease, and the carrying capacity of the environment in which they live. The size of each population in relation to the other populations of the ecosystem is also determined by the interactions of that population with other populations and with the environment. A species with a broad set of tolerances of environmental conditions and a broad ability to interact with other species will be relatively common over a large area. In contrast, a species with a narrow set of tolerances and a very specialized role in the system will be common only locally.

Depending on the actual set of adaptive traits of each species, the outcome of its interaction with other species will vary. When the adaptations of two species are very similar, and resources are insufficient to maintain populations of both, *competition* can occur. One species can begin to dominate another through the removal of essential materials from the environment. In other cases, a species can add materials to the environment, modifying conditions that aid its own ability to be dominant to the detriment of others. Some species have developed ways of interacting with each other that can be of benefit to them both, leading to relationships of *mutualism*, where resources are shared or partitioned (the importance of mutualisms in agroecology is discussed in Chapter 15). In natural ecosystems, selection through time has tended to result in the most complex structure biologically possible within the limits set by the environment, permitting the establishment and maintenance of dynamic populations of organisms.

## ECOSYSTEM CHANGE

Ecosystems are in a constant state of dynamic change. Organisms are coming into existence and dying, matter is being cycled through the component parts of the system, populations are growing and shrinking, and the spatial arrangement of organisms is shifting. Despite this internal dynamism, however, ecosystems are remarkably stable in their overall structure and functioning. This stability is due in part to ecosystems' complexity and species diversity.

One aspect of ecosystem stability, as discussed earlier in terms of communities, is the observed ability of ecosystems to either resist change that is introduced by disturbance, or to recover from disturbance after it happens. The recovery of a system following disturbance, a process called *succession*, eventually allows the reestablishment of an ecosystem similar to that which occurred before the disturbance. This "end point" of succession is called the *climax* state of the ecosystem. As long as disturbance is not too intense or frequent, the structure and function that characterized an ecosystem before perturbation is reestablished, even if the community of organisms that eventually regains dominance may be slightly different.

Nevertheless, ecosystems do not develop toward or enter into a steady state. Instead, due to constant natural disturbance, they remain dynamic and flexible, resilient in the face of perturbing forces. Overall stability combined with dynamic change is often captured in the concept of *dynamic equilibrium*. The dynamic equilibrium of ecosystems is of considerable importance in an agricultural setting. It permits the establishment of an ecological "balance," functioning on the basis of sustained resource use, which can be maintained indefinitely despite ongoing and regular change in the form of harvest, soil cultivation, and replanting.

## AGROECOSYSTEMS

Human manipulation and alteration of ecosystems for the purpose of establishing agricultural production makes agroecosystems very different from natural ecosystems. At the same time, however, the processes, structures, and characteristics of natural ecosystems can be observed in agroecosystems.

### NATURAL ECOSYSTEMS AND AGROECOSYSTEMS COMPARED

A natural ecosystem and an agroecosystem are diagrammed, respectively, in Figure 2.6 and Figure 2.7. In both figures, flows of energy are shown as solid lines and movement of nutrients is shown with dashed lines.

A comparison of Figure 2.6 and Figure 2.7 reveals that agroecosystems differ from natural ecosystems in several key respects.

*Energy Flow:* Energy flow in agroecosystems is altered greatly by human interference. Inputs are derived from primarily human sources and are often not self-sustaining. Thus agroecosystems become open systems where considerable energy is directed out of the system at the time of each harvest, rather than stored in biomass, which could otherwise accumulate within the system.

*Nutrient Cycling:* Recycling of nutrients is minimal in most agroecosystems and considerable quantities are lost from the system with the harvest or as a result of leaching or erosion due to a great reduction in permanent biomass levels held within the system. The frequent exposure of bare soil between crop plants and, temporally, between cropping seasons, also creates "leaks" of nutrients from the system. Farmers have recently come to rely heavily upon petroleum-based nutrient inputs to replace these losses.

*Population Regulating Mechanisms:* Due to the simplification of the environment and a reduction in trophic interactions, populations of crop plants or animals in agroecosystems are rarely self-reproducing or self-regulating. Human inputs in the form of seed or control agents, often dependent on large energy subsidies, determine population sizes. Biological diversity is reduced, trophic structures tend to become simplified, and many niches are left unoccupied. The danger of catastrophic pest or disease outbreak is high, despite the intensive human interference.

*Stability:* Due to their reduced structural and functional diversity in relation to natural ecosystems, agroecosystems have much less resilience than natural ecosystems. A focus on harvest outputs upsets any equilibrium that is established, and the system can only be sustained if outside interference — in the form of human labor and external human inputs — is maintained.

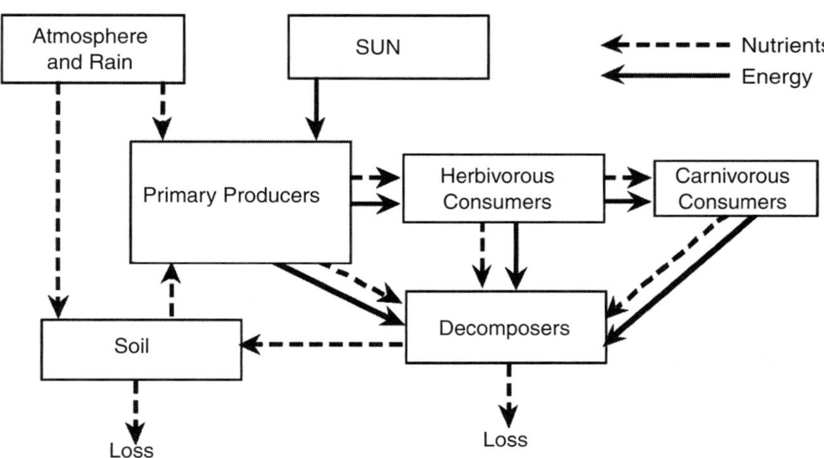


**FIGURE 2.6 Functional components of a natural ecosystem.** The components labeled "Atmosphere and Rain" and "Sun" are outside any specific system and provide essential natural inputs.

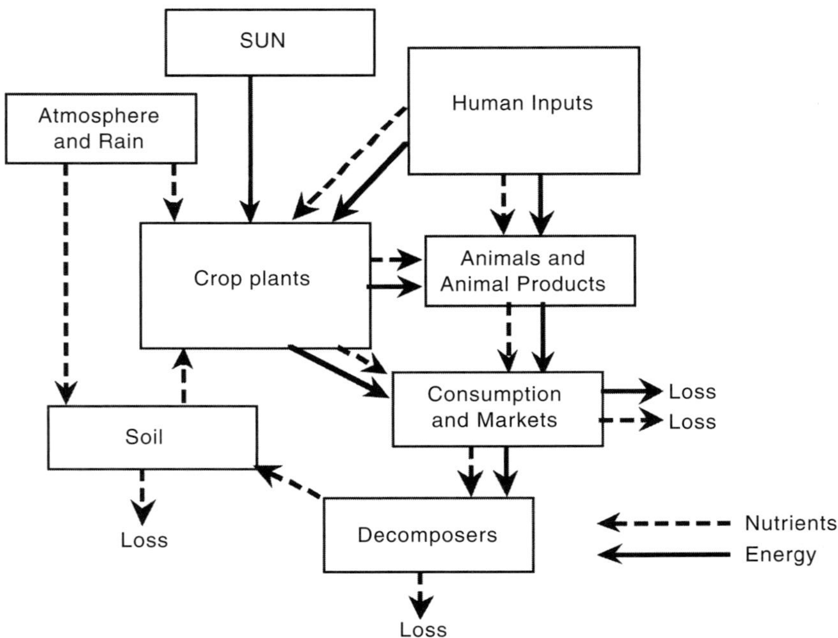


**FIGURE 2.7 Functional components of an agroecosystem.** In addition to the natural inputs provided by the atmosphere and the sun, an agroecosystem has a whole set of human inputs that come from outside the system. An agroecosystem also has a set of outputs, labeled here as "Consumption and Markets."

**TABLE 2.2**
**Important Structural and Functional Differences Between Natural Ecosystems and Agroecosystems**

| | Natural Ecosystems | Agroecosystems |
|---|---|---|
| Net productivity | Medium | High |
| Trophic interactions | Complex | Simple, linear |
| Species diversity | High | Low |
| Genetic diversity | High | Low |
| Nutrient cycles | Closed | Open |
| Stability (resilience) | High | Low |
| Human control | Independent | Dependent |
| Temporal permanence | Long | Short |
| Habitat heterogeneity | Complex | Simple |

*Source:* Odum, E. P. 1969. *Science* 164: 262–270.

The key ecological differences between natural ecosystems and agroecosystems are summarized in Table 2.2.

Although sharp contrasts have been drawn between natural ecosystems and agroecosystems, actual systems of both types exist on a continuum. On one side of the continuum, few 'natural' ecosystems are truly natural in the sense of being completely independent of human influence; on the other side, agroecosystems can vary greatly in their need for human interference and inputs. Indeed, through application of the concepts presented in this text, agroecosystems can be designed that come close to resembling natural ecosystems in terms of such characteristics as species diversity, nutrient cycling, and habitat heterogeneity.

## The Agroecosystem as a Unit of Analysis

We have so far described agroecosystems conceptually; it remains to explain what they are physically. In other words, what is the thing we are talking about when we discuss the management of an agroecosystem? This is first of all an issue of spatial boundaries. The spatial limits of an agroecosystem in the abstract, like those of an ecosystem, are somewhat arbitrary. In practice, however, an "agroecosystem" is generally equivalent to an individual farm, although it could just as easily be a single farm field or a grouping of adjacent farms.

Another issue involves the relationship between an abstract or concrete agroecosystem and its relationship and connection to the surrounding social and natural worlds. By its very nature, an agroecosystem is enmeshed in both. A web of connections spreads out from every agroecosystem into human society and natural ecosystems. Coffee drinkers in Seattle are connected to coffee-producing agroecosystems in Costa Rica; the Siberian taiga may experience impacts from conventional corn production systems in the U.S.

In practical terms, however, we must distinguish between what is external to an agroecosystem and what is internal. This distinction becomes necessary when analyzing agroecosystem inputs, since something cannot be an input unless it comes from outside the system. The convention followed in this text is to use an agroecosystem's spatial boundary (explicit or implicit) as the dividing line between internal and external. In terms of inputs supplied by humans, therefore, any substance or energy source from outside the spatial boundaries of the system is an *external human input*. Even though the word *external* is redundant with *input*, it is retained in this phrase to emphasize off-the-farm origins. Typical external human inputs include pesticides, inorganic fertilizers, hybrid seed, fossil fuels used to run tractors, the tractors themselves, most kinds of irrigation water, and human labor supplied by nonfarm residents. There are also natural inputs, the most important of which are solar radiation, precipitation, wind, sediments deposited by flooding, and plant propagules.

### Sustainable Agroecosystems

The challenge in creating sustainable agroecosystems is one of achieving natural ecosystem-like characteristics while maintaining a harvest output. Working toward sustainability, the manager of any particular agroecosystem strives as much as possible to use the ecosystem concept in his or her design and management. Energy flow can be designed to depend less on nonrenewable sources, and a better balance achieved between the energy used to maintain the internal processes of the system and that which is available for export as harvestable goods. The farmer can strive to develop and maintain nutrient cycles that are as "closed" as possible, to lower nutrient losses from the system, and to search for sustainable ways to return exported nutrients to the farm. Population regulation mechanisms can depend more on system-level resistance to pests, through an array of mechanisms that range from increasing habitat diversity to ensuring the presence of natural enemies and antagonists. Finally, an agroecosystem that incorporates the natural ecosystem qualities of resilience, stability, productivity, and balance will better ensure the maintenance of the dynamic equilibrium necessary to establish an ecological basis for sustainability. As the use of external human inputs for control of agroecosystem processes is reduced, we can expect a shift from systems dependent on synthetic inputs to systems designed to make use of natural ecosystem processes and interactions and materials derived from within the system.

### Agroecosystems in Context: The Food System

Agroecology finds its most immediate applications at the farm or agroecosystem level, where it can effectively deal with production, short-term enterprise economics, and environmental impacts in the immediate vicinity of the farm. But each farm or agroecosystem is part of a much larger system, a global network of food production, distribution, and consumption called the *food system*.

Sustainability in agriculture can only come from understanding the interaction of all components of the food system. Therefore, this text lays the groundwork for developing a food-system perspective from which to view all questions of agricultural sustainability. This perspective pays attention as much to the people in agroecosystems as it does to the ecological conditions on the farm. It takes into account the large amounts of energy and materials that are integral to the processing, transportation, and marketing that take place in the human "food chain." It pays attention to the equity issues of hunger, *food security*, and access to good nutrition and diet. It weighs the impacts of globalization in the marketplace and in farm communities, and sees producers and consumers as actively connected parts of a single system.

The agroecosystem concept and the science of agroecology provide a foundation for examining and understanding the interactions and relationships among the diverse components of the food system (Francis et al., 2003). A grounding in ecosystem thinking — wherein a complex web of interacting and independent parts contribute to the emergence of a sustainable whole — allows a framework for integrating social, economic, political, and ecological perspectives to take shape. It is the goal of the chapters ahead to establish this grounding in ecological thinking, apply it to agricultural systems, and then to broaden the scope of agroecology to include all components of the food system.

## FOOD FOR THOUGHT

1. What kinds of changes need to be made in the design and management of agriculture so that we can come closer to farming in "nature's image"?
2. It seems that for modern agriculture to be sustainable, it has to solve the problem of how to return nutrients to the farms that they come from. What are some ways this might be done in your own community?
3. The concept of ecosystem stability is one that is currently under much discussion in ecology. Some ecologists claim that there is no such thing as stability in ecosystems, since change is constant and disturbance inevitable. Yet in agroecology, we strive for stability of agroecosystem structure and function. How is the concept of stability being applied differently in these different contexts?
4. As a consumer, how do your choices affect the global food system?

## INTERNET RESOURCES

Agroecology in Action
www.agroeco.org
A website dedicated to demonstrating the many and varied ways to apply agroecology, with special emphasis on issues in Latin America.

Center for Agroecology and Sustainable Food Systems
zzyx.ucsc.edu/casfs
The Center for Agroecology & Sustainable Food Systems is a research, education, and public service program at the University of California, Santa Cruz, dedicated to increasing ecological sustainability and social justice in the food and agriculture system.

Ecology and Society
www.ecologyandsociety.org
A journal of integrative science for resilience and sustainability

University of California Santa Cruz Agroecology Program.

www.agroecology.org
This website is an information resource for developing sustainable agroecosystems, emphasizing international training, research, and application of agroecological science to solving real world problems.

## RECOMMENDED READING

Altieri, M.A. 1995. *Agroecology: The Science of Sustainable Agriculture*. 2nd ed. Westview Press: Boulder, CO. A pioneering book on the foundations of agroecology, with emphasis on case studies and farming systems from around the world.

Carroll, C.R., J.H. Vandermeer, and P.M. Rosset. 1990. *Agroecology*. McGraw-Hill: New York. An edited overview of agroecology that introduces the reader to many of the main currents of thought in the field in an interdisciplinary context.

Cox, G.W. and M.D. Atkins. 1979. *Agricultural Ecology*. W. H. Freeman: San Francisco. A groundbreaking work that points out the ecological impacts of agriculture and the need for an ecological approach to solving the problems they create.

Daubenmire, R.F. 1974. *Plants and Environment*. 3rd ed. John Wiley and Sons: New York. The primary work in the area of autecology, emphasizing the relationship between an individual plant and the factors of the environment in which it must develop.

Gliessman, S.R. 1990. *Agroecology: Researching the Ecological Basis for Sustainable Agriculture*. Ecological Studies Series #78. Springer-Verlag: New York. A survey of research approaches in the search for the ecological foundations for sustainable agroecosystem design and management.

Gurevitch, J., S.M. Scheiner, and G.A. Fox. 2002. *The Ecology of Plants*. Sinauer Associates, Inc.: Sunderland, MA. A text focusing on the interactions between plants and their environments, over a range of scales.

Golley, F.B. 1993. *A History of the Ecosystem Concept in Ecology*. Yale University Press: New Haven, Connecticut. The essential review of how the ecosystem concept was developed and how it has been applied as a central concept in ecology.

Lowrance, R., B.R. Stinner, and G.J. House. 1984. *Agricultural Ecosystems: Unifying Concepts*. John Wiley and Sons: New York. A conceptual approach to applying ecological concepts to the study of agricultural systems.

Molles, M.C. 2004. *Ecology: Principles and Applications*. 3rd ed. McGraw-Hill. A good introductory ecology text.

Odum, E.P. 1997. *Ecology: A Bridge Between Science and Society*. Sinauer Associates: Sunderland, MA. An introduc- tory text covering the principles of modern ecology as they relate to the threat to Earth's life-support systems.

Pretty, J. (ed.) 2005. *The Earthscan Reader in Sustainable Agriculture*. Earthscan/James and James: London. Founded on ecological principles, this remarkable edited book brings together experts who present a vision for a sustainable agriculture.

Ricklefs, R.E. 2000. *The Economy of Nature*. 5th ed. W. H. Freeman and Company: New York. A very complete textbook of ecology for the student committed to understanding the way nature works.

Smith, R.L. T.M. and Smith. 2005. *Elements of Ecology*. 6th ed. Benjamin Cummings: San Francisco. A commonly used textbook of ecology for the serious student in biology or environmental studies.

# Section II

## Plants and Environmental Factors

Even in the simplest of agroecosystems, there are complex relationships among crop plants, non-crop plants, animals, and soil microorganisms, and between each of these types of organisms and the physical environment. Before attempting to understand all these relationships and interactions at their full level of complexity, it is helpful to study agroecosystems from a more focused perspective — that of the individual crop plant in relation to its environment. Such a perspective is termed *autecological*. We restrict our attention to plants because they, as producers, make up the foundation of all agroecosystems, even those incorporating animals.

Autecological study of agroecosystems begins by breaking down the environment into individual *factors* and exploring how each factor affects the crop plant. Consistent with this approach, the core chapters in this section are each devoted to a single environmental factor of importance in agroecosystems. Each chapter describes how its factor functions in time and space, and then gives examples of how farmers have learned to either accommodate their crops to this factor, or take advantage of it to improve the sustainability of the agroecosystem.

These chapters are preceded by a chapter that reviews the basic structure and function of the plant itself, providing a basis for understanding its responses. The section concludes with a chapter explaining how the separate factors must be viewed as parts of a whole dynamic system.

**A young corn plant emerging through the organic debris left after the burning of fallow second growth vegetation in Tabasco, Mexico.** This plant will respond in different ways to the environmental conditions and factors it encounters during its life cycle.

# 3 The Plant

The design and management of sustainable agroecosystems has important foundations in our understanding of how individual plants grow, develop, and eventually become the plant matter we use, consume, or feed to our animals. This chapter reviews some of the more important plant physiological processes that allow a plant to live, convert sunlight into chemical energy, and store that energy in parts of the plant and in forms we can use. The chapter also reviews some of the principal nutritional needs of plants. Finally, by way of introduction to the rest of the chapters in Section II, the chapter reviews some of the most important concepts and terms used to describe the ways individual plants respond and adapt to the range of environmental factors we will be examining.

## PLANT NUTRITION

Plants are autotrophic (self-nourishing) organisms by virtue of their ability to synthesize carbohydrates using only water, carbon dioxide, and energy from the sun. Photosynthesis, the process by which this energy capture takes place, is thus the foundation of plant nutrition. Yet manufacturing carbohydrates is just part of plant growth and development. An array of essential nutrients, along with water, are needed to form the complex carbohydrates, amino acids, and proteins that make up plant tissue and serve important functions in plants' life processes.

### PHOTOSYNTHESIS

Through the process of photosynthesis, plants convert solar energy into chemical energy stored in the chemical bonds of sugar molecules. Since this energy-trapping process is so important for plant growth and survival, and is what makes plants useful to humans as crops, it is important to understand how photosynthesis works.

The descriptions of the processes of photosynthesis that follow are very simplified. For our purposes, it is more important to understand the agroecological consequences of the different types of photosynthesis than to know their actual chemical pathways. However, if a more detailed explanation is desired, the reader is advised to consult a plant physiology text.

As a whole, the process of photosynthesis is the solar-energy-driven production of glucose from water and carbon dioxide, as summarized in this simple equation:

$$6CO_2 + 12H_2O + \text{light energy} \longrightarrow C_6H_{12}O_6 + 6O_2 + 6H_2O$$

Photosynthesis is actually made up of two distinct processes, each with multiple steps. These two processes, or stages, are called the *light reactions* and the *dark reactions*.

The light reactions function to convert light energy into chemical energy in the form of adenosine triphosphate (ATP) and a compound called oxidized from of nicotinamide adenine dinucleotide phosphate (NADPH). These reactions use water and give off oxygen. The dark reactions (which take place independently of light) take carbon atoms from carbon dioxide in the atmosphere and use them to form organic compounds; this process is called *carbon fixation* and is driven by the ATP and NADPH produced by the light reactions. The direct end product of photosynthesis, often called *photosynthate*, is made up mainly of the simple sugar glucose. Glucose serves as an energy source for growth and metabolism in both plants and animals, because it is readily converted back to chemical energy (ATP) and carbon dioxide by the process of respiration. Glucose is also the building block for many other organic compounds in plants. These compounds include cellulose, the plant's main structural material, and starch, a storage form of glucose (Figure 3.1).

From an agroecological perspective, it is important to understand how photosynthesis can be limited. Temperature and water availability are two important factors. If temperatures are too high or moisture stress too great during the day, the openings in the leaf surface through which carbon dioxide passes begin to close. As a result of the closing of these openings — called *stomata* — carbon dioxide becomes limiting, slowing down the photosynthetic process. When the internal concentration of $CO_2$ in the leaf goes below a critical limiting concentration, the plant reaches the so-called *$CO_2$ compensation point*, where photosynthesis equals respiration, yielding no net energy gain by the plant. To make matters worse, the closing of the stomates under water or heat stress also eliminates the leaf's evaporative cooling process and increases leaf $O_2$ concentration. These conditions stimulate the energetically wasteful process of *photorespiration*, in which $O_2$ is substituted for $CO_2$ in the dark reactions of photosynthesis, producing useless products that require further energy to metabolize.

Some kinds of plants have evolved different ways of fixing carbon that reduce photorespiration. Their alternate forms of carbon fixation constitute distinct photosynthetic pathways. Altogether, three types of photosynthesis are known to exist. Each has advantages under certain conditions and disadvantages in others.

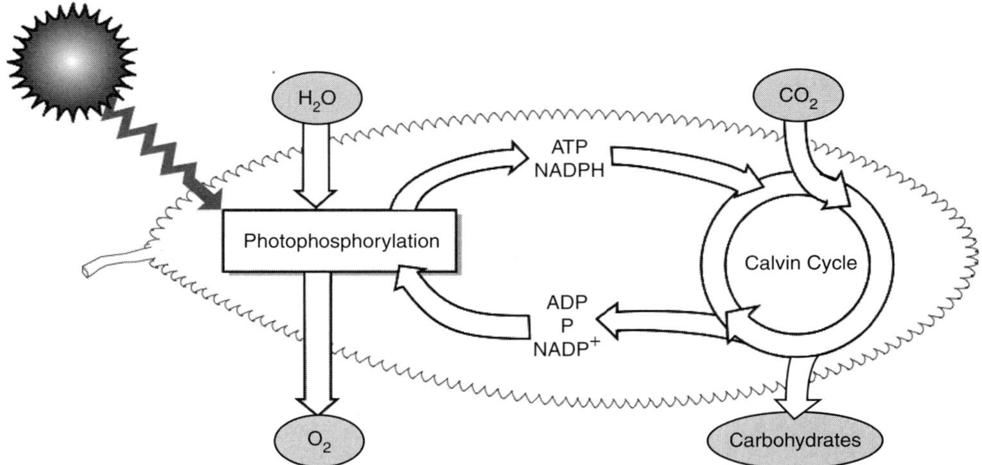


**FIGURE 3.1 Basic processes of photosynthesis.** Photophosphorylation is another name for what occurs during the light reactions; the Calvin Cycle is the basis of the dark reactions.

### C3 Photosynthesis

The most widespread type of photosynthesis is known as C3 photosynthesis. The name comes from the fact that the first stable compound formed in the dark reactions is a three-carbon compound. In plants that use this pathway, carbon dioxide is taken in during the day through open stomata and used in the dark reactions to form glucose.

C3 photosynthesis plants do well under relatively cool conditions, since their optimum temperature for photosynthesis is relatively low (Table 3.1). However, because their stomata must be open during the day to take in carbon dioxide, C3 plants are subject to photosynthetic limitation during times of heat or drought stress: the closure of the stomata to prevent moisture loss also limits the intake of carbon dioxide and increases photorespiration. Common crops that use C3 photosynthesis are beans, squash, and tomatoes.

**TABLE 3.1**
**Comparison of the Three Photosynthetic Pathways**

| | C3 | C4 | CAM |
|---|---|---|---|
| Light saturation point (foot-candles) | 3000–6000 | 8000–10,000 | ? |
| Optimum temperature (°C) | 15–30 | 30–45 | 30–35 |
| $CO_2$ compensation point (ppm of $CO_2$) | 30–70 | 0–10 | 0–4 |
| Maximum photosynthetic rate (mg $CO_2/dm^2/h$) | 15–35 | 30–45 | 3–13 |
| Maximum growth rate ($g/dm^2/d$) | 1 | 4 | 0.02 |
| Photorespiration | High | Low | Low |
| Stomata behavior | Open day, closed night | Open or closed day, closed night | Closed day, open night |

*Source:* Larcher, W. 1980. *Physiological Plant Ecology.* Springer-Verlag: New York; Larcher, W. 1995. *Physiological Plant Ecology.* 3rd ed. Springer: New York; Etherington, J. R. 1995. *Environment and Plant Ecology.* 3rd ed. John Wiley and Sons: New York.

### C4 Photosynthesis

A more recently discovered form of photosynthesis is known as the C4 type. In this system, $CO_2$ is incorporated into four-carbon compounds before it enters the dark reactions. This initial carbon-fixing process takes place in special chlorophyll-containing cells in the leaf. The four-carbon compound is transported to special cells known as bundle sheaths, clustered around veins in the leaves, where enzymes break loose the extra carbon as $CO_2$. The $CO_2$ is then used to form the three-carbon compounds used in the dark reactions, just as in C3 photosynthesis.

The C4 pathway allows carbon fixation to occur at much lower concentrations of $CO_2$ than does the C3 pathway. This enables photosynthesis to take place while the stomata are closed, with $CO_2$ liberated by internal respiration being captured rather than $CO_2$ from outside air. The C4 pathway also prevents photorespiration from occurring because it makes it much more difficult for $O_2$ to compete with $CO_2$ in the dark reactions. Thus, photosynthesis in C4 plants can occur under conditions of moisture and temperature stress, when photosynthesis in C3 plants would be limited. At the same time, C4 plants usually have a higher optimum temperature for photosynthesis.

C4 plants therefore use less moisture during times of high photosynthetic potential, and under warm and dry conditions have higher net photosynthesis and higher biomass accumulation than C3 plants (Table 3.2). C4 photosynthesis

**TABLE 3.2**
**Comparison of Net Photosynthetic Rates among C3 and C4 Plants**

| Crop Type | Net Photosynthetic Rate (mg $CO_2/dm^2$ leaf area/h)[a] |
|---|---|
| **C3 plants** | |
| Spinach | 16 |
| Tobacco | 16–21 |
| Wheat | 17–31 |
| Rice | 12–30 |
| Bean | 12–17 |
| **C4 plants** | |
| Corn | 46–63 |
| Sugar cane | 42–49 |
| Sorghum | 55 |
| Bermuda grass | 35–43 |
| Pigweed (*Amaranthus*) | 58 |

[a] Determined under high light intensity and warm temperatures (20–30°C).

*Source:* Zelitch, I. 1971. *Photosynthesis, Photorespiration, and Plant Productivity.* Academic Press: New York; Larcher, W. 1980. *Physiological Plant Ecology.* Springer-Verlag: New York.

involves an extra biochemical step, but under conditions of intense direct sunlight, warmer temperature, and moisture stress, it provides a distinct advantage.

Some well known crops that use C4 photosynthesis are corn, sorghum, and sugarcane. A lesser-known C4 crop is amaranth. C4 plants are more common in tropical areas, especially the drier tropics. Plants that originated in drier desert regions or grassland communities of warm temperate and tropical climates are more likely to be C4 plants.

### Crassulacean Acid Metabolism Photosynthesis

A third type of photosynthesis is called Crassulacean acid metabolism (CAM) photosynthesis. It is similar to C4 photosynthesis. During the night, while the stomata can be open without causing the loss of undue amounts of moisture, carbon dioxide is taken in and the four-carbon compound malate is formed and stored in cellular organelles called vacuoles. The stored malate then serves as a source of $CO_2$ during the day to supply the dark reactions. Plants using CAM photosynthesis can keep their stomata closed during the day, taking in all the $CO_2$ they need during the night. As would be expected, CAM plants are common in hot and dry environments, such as deserts; they include many succulents and cactus. Bromeliads that live as epiphytes (plants attached to other plants and not rooted in soil) are also CAM plants; their habitat in the canopy of rainforests is much drier than the rest of the rainforest community. An important crop plant using CAM photosynthesis is pineapple, a member of the Bromeliaceae.

### The Photosynthetic Pathways Compared

A comparison of the different photosynthetic pathways is presented in Table 3.1. The different arrangements of chloroplasts within the leaves of each type are correlated with different responses to light, temperature, and water. C3 plants tend to have their peak rate of photosynthesis at moderate light intensities and temperatures, while actually being inhibited by excess light exposure and high temperatures. C4 plants are better adapted to high light and temperature conditions, and with the ability to close stomata during daylight hours in response to high temperature and evaporative stress, they can use water more efficiently under these conditions. CAM plants can withstand the most consistently hot and dry conditions, keeping stomata closed during daylight hours, but they sacrifice growth and photosynthetic rates in exchange for tolerance of extreme conditions.

Despite the greater photosynthetic efficiency of C4 plants, C3 plants such as rice and wheat are responsible for the great bulk of world food production. The superiority of C4 photosynthesis makes a difference only when the ability of the crop to convert light into biomass is the sole limiting factor, a situation that seldom occurs in the field.

## Carbon Partitioning

The carbon compounds produced by photosynthesis play critical roles in plant growth and respiration because of their dual role as an energy source and as carbon skeletons for building other organic compounds. How a plant distributes the carbon compounds derived from photosynthesis and allocates them to different physiological processes and plant parts is described by the term *carbon partitioning*. Since we grow crops for their ability to produce harvestable biomass, carbon partitioning is of considerable agricultural interest.

Although photosynthesis has an efficiency of energy capture of about 20%, the process of converting photosynthate into biomass has an efficiency that rarely exceeds 2%. This efficiency is low mainly because internal respiration (oxidation of photosynthate for cell maintenance) uses up much of the photosynthate and because photorespiration limits photosynthetic output when photosynthetic potential is highest. Much research aimed at improving crop yield has focused on increasing the efficiency of photosynthetic carbon fixation, but this work has been wholly unsuccessful. The primary basis for increasing crop yield through plant breeding, both traditional and modern, has been the enhancement of harvested biomass relative to total plant biomass.

Since the ability of plants to create biomass is limited, how they partition the fixed carbon they do create is of paramount importance in agriculture. Humans select plants that shunt more photosynthate to the part of the crop that is to be harvested, at the expense of other plant parts.

The harvestable or harvested portion of most crop plants usually has limited photosynthetic capacity itself, hence yields depend a great deal on carbohydrate that is transported through phloem cells from photosynthetically active parts of the plants to the harvestable parts (Figure 3.2).

In ecological terms, we often refer to carbon partitioning as a "source, path, and sink" phenomena. The source is usually the leaf, the chloroplasts in particular. Much detailed research has been done on the physiology and biochemistry of the actual transfer of carbon out of the chloroplast and into transport paths. A complex set of chemical locators and enzymes are active in this process. Once in the phloem, carbon then moves through the stem to grain, flowers, fruits, tubers, or other parts, which are the sinks. At this point there is phloem "unloading" and sink uptake. The actual transfer from vascular strands to sink tissue is often based on a sugar concentration gradient.

The products of photosynthesis are compounds of carbon, oxygen, and hydrogen that make up an average of 90% of plant dry matter. Therefore, there is a close relationship between whole plant photosynthesis and whole plant productivity. Overall photosynthetic rates are related to rates per unit leaf area, as well as to the production of new leaf area, but they are also dependent on the rate of transfer from source to sink. Carbon is kept in the area of leaf development while new leaves are forming; only after all leaves are formed can the transfer to other sinks take place. After the canopy closes, crop photosynthesis and growth depend mainly on net $CO_2$ fixation per unit leaf area.

Over the growing season, the various sinks of the plant compete with each other for the supply of fixed carbon produced by the leaves, with the result that some parts of the plant accumulate more biomass than others. The mechanisms regulating this partitioning of photosynthate within the plant are not well understood, though it is clear that the process is dynamic and related to both environmental conditions and the genetically determined developmental patterns of the plant. Ways of modifying carbon partitioning in crop plants are being explored by researchers; one example involves the development of perennial grain crops, where the challenge is to balance the partitioning of carbon between the vegetative body of the perennial plant (especially the roots and stems) and the grain.

## Nutritional Needs

Photosynthesis provides a plant with a large portion of its nutritional needs — energy, and carbon and oxygen for building important structural and functional compounds. Together with hydrogen — derived from the water that enters plant roots as a result of transpiration — carbon and oxygen make up approximately 95% of the average plant's fresh weight.

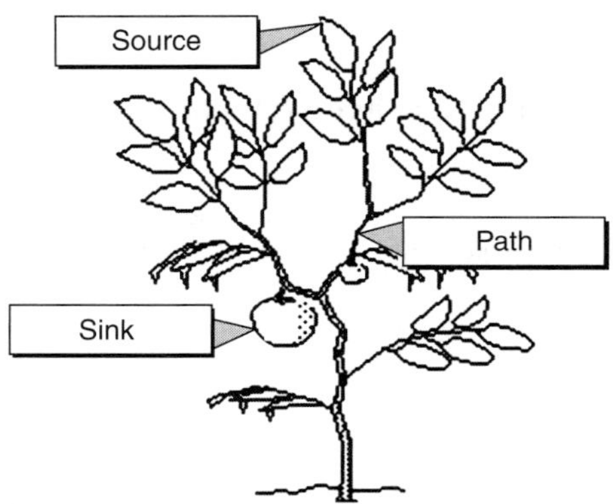


**FIGURE 3.2 Carbon partitioning.**

The elements that make up the other 5% of living plant matter must come from somewhere else — namely the soil. These elements are plants' essential nutrients. They are needed to form the structures of the plant, the nucleic acids directing various plant processes, and the enzymes and catalysts regulating plant metabolism. They also help maintain internal osmotic balance and have a role in the absorption of ions from the soil solution. If an essential nutrient is not available in adequate supply, the plant suffers and does not develop properly. In agriculture we have learned how to adjust the supply of these nutrients in the soil to meet the needs of our crops.

The three nutrients that are required in relatively large amounts, and have played such important roles as inorganic fertilizers in agriculture, are nitrogen, phosphorus, and potassium. These are classified as macronutrients. Plants vary in the actual amounts of these nutrients they require. Since each plant variety has become adapted to different habitats with different environmental conditions, it makes sense for there to be such variation in nutrient requirements. A review of some of this nutritional variation can tell us a lot about proper crop selection and fertility management.

### Nitrogen

Nitrogen is needed in large amounts by plants, but at the same time is the most universally deficient nutrient. It occurs in every amino acid, and as a result is a major component of proteins. Nitrogen is therefore involved in some way with up to 50% of dry plant biomass. It is required in enzyme synthesis, with a deficiency affecting almost every enzymatic reaction. Since nitrogen forms part of chlorophyll and is required in its synthesis, it is no wonder that nitrogen-deficient plants show the yellowing that is indicative of limiting amounts of this nutrient in the soil. Adequate supplies of nitrogen are also needed for normal flowering and fruit set in all plant species. Plants commonly have 1 to 2% nitrogen as a proportion of dry weight, but contents above 5% are not uncommon.

Except for nitrogen that is captured directly from the air by symbiotic microorganisms that live in the roots of most members of the Fabaceae and a few other plant families and passed on to the host plants in an available form, most plants obtain their nitrogen from ion exchange with the soil solution as $NO_3^-$ or from $NH_4^+$ adsorbed to humus or clay minerals. Available forms of nitrogen in the soil are generally kept at low levels by rapid uptake of nitrogen when it is available coupled with nitrogen's high potential for leaching loss with rainfall or irrigation percolation.

### Phosphorus

Phosphorus is an important component of nucleic acids, nucleoproteins, phytin, phospholipids, ATP, and several other types of phosphorylated compounds including some sugars. Phosphorus is built into the DNA of chromosomes and the RNA of the nucleus and ribosomes. Cell membranes depend on phospholipids for the regulation of movement of materials in and out of the cells and organelles. Phosphorus in the form of phosphates ($PO_4$+) occurs in certain enzymes that catalyze metabolic reactions. Sugar metabolism in plants, for example, depends on phosphoglucomutase. Phosphorus also occurs in primary cell walls in the form of enzymes that affect cell wall permeability. The initial reactions of photosynthesis also involve phosphorus; it is found in the five-carbon sugar with which $CO_2$ initially reacts.

Phosphorus is absorbed as phosphates from the soil solution through plant roots. Phosphates in solution are readily available and taken up by plants, but except in soils that are derived from parent materials high in phosphorus or where phosphorus levels have built up over time in response to many years of fertilization, available phosphorus in most soils is quite low. Plants will opportunistically take up large amounts of this nutrient when it is available, accumulating about 0.25% of dry weight, but are quick to show signs of deficiency when it is lacking. Leaves take on a bluish cast or remain dark green, and purple pigments (anthocyanins) become prominent on the underside of the leaves and along the veins or near the leaf tip. Root and fruit development are severely restricted when phosphorus is limiting.

### Potassium

Potassium is not a structural component of the plant, nor a component in enzymes or proteins. Its function seems to be primarily regulatory: it is involved, for example, in osmoregulation (stomatal movement) and as a cofactor for many enzyme systems. We know a lot about where potassium occurs in the plant, but much less about what it actually does. Most metabolic processes that have been studied are affected by potassium. In protein metabolism, for example, it appears that potassium activates certain enzymes that are responsible for peptide bond synthesis and the incorporation of amino acids into protein. Potassium needs to be present for the formation of starches and sugars, as well as for their later transport throughout the plant. This nutrient has been shown to be needed for cell division and growth, and is in some way linked to cell permeability and hydration. Plants show better resistance to disease and environmental stress when potassium supplies are adequate.

Plants obtain potassium in the form of the cation $K^+$, taking it in through the roots as exchangeable ions from adsorption sites in the soil matrix or from a dissolved form in the soil solution. When potassium is deficient, plants primarily show disruptions in water balance; these include drying tips or curled leaf edges, and sometimes a higher predominance of root rot. Potassium is usually quite abundant in soils, with plant tissues being made up of 1 to 2% potassium by dry weight under optimum conditions, but excessive removal through harvest or soil leaching can lead to potassium deficiency.

### Other Macronutrients

Three other nutrients — calcium (Ca), magnesium (Mg), and sulfur (S) — are also considered to be macronutrients, but this classification is more a function of the relatively high levels in which they accumulate in plant tissue and less because of their importance in different plant structures or processes. This is not to say that they do not play valuable roles, because when any of these nutrients are deficient in the soil, plant development suffers and symptoms of deficiency show up quickly. Calcium and magnesium are readily absorbed by plant roots through cation exchange (as $Ca^{2+}$ and $Mg^{2+}$), but sulfur is taken up sparingly as an anion ($SO_4^{2-}$) from organically bound sites in the soil or upon dissociation of sulfates of Ca, Mg, or Na.

### Micronutrients

Iron (Fe), copper (Cu), zinc (Zn), manganese (Mn), molybdenum (Mo), boron (B), and chlorine (Cl) make up what are called the micronutrients or the trace elements. Each one plays some vital role in plants, but usually in extremely small quantities. In fact, most of these elements are toxic to plants when they occur in the soil in large quantities. All are taken up from the soil solution through ion exchange at the root surface.

The role that each of the micronutrients plays in plants' life processes is outlined in Table 3.3. As one would imagine, any of the important physiological processes listed could be inhibited or altered by a deficiency of the micronutrient concerned. Many inorganic fertilizers carry small quantities of these elements as contaminants, and mixtures of trace elements are now commonly added

**TABLE 3.3**
**Micronutrients and the Processes in Which They are Involved**

| Nutrient | Processes |
|---|---|
| Boron (B) | Carbohydrate transport and metabolism, phenol metabolism, activation of growth regulators |
| Chlorine (Cl) | Cell hydration, activation of enzymes in photosynthesis |
| Copper (Cu) | Basal metabolism, nitrogen metabolism, secondary metabolism |
| Iron (Fe) | Chlorophyll synthesis, enzymes for electron transport |
| Manganese (Mn) | Basal metabolism, stabilization of chloroplast structure, nitrogen metabolism |
| Molybdenum (Mo) | Nitrogen fixation, phosphorus metabolism, iron absorption and translocation |
| Zinc (Zn) | Chlorophyll formation, enzyme activation, basal metabolism, protein breakdown, hormone biosynthesis |

*Source:* Adapted from Treschow, M. 1970. *Environment and Plant Response.* McGraw-Hill: New York.

to soils that have undergone a long period of conventional management. Organic fertilizers, especially those made from composted plant material and manure, are rich in micronutrients.

#### Transpiration

All of a plant's life processes, including photosynthesis, carbon partitioning, and metabolism, are dependent on the continual flow of water from the soil to the atmosphere along a pathway that extends from the soil, into the roots, up the stem to the leaves, and out of the leaves through the stomata. This flow process is called *transpiration.*

Water loss from the leaves creates a concentration gradient, or a lower leaf water potential, that then through capillarity moves more water into the plant and to the leaves to replace the loss. The actual amount of water that is chemically bound in plant tissues or that is actively involved in processes such as photosynthesis is very small in proportion to the transpirational loss of water on a daily basis. Water movement through plants is very important in nutrient cycles and under conditions of limited water availability in the soil, as we will see in later chapters.

## THE PLANT IN ITS INTERACTION WITH THE ENVIRONMENT

Each of the physiological processes described above allow the plant to respond to and survive in the environment in which it lives. An understanding of the ways individual plants and their physiology are impacted by different factors of the environment is an essential component in the design and management of sustainable cropping systems.

The ecological study of individual plant response to the diverse factors of the environment — termed autecology or physiological ecology in the pure sense and crop ecology in the applied sense — is therefore a foundation of agroecological understanding. Some of the conceptual basis of autecology is reviewed in the next section. Each factor of the environment and its effects on crop plants is then explored in a separate chapter in preparation for expanding our view to the agroecosystem level.

#### A Plant's Place in the Environment

Each species occupies a particular place in the ecosystem, known as the *habitat*, that is characterized by a particular set of environmental conditions that includes the interaction of the species with the other species in the habitat. Within its habitat, the species carries out a particular ecological role or function, known as the *ecological niche* of that species. For example, redwoods (*Sequoia sempervirens*) occupy a specific habitat on the north coast of California characterized by a moderating maritime climate and the occurrence of summer fog that compensates for a lack of rainfall during this time. At the same time, redwoods occupy the ecological niche of autotrophic producers capable of modifying the microclimate under their emergent canopies and being the dominant species in their community.

#### Responses to Factors of the Environment

Every plant during its lifespan goes through distinct stages of development, including germination of the seed, initial establishment, growth, flowering, and dispersal of seed. Each of these stages involves some kind of physiological change, or *response*, in the plant. Most plant responses are tied directly to environmental conditions.

### Triggered Responses

Many plant responses are triggered by some external stimulus. They come about as a result of a certain condition, but that external condition does not have to be maintained in order for the response to continue. For example, tobacco seed requires exposure to light in order to germinate, but that exposure need only last for a fraction of a second. After a brief exposure to light, the seed will germinate even if it is planted in total darkness.

### Dependent Responses

Some plant responses depend on the continued presence of a particular external condition. The response is both induced and maintained by the condition. The production of leaves on the spiny stems of ocotillo (*Fouquieria splendens*) in the Sonoran desert is an example of this type of response. Within a day or two after significant rainfall, leaves appear on the stems; as long as moisture levels are sufficient in the soil, the leaves are retained, but immediately upon reaching the wilting point the leaves are dropped.

### Independent Responses

Finally, certain responses in plants occur regardless of conditions in the immediate environment and are the result of some internally controlled, physiologically determined set of factors. For example, a corn plant begins to flower because a particular stage in growth and development has been achieved. External conditions may force later or earlier flowering by affecting growth, but the actual shift to flowering is internally controlled.

## Limits and Tolerances

The ability of an individual species to occupy its particular habitat is the result of a set of adaptations that have evolved over time for that species. These adaptations allow the plant to cope with certain levels of moisture availability, temperature, light, wind, and other conditions. For each of the factors that delimit the habitat for the species, there is a maximum level of tolerance and a minimum level of tolerance beyond which that species cannot cope. Between these two extremes there is an optimum at which the species performs or functions the best. For example, the tropical plant banana has a mean monthly temperature optimum of 27°C; above 50°C banana trees suffer sunscorch and stop growing; below 21°C growth is checked by reduction in leaf production and delayed shooting of the bunches.

A particular species' range of tolerance limits and optimum for a factor of the environment is ultimately the result of how that factor affects each of the physiological processes of the plant. A species' tolerance of a range of temperatures, for example, is linked to how temperature affects photosynthesis, transpiration, and other physiological processes of the plant. When all of the abiotic and biotic factors of the environment are entered into the tolerance equation, the full range of a species' adaptability becomes apparent. An individual's habitat and niche become fully integrated (Figure 3.3).

A species with a broad set of tolerances of environmental conditions (known as a *generalist*) and a broad ability to interact with other species (often referred to as a species with a broad niche or the capability of considerable niche overlap) will be more common over a larger area. In contrast, a species with a narrow set of tolerances and a very specialized niche (a *specialist*), will be less common over larger areas and only seen as common at a very localized level. Redwood sorrel (*Oxalis oregana*), an ecological specialist, can form dense stands in which it is the locally dominant plant, but it is restricted to the specific conditions encountered in the partially shaded understory of a redwood forest. If the shade is too dense, photosynthetic activity is not great enough to meet the plant's respiratory needs, and if the sun is too intense, sorrel is unable to tolerate the desiccating effects of direct solar radiation. Sorrel's optimum level of light is intermediate to these two extremes.

In summary, each individual plant species occurs in a particular habitat as a result of the development over time of a particular set of adaptive responses to the environment in which it lives. The species' limits of tolerance restrict individuals of that species to a particular habitat,

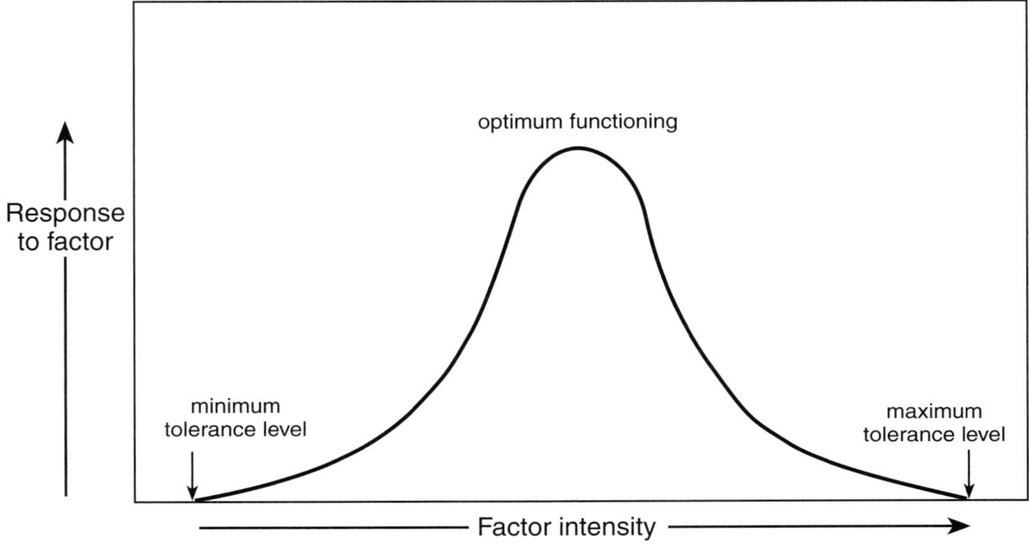


**FIGURE 3.3 A plant's range of tolerance for an environmental factor.**

within which interactions with other species occur. This is the case in both agroecosystems and natural ecosystems. How each plant in an agroecosystem performs will depend on how each factor of the environment impacts it. We will explore these factors in detail in the following chapters.

## FOOD FOR THOUGHT

1. How might the different forms of photosynthesis that occur in plants have come about? What specific conditions of the environment would select for each type and how might we use this knowledge in agriculture?
2. What would you consider to be "balanced plant nutrition" and how would you try to maintain it in an agroecosystem setting?
3. Why does a plant partition carbon to different parts of the plant structure?
4. How many factors need to be included to be able to thoroughly understand the full range of conditions that determine an individual plant's habitat?
5. How is plant nutrition affected by the shift from easily soluble synthetic fertilizers to more complex organic soil amendments, as commonly applied in organic farming systems?

## RECOMMENDED READING

Campbell, Neil and J.B. Reece. 2001. *Biology.* 6th ed. Menlo Park: Benjamin Cummings. One of the most complete and best-respected textbooks on general biology.

Epstein, E. and A.J. Bloom. 2004. *Mineral Nutrition of Plants: Principles and Perspectives.* 2nd ed. Sunderland, MA: Sinauer Associates. A detailed work on the important field of plant nutrition. The authors trace the movement of nutrients from soil to roots and throughout the plant, providing details on physiology and metabolism.

Hall, A.E. 2001. *Crop Responses to Environment.* Boca Raton, FL: CRC Press. Presents the principles, theories, and experimental observations concerning plant responses to the environment, with specific reference to crop cultivars and management.

Lambers, H., F.S. III Chapin, and T.L. Pons. 1998. *Plant Physiological Ecology.* New York: Springer-Verlag. An in-depth analysis of the mechanisms underlying plant physiological ecology, including biochemistry, biophysics, molecular biology, and whole-plant physiology.

Larcher, W. 2003. *Physiological Plant Ecology.* 3rd ed. Berlin: Springer-Verlag. A well-known textbook that focuses on the science of plant function in interaction with the environment.

Loomis, R.S. and D.J. Connor. 1992. *Crop Ecology: Productivity and Management in Agricultural Systems.* Cambridge: Cambridge University Press. A textbook that emphasizes physiological ecology and how to adjust the crop environment to meet the needs of the crop plant.

Taiz, L. and E. Zeiger. 2002. *Plant Physiology.* 3rd ed. Sunderland, MA: Sinauer Associates. A very thorough review of the field of plant physiology; it balances chemical and molecular specificity with the broader ecological applications.

Wilkinson, R.E. (ed.) 2000. *Plant–Environment Interaction.* 2nd ed. New York: Marcel Dekker. A comprehensive presentation of plant responses to changing environments, with a focus on how stress factors influence plant survival.

# 4 Light

Light from the sun is the primary source of energy for ecosystems. It is captured by plants through photosynthesis and its energy is stored in the chemical bonds of organic compounds. Sunlight also drives the earth's weather: light energy transformed into heat affects rainfall patterns, surface temperature, wind, and humidity. The way these factors of the environment are distributed over the face of the earth determines climate and is of considerable importance in agriculture. All these light-related factors will be reviewed in more detail in subsequent chapters.

This chapter focuses on the light environment as it directly affects agroecosystems. The light environment includes that portion of the electromagnetic spectrum from the invisible ultraviolet (UV), through the visible light spectrum to the invisible infrared (IR). This chapter also discusses how the light environment can be managed to more efficiently channel this renewable source of energy through the system, use it to maintain the many and diverse functions of the system, and ultimately convert part of it into sustainable harvests.

## SOLAR RADIATION

The energy the earth receives from the sun arrives in the form of electromagnetic waves varying in length from less than 0.001 nanometers (nm) to more than 1000 million nm. This energy makes up what is known as the electromagnetic spectrum. The portion of the electromagnetic spectrum between about 1 and 1 million nm is considered to be light, although not all of it is visible. Light with a wavelength between 1 and 390 nm is UV light. Visible light is the next component, made up of light with wavelengths between 400 and 760 nm. Light with a wavelength longer than 760 nm and shorter than 1,000,00 nm is known as IR light, and like UV light is invisible to the eye; when the wavelength of IR light extends beyond 3000 nm, however, it is sensed as heat. Figure 4.1 shows how the electromagnetic spectrum is divided into types of energy.

## THE ATMOSPHERE AS FILTER AND REFLECTOR

When light first arrives from the sun at the outer edge of the earth's atmosphere, it is comprised of approximately 10% UV light, 50% visible light, and 40% IR light or heat energy. As this light interacts with the earth's atmosphere, several things can happen to it, as shown in Figure 4.2.

Some light is *dispersed* or scattered — its path toward the surface is altered due to interference from molecules in the atmosphere, but its wavelength is not changed in the process. Most dispersed light reaches the surface, but in the process gives the atmosphere its unique blue color. Some light is *reflected* off the atmosphere back out into space; its wavelength is also unchanged in the process. Finally, some light is *absorbed* by water, dust, smoke, ozone, carbon dioxide, or other gases in the atmosphere. The absorbed energy is stored for a period of time, and then *reradiated* as longer-wave heat energy. Almost all UV light with a wavelength of 300 nm or less is absorbed by the earth's atmosphere before it strikes the surface. (UV light with a wavelength below 200 nm is potentially lethal to living organisms.) The light that is not reflected off the atmosphere or absorbed is *transmitted* and reaches the surface. This energy is mostly visible light, but also includes some UV light and IR light.

At the earth's surface, this transmitted light is absorbed by soil, water, or organisms. Some of the absorbed energy is reflected back into the atmosphere, and some is reradiated as heat. It is the absorption of visible light by plants and its role in photosynthesis that concerns us in this chapter.

## THE ECOLOGICAL SIGNIFICANCE OF LIGHT ON EARTH

All wavelengths of light that reach the earth's surface have significance for the living organisms that occupy the planet. Over evolutionary time, organisms have developed different adaptations for accommodating themselves to the various spectra. These adaptations vary from active energy capture to deliberate avoidance of solar energy exposure.

### UV LIGHT

Despite the fact that UV light cannot be seen, it can be very active in certain chemical reactions in plants. Together with the shorter wavelengths of visible light, UV tends to promote the formation of plant pigments known as anthocyanins, and can be involved in the inactivation of certain hormonal systems important for stem elongation and phototropism.

## OZONE DEPLETION

Only about 1% of the UV light entering the earth's outer atmosphere actually reaches the surface. The rest is absorbed by a layer of ozone gas high in the atmosphere. Organisms today are completely dependent on the screening effect of the ozone, because most have no means for protecting themselves against the harmful effects of UV, which include burning, cancer, and lethal mutations.

When ultraviolet light strikes an ozone molecule ($O_3$), the ozone is split apart and the energy of the UV light is absorbed. An oxygen molecule ($O_2$) and a free oxygen atom, called a free radical, are created. The oxygen free radical is extremely reactive, however, and readily combines with an oxygen molecule to reform a molecule of ozone. When this reaction occurs, energy is released in the form of heat. Thus, absorption of UV light in the ozone layer involves the continual destruction and creation of ozone, and the transformation of UV light into heat energy (infrared light). There are enough ozone molecules in the ozone layer to intercept nearly all the UV light passing through it.

The ozone layer lies in the outer stratosphere, beginning about 20 km above sea level and extending for another 30 km out toward space. The stratosphere is well above the thick, turbulent region of the atmosphere responsible for our weather, far removed from most human activities and surface sources of pollution. Nevertheless, human beings do have an effect on the ozone.

For many decades, we have produced artificial gases called chlorofluorocarbons (CFCs) to use as coolants for refrigerators and air conditioners, as spray can propellants, and for making plastic foam. These gases have been freely released into the atmosphere, and they have leaked from cooling systems. Once they enter the atmosphere, they slowly migrate into the stratosphere.

In the stratosphere, UV light bombards the CFC molecules, eventually breaking a chlorine atom off each one in the form of a chlorine free radical. The chlorine free radicals formed through this process of photodissociation attack and destroy ozone molecules, forming chlorine oxide (ClO) and molecular oxygen ($O_2$).

$$Cl^- + O_3 \rightarrow ClO + O_2$$

The chlorine oxide thus formed has the ability to react with and destroy ozone as well.

$$ClO + O_3 \rightarrow ClO_2 + O_2$$

Worse, each chlorine oxide molecule can also react with one of the oxygen free radicals constantly being generated by the absorption of UV by ozone, preventing the oxygen free radical from reforming ozone, and regenerating the chlorine free radical!

$$ClO + O^- \rightarrow Cl^- + O_2$$

Because the chlorine free radical can be regenerated, a single one, according to estimates, can destroy up to 100,000 ozone molecules before it reacts with an ozone molecule to form the relatively inactive chlorine dioxide ($ClO_2$).

CFCs are not the only ozone-destroying compounds that humans release into the atmosphere. Besides other chlorine-containing compounds such as carbon tetrachloride and methyl chloroform, there are bromine-containing compounds such as methyl bromide, an agricultural chemical used to fumigate and sterilize soil before the planting of certain crops, such as strawberries. All of these substances, collectively called halocarbons, affect ozone in much the same way, although bromine free radicals are even more reactive than chlorine, and so more destructive.

In 1987, the Montreal Protocol on Substances that Deplete the Ozone Layer, now signed and ratified by nearly all countries, called for most ozone-depleting chemicals to be phased out by 2000. Most countries have adhered to its provisions, and the production and release of ozone-depleting chemicals has decreased. As a result, increases in the amounts of these chemicals in the stratosphere appear to have leveled off. Nevertheless, significant depletion of atmospheric ozone continues to be observed every year, and the latest measurements show a 6 to 14% increase in UV irradiation since the early 1980s at a variety of sites around the world (WMO, 2003).

Atmospheric scientists have been measuring the ozone layer since the 1970s. Although the concentration of ozone in the stratosphere varies naturally from year to year, a marked seasonal depletion has been observed since at least 1984, when a summer "hole" in the ozone layer over Antarctica was first detected.

It is difficult to predict how much the ozone layer will be depleted in the future. The World Meteorological Organization predicts that if the Montreal Protocol is adhered to fully, the ozone layer could be restored to its pre-1980 levels by the middle of the 21st Century (WMO, 2003). In the meantime, enhanced UV irradiation will be a fact of life, and no one knows for sure that UV irradiation will not continue to increase.

There is disagreement in the scientific community about the consequences of increased UV exposure. Certainly, even slight increases in exposure to UV can be harmful to plants. Too much UV can damage leaf cells, inhibit photosynthesis and growth, and promote mutations. Different crop plants have varying levels of sensitivity to increased UV exposure, but even if only a few crops suffer decreases in yield, the effect on worldwide food production could be dramatic. In addition, if UV radiation increases significantly, agriculture all over the world could be threatened, in addition to natural terrestrial and marine ecosystems.

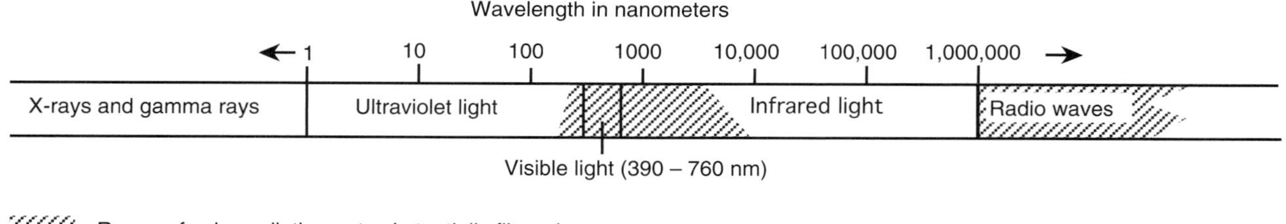


**FIGURE 4.1 The electromagnetic spectrum.** The sun emits the full spectrum of electromagnetic energy, but the atmosphere reflects and filters out most of the shortwave radiation, much of the infrared, and the longest wavelength radio waves. A relatively narrow band of energy centered on the visible light spectrum reaches the earth's surface mostly unimpeded.

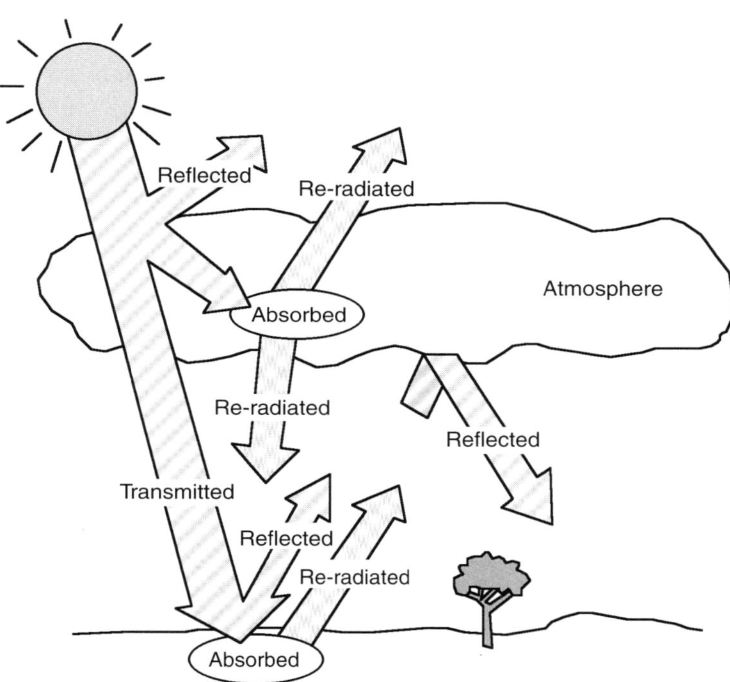


**FIGURE 4.2 The fate of light upon reaching the earth.** Transmitted light from the sun is mostly in the visible-light range; reradiated energy is mostly in the infrared range.

In general, however, because UV radiation can be harmful to plant tissues and because the overall level of UV energy reaching the surface is greatly reduced, plants have not developed many adaptations for its use. Instead UV radiation is mostly avoided: the opaque epidermis of most plants keeps most harmful UV radiation from entering sensitive tissue or cells. Reduction of the ozone layer of the upper atmosphere is cause for concern because of the potentially negative effects that excess UV radiation can cause in both plants and animals.

### Photosynthetically Active Radiation

The light energy in the visible spectrum is of greatest importance in agroecosystems. Depending on local climatic conditions, it forms 40 to 60% of the total energy of solar radiation reaching the earth's surface. Also known as photosynthetically active radiation (PAR), this is the light with wavelengths between 400 and 760 nm. Green plants will not grow without a combination of most of the wavelengths of light in the visible spectrum.

Not all the light in this spectrum is of equal value in photosynthesis, however. The photoreceptors in chlorophyll are most absorptive of violet-blue and orange-red light; green and yellow light is not as useful. Since chlorophyll cannot absorb green light very well, most of it is reflected back, making plants appear green. Figure 4.3 shows how the absorbance of chlorophyll varies with wavelength. The wavelengths of light that chlorophyll absorbs best correspond roughly to the wavelengths at which photosynthesis is most efficient.

### Infrared Light

Infrared light energy with a wavelength from 800 to 3000 nm sometimes referred to as the near IR range — has an important role in influencing the hormones involved in germination, a plant's responses to changes to daylength, and other plant processes. In the range beyond 3000 nm, IR light becomes heat, and different ecological impacts are evident. (Temperature as an ecological factor is discussed in the next chapter.)

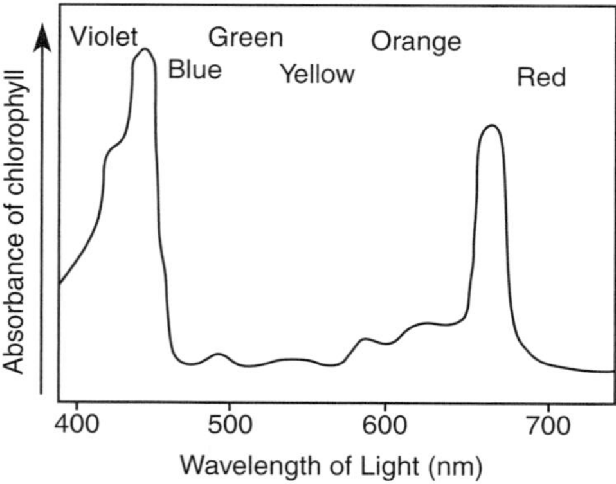


**FIGURE 4.3 Absorbance of chlorophyll in relation to the wavelength of light.** Chlorophyll absorbs mostly violet-blue and orange-red light; thus leaves reflect green and yellow light.

## CHARACTERISTICS OF VISIBLE LIGHT EXPOSURE

Light energy in the visible or PAR range is converted by photosynthesis into chemical energy, and eventually into the biomass that drives the rest of the agroecosystem, including the part we harvest for our own use. To increase the efficiency of this process, it is important to understand how the light to which plants are exposed can vary.

### Quality

Visible light can vary in the relative amounts of the colors that make it up — this is referred to as the light's quality. The largest proportion of direct sunlight at the earth's surface is at the center of the visible-light spectrum, dropping off slightly at both the violet and red ends. The diffuse light from the sky — what occurs in the shade of a building — is relatively higher in blue and violet light. Since different portions of the visible light spectrum can be used for photosynthesis more efficiently than others, light quality can have an important effect on photosynthetic efficiency.

A number of factors can cause light quality to vary. In the interior of some cropping systems, for example, canopy species remove most of the red and blue light, leaving primarily transmitted green and far red light. Light quality can therefore become a limiting factor for plants under the canopy, even though the total amount of light may appear to be adequate.

### Intensity

The total energy content of all the light in the PAR range that reaches a leaf surface is the intensity of that light. Light intensity can be expressed in a variety of energy units, but the most common are the langley (calories per $cm^2$), the watt (Joules per second), and the Einstein ($6 \times 10^{23}$ photons). All of these units of measure express the amount of energy falling on a surface over some time period. At very high light intensities, photosynthetic pigments become saturated, meaning that additional light does not effectively increase the rate of photosynthesis. This level of light intensity is called the *saturation point*. Excessive light can lead to degradation of chlorophyll pigments and even cause harm to plant tissue. At the other extreme, low levels of light can bring a plant to the *light compensation point*, or the level of light intensity where the amount of photosynthate produced is equal to the amount needed for respiration. When the light intensity goes below the compensation point, the energy balance for the plant is negative. If the negative balance is not offset by a time period of active photosynthesis and energy gain, the plant may die.

### Duration

The length of time that leaf surfaces are exposed to sunlight each day can impact photosynthetic rates as well as longer-term plant growth and development. Duration of light exposure is also an important variable in how light intensity or quality can affect a plant. Exposure to excessive levels of light for a short time, for example, can be tolerated, whereas a longer period of exposure can be damaging. Or a short period of intensive light, allowing the plant to produce an excess of photosynthate, can then allow for tolerance of a longer period below the light compensation point.

The total number of hours of daylight — the *photoperiod* — is also an important aspect of the duration of light exposure. A variety of plant responses, as will be discussed in detail below, have specific chemical triggers or control mechanisms that can be activated or deactivated depending on the number of hours of daylight, or in some cases, the number of dark hours without sunlight.

## DETERMINANTS OF VARIATIONS IN THE LIGHT ENVIRONMENT

The quality and quantity of light received by a plant in a specific location and the duration of its exposure to light are a function of several important factors including (1) seasonality, (2) latitude, (3) altitude, (4) topography, (5) air quality, and (6) the structure of the vegetation canopy.

### Seasonality

Except at the equator, daylight hours are longest during the summer and shortest in the winter, reaching their extremes at the corresponding solstice. Since the angle of the sun in relation to the surface is much lower toward the poles during the winter, the sunlight that is available has to pass through more atmospheres before it reaches the plant, making that sunlight much less intense. Therefore, both intensity and duration of light are affected by seasonality. Many plants have adapted to the seasonal variations in day length and light intensity through the selection of adaptations that either prepare the plant for the upcoming winter or get it ready to take advantage of more optimal conditions for growth and development as spring progresses into summer. The timing of many agricultural activities — planting and pruning — correspond to the changing hours of daylight at specific times of the year.

### Latitude

The closer to either of the poles, the greater the seasonal variation in daylength. Above the arctic circle, 24-h periods of daylight in the summer are balanced by 24-h periods of night in the winter. Near the equator, the constancy of 12-h d throughout the year makes for a light environment that promotes year-round high net primary productivity and permits an agriculture that is characterized by either multiple plantings during the annual crop calendar or an abundance of perennial crops that provide a mixture or succession of harvests throughout the year.

### Altitude

As elevation increases, light intensity also increases because the thinner atmosphere absorbs and disperses less light. Plants growing at higher elevations, therefore, are more subject to conditions of light saturation and face greater danger of chlorophyll degradation than plants at sea level. Many high-elevation plants have evolved reflective coloration or protective hairs or scales on leaf cuticles to reduce the amount of light penetrating the leaves.

### Topography

The slope and direction of the soil surface can create localized variations in the intensity and duration of exposure to sunlight. Although the temperature effects of this variation may be of greater significance, steep slopes facing the poles can receive significantly lower, direct insolation than other sites. Slope orientation usually becomes more important during the winter months, when a hillside or other topographic feature can cast a shadow over the vegetation. In farming systems, minor topographic variation can create subtle differences in microclimate that affect plant development, especially when plants are still very small (Figure 4.4).

### Air Quality

Suspended materials in the atmosphere can have a significant screening effect. Smoke, dust, and other pollutants, either natural or human-produced, can greatly interfere with photosynthetic activity, either by reducing the amount of light energy that reaches the leaf or by coating the leaf and cutting down the amount of light that penetrates the cuticle. Such air quality problems are usually most common in and around urban or industrial regions, but poor air quality associated with agricultural activities such as burning and soil disturbance can also occur. Greenhouse horticulture is particularly affected by deposition of particulates from dirty air; even when glass is clean it reduces light passage by about 13% (Figure 4.5).

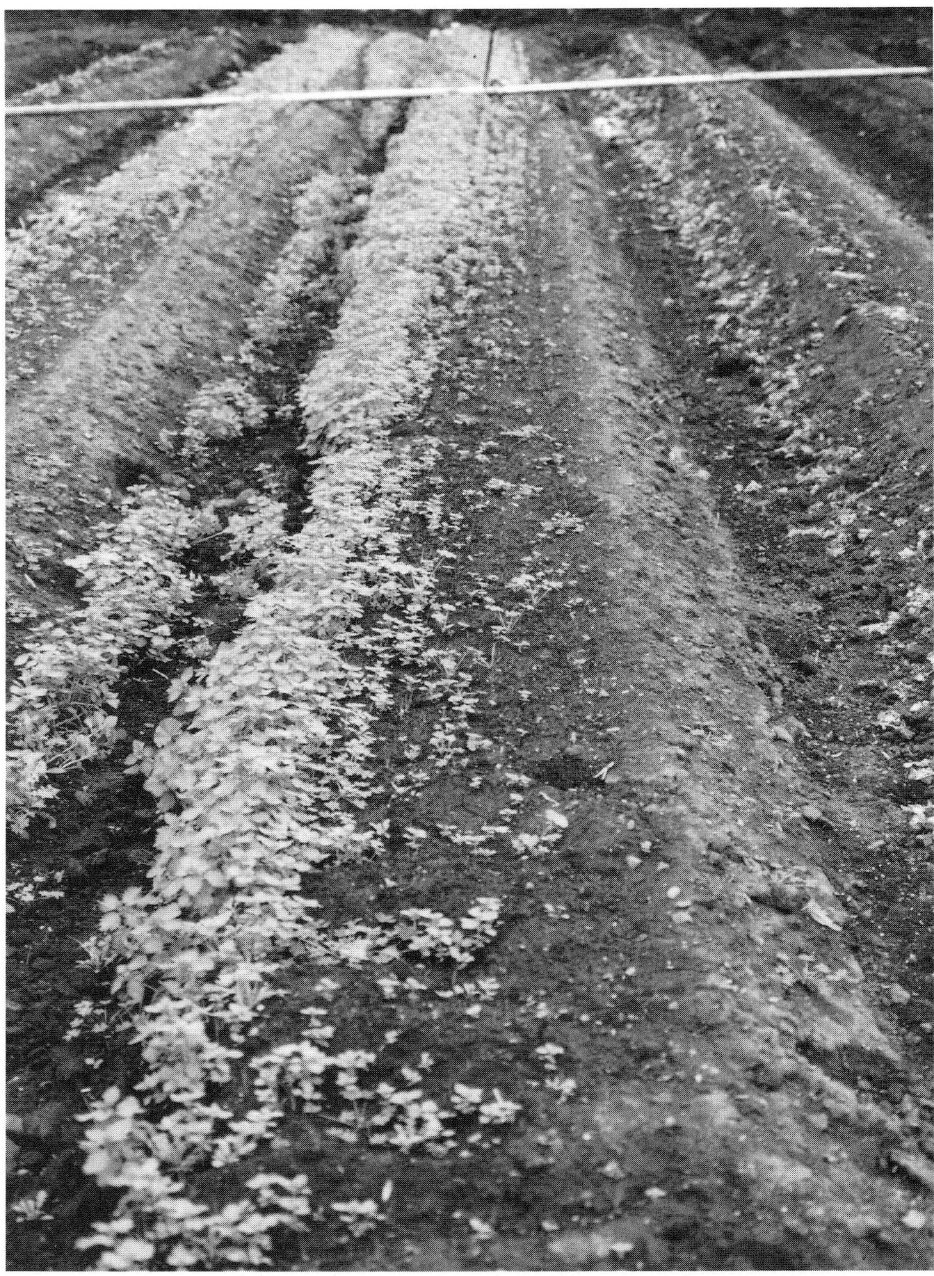

**FIGURE 4.4 Concentrated weed growth on the north-facing side of a furrow.** Because this side of the furrow received less light than the south-facing side, it remained cooler and moister, favoring the development of these particular weeds.

## Vegetation Canopy Structure

The average leaf allows the transmission of about 10% of the light that strikes its surface. Depending on the structure of the canopy of the vegetation, leaves will overlap one another to a greater or lesser extent, adding to the density of the canopy and reducing both the quantity and quality of light that eventually reaches the soil surface. At the same time, however, considerable sunlight may pass between leaves or through the spaces that become available between leaves as wind moves the canopy and as the sun moves across the sky. Some of this additional light enters as diffused sidelighting (sky light), and other light enters directly from the sun and forms sunflecks (small, usually mobile spots of unobstructed light). From an agricultural perspective, it is important to understand how light varies inside of the vegetative canopy, especially when dealing with diverse intercropped systems, agroforestry systems, and even the management of noncrop plant species in the interior of a cropping system.

**FIGURE 4.5 Smog in the Valley of Mexico.** The high level of air pollution in this mountain-ringed valley impacts light quality at ground level. One of the peaks of Volcán Ixtacihuatl extends above the smog.

The *relative rate of light transmission* of a canopy is expressed as the average amount of light that is able to penetrate the canopy as a percentage of the total incident light available at the top of the canopy or on the surface of an adjacent area free of vegetation. Since we also know that the change in average light penetration depends on the density of the foliage and arrangement of leaves, another way of determining the potential for light absorption of a particular canopy is to measure *leaf-area index* (LAI). This is done by calculating the total surface area of leaves above a certain area of ground; since the units for both are identical ($m^2$), LAI becomes a unitless measure of the amount of cover. If the LAI is determined to be 3.5, for example, the given area is covered by the equivalent of 3.5 layers of leaves in the canopy, implying that light will have to travel through that many layers before reaching the ground. The height of each layer, however, is an important determinant of the sequential reduction of light as it travels through the canopy.

Not only is the more obvious measure of total light intensity reduced as we enter deeper into the vegetative cover, but also the quality of that light changes as well. The "light of shade" usually has a very low amount of red and blue light, and a relatively high amount of green and IR light. This effect is particularly pronounced under broad-leaved evergreen canopies. Conifer forests, on the other hand, have much more red and blue light at the forest floor because of the structure of the leaves (needles) and the fact that they are much more reflective rather than absorbing and transmitting of visible light.

Given the extreme variations in canopy structure among natural vegetations and cropping systems, light levels inside canopies are highly variable as well. They can range from only a few percent of full sunlight at soil level in a dense forest to nearly 100% of full sunlight in a cropping system in the early stages of crop development. The light intensity in a fully mature cotton crop is reduced to 30% of full sunlight at a point halfway between canopy top and soil surface, and is less than 5% of full sunlight at the soil surface. The ways in which a squash crop, a corn crop, and a corn/squash intercrop modify the light environment under their canopies is illustrated in Figure 4.6.

## PHOTOSYNTHETIC RATE

Once light is absorbed by the leaf and activates the processes in the chloroplast that eventually lead to the production of energy-rich sugars, differences in the actual

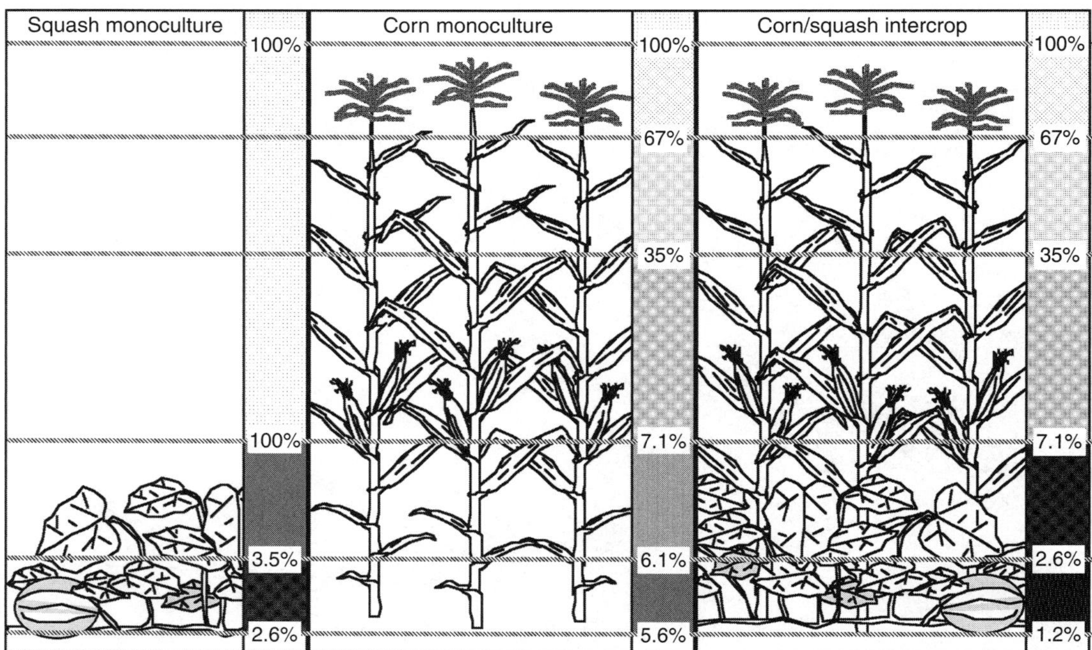


**FIGURE 4.6 Light attenuation under the canopy of a squash monoculture, a corn monoculture, and a corn/squash intercrop.** The data for each crop show the percentage of full sunlight remaining at each of six horizontal levels. (From Fujiyoshi, 1997)

rate of photosynthesis become important. Photosynthetic rate is primarily determined by three different sets of factors: (1) the plant's developmental stage (discussed in the next section), (2) the environmental conditions surrounding the plant, including the light environment, and (3) the type of photosynthetic pathway (C3, C4, or CAM) used by the plant. It is important to know what determines variations in photosynthetic rate when managing the light environment in agroecosystems.

## Photosynthetic Efficiency and Factors of the Environment

Like any plant response, photosynthesis is greatly affected by environmental conditions. These include temperature, light intensity, light quality, duration of light exposure, availability of carbon dioxide, availability of moisture, and wind. For each of these factors, a plant has maximum and minimum tolerances, as well as an optimum condition making photosynthesis most effective. The effects of these factors will be dealt with in more detail in later chapters.

In general it can be said that much of an individual plant's structure and function has evolved over time for photosynthetic efficiency. But despite a host of adaptations, from leaf structure to chemical pathways, only a small percentage of available solar energy is captured by the process. Most leaves reach saturation at only about 20% of full sunlight. Of the solar energy absorbed by leaves, only about 20% gets converted to chemical energy in sugar molecules. This gives photosynthesis a theoretical efficiency of about 4%, which can be lowered even more as carbon dioxide around the leaf is depleted. In addition, only part of the energy in photosynthate is actually converted to biomass, reducing the efficiency of the entire process between 1 and 3%. Since we have yet to find ways of altering the photosynthetic process itself, it becomes most important to try to maintain environmental conditions as close to optimum as possible, as well as to select crop plants with the appropriate pathway for a particular environment.

## Differences in Photosynthetic Pathways

The research that has helped us understand the different types of photosynthetic pathways and their conditions of optimum functioning has also helped us refine our selection of crops for different locations. The higher photosynthetic rates, virtual lack of photorespiration, and morphological adaptations (bundle sheaths) in C4 plants combine to give these plants an advantage under conditions of high light intensity and warm temperatures. These two conditions often occur in moisture-limited situations as well. Therefore, even under moisture stress and accompanying stomatal closure, C4 plants can continue to photosynthesize through the scavenging of internally-produced carbon dioxide and because of an ability to maintain the process even at low compensation points for carbon dioxide. C4 plants, however, are somewhat restricted to these conditions of high light intensity and warmth. C3 plants have a much wider distribution and a better ability to function under conditions of lower temperatures, shading, and climatic variation. Researchers have recently shown that when C3 and C4 crops (oats and peas, for example) are grown together in the same cropping system, the complementarity in light needs helps produce a yield advantage for the mixture (Kwabiah, 2004). Rotations of C3 and C4 crops can also respond to changing light conditions that occur seasonally.

### Measurement of Photosynthetic Rate

The measurement of photosynthetic rates in the field allows us to monitor the efficiency of energy capture in various crops. The most accurate measurement is of actual gas exchange by the plant. An individual leaf, plant part, or whole plant is enclosed in a transparent chamber where conditions are monitored and maintained as close to ambient conditions as possible. Air is passed through the chamber and into an IR gas analyzer so that changes in carbon dioxide content caused by the photosynthesis–respiration balance can be determined.

The other form of measurement is based on the weight gain in dry biomass by the whole plant or the determination of the correlation between weight gain of specific plant parts and the whole plant over time. For an annual plant that begins as a seed and completes its life cycle in a single season, net photosynthetic activity is directly related to the dry weight of the plant at harvest. For perennials, some part of the plant has to be harvested, and by using models of whole plant development and biomass distribution, approximate values of net photosynthetic activity can be determined. The LAI described earlier can also be used to estimate the potential leaf area available for photosynthesis in a crop system, and then based on our knowledge of approximate photosynthetic rates for individual plants or plant parts, estimates of the photosynthetic rate for the whole system can be made.

## OTHER FORMS OF RESPONSE TO LIGHT

Plants respond to light in other ways besides using light to produce energy-rich sugars. Light has an influence on the plant from germination of the seed to its production of new seeds.

### Germination

The seeds of many plants require light to germinate; when buried beneath the soil they do poorly. A single, brief exposure to light, however, such as during cultivation when a weed seed is brought to the surface but immediately buried again as the soil is turned, can be enough to induce germination. Other seeds need repeated exposure, or even constant exposure to the light in order to germinate. Lettuce is perhaps one of the best known examples of such a crop species — without light exposure, germination is reduced by 70% or more. The seeds of other plants, such as those of many of the cucurbits, have the opposite requirement: the seed must be buried fully in order to germinate because light actually inhibits germination. In all of these cases, a light-sensitive hormone controls the response.

### Growth and Development

Once a seed germinates, the newly emerged plant begins the process of growth and development. At any stage in the process, light intensity or duration of light exposure can control the plant's response, either as a stimulus for the response or as a limiting factor.

#### Establishment

Early seedling establishment can be very much affected by light levels, especially when seed germination or seedling establishment takes place under the canopy of already established plants. Some seedlings are less shade-tolerant than others, and have more difficulty establishing when there is a lack of sufficient light to maintain further plant development. An example of the importance of differences in shade tolerance is seen in the comparison of seedlings of white pine and sugar maple in forests of the eastern U.S. White pine seedlings experience a photosynthetic deficit at 10% of full sunlight and sugar maple seedling reach it at 3%. This difference in light compensation point means that sugar maple is more shade tolerant than white pine, so in a dense forest with light levels consistently below 10%, only sugar maple seedlings will reproduce. The greater shade tolerance of sugar maple can be an important factor in forest succession. After logging, pines establish first, but as the forest closes in and shade deepens, sugar maples begin to establish and eventually replace the pines.

#### Plant Growth

When a plant is surrounded by other plants, the amount of light reaching its leaves can become limiting and competition for light begins to occur. Competition for light is especially likely in same-species plant populations or in plant communities made up of very similar species with very similar light needs. Stem and leaf growth can be severely limited if competition reaches the point where a plant is completely shaded by its neighbors. If some part of the plant is able to emerge from the shade and reach full sunlight, photosynthesis in that part may be able to compensate for the shading occurring over the rest of the plant and permit adequate development.

Many plants develop anatomically different leaves depending on the level of shading or sun. Shade leaves are thinner and have larger surface per unit weight, a thinner epidermis, less photosynthetic pigment, spongier leaf structure, but more stomata than sun leaves. Interestingly, shade leaves often appear to be adapted to the lower light environment, being able to photosynthesize above the compensation point due in part to the larger surface area for light capture. But it is important that shade leaves be protected from the harmful effects of too much light.

### Phototropism

Light can induce a plant to synthesize chlorophyll and anthocyanins, which stimulate growth in certain plant parts such as the leaf petiole or the flower peduncle, causing the phenomenon of growing toward or away from light. In some cases, this growth pattern is triggered by a hormone that is activated by blue light. Leaves can be oriented toward the sun to capture more light, or away from the sun in high light environments. Sunflowers receive their name from the characteristic orientation of the disc of the inflorescence toward the morning sun.

### Photoperiod

Because the earth is tilted on its axis, the relative proportion of daylight and nighttime hours varies from one time of year to another. Because of the correlation of hours of light or dark with other climatic factors, especially temperature, plants have developed adaptive responses to the changing light/dark regimes over time. Important processes such as flowering, seed germination, leaf drop, and pigmentation changes are examples. A pigment in plants known as phytochrome is the major photoreceptive agent responsible for regulating these responses.

The phytochrome pigment has two forms; one form has an absorption peak for red light with a wavelength of 660 nm, the other has an absorption peak for far-red light with a wavelength of 730 nm. In daylight, the red light form is rapidly converted to the far-red form, and in the dark, the far-red form slowly converts back to the red form. The far-red phytochrome is biologically active and responsible for the basic responses of plants to the number of hours of light or darkness.

In the morning, after only a few minutes of light exposure, the far-red phytochrome becomes the dominant form and remains so throughout the day. This dominance is maintained into the night as well, since the conversion back to red phytochrome during darkness is slow. Therefore, when the length of the night is relatively short, there is insufficient time for enough far-red phytochrome to convert to the red form, and the far-red form stays dominant. However, as the number of hours of darkness increases, a point is reached at which night is long enough to allow a shift of dominance to the red form. Even when this period of red dominance is short, changes occur in the plant's response.

In chrysanthemums, for example, the end of the far-red phytochrome's continual dominance in autumn triggers the growth of flowering buds. This type of response is known as a "short-day" response, even though the actual response is activated by the longer nighttime hours. The importance of the dark period is accentuated by the fact that even a short period of artificial light in the middle of the night for greenhouse-raised mums allows for the conversion of enough far-red phytochrome to suppress flowering.

Strawberries have the opposite type of response. In the spring, shorter nights allow the far-red phytochrome to regain continual dominance, causing a shift from vegetative production to flower production. Plants with this kind of response are called "long-day" plants, even though it is shorter nights that actually trigger the change. So-called day-neutral varieties of strawberries have been developed to extend flowering later into the summer and early fall when normal strawberries undergo the shift to vegetative growth characteristic of long-day plants.

### Production of the Harvestable Portion of the Plant

The conditions of the light environment have a crucial role in the production of the part of the plant that we intend to harvest. In general, crop plants have been selected to shunt a great deal of photosynthate to the portions of the plant that are harvested. In other words, the harvested portions are major "sinks" in carbon partitioning. Nevertheless, the ability of the plant to produce the desired amount of biomass in its harvested parts is dependent on the conditions of its light environment. By understanding the complex relationships between plant response and light quantity, quality, and duration of exposure as discussed above, the light environment can be manipulated and plants selected in order to optimize output from the agroecosystem.

## MANAGING THE LIGHT ENVIRONMENT IN AGROECOSYSTEMS

There are two main approaches to managing the light environment of an agroecosystem. Where light is generally not a limiting factor, management is oriented toward accommodating the system to the excess of light that can occur; where light is more likely to be a limiting factor, the focus is on how to make enough light available for all the plants present in the system.

Regions where light is not a limiting factor are generally dry regions. In these locations, the key issue in determining the structure of the vegetation and the organization of a cropping system is usually the availability of water, not light. Plants are usually more separated from each other, light relations are of less importance since there is usually an overabundance of solar energy, and many organisms must display adaptations for "avoidance" of light rather than capture. Leaves are often vertically oriented to avoid direct exposure to light, they have less chlorophyll content so as to absorb less light energy and thus less heat, and contain higher proportions of red pigments so as to reflect the red light normally absorbed in photosynthesis.

Light is more likely to be a limiting factor in humid regions. Both natural vegetation and agroecosystems in humid areas are much more layered or stratified, with both light quantity and quality being altered as light passes through those layers on its way to the soil surface. In these regions, the management of light can be an important factor in optimizing the productivity of agroecosystems. The more stratified the vegetation structure, the greater the challenges for light management. In forestry and agroforestry systems, for example, the seedlings of the canopy species often do not germinate well in the shaded environment of the forest floor, a factor that must be taken into account in managing the diversity of the system.

### Crop Selection

One aspect of managing the light environment is to match the availability of light in the system to the plants' response to light. The light requirements of plants, as well as their tolerances, are important factors in the crop selection process.

The type of photosynthetic pathway of the crop plants is the most basic determinant of light requirements. As discussed previously, plants with C4-type photosynthesis require high light intensity and long duration of light exposure to produce optimally, in addition to not being well adapted to areas with cooler, moister conditions, especially cooler nighttime conditions. In contrast, many C3 plants will not grow well in the same light conditions favored by C4 plants.

In central coastal California, for example, where the adjacent cold ocean currents normally keep summer nighttime temperatures at low to moderate levels and produce regular morning fog, C4 crops such as sweet corn are very slow to develop and rarely obtain the yields or sweetness of the ears grown in plantings in the interior valleys of the state just 50 mi to the east. In contrast, many C3 crops such as lettuce grow very well in the coastal climate.

Sugar cane is good example of a C4 crop requiring high light intensity. When planted in areas with adequate light and moisture, this C4 crop achieves one of the highest rates of photosynthetic efficiency known for crop plants. Variety selection, row arrangement, planting density, fertility management, and other factors have been combined with the 4% conversion rate of PAR to biomass to produce some of the highest net dry matter returns known for a cropping system (up to 78 t dry matter/ha/yr).

Even within crops of the same photosynthetic pathway, crop selections can be made. Different light compensation points, for example, could determine which crops to select for shadier environments.

### Cropping Diversity and Canopy Structure

The light environment in the interior of a cropping system varies considerably. Cropping systems can be designed to create regions in the system where the light environment is most appropriate for a particular crop. In the tropics, for example, farmers make full use of the altered light environment under the canopy of trees to grow crops such as coffee, cacao, and vanilla. Cacao and vanilla plants do not tolerate direct sun for any appreciable amount of time, and often they need to have the shade-producing canopy in place before they can be planted. Only recently have varieties of coffee been developed that can be planted in direct sunlight.

In mixtures of annual crops, the light environment within the canopy of the system changes as the crop system matures, with LAI and light intensity at different levels undergoing considerable variation over time. Farmers have learned to take advantage of these changing conditions. A well-known example is the traditional corn–bean–squash intercrop of Mesoamerica. In a particular form of this multiple cropping system in southeastern Mexico (Amador and Gliessman, 1990), all three crops are planted at the same time, hence each encounters a very similar light environment when they first emerge. But the corn component of the system soon dominates the canopy structure, casting shade on the beans and squash below. As the corn canopy closes, beans occupy the lower half to two thirds of the corn stalk by climbing up the corn stalk. The squash is confined to the darker understory, itself casting yet a deeper shade on the soil surface and aiding in weed control within the cropping system (Gliessman, 1988). Although both the beans and squash receive less-than-optimal light exposure, they both receive enough to produce adequately and do not interfere with the very high light needs of the corn. Corn is a C4 crop, and beans and squash are C3. Such an agroecosystem is evidence that crops of different photosynthetic pathways can be combined in intercropping systems, and research aimed in this direction could certainly come up with more.

Diverse home garden agroforestry systems are perhaps the most complex examples of the management of the light environment in agroecosystems (Mendez, 2000, Nair, 2001); they are discussed in much more detail in Chapter 17. Their high LAI (3.5 to 5.0), diversity of distribution of the canopy layers, high light absorbance by the foliage (90 to 95%), and patchy horizontal structure due to either successional development or intentional human intervention make for a highly diverse light environment that promotes one of the correspondingly highest plant species diversities known for an agroecosystem. Much needs to be known about the specific light requirements and tolerances of each component of such a system.

A study of the light environments of the nine different agroecosystems in Mexico and Costa Rica provides some impression of the possible variation in the structure and characteristics of light environments. The data from this study are presented in Table 4.1.

In general, the polycultures in the study were more effective at intercepting light than the monocultures, although the sweet potato monoculture, with its broad leaves, intercepted light as effectively as the home garden and the shaded coffee system. These mixed results point out the difficulty of determining a system's efficiency of light use. Simply measuring vegetative cover, LAI, and the transmission of light to the surface does not by itself elucidate how light is used by the components of the system. Nor does it show how a well-designed system can create a light environment that meets the needs of a diversity of different plants at the same time.

**TABLE 4.1**
**Measures of the Light Environment in a Range of Agroecosystems and Natural Ecosystems in Costa Rica and Mexico**

| | Species | LAI | Cover (%) | Transmission (%) |
|---|---|---|---|---|
| 2-month-old corn monoculture, conventionally managed | 7 | 1.0 | 56 | 35 |
| 3.5-month-old corn monoculture, traditionally managed | 20 | 2.6 | 88 | 12 |
| Sweet potato, weeded and treated with insecticide | 8 | 2.9 | 100 | 11 |
| 2.5-year-old intercrop of cacao, plantain, and the native timber tree *Cordia alliodora* | 4 | 3.4 | 84 | 13 |
| Old wooded home garden containing a diverse mixture of useful plants | 18 | 3.9 | 100 | 10 |
| Coffee plantation with an overstory of erythrina trees | 7 | 4.0 | 96 | 4 |
| Plots planted with useful plants to mimic natural succession, 11 months after clearing | 27 | 4.2 | 98 | 7 |
| Gmelina plantation (trees grown for timber and pulp intercropped with beans and corn) | 8 | 5.1 | 98 | 2 |
| Plots undergoing natural succession, 11 months after clearing | 35 | 5.1 | 96 | <1 |

*Source:* Ewel et al., 1982. *Agro-Ecosystems* 7: 305–326.

## Temporal Management

Over time, the light environment in an agroecosystem changes. One type of change results from the growth of the plants in the system, and another from seasonal changes. Both kinds of changes can be taken advantage of, modified, or used as cues for initiating specific techniques.

One kind of temporal management that takes advantage of the changes in the light environment that occur as a crop matures is the "over-sowing" of one crop into another. This is done, for example, to produce an oat/legume hay crop: instead of sowing the oats, harvesting the oats, and then planting the legume cover crop (such as clover or vetch), the seed of the legume can be sown when the oats reach a particular stage of development and the light environment is most conducive to the establishment of the legume. Specifically, the legume is planted just before the heads of oats begin to form, when light levels at 3 in. above the soil are about 40% of full sunlight. Clover seems to establish best around 50% of full sunlight, so over-seeding that occurs just before heads start to form gets the legume off to a good start. After the oats are harvested, the light levels reaching the established clover plants approach once again those of full sunlight, promoting the rapid growth of this species as a nitrogen-fixing cover crop (Figure 4.7).

Management of seasonal variations in light is common in perennial and agroforestry systems. Coffee systems in Costa Rica — the subject of considerable applied shade management research — offer a good example of this form of temporal light management (Lagemann and Heuveldop, 1982; Bellow and Nair, 2003). As discussed previously, coffee is typically grown under the shade of trees, often species of the leguminous genus *Erythrina*. Although coffee is a very shade-tolerant plant, it suffers when shade becomes too dense. This is especially true during the wet-season time of the year, when relative humidity inside the coffee cropping system stays close to 100% most of the time, promoting fungal diseases that can cause coffee defoliation and fruit drop. Therefore, a common practice is to heavily prune the shade trees at the beginning of the wet season (during June) in order to allow more light into the interior, promoting drier conditions and hence a reduced chance of disease. The greater cloud cover during the wet season lessens the need for shade over the coffee. Close to the end of the wet season (usually November or December) another less intensive pruning occurs that opens up the canopy of the plantation again, possibly promoting not only the development of flower buds that open later in the dry season, but also stimulating the turnover of nitrogen-rich biomass that aids the more rapid growth of the coffee plants during this period (Figure 4.8).

**FIGURE 4.7 Over-sown clover plants exposed at the early July harvesting of the overstory oat crop at the Rodale Research Farm, Kutztown, Pennsylvania.** The clover will be ready to harvest for forage or incorporated as a green manure crop in less than 2 months.

## Carbon Partitioning and Sustainability

As was discussed in Chapter 2, a relatively small percentage of the carbon that gets fixed by photosynthesis into carbohydrate form eventually gets transformed into biomass. For agriculture, it is the portion of that biomass that finds its "sink" in the form of harvestable, consumable, and/or marketable organic matter that is of greatest importance. All of the discussions of how the light environment can be managed to increase the size of this sink must also take into consideration what the long-term impacts might be of harvesting and removing this biomass from the agroecosystem.

The experience of corn farmers in Puebla, Mexico offers an interesting example of how increasing the proportion of carbon partitioned into harvestable material is not necessarily positive. Many of the small traditional farmers of the region switched to higher-yielding "green revolution" corn varieties in the late 1960s and early 1970. These varieties had been bred to produce more grain at the expense of biomass normally stored in other parts of the plant — especially the stems and leaves. After planting these varieties for a few years, the farmers went back to using their traditional varieties of corn. Since these farmers used animals so extensively in their farming systems (especially for cultivation and transport), and since corn stover was an important supplemental feed for the animals, the great reduction in stems and leaves from the new varieties did not allow the production of adequate animal feed. In this case, concentrating the carbon sink in grain did not take into account the sustainability of all parts of the agroecosystem.

The same process may be going on with other crops. Traditional rice varieties, for example, store over 90% of their carbon in leaves, stems, and roots, whereas new varieties have raised the portion of carbon stored in grain to well over 20% (Gliessman and Amador, 1980). In cultures where rice straw plays important roles elsewhere in the agroecosystem, such as for building material, fuel, and feed for animals, human needs would dictate the need for care in transitioning to varieties that sacrifice some forms of biomass for rice grain. Within the agroecosystem itself, we must also understand the possible impacts of this "loss" of organic matter on such ecological components as soil organic matter maintenance, soil aggregate stability, biological activity in the soil, and nutrient inputs that are essential for the long-term sustainability of the agroecosystem.

**FIGURE 4.8 Pruned shade trees in a coffee plantation in Turrialba, Costa Rica.** The common shade trees (Erythrina poeppigina) are heavily pruned at the beginning of the wet season to open up the coffee plantation to better light penetration during the more cloudy and rainy time of the year.

## FUTURE RESEARCH

Much work needs to be done on managing the light environment in agroecosystems. We have recently learned a lot about photosynthetic pathways, carbon partitioning, and how to raise the yield of harvestable biomass from cropping systems. But we need also to understand that agroecosystem management requires that we return as much organic matter to the system, especially to the soil, as we remove from it. The energy that is captured from the sun must contribute as much to long-term agroecosystem sustainability as it does to short-term harvests. Research on how to balance these needs is key to developing the sustainable agroecosystems of the future.

## FOOD FOR THOUGHT

1. What are the basic differences between too much light and too little light in terms of plant response? What are some of the ways of compensating for either extreme in the design of an agroecosystem?
2. Our understanding of the different types of photosynthetic pathways in plants has come mostly from basic laboratory research, but this knowledge has helped considerably in the management of the light environment in agroecosystems. What other basic research questions, greatly isolated from the field, might be of great potential significance for sustainability?
3. What are some of the most significant ways by which humans and human activities are impacting the light environment? What might the consequences be for agriculture in the future?
4. Light energy is considered to be one of our most available and easily used sources of renewable energy. What are some of the factors that have slowed the development of better ways to take advantage of this energy source in agriculture?

## INTERNET RESOURCES

Centre for Atmospheric Research: Ozone Hole Tour
www.atm.ch.cam.ac.uk/tour

International Society of Photosynthesis Research
www.photosynthesisresearch.org

## RECOMMENDED READING

Bainbridge, R., G.C. Evans, and O. Rackham. 1968. *Light as an Ecological Factor.* Blackwell Scientific: Oxford. Proceedings of an international symposium that covers a wide range of topics related to light as an important factor in the environment.

Evans, G.C., R. Bainbridge, and O. Rackham. 1975. *Light as an Ecological Factor: II.* Blackwell Scientific: Oxford. A follow-up to the symposium held in 1968, with a broader range of topics covered.

Hall, D.O. and K.K. Rao. 1999. *Photosynthesis.* 6th ed. Cambridge University Press: New York. An excellent introductory textbook on the photosynthetic process at both the macro- and molecular-level, with a special focus on the role of photosynthesis as a source of food and fuel.

Lawlor, D.W. 2001. *Photosynthesis: Molecular, Physiological and Environmental Processes.* 3rd ed. BIOS Scientific Publishers/Springer-Verlag: New York. This updated edition provides a comprehensive review of photosynthesis and an introduction to the existing scientific literature. It incorporates many recent research advances, especially in the areas of the molecular basis of photosynthesis and the effects of environmental change. It provides a good basis for those interested in the ecological and environmental factors related to photosynthesis.

# 5 Temperature

The effect of temperature on the growth and development of plants and animals is known and easily demonstrated. Each organism has certain limits of tolerance for high and low temperatures, determined by its particular adaptations for temperature extremes. Each organism also has an optimum temperature range, which can vary depending on the stages of development. Because of their different reactions to temperature, papayas are not planted in the cool coastal temperate environment of the Monterey Bay of California, and apples would not do well if planted in the humid tropical lowlands of Tabasco, Mexico.

Thus the temperature range and degree of temperature fluctuation in an area can set limits on the crop species and cultivars that a farmer can grow, and can cause variations in quality and average yield for the crops that are grown. It is necessary to consider the temperature factor in selecting crops that are appropriate to the range of temperature conditions that might occur from day to day, between day and night, and from season to season. Aboveground temperatures are as important as those below ground.

When we measure the temperature of the air, soil, or water, we are measuring the heat flow. In order to fully understand temperature as a factor, it is useful to think of this heat flow as part of the energy budget of the ecosystem, the basis of which is solar energy.

## THE SUN AS THE SOURCE OF HEAT ENERGY ON EARTH

The energy flowing from the sun is predominantly short-wave radiation, usually thought of as light energy made up of both visible and invisible spectra. Recall that the fate of this energy once it reaches the atmosphere of the earth was discussed in the previous chapter and diagrammed in Figure 4.1. Incoming solar radiation is either reflected, dispersed, or absorbed by the atmosphere and its contents. Reflected and dispersed energy is little changed, but absorbed energy is converted to a long-wave form of energy manifested as heat. Similarly, the short-wave energy that reaches the earth's surface is either reflected or absorbed. The absorption process at the surface, by which short-wave light energy is converted into long-wave heat energy, is known as *insolation*. Heat formed by insolation can be stored in the surface, or reradiated back into the atmosphere. Some of the heat reradiated into the atmosphere can also be reflected back to the surface.

As a result of these processes, heat energy is trapped at and near the earth's surface, and the temperature there remains relatively high compared to the extreme cold of the outer space. Overall, this warming process is termed "the greenhouse effect."

Temperatures at the earth's surface vary from place to place, from night to day, and from summer to winter; nevertheless, an overall equilibrium is maintained between the heat energy gained by the earth and its atmosphere, and the heat energy lost. This balance between heating and cooling is represented in the following equation:

$$S(1-\alpha) + L_d - L_u \pm H_{air} \pm H_{evap} \pm H_{soil} = 0$$

where

$S$ = solar gain

$\alpha$ = the albedo of the earth's surface (with a value between 0 and 1)

$L_d$ = the flux of long-wave heat energy to the surface

$L_u$ = the flux of long-wave heat energy away from the surface

$H$ = the gain or loss of heat energy from air, soil, and water (evap).

This equilibrium is currently undergoing a shift in response to human-induced changes in the atmosphere. These changes include a rise in carbon dioxide levels from the combustion of fossil fuels. As more carbon dioxide and other "greenhouse gases" are added to the atmosphere, more heat is trapped between the atmosphere and the surface. Many scientists are concerned about the possible impacts on agriculture by a global rise in temperature.

## CAUSES AND CONSEQUENCES OF CURRENT CLIMATE CHANGE

The global climate is changing rapidly, and human activities since the beginning of the industrial age are clearly a major, if not primary, causal factor. Our power plants, factories, and automobiles release into the atmosphere huge amounts of carbon dioxide and other gases that trap solar radiation in the earth's atmosphere. So far, the cumulative effect of more than a century of intensive fossil fuel burning is a slight increase in the average global surface temperature over the last 50 years. Due to the vast complexity of global atmospheric dynamics, it is difficult for scientists to predict with any certainty what will occur in the future, but there is substantial concern in the scientific community about the consequences of continuing to dump carbon dioxide and other greenhouse gases into the atmosphere.

The amount of carbon in the atmosphere has increased about 30% since the beginning of the industrial age. This increase is mainly due to the burning of fossil fuels in industrial manuzacturing and energy production, and due to deforestation. Deforestation is doubly detrimental because not only is the cleared vegetation usually burned, releasing more carbon, but also plants that had previously taken up carbon dioxide are lost.

Although modern agricultural practices directly account for only a small portion of the yearly input of greenhouses gases into the atmosphere, they are an indirect cause of far more. The clearing of forests for agricultural purposes (including grazing), for example, is a significant cause of deforestation. In addition, fossil fuels are burned to produce the energy needed for the synthesis of pesticides and fertilizers, and the shipping of agricultural products around the world requires more fossil fuel consumption.

On average, the earth is now 0.6°C hotter than it was 100 years ago. Many in the scientific community are worried that global temperatures are likely to continue to rise, and that the effects may be extremely serious. Already, the average snow cover on the earth has decreased about 10% since the 1950s, and the spring–summer extent of artic sea ice has decreased about 15%. Current projections from climate change models are for global temperatures to rise another 2.7 to 4.4°C by 2100 (CGER, 2001). Scientists are particularly concerned about positive feedback mechanisms that may rapidly accelerate global warming once a "tipping point" is reached. In the arctic, for example, less snow cover means less reflection of light and more absorption as heat, which causes even less snow cover, loss of permafrost, and more melting, which in turn causes more heat absorption.

Studies suggest that a warming climate will cause more local climate extremes such as floods and droughts. While some areas might receive increased rainfall and warmer temperatures beneficial to agriculture, atmospheric models indicate that other regions, including South and Southeast Asia, Latin America, and sub-Saharan Africa, would likely suffer from the increased heat and disrupted rainfall. Another concern is that prime agricultural land in low-lying coastal regions throughout the world will be flooded by the continued melting of the polar ice caps and the rise in water levels associated with the expansion of water as it warms.

Although human-induced climate change is certainly real, there is still much debate in the scientific community about its severity and consequences, in part because of uncertainties in the rate of temperature increases and the extreme complexity and dynamism of the atmosphere and its interactions with the earth's surface and the oceans. Regardless of the actual amount of change, however, the rates are great cause for concern and reason for changing our current habits of fossil fuel consumption.

## PATTERNS OF TEMPERATURE VARIATION ON THE EARTH'S SURFACE

There are several ecological aspects to temperature distribution that are useful for understanding the variation and dynamics of temperature conditions at the surface. We need to know this information, first of all, not only to make the proper selections of our crop types, but also to adapt agroecosystems to temperature conditions and to alter these conditions where possible.

Temperature variation occurs at the largest scale when we consider world climates, made up of the seasonal patterns of temperature, rainfall, wind, and relative humidity. At the other end of the scale, important variation also occurs at the micro level when we consider the temperature conditions inside a crop canopy or those just below the surface of the soil.

### LATITUDINAL VARIATION

The amount of solar radiation actually absorbed by the surface over a particular period of time is affected greatly by latitude. At or near the equator, incoming radiation strikes the earth's surface at a vertical angle. At increasing distances from the equator, however, the sun's rays strike the surface at an increasingly shallow angle. As this angle becomes shallower, the same amount of incoming solar radiation is spread over a larger and larger area of the

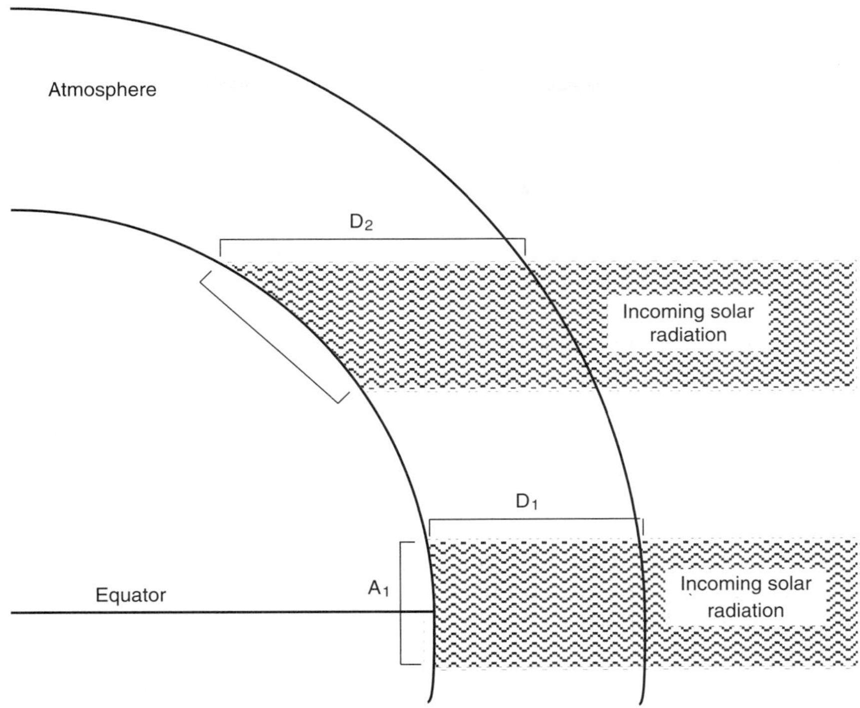


**FIGURE 5.1 The effect of latitude on solar gain.** The higher the latitude, the greater the distance that solar radiation must travel through the atmosphere ($D_2 > D_1$) and the greater the surface area over which a certain amount of solar radiation is spread ($A_2 > A_1$).

Earth's surface, as shown in Figure 5.1. In addition, the sun's rays must pass through an increasingly thick atmospheric layer at higher latitudes, resulting in a loss of energy to reflection and scattering by materials in the atmosphere, such as water droplets and dust. The overall effect is a regular decline in the intensity of solar radiation per square unit of surface as one moves away from the equator. This latitudinal variation in solar gain is one of the major causes of latitudinal variations in temperature.

## Altitudinal Variation

At any latitude, as altitude increases, temperature decreases. On the average, for each 100 m of elevation gained, ambient temperature drops approximately 0.5°C. In locations where increased cloud cover during the day is associated with this elevation gain, temperature differences can be even greater due to reduced solar gain. At the same time, the increasing thinness of the atmosphere at higher altitude results in a greater loss of heat from both the soil surface and the air just above it by reradiation at night. This phenomenon contributes significantly to lower nighttime temperatures at elevations much above the sea level. In mountainous regions at high elevations in the tropics (above 3000 m) and at progressively lower elevations as one moves toward the poles, reradiation at night is so intense that wintertime temperature conditions are encountered almost every night when the sky is clear.

## Seasonal Variation

Seasonal differences in temperatures over the surface of the earth are the result of changes in the orientation of the earth in relation to the sun as it revolves around the sun on its tilted axis. Through the course of the year, a belt of maximum solar gain or insolation moves back and forth across the equator in relation to the angle of incidence of the sun's rays and the length of the day. Longer days lead to more solar gain. This swing in insolation is the direct cause of a seasonal swing in temperature. The degree of seasonal variation in average temperatures increases with increasing distance from the equator.

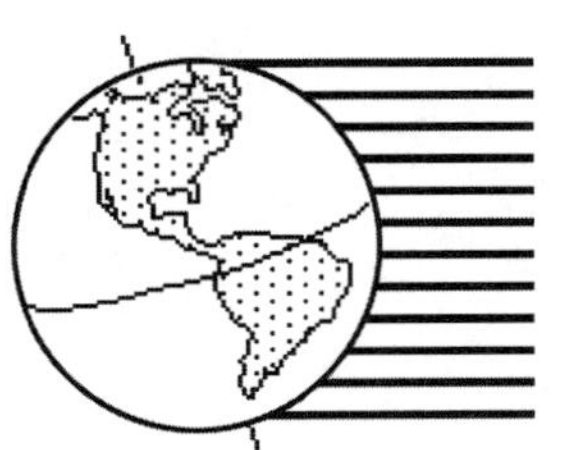

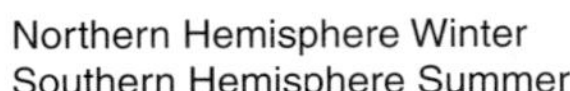


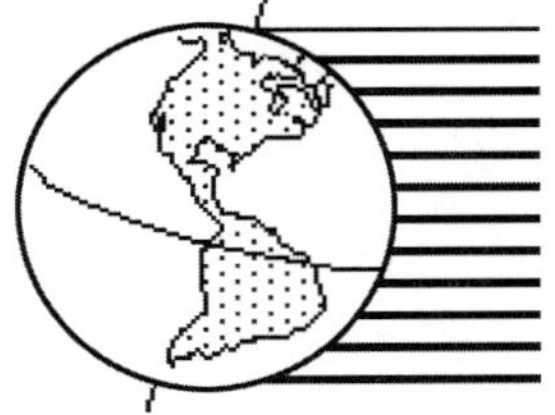


**FIGURE 5.2 Seasonal variation in the sun's angle of incidence.** The tilt toward the sun that occurs in summer increases both the length of the day and the intensity of solar radiation striking the ground.

### Maritime vs. Continental Influence

Large bodies of water, especially the oceans, greatly affect the temperature of adjacent landmasses. Because water reflects a larger proportion of insolation in relation to land and loses heat readily through surface evaporation, it has a high specific heat, and readily mixes layers vertically; the temperature of large bodies of water is slower to change than that of landmasses. Land heats up more during the summer because all the absorbed heat stays in the surface horizon and the atmosphere close to that surface, and it cools to a lower temperature during the winter because of reradiation and heat loss. Water masses are, therefore, moderators of broad fluctuation in temperature, tending to lower temperatures in the summer and to raise temperatures in the winter. This water- or marine-mediated effect on temperature is called a *maritime influence*, in contrast to the more widely fluctuating variations in temperature encountered at a distance from water under a *continental influence*. Maritime influences help create the unique Mediterranean climates of such places as coastal California and Chile, where nearby upwelling cold currents accentuate the moderating influences during the dry summer season.

### Topographic Variation

Slope orientation and topography introduce variation in temperature as well, especially at the local level. For example, slopes that face toward the sun as a result of the inclination of the earth on its axis experience more solar gain, especially in the winter months. Hence, an equator-facing slope is significantly warmer than a pole-facing slope — all other factors being equal — and offers unique microclimates for crop management.

Valleys surrounded by mountain slopes create unique microclimates as well. In many parts of the world, air that moves downslope due to winds or pressure differences can rapidly expand and heat up as it descends — a process known as catabatic warming. (The wind associated with this phenomenon will be discussed in Chapter 7.) As the air is warmed, its ability to hold moisture in vapor form (relative humidity) goes up, increasing the evaporative potential of the warmer air.

Valleys are subject to nighttime microclimate variation as well. On the higher elevation slopes above a valley, reradiation occurs more rapidly; since the cooled air that results is heavier than the warmer air below, the cooler air begins to flow downslope — a phenomenon

**FIGURE 5.3 Lettuce grown year-round in a temperate maritime climate.** Cooling summer fog and the warming effect of the nearby ocean in the winter permit year-round vegetable and fruit production on the central coast of California.

called *cold air drainage*. Often this cooler air passes under warmer air, pushing the warmer air above it and forming an *inversion*, in which a warmer layer of air becomes "sandwiched" between two layers of colder air. In some locations, the cold pocket of air can lead to frost formation and plant damage, whereas the warm air inversion just above it stays significantly warmer. This pattern of local temperature variation is illustrated in Figure 5.5. The planting of frost-sensitive citrus between 500 and 1000 ft elevation on the lower slopes of the foothills of the Sierra Nevada Mountains of the Central Valley of California is a good example of how farmers have learned to take advantage of a wintertime inversion layer of warmer air that is forced up by the drainage of colder air into a valley floor below.

## RESPONSES OF PLANTS TO TEMPERATURE

All physiological processes in plants — including germination, flowering, growth, photosynthesis, and respiration — have limits of tolerance for temperature extremes, and a relatively narrow temperature range at which functioning is optimized. Thus the temperature regime to which a plant is exposed is ultimately connected to its yield potential. For example, temperature conditions may allow a plant to establish and grow, but then a sudden change in the weather (e.g., a cold spell) might prevent it from flowering and setting fruit and producing seed.

Farmers must carefully adapt their practices to the local temperature regime, taking into account diurnal variations, seasonal variations, moderating influences,

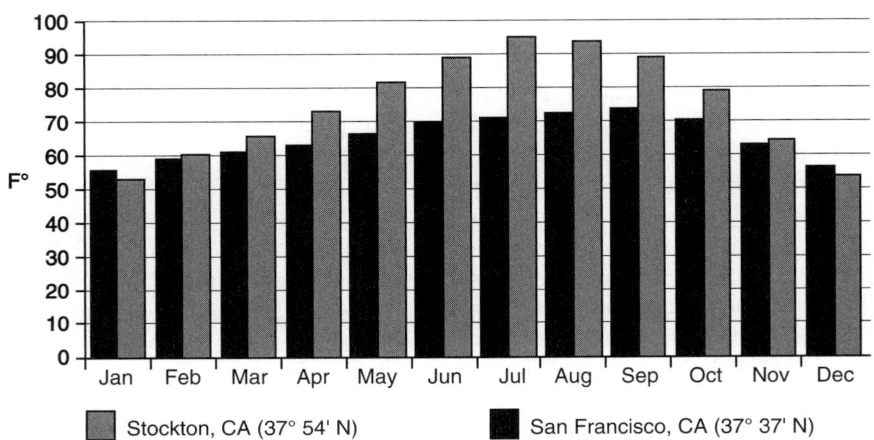


**FIGURE 5.4 Monthly average daily high temperatures at San Francisco, and Stockton, CA.** Both cities are at nearly the same latitude and elevation, but coastal San Francisco has a maritime climate, and Stockton, 100 km to the east, is under more of a continental influence. (Data from Conway and Liston, 1990. *The Weather Handbook*. Conway Data: Atlanta.)

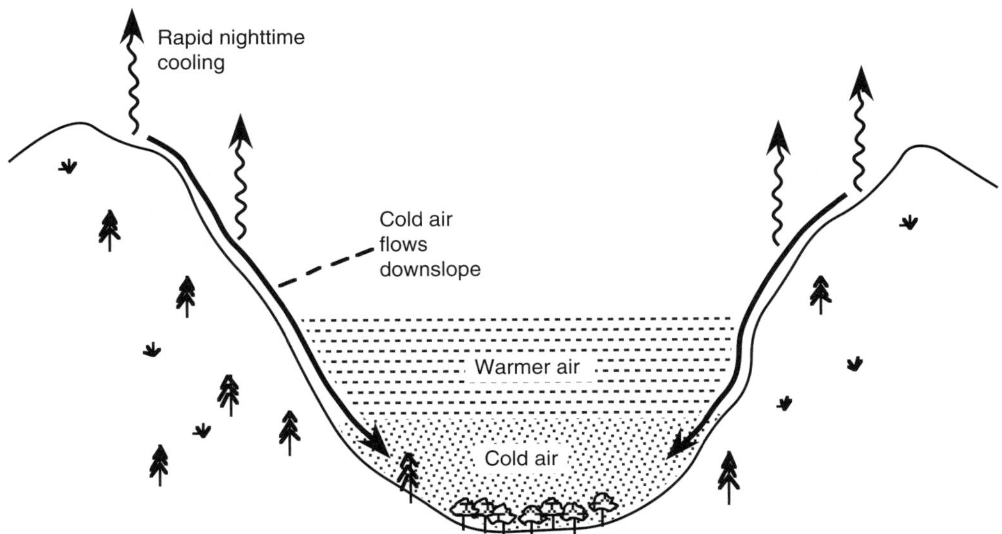


**FIGURE 5.5 Cold air drainage and inversion layer.** Cold air can drain into valley bottoms at night and pool beneath a layer of warmer air.

microclimate, other temperature-related factors, and the particular temperature responses of specific crops. In California, for example, farmers shift to cool-season varieties of crops such as broccoli for winter planting, plant cover crops during the wet and cool time of the year when many vegetable crops would not do well, plant avocado trees close to the coast in areas that are frost-free because of the maritime influence, and plant lettuce during the winter in the interior desert valleys of southern California. Other farming regions offer similar examples.

Temperature can also be used as a tool to cause desired changes in plants. For example, farmers in central coastal California chill strawberry transplants in order to induce vegetative growth and good crown development.

## Adaptations to Temperature Extremes

Natural ecosystems are made up of plants and animals that have been "screened" by natural selection. Periodic temperature extremes is one of the factors that have eliminated those species that are not tolerant of local conditions. Therefore, we can expect the temperature range tolerances of the species of local natural systems to give us an indication of the temperature extremes we might expect when we try to farm in an area. Recognizing these indicators, as well as selecting for adaptations to extremes in our crop species, can help in the development of farming systems that lower the risk associated with the natural variability in temperature extremes.

### Heat

The effects of high temperatures on crops are the result of a complex interaction between evaporative water loss, changes in internal water status, and changes in other physiological processes. Heat stress causes a decline in metabolic activity, which is thought to come from the inactivation of enzymes and other proteins. Heat also raises the rate of respiration, which can eventually overtake the rate of photosynthesis, halting plant growth, and ultimately killing plant tissue.

Plants, native to temperate areas, generally have lower limits to temperature stress than plants of more tropical areas. In all cases, though, leaf functions become impaired at about 42°C, and lethal temperatures for active leaf tissue are reached in the range of 50 to 60°C.

Common morphological adaptations of plants to excess heat include the following:

- a high $CO_2$ compensation point for the photosynthesis/respiration ratio, often aided by changes in leaf structure
- white or gray leaves that reflect light and thus absorb less heat
- hairs (pubescence) on the leaves that insulate leaf tissue
- small leaves with less surface area exposed to sunlight
- leaves with a lower surface-to-volume ratio for gaining less heat
- vertical orientation of leaves to reduce heat gain
- more extensive roots, or a greater root-to-shoot ratio, for absorbing more water to offset water loss from the leaves or to maintain more water intake relative to leaf area
- thick, corky, or fibrous bark that insulates the cambium and phloem in the plant trunk
- lower moisture content of the protoplasm and higher osmotic concentration of the living tissue

These characters can be incorporated into farming systems where water availability is limited and temperatures are high, either through the use of crop plants with these characters, or the breeding of varieties that show them.

### Cold

When temperatures drop below the minimum required for growth, a plant can become dormant, even though metabolic activity may slowly continue. Chlorosis may occur, followed eventually by death of the tissue. Death at low temperature is due to protein precipitation (which can occur at temperatures above freezing), the drawing of water out of protoplasm when intercellular water freezes, and the formation of damaging ice crystals inside the protoplasm itself.

Resistance to extremes of cold depend greatly on the degree and duration of the low temperature, how quickly the cold temperature comes about, and the complex of environmental conditions that the plant may have undergone before the cold event. Some specific structural adaptations provide resistance as well, such as coverings of wax or pubescence that allow leaves to endure extended cold without freezing the interior tissue, or the presence of smaller cells in the leaf that resist freezing.

Temporary cold hardiness can be induced in some plants by short-term exposure to temperatures, a few degrees above freezing or withholding water for a few days. Such plants undergo *hardening*, giving them limited resistance to extreme cold when it occurs. Greenhouse-grown seedlings can be hardened to cold by exposing them to cooler temperatures in a shade house and cutting back on irrigation for a few days before transplanting to the field.

Many plants are adapted to extreme cold through mechanisms that allow them to avoid cold. Deciduous perennial shrubs or trees that lose their leaves and go dormant during the cold period, bulbous plants that die back to the belowground plant parts, and annuals that complete their life cycle and produce seeds, are all examples of plants avoiding cold.

### Thermoperiod in Plants

Some plants need daily variation in temperature for optimal growth or development. In a classic paper in ecophysiology (Went, 1944), it was demonstrated that tomato plants grown with equal day and night temperatures did not develop as well as tomato plants grown with normal day temperatures and lower night temperatures. This response occurs when the optimal temperature for growth — which takes place mostly at night — is substantially different from the optimal temperature for photosynthesis — which takes place during the day.

Diurnal variation in temperature is encountered by plants in many natural ecosystems and open-field agroecosystems, but in very controlled agroecosystems such as greenhouses, the diurnal temperature variation is much less pronounced. In other situations, plants from climates with cool nights do not do as well in regions with relatively constant day and night temperatures, such as the humid tropics or in temperate continental regions during the summertime.

### Vernalization

Some plants need to undergo a period of cold, called *vernalization*, before certain developmental processes can take place. For example, in the California grasslands, many native herbaceous species will not germinate until after a cold spell of several days duration, even though rainfall may have already occurred. Since the timing of the first rain of the season in this area is highly variable and early rain is usually followed by a very dry spell before more consistent precipitation begins, if germination were to occur with the initial rainfall, most of the new seedlings would probably not survive. There is thus a selective advantage to delaying germination until after vernalization has occurred.

Many agricultural and horticultural plants respond to vernalization. Lily bulbs, for example, are treated with cold at the appropriate time before planting so that they can be blooming for Easter in north temperate areas. In other cases, seeds of crops are treated with cold before planting in order to ensure more uniform germination.

## MICROCLIMATE AND AGRICULTURE

Temperature has thus been discussed as a factor of climate. Climate is made up of the fairly predictable, but highly variable, patterns in atmospheric conditions that occur over the long term in a certain geographic area. Climatology, or the study of climatic patterns, can tell us what the average temperatures for any particular part of the earth might be, and the degree of variation from the average that can be expected. There is little chance in the near future that humans will be able to intentionally modify climate on any kind of large scale. This is especially true for temperature. The large-scale aspects of climate, such as cold fronts, windstorms, and rainfall patterns, are best dealt with by selecting crops adapted to the range of climatic conditions that are expected.

But at the level of the individual crop organism or crop field, there is an aspect of climate that can be managed — the *microclimate*. Microclimate is the localized conditions of temperature, humidity, and atmosphere in the immediate vicinity of an organism. According to some definitions, the microclimate is made up of the conditions in a zone, four times the height of the organism being considered. Although microclimate includes factors other than temperature, farmers are most likely to be concerned with temperature when modifying microclimate or taking advantage of microclimatic variations.

### Microclimatic Profile

Within a cropping system, the conditions of temperature, moisture, light, wind, and atmospheric quality vary with specific location. Conditions just above the canopy of the cropping system can be very different from those in the interior, at the soil surface, and below the soil into the root zone. The specific microclimatic conditions along a vertical transect within a cropping system form what is called the microclimatic profile of the system. Both the structure of the system and the activities of the component parts have impact on the microclimatic profile. The profile also changes as the component plant species develop.

Table 5.1 shows the microclimatic profile of a corn, bean, and squash intercropping system in a schematic form, with each factor measured in relative terms through five layers of the canopy. In such a system, the microclimatic profile is very different at each stage of development, from early germination to full growth.

The belowground microclimate profile is also important; it extends from the soil surface to a small distance below the deepest roots of the crop plants. Under certain conditions, the soil and atmospheric microclimates of a crop may be so different as to cause problems for the crop. For example, warm wind currents when the soil is very cold can cause desiccation of the aboveground part of the plant since the roots are unable to absorb water fast enough to offset water loss.

**TABLE 5.1**
**Schematic microclimatic profile of a mature corn/bean/squash intercrop system, showing relative levels of five factors at each layer in the canopy at midday**

| | Temperature | Wind speed | Water vapor | Light | $CO_2$ |
|---|---|---|---|---|---|
| above corn canopy | medium level | highest level | lowest level | highest level | high level |
| upper corn canopy | highest level | high level | low level | medium level | lowest level |
| mid-interior | high level | medium level | medium level | low level | lowest level |
| below squash leaves | low level | low level | high level | lowest level | high level |
| soil surface | lowest level | lowest level | highest level | lowest level | highest level |

Legend: highest level; high level; medium level; low level; lowest level

Source: adapted in part from Montieth 1973

## Modifying the Temperature Microclimate

Through appropriate design and management, the microclimate of a system can be modified. Such modification is especially important if the goal of the farmer is to create or maintain microclimatic conditions that favor the sustainability of the cropping system. If this is the case, each modification must be evaluated as much as for its contribution to short-term yield and market return as for its contribution to the longer-term sustainability of the system.

Although microclimate includes many factors, its modification is often focused specifically on temperature. Practices and techniques used to modify the temperature microclimate are described below. Although modification of temperature is the main purpose of these practices, they will also impact other factors of the microclimate, such as humidity and light.

### Canopy Vegetation

Trees or other tall plants that create a canopy over the other plants in a system can greatly modify the temperature conditions under the canopy. Shade from the canopy reduces solar gain at the surface of the soil, as well as helps the soil retain moisture. Agroforestry systems in the tropics are a good example of this kind of practice.

The data from a study in Tabasco, Mexico (Gliessman, 1978c) clearly show the temperature-modifying effects of trees. In this study, the temperature microclimate of a tree-covered cacao orchard was compared with that of a nearby open grass pasture. As shown in Figure 5.6, temperature changes over a 24-h period at various levels in the cacao plantation were much more moderate than they were at the same levels in the pasture system. The pasture system became warmer during

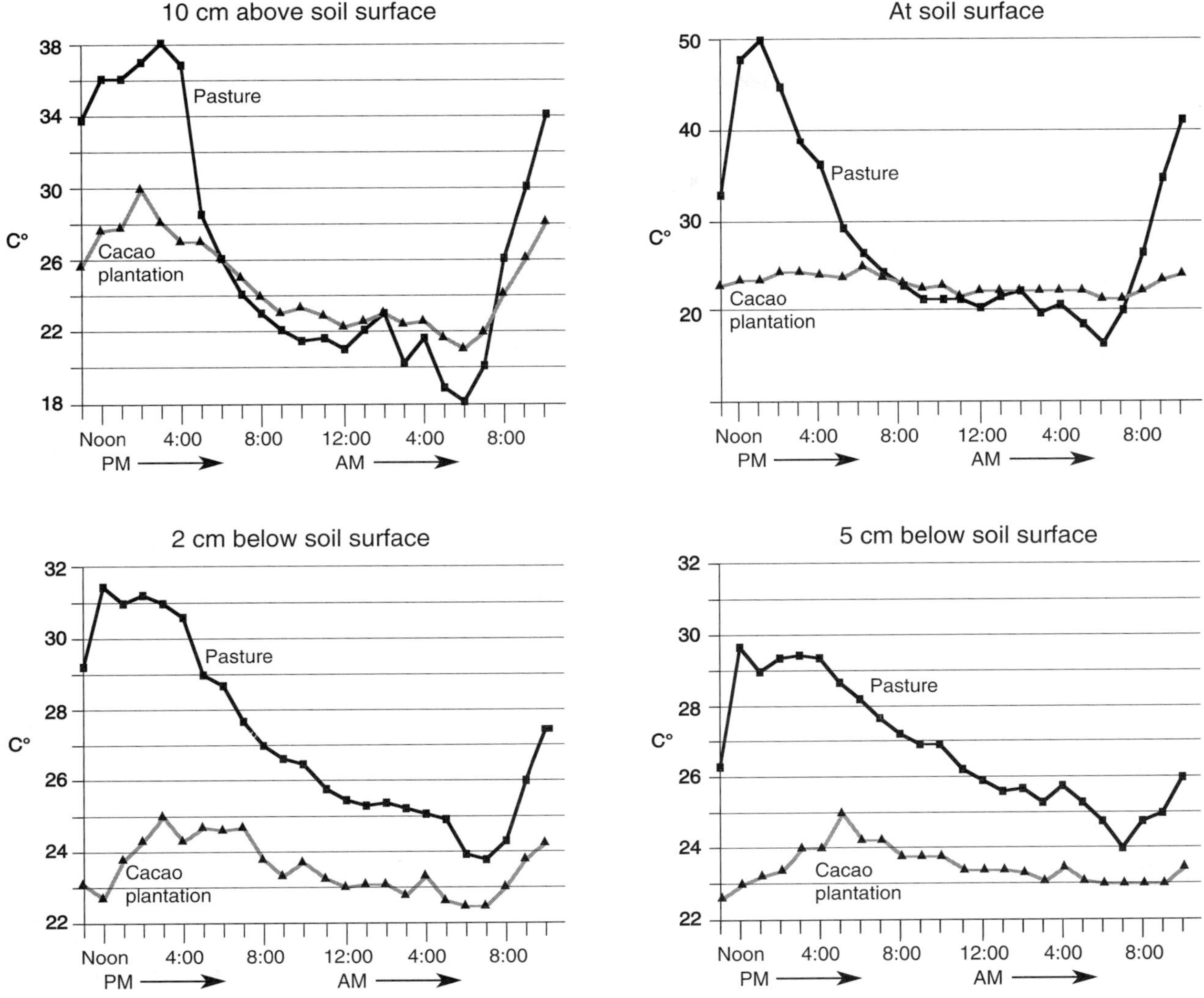


**FIGURE 5.6 Temperature changes over a 24-h period at four different levels in an open pasture and in a tree-covered cacao plantation in Tabasco, Mexico.** The presence of trees in the cacao system moderates temperature changes at all levels, keeps belowground temperatures lower than those in open pasture, and keeps aboveground temperatures higher at night. A similar pattern is shown for relative humidity: in the pasture system, humidity fluctuates more over a 24-h period than it does in the cacao system. Note that the scales on the vertical axes are not all identical (Data from Gliessman S.R., 1978c. Colegio Superior de Agricultura Tropical).

the day than the cacao system, and became colder aboveground during the night.

## Nonliving Canopies

Other means of creating a canopy for a cropping system are possible as well. Floating row covers of nylon fiber, for example, have been used over organic strawberries in California during the early winter season in an attempt to allow more insolation of the soil surface below, yet provide a localized greenhouse effect for reradiated heat given off from the soil surface. Figure 5.7 shows the results of one study of this practice, in which temperatures in the upper 5 cm of the soil were significantly raised during the critical root and crown development period for the strawberry plant (Gliessman et al., 1996).

There has also been considerable research and practical experimentation in the use of "hoop houses" or plastic tunnels for vegetable production in California, Spain, and elsewhere (Illic, 1989). Wire or plastic hoops are placed over planted beds in the field, and then covered with plastic or cloth. The localized greenhouse effect of these structures traps and holds additional heat during the day, and the covering reduces heat loss during the night. Hoop houses can allow for the earlier planting of warm-weather crops such as tomatoes or peppers, or the extension of the cropping season into the fall or early

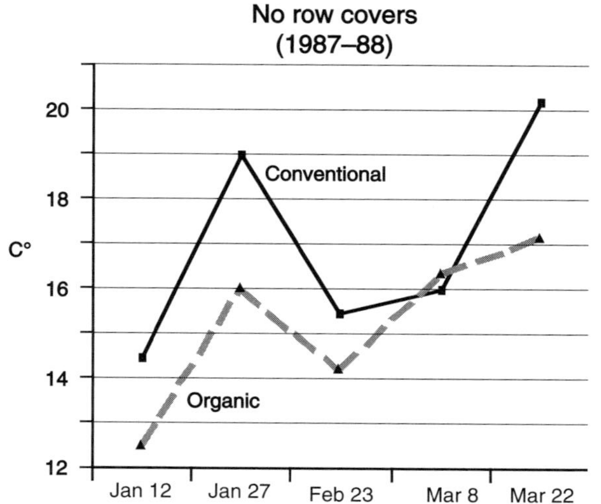


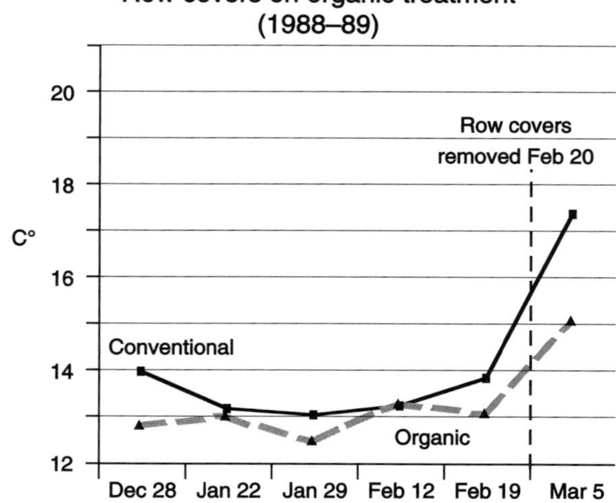


**FIGURE 5.7 Effect of floating row covers on soil temperature in an organic strawberry system.** When strawberries are grown under conventional methods, it is possible to use clear plastic as a soil-temperature-elevating soil covering during the winter, because weeds have been killed by prior soil fumigation. In organically-grown strawberries, black plastic must be used instead to prevent weed growth. Black plastic, however, is less efficient than clear plastic in raising the soil temperature, as shown in the left-hand graph. In an attempt to compensate for this difference, nylon floating row covers were placed over the organic strawberries during the second year of the study. As shown in the right-hand graph, the row covers were successful in narrowing the soil temperature differences between the conventional and organic treatments during the period the covers remained on the beds. (Data from Gliessman et al., 1996. *California Agriculture* 50: 24–31.)

winter where light frost becomes possible. Due to their high cost, these structures are mostly restricted to use with higher value crops (Figure 5.8).

## Soil Surface Cover

Changes in the soil temperature microclimate can be induced by covering the surface of the soil. Growing a cover crop is one well-recognized method of modifying soil temperature. The cover crop shades the soil, hence lowering soil temperatures, and has additional positive impacts on soil organic matter content, weed seed germination, and moisture conservation. When a cover crop is planted in between active crop plants, it is often called a *living mulch*. A living mulch can change the albedo of the soil surface, making it less reflective and raising the temperature of the air immediately above the crop. A living mulch can also have the opposite effect on temperature by increasing evaporation off of the vegetation.

Nonliving mulches, of either organic or inorganic materials, can change the temperature microclimate as well; their effect depends on the color, texture, and thickness of the material. Straw from crops such as wheat, oats, and barley are commonly used for a dry mulch, as are many other kinds of crop residues or grasses gathered from fallow fields, gardens, or nearby noncrop areas. Aquatic plants such as water hyacinth (*Eichornia crassipes*) or duckweed (*Lemna* spp.), usually considered a problem in waterways, especially in tropical areas, can be pulled from the water and applied as mulch. Plant-derived mulches eventually get incorporated into the soil, benefiting soil organic matter content. In recent times, some nonplant mulching materials have become popular; these include newspaper, cloth, and plastic sheeting. Specialized horticultural papers have been developed that biodegrade after a period of time and can be worked back into the soil.

A practice with effects similar to those of adding a mulch is to let a mulch accumulate naturally. This is accomplished through the use of a no-till system. Crop residues are left on the soil surface, forming a mulch that modifies the temperature of the soil and prevents moisture loss.

A final kind of practice is to change the color of the soil surface; to alter its albedo and the amount of solar energy it absorbs. Burning crop residue is one way of doing this. Residue burned to carbon black will absorb a greater amount of heat, and residue burned to ash white will absorb less heat.

## Greenhouses and Shade Houses

Shade houses and greenhouses are now common ways of modifying the temperature environment at the microclimatic level. Shade houses block a portion of incoming solar radiation, lowering solar gain and temperature.

**FIGURE 5.8 Hoop houses protecting frost-sensitive crops.** The hoop-house coverings, acting as a nonliving canopy, are put in place at the end of the day to trap heat and reduce nighttime heat loss; in the morning they are removed to allow light to reach the crop. Frost is still visible on the ground just outside the shadow of the center hoop house.

Greenhouses, on the other hand, are more often used to conserve or trap heat. Light energy penetrates the glass or plastic cover on a greenhouse, and inside it is absorbed and reradiated as long-wave heat energy. The reradiated energy then becomes trapped inside the greenhouse. During extended cold or cloudy periods, growers can heat the interiors of their greenhouses from many different sources. Recirculating hot water is often used to heat the floors of greenhouses, or at least provide heat on benches in the houses for germination or early plant development.

At certain times of the year or in particular climate zones, excess heat can be trapped in a greenhouse, requiring venting and air cooling. Another way of reducing greenhouse temperatures is to block some of the incoming solar radiation with shade cloth or other materials. Sophisticated greenhouse management now employs computer technology and automation to achieve remarkable levels of microclimate control.

### Methods of Preventing Frost Damage

In more temperate regions of the world, especially at higher elevations and latitudes, frost damage early or late in the growing season may be a constant danger. Mulching and row covers are important ways of providing some frost protection, but other means exist as well (Figure 5.9).

Raising soil moisture with irrigation when frost is expected may help raise temperatures close to the ground because evaporation of the moisture transfers heat from the soil to the evaporated water vapor, which then surrounds the crop plants. The increased atmospheric moisture itself provides some protection for the plants.

In low-lying areas subject to cold air drainage at night, farmers have long employed relatively simple means of raising the temperature to few degrees necessary to avoid frost damage. One technique is smudging, in which some kind of fuel — diesel fuel, garbage, old tires, or plant material — is burned to generate heat-trapping smoke or to create enough air turbulence to keep cold air from settling in depressions during a calm night. Recent concerns about health hazards and air pollution have reduced the use of smudging, however, and prompted farmers to use large fans to keep the air moving in frost-prone areas. Obviously, such techniques work only under certain conditions and when a few degrees of temperature difference will matter.

## Temperature and Sustainability

Temperature is a factor of considerable agroecological importance. Part of dealing with this factor is understanding local climatic and weather patterns and how larger-scale patterns may affect them. Another part is knowing how to control and modify the microclimate. Farmers have employed techniques for modifying the microclimate for a long time, and modern scientific knowledge has provided many new ones. Yet, agriculture still faces the challenge of finding more and better ways to design agroecosystems that modify microclimate themselves rather than relying on costly and often nonrenewable external inputs.

**FIGURE 5.9 Precise microclimate control in a greenhouse.** Hot water circulating in tubing below germination trays maintains warm soil temperatures for vegetable seedlings destined for early season transplanting.

## FOOD FOR THOUGHT

1. Describe several examples of farmers being able to grow crops in an area subject to temperature extremes greater than the normal tolerance levels for the particular crop species. What is the ecological basis for success in such situations?
2. What are some examples of food crops you now consume during a time of the year when temperature regimes in your local region would normally not allow them to be grown?
3. How might global climate change alter our patterns of food production and consumption?
4. Although we probably will never be able to intentionally control temperature conditions at the climatic level, we can manage temperature at the microclimate level. How it is possible to modify the microclimate to extend the growing season for a crop? How it is possible to allow planting earlier in the season? How it is possible to allow planting at a higher elevation? How it is possible to protect a crop from excessively high temperatures?

## INTERNET RESOURCES

Global Climate Change Research Reporter
www.exploratorium.edu/climate

Pew Center on Global Climate Change
www.pewclimate.org

Intergovernmental Panel on Climate Change
www.ipcc.ch

NASA GISS Surface Temperature Analysis (GISTEMP)
data.giss.nasa.gov/gistemp

Western Regional Climate Center
www.wrcc.dri.edu

## RECOMMENDED READING

Bonan, G.G. 2002. *Ecological Climatology: Concepts and Applications*. Cambridge University Press. This book integrates the perspectives of atmospheric science and ecology to describe and analyze climatic impacts on natural and managed ecosystems. In turn, it discusses the feedback mechanisms on climate from the use and management of land by people. The book includes detailed information on the science of climatology, as well as specific chapters on the interactions between climate and terrestrial ecosystems, including agroecosystems and urban ecosystems.

Geiger, R. 1965. *The Climate near the Ground*. Harvard University Press. Cambridge, Mass. The most thorough treatment of the field of micrometeorology, or the study of the microclimate within 2 m of the surface, where most crop organisms live.

Hellmers, H. and I. Warrington. 1982. Temperature and Plant Productivity. In: Recheigl Jr., M. (ed.) *Handbook of Agricultural Productivity.* Vol. 1. CRC Press: Boca Raton, Florida. pp. 11–21. A review of the complex relationships between temperature and plant growth and development, with a particular focus on crop plants.

Oliver, J.E. and J.D. Hidore. 2002. *Climatology: An Atmospheric Science*. Prentice Hall. A textbook on climate patterns, processes, and dynamics.

Reddy, K.R. and H.F. Hodges. (eds.) 2000. *Climate Change and Global Crop Productivity.* Oxford University Press & CABI. An edited volume by leading international experts, which presents a comprehensive examination of the potential effects of climate change on agricultural systems around the world. It includes chapters focusing on specific crops, agroecosystems and agroecological processes, mitigation strategies, and socioeconomic impacts.

Rosenzweig, C. and D. Hillel. 1999. *Climate Change and the Global Harvest: Potential Impacts of the Greenhouse Effect on Agriculture*. Oxford University Press. One of the most authoritative analyses of the potential impacts of climate change on agriculture.

# 6 Humidity and Rainfall

A place's natural vegetation is usually a reliable indicator of its rainfall regime. Deserts, with their sparse, slow-growing vegetation, tell the observer that the local annual rainfall is minimal. The lush vegetative growth of tropical and temperate rainforests points to abundant rainfall through at least most of the year. Rainfall amounts and vegetation have this direct relationship because for most terrestrial ecosystems, water is the most important limiting factor.

Water is also a primary limiting factor in agroecosystems. Agriculture can be practiced only where there is adequate rainfall or where it is possible to overcome, through irrigation, the limits imposed by a dry climate.

In this chapter, we discuss the agroecological significance of water in the atmosphere, both as humidity and as precipitation. Despite this focus, the reader should keep in mind that water in the atmosphere is only one aspect of a larger set of environmental factors affecting plants — those involving the atmosphere as a whole. Patterns of movement and change in the atmosphere influence not only rainfall patterns but also wind and variations in temperature. Combined atmospheric factors make up climate (when we are referring to the annual average conditions) and weather (when we are referring to the climatic conditions at one moment in time).

## WATER VAPOR IN THE ATMOSPHERE

Water can exist in the atmosphere in a gaseous form (as water vapor) or in a liquid form (as droplets). At constant pressure, the amount of water vapor that air can hold before it becomes saturated, and its water vapor begins to condense and form droplets, is dependent on temperature. As the temperature of the air goes down, the amount of water that can be held in vapor form goes down as well. Because of this dependence on temperature, humidity — the amount of moisture in the air — is usually measured in relative terms rather than according to the absolute amount of moisture in the air. *Relative humidity* is the ratio of the water-vapor content of the air to the amount of water vapor the air can hold at that temperature. At a relative humidity of 50%, for example, the air is holding 50% of the water vapor it could hold at that temperature. When the relative humidity is 100%, the air is saturated with water vapor, and water vapor condenses to form mist, fog, and clouds.

Relative humidity can change as a result of either changes in the absolute amount of water vapor or changes in temperature. If the absolute amount of water vapor in the air is high, small variations in temperature can greatly influence relative humidity. A drop of a few degrees in temperature in the evening or morning hours, for example, can push the relative humidity to 100%. Once relative humidity reaches 100%, water vapor begins to condense into water droplets, and shows up as dew. The temperature at which this condensation begins to occur is called the *dew point*.

In natural systems, the interaction of temperature and the air's moisture content can be a very important factor in determining the structure of an ecosystem. The redwood forest community along the coast of California is a good example. Cold ocean currents condense the moisture-laden air over the ocean, forming fog. The occurrence of fog almost every night during the dry summer months compensates for the lack of rainfall and is believed to be the main reason that redwoods still exist where they do. Some studies estimate that fog and dew add at least an extra 10% to the effective total of rainfall for redwood regions.

For similar reasons, humidity can affect agroecosystems. Crops grown in the redwood forest region, for example, may benefit from the extra moisture that fog and dew provide; farmers of crops such as Brussels sprouts, lettuce, and artichokes use less water as a result.

## PRECIPITATION

Although dew and fog can contribute significant quantities of moisture to some regions, the primary (natural) source of water for agroecosystems is precipitation, usually in the form of rain or snow. Precipitation contributes moisture to the soil directly, and in irrigated agroecosystems, it does so indirectly by being the ultimate source of most irrigation water.

### THE HYDROLOGICAL CYCLE

Precipitation is part of the *hydrological cycle*, a global process moving water from the earth's surface to the atmosphere and back to the earth. A diagram of the hydrological cycle is presented in Figure 6.1. The core of the hydrological cycle is made up of the two basic physical processes of evaporation and condensation. Evaporation occurs at the earth's surface, as water evaporates from soil, bodies of

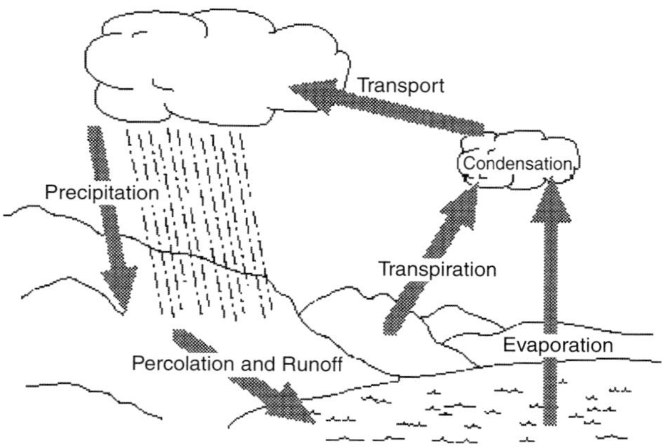


**FIGURE 6.1 The hydrological cycle.**

water, and other wet surfaces. Evaporation of water from inside the bodies of plants also occurs on the surface of leaves. This kind of evaporation, called transpiration, is part of the mechanism, by which plants draw water from the soil into their roots (Chapter 3). Evaporation from all these sources is collectively termed "evapotranspiration."

When the absolute amount of water vapor in the air is sufficient to approach or exceed 100% relative humidity, condensation begins to occur. Small water droplets form and aggregate to create clouds. Precipitation occurs when droplets of water in clouds become heavy enough to fall. This usually happens when the moisture-containing air rises (by being forced up a mountain by winds or rising on currents of warm air) and begins to cool. As the air cools, its ability to hold moisture in vapor form or as very small cloud droplets begins to decrease, resulting in more condensation and aggregation of droplets. This cooling and condensing process is called adiabatic cooling. The precipitation formed by adiabatic cooling falls to earth, enters watersheds or the ocean, and eventually returns to the atmosphere.

## TYPES OF RAINFALL

The precipitation part of the hydrological cycle is highly variable. Masses of moisture-laden air are constantly being moved over the earth's surface by the complex movements of the atmosphere. Rainfall (and other forms of precipitation) occurs locally in different ways depending on latitude, season, temperature, topography, and the movement of the air masses. In general, however, rainfall can be classified into three types depending on the mechanism that produces the adiabatic cooling of the moist air mass.

### Convective Rainfall

Convective rainfall occurs when high levels of solar gain heat the air close to the ground, causing it to rise rapidly, cool, and condense the moisture it contains. Often the rising air draws moisture-laden air from some distant source such as a lake, gulf, or ocean. The rain associated with summer thunderclouds is an example of convective rainfall. High winds, and even tornadoes, can accompany these storms, as can lightning and localized fires. In many regions, such as the American Midwest, agroecosystems are dependent on this type of rainfall, at least at certain times of the year. Traditional Hopi agriculture in the southwest of the U.S. is completely dependent on convective rainfall with the torrent, which often accompanies these storms being channeled down, washes from the mountains and then spreads out over planted fields at the mouths of the canyons.

### Orographic Rainfall

Orographic rainfall occurs when a moisture-laden air mass meets a mountain range that forces it up into the cooler layers of the atmosphere. Such precipitation occurs on the western flanks of California's Sierra Nevada — as rain in the foothills and as snow at the higher elevations. This precipitation is an important replenisher of streams and aquifers, which later become sources of irrigation water downstream in drier locations. Agriculture in a region such as the Great Central Valley of California would not be possible without orographic precipitation in nearby mountains.

### Cyclonic Rainfall

This type of rainfall is associated with areas of low atmospheric pressure that form over the ocean. Warm, moisture-laden air rises, creating a low-pressure area. As this air rises, it cools, forms precipitation, and then falls back toward the ocean surface where it can collect more moisture. In addition, the air currents of this self-perpetuating system begin to revolve counter-clockwise around the low-pressure area, and the entire system begins to move. The revolving air currents form the characteristic cyclonic storms and frontal systems, which we can see on weather maps. When one of these cyclonic systems moves ashore, the moisture-laden air masses may be forced up against mountain masses, creating rainfall with both orographic and cyclonic causes (Figure 6.2).

**FIGURE 6.2 A cyclonic storm system over the eastern Pacific.**

## DESCRIBING RAINFALL PATTERNS

Each region of the earth has its characteristic patterns of precipitation. The total amount of precipitation received in a typical year, its distribution throughout the year, the intensity and duration of precipitation events, and the regularity and predictability of the precipitation patterns are all important determinants of the opportunities for, and constraints upon, agriculture in a particular region.

Below, these facets of rainfall patterns are described using rainfall data collected by the author from the Cuyama Valley, California. These data are shown in Table 6.1.

- **Average total annual rainfall.** The total amount of precipitation that falls in an area during an average year is a good indicator of the moistness of that area's climate. From an ecological perspective, however, it is also important to know how much variability there can be in this rainfall amount from one year to the next. Extremes at either end of the average can have significant negative impact on an agricultural system, even if that extreme only occurs rarely. Table 6.1 shows that in the Cuyama Valley the annual total is highly variable: during the data collection period there were three drought years, 3 years of near-normal precipitation, one wet year, and one excessively wet year (associated with El Niño patterns in the Pacific Ocean).
- **Distribution and periodicity.** This refers to how rainfall is spread out through the year, both on average and during a specific year. In many parts of the world, rainfall is distributed in such a way as to create predictable wet and dry periods; the Cuyama Valley, where precipitation is largely confined to the period from October to May, is a good example. Within this overall climatic distribution pattern, however, rainfall is often distributed differently each year: if the data for the Cuyama Valley were graphed, for example, the peaks and valleys for each year would not correspond, and some years, such as 2004 to 2005, would show much more evenly distributed rainfall than others.
- **Intensity and duration.** The absolute amount of rainfall in a long time period such as a month or even a day does not fully describe the ecological relevance of the rainfall. How intense the rainfall is, and for what length of time that rainfall occurs, are important aspects. Two inches of rainfall in less than an hour can have very different ecological impacts than a 2-in. rain spread over 24 h. For example, of the 12.66 in. of rainfall recorded during February 1998 in the Cuyama Valley, over 8 in. fell in one 3-h rainfall event, with associated excessive runoff and flooding.
- **Availability.** It is also important to know how much of the rainfall becomes available as soil moisture. Does it penetrate into the root zone? What were the weather conditions immediately following the rainfall event? What was the temperature and what were the wind conditions?
- **Predictability.** Every region has a characteristic degree of variability in its rainfall patterns. The higher the variability, the less predictable is the rainfall for any particular time period.

**TABLE 6.1**
**Monthly and Seasonal Rainfall Totals in Inches at Cottonwood Canyon, Cuyama Valley, Santa Barbara County, California**

| Season | Sept | Oct | Nov | Dec | Jan | Feb | Mar | Apr | May | Total |
|---|---|---|---|---|---|---|---|---|---|---|
| 1996–1997 | 0.0 | 2.3 | 2.12 | 4.31 | 5.6 | 0.37 | 0.0 | 0.0 | 0.0 | 14.7 |
| 1997–1998 | 0.2 | 0.1 | 3.65 | 4.93 | 6.75 | 12.66 | 3.76 | 1.78 | 1.82 | 35.65 |
| 1998–1999 | 1.43 | 0.18 | 0.87 | 0.93 | 0.23 | 3.4 | 2.29 | 0.85 | 0.0 | 10.18 |
| 1999–2000 | 0.0 | 0.0 | 0.9 | 0.04 | 1.91 | 2.99 | 4.85 | 2.6 | 0.18 | 13.46 |
| 2000–2001 | 0.0 | 1.06 | 0.02 | 0.17 | 5.32 | 5.05 | 5.6 | 2.35 | 0.0 | 19.52 |
| 2001–2002 | 0.5[a] | 0.58 | 2.4 | 2.54 | 0.08 | 0.8 | 0.87 | 0.03 | 0.2 | 8.2 |
| 2002–2003 | 0.0 | 0.0 | 3.73 | 2.06 | 2.28 | 1.64 | 2.3 | 0.95 | 1.2 | 14.16 |
| 2003–2004 | 0.8[b] | 0.45 | 0.44 | 1.88 | 0.42 | 1.98 | 2.90 | 0.1 | 0.0 | 9.05 |
| 2004–2005 | 0.0 | 4.25 | 0.06 | 4.32 | 7.06 | 2.25 | 2.30 | 0.66 | 0.75 | 21.65 |
| Averages | 0.33 | 0.99 | 1.58 | 2.35 | 3.29 | 3.46 | 2.76 | 1.04 | 0.46 | 16.29 |

Rainfall from June to August is usually negligible.

[a] Includes 0.5 in. from July.

[b] Includes 0.88 from late July/early August.

## ACID PRECIPITATION

Rainfall is the lifeblood of both natural ecosystems and most agroecosystems. Yet in many areas of the world, it is poisoning the very systems it supports. The rain (and snow) that falls in these areas is acidic enough to damage crops and forests, kill fish and other aquatic organisms, and acidify the soil.

Acid precipitation is just one of the many consequences of humans' pollution of the atmosphere. The burning of fossil fuels in automobiles and in power plants dumps large amounts of nitrogen oxides and sulfur oxides into the atmosphere near urban and industrial areas. These compounds are called acid precursors; they easily combine with atmospheric water to form nitric acid and sulfuric acid. These acids are then dissolved into the naturally occurring water droplets of the atmosphere. After drifting for some distance, the droplets eventually fall as acid precipitation. (Oxides of nitrogen and sulfur can also form nitrates and sulfates in the atmosphere and "rain" down as solid particles; when they combine with water, these particles are converted to acids and have the same effect as acid rain.)

Rainfall in an unpolluted environment is naturally slightly acidic — whereas pure water has a neutral pH of 7.0, the pH of natural rainfall is about 5.7. This normal acidity is the result of atmospheric carbon dioxide dissolving in water droplets in clouds and forming weak carbonic acid. Thus acid precipitation is a problem only where human-produced acid precursors lower the pH of the precipitation to less than 5.7. Where this will occur depends on rainfall and wind patterns and on the location of significant anthropogenic sources of acid precursors. Prevailing wind patterns, for example, tend to carry acid precursors from urban areas and power plants in the northeastern U.S. to the Adirondack Mountains of New York state. There, the rainfall has an average pH of about 4.1, and has been measured as low as 2.3. Other areas where acid precipitation is a problem include northern Europe, much of the eastern U.S., southeastern Canada, and parts of southern California. The distribution of acid rainfall, however, is extremely variable, and almost any area can receive it.

Acid precipitation has been shown to have many deleterious effects. Aquatic ecosystems are particularly vulnerable; years of acid precipitation has acidified many of the lakes in mountainous areas of the eastern U.S. and Canada and left them virtually lifeless. Acid rain also damages forests: it harms needles and leaves, impairs seed germination, and erodes protective waxes from leaves.

In the U.S., some progress has been made in reducing sulfur and nitrogen oxide emissions from power plants since 1980, leading to small decreases in acid deposition and a reduction in the acidity of many lakes and streams. Acid precipitation continues to occur, however, and further reductions in acid-precursor emissions will be necessary for already-damaged ecosystems to recover (NAPAP, 2005). Acid precipitation continues to be a serious problem in many parts of the developing world, particularly China and other parts of Asia.

The extent to which acid precipitation damages agroecosystems is difficult to assess. Some studies have shown decreased crop productivity and inhibition of the dark reactions of photosynthesis. Other studies have documented damage to leaves and buds and the leaching of calcium from leaves. Where the soil lacks the ability to neutralize acids, acid precipitation has caused acidification of the soil and changes in nutrient availability. Although the distribution of acid precipitation varies, and different crops and soils have different levels of sensitivity, acid precipitation is a global problem with the potential for significant direct and indirect effects on agriculture.

> The rainfall data in Table 6.1 show that the Cuyama Valley has fairly high variability, for example. Based on these data, a farmer could not count on there being at least 1 in. of rain in April, even though the 6-years average for that month is 1.04 in.

Additional aspects of rainfall may be relevant from an agroecological perspective as well. For example, it may be important to know how much moisture was in the soil when rainfall occurred, as well as the stage of crop development. In the Paso Robles and Santa Maria regions of California, for example, two storms with total rainfall of about 1.5 in. occurred during the first 2 weeks of September in 1998. Since most grapes were still on the vine at this time, the rains damaged the crop (in most years, significant rainfall does not occur until early November, after the grapes have been harvested).

## RAINFED AGROECOSYSTEMS

Agriculture in most of the world is carried out using natural precipitation to meet the water needs of crops. These *rainfed agroecosystems* must adjust to the distribution, intensity, and variability of the rainfall that is characteristic of the local climate. The challenge is either to maintain a balance between precipitation (P) and potential evapotranspiration (PET) by manipulating evapotranspiration, or to somehow work around a water deficit (P–PET < 0) or a water surplus (P–PET > 0).

Several examples of how agroecosystems function within the constraints of local rainfall regimes are presented below, providing another way of examining the aspects of sustainability inherent in farming approaches that work with ecological conditions rather than striving for their alteration or control. These examples were chosen to cover the range from very wet to very dry rainfed agriculture. The aspects of managing moisture once it gets into the soil will be described in more detail in Chapter 9.

### Agroecosystems Adapted to a Long Wet Season

In very humid regions with extended rainfall, farmers are concerned more with excess water than with water deficits. Frequent and heavy rainfall creates problems of waterlogging, root diseases, nutrient leaching, abundant weed growth, and complications for most farming operations. Even wetland-adapted crops such as rice or taro are difficult to manage in regions with a long wet season. Conventional approaches to excess precipitation most often look to some type of major habitat modification such as drainage projects and flood control. An agroecological approach to an extended wet season, in contrast, looks for ways to accommodate the system to the excess moisture.

A very interesting and productive use of land that is flooded for the entire wet season is seen in Tabasco, Mexico (Gliessman, 1992a). This region receives more than 3000 mm of rainfall distributed over a long wet season that extends from May until February of the next year. The staple local crop of corn is planted on higher ground around wetlands that are shallowly flooded during most of the year. In March, however, the drop in rainfall permits the planting of another corn crop. Low-lying areas dry out enough for the soil surface to become exposed. Farmers follow the receding water line with this special corn planting, known locally as the March planting or *marceño*.

During much of the year, constant rainfall keeps the low areas inundated to a depth that ranges from a few centimeters to as much as a meter. The marsh vegetation that densely covers these areas during the wet season is felled quickly with machetes as the water level recedes. A very dense, 10 to 20 cm mat of organic matter is produced by this process. Seed is planted into holes made with a pointed stick driven into the mat. About a week after the sowing, fire is used to burn part of the organic mat, as well as to kill back any weed seedlings or sprouts of the marsh plants. The burning must be timed so as to burn only the dry leaves on top of the mat and not the moist lower layers or the soil. The corn seed, planted 10 to 15 cm below the surface of the soil, is not harmed by the fire. Local short-cycle varieties of corn (2 to 3 months from planting to harvest) are most frequently used. The practice of using seed from the previous harvest for the subsequent planting favors the use of local varieties, rather than the purchase of hybrid or "improved" seed produced at distant locations. The name of one corn variety — *mejen*, from a Maya word meaning "precocious" or "early maturing" — shows the link to the past that this system may have (Figure 6.3).

The corn grows very quickly in this system, and when fire is not used excessively and flooding is allowed to occur every year, weeding is usually not necessary. After about two and a half months of growth, the mature corn stalks are "doubled-over" just below the corn ear, facilitating final drying of the grain for another 2 to 4 weeks before harvest. Yields of 4 to 5 t/ha of dry grain are common, with some yields reaching 10 t/ha. This is many times the average yield of 1 to 1.5 t/ha for mechanized production on lands that have been cleared and drained in the same region. These greater yields are obtained at a fraction of the input costs and labor invested in mechanized production systems (Amador, 1980).

Following the harvest, all crop and noncrop residues end up on the soil surface. This contributes to a key element in the productivity of the system — maintenance of organic matter in the soil. Soil profiles demonstrate the presence of a thick, organic-rich soil to a depth of 30 to 40 cm below the surface. During the 9-month inundation, organic matter produced by the marsh plants or left by the previous cropping cycle is incorporated into the soil and conserved in the anoxic conditions under water. In addition, nutrient minerals that enter the system with surface drainage are captured by the highly productive aquatic sector of the ecosystem. These factors result in the formation of a soil that has organic matter levels over 30%, total nitrogen as high as 3%, and high levels of other important plant nutrients. The key element in the management of this system, then, is the way in which inundation during the wet season is taken advantage of. When the system is drained artificially in an attempt to extend the cropping season, the organic layer in the soil can be reduced to 5 cm in less than 2 years, and yields drop dramatically.

### Agroecosystems Adapted to Alternating Wet–Dry Seasons in the Tropics

Many parts of the world have a monsoon-type climate in which average annual rainfall is relatively high, but nearly all the rain falls during a wet season of medium length. Farmers in these areas have to deal with excess rainfall at one time, and a lack of rainfall at another.

A very interesting and productive agroecosystem in such an alternating rainfall regime has been observed in the state of Tlaxcala, Mexico (González, 1986; Anaya, et al., 1987; Crews & Gliessman, 1991; Wilken, 1969). In an area known as the Puebla Basin, a triangular flood plain of about 290 km$^2$ is formed where the Atoyac and Zahuapan Rivers meet in the southern part of the state.

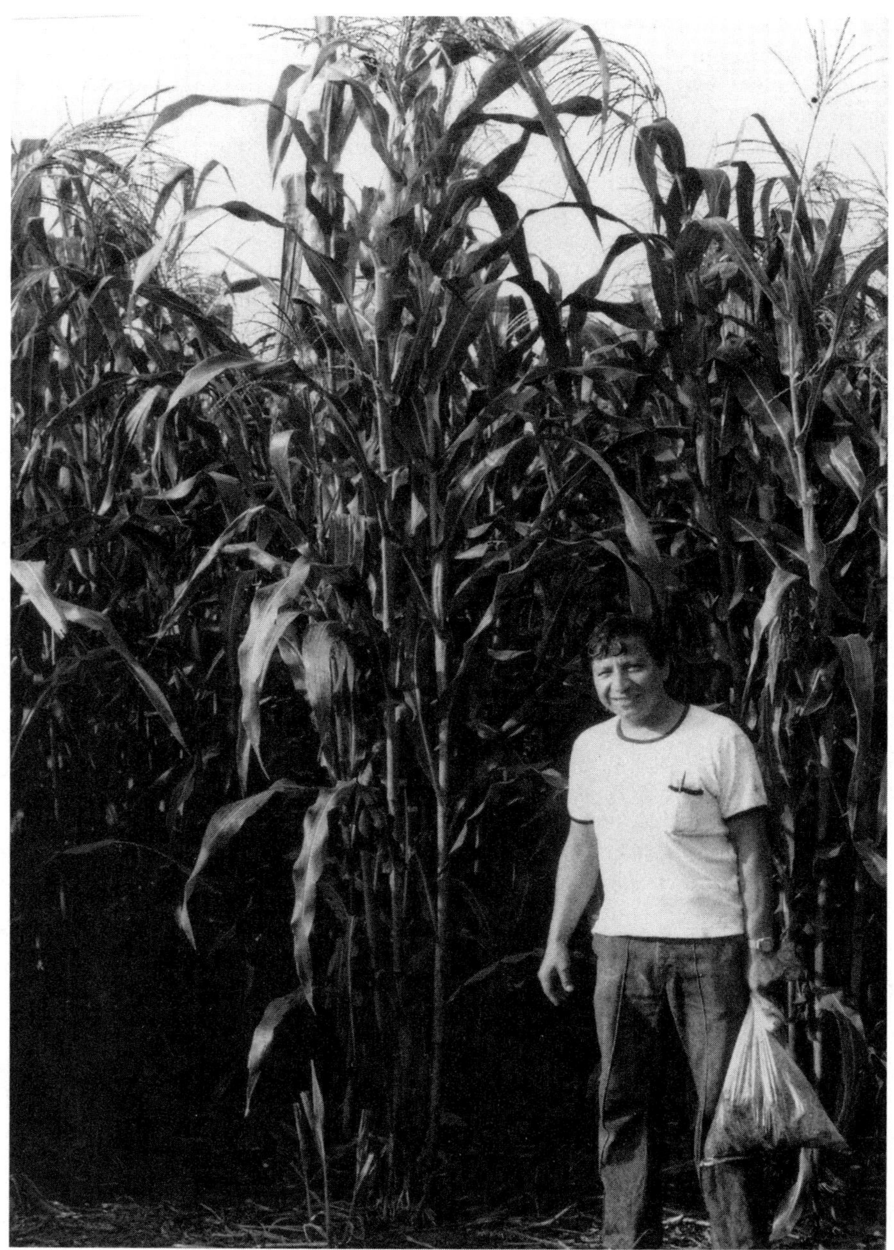

**FIGURE 6.3 The local variety of corn called *mején* close to maturity 10 weeks following planting in Cárdenas, Tabasco, Mexico.** This site is a wetland normally flooded for 8 to 9 months out of the year.

Average annual rainfall is about 700 mm. A large part of the basin floor has a water table less than three feet below the surface during much of the year, with soils that are poorly drained and swampy. In order to make such land agriculturally productive, most present-day agronomists would probably recommend draining the region so that large-scale mechanized cropping practices could be introduced. But the local, traditional cropping systems provide an alternative that makes use of the high water table and rainfall distribution in the watershed (Figure 6.4).

Using a system that is prehispanic in origin, raised platforms (locally called *camellones*) have been constructed from soil excavated from their borders, creating a system of platforms and canals (called *zanjas*). Individual platforms are 15 to 30 m wide, 2 to 3 m high, and 150 to 300 m long. A diverse mixture of crops are grown on the platforms, including intercropped maize, beans, and squash, vegetables, alfalfa, and other annuals. Crop rotations with legumes such as alfalfa or fava beans help maintain soil fertility, and the crop mixtures themselves help in weed control. Soil fertility is also maintained with

**FIGURE 6.4 A camellón (raised field) near Ixtauixtla, Tlaxcala, Mexico.** The field is planted with rotational strips of alfalfa and intercropped corn and beans; alder trees mark the edge of the canals dug to raise the field. The compost in the foreground is used as a fertilizer.

frequent applications of composted animal manures and crop residues. Much of the feed for the animals comes from alfalfa grown on the platforms, or from residues of other crops that cannot be directly consumed by humans (e.g., corn stalks). Supplemental feed for animals is derived from the noncrop vegetation (i.e., weeds) that is selectively removed from the crop area, or periodic harvests that are made of the ruderals and natives that grow either along the canals or directly in them as aquatic species. This latter source of feed can constitute a very significant component of livestock diets during the dry season.

A very important aspect of this traditional agroecosystem is the management of the complex set of canals. Besides originally serving as a primary source of soil for raising the platform surfaces, they also serve as a major reservoir of water during the dry season. Organic matter accumulates in the canals as aquatic plants die, leaves from trees along the canal borders fall into the water, and even weeds from the crop field are thrown into the canals. Soil from the surrounding hillsides and the platforms is also washed into the canals by the heavy wet-season rains. Every 2 to 3 years, the canals are cleaned of the accumulated soil and muck, with the excavated materials being applied as a nutrient-rich top dressing on the platforms.

The canals thus play a very important role in the sustainability of this agroecosystem. They function as a nutrient "sink" for the farmer, and are managed in ways that permit the capture of as much organic material as possible. Supplemental irrigation water can be taken from them in the dry season, and the plants rely greatly on moisture that moves upward through the soil from the water table by capillarity. The raised platforms provide suitable planting surface even during the peak of the rains. Water levels in the canals are controlled by an intricate system of interconnected canals that eventually lead to the rivers of the basin, but flow in the canals is very limited. Farmers often block the flow of canals along their fields during the dry season in order to maintain a higher water table, and even in the wet season, water flow out of system is minimal. Only at times of excessive rainfall, do appreciable quantities of water drain from the area. Rainfall is both an input and a tool in the management of the system, and permits year-round cropping.

### Agroecosystems Adapted to Seasonal Rainfall

Outside of the wet tropics, a common rainfall regime is one in which one or more wet seasons are interspersed with relatively long dry seasons. In these areas, crops are often planted at the beginning of the rainy season, grow and develop while there is moisture in the soil, and become ready to harvest at the end of the wet season or the beginning of the dry season.

This kind of wet-season cropping takes many forms. In much of the midwestern heartland of the U.S., for example, spring wheat, corn, and soybeans are planted in the late spring and depend on convective summer rainfall to develop. In Mediterranean climates around the world, the mild, wet winters and dry summers are appropriate for grain crops such as oats, barley, and rye grown in winter, with the land being left fallow or grazed during the summer unless irrigation can be provided.

A seasonally rainfed cropping system of considerable importance is the Mesoamerican corn/bean/squash polyculture system. Adapted to a wide range of rainfall intensities and amounts, this intercropping system is found throughout much of Latin America (Pinchinat, et al., 1976; Davis, et al., 1986; Laing, et al., 1984). These three crops are planted in many different arrangements, sequences, and patterns, sometimes only two of them together, and at other times all three. But regardless of the combination, it is the arrival of the rainy season that determines planting.

If shifting cultivation practices are used, clearing and burning takes place during the dry season. Sometimes farmers wait to burn until after the first rains of the wet season dampen the lower layers of the slash. Since these first rains are most often interspersed with periods of sun, the upper layer of the slash is dried enough between rains to carry a fire, while the newly acquired moisture below prevents excessive heat from reaching the soil. Crop seed is then planted into a mulch made up of nutrient-rich ash and a protective layer of unburned organic matter. This practice achieves the dual goals of nutrient supply and soil erosion protection. Soil protection is important in many areas where this crop system is used, since the early rains of the season occur most often as intense, convective showers.

Once the rains begin, crop seeds germinate and develop quickly, covering the soil and protecting it against the continued rains. The amount of time it takes for the crop to mature (from 4 months to 6 months) depends on the length of the wet season.

In areas such as the wet lowlands of Tabasco, Mexico, two corn crops can be planted because the wet season is longer and characterized by a bimodal distribution, with one rainfall peak in June/July and another in September/October. One crop is planted in May at the beginning of the wet season, with fire being used to clear the slash, and the crop (called *milpa de año*) being harvested in September. The second crop (called *tolnalmil*) is planted just following the second rainfall peak in late October or November for harvest at the beginning of the dry season in late February. The second crop depends greatly on the presence of residual soil moisture extending into the dry season, and since the crop is planted during the wet season, any slash on the surface at planting is not burned. Different local varieties of corn are used in each planting system.

## DRYLAND FARMING

In many parts of the world, rainfall during the cropping season does not meet the needs of the crop, either because the area does not receive enough rainfall to offset the moisture lost through evapotranspiration, or because the cropping cycle does not coincide with the wet season. The type of agriculture developed in such climates — when irrigation is not an option — is termed "dryland agriculture or *dry farming*."

Dryland agriculture is defined as crop production without irrigation in semiarid regions of the world, where annual rainfall is mostly between 250 and 500 mm. But total rainfall is only one influence on dryland agriculture; annual and seasonal variations in temperature and the type and distribution of rainfall are key factors as well. The traditional agriculture in most dryland regions is pastoral in nature, with cultivated crops limited to small areas farmed by hand tools or animal power. Today, mechanization has added a new dimension to dry farming, but the types of tillage, seeding management, and harvest procedures remain much the same. In many countries, hand labor still plays a major role.

The most important aspects of dry farming are (1) the use of some type of cultivation system that promotes the penetration of rainwater into the soil profile and its storage there, and (2) the frequent use of summer fallows or rest seasons to allow replenishment of water reserves depleted by cropping. Other practices can be important as well. Cultivation of the surface soil during the cropping cycle is used to control potential water-using weeds and to create a "dust mulch" of pulverized surface soil that reduces the proportion of large pores and therefore reduces evaporation. Drought-resistant cultivars are often planted to reduce moisture use. Altogether, these practices allow a much higher proportion of the moisture from rainfall to be channeled through the crop rather than to pass from the soil to the atmosphere.

The most highly developed modern dryland agricultural systems, at least in terms of intensive management and technology, are in Australia, Canada, and the U.S.

In all of these regions, grain crops are the primary focus. In Australia, however, wheat in rotation with grazing, especially for sheep and wool production, has led to the development of unique systems where a grain crop is grown alternately with pasture. Pasture actually allows for the replenishment of moisture reserves necessary to produce a grain crop.

A unique example of dry farming occurs in coastal central California, where several vegetable crops are planted, either from transplanted seedlings or direct seeding, at the beginning of the dry Mediterranean summer in May. Rarely does rainfall occur in summer in this climate; so these vegetable crops must rely solely on the moisture reserves stored in the soil. Tomatoes seem to be a crop that is particularly well suited to this system. Tomato seedlings are planted deeply into moist soil in May, with no irrigation applied. Cultivation of the soil surface maintains a weed-free dust mulch, and because the soil surface is dry and no rain occurs during the growing season, the plants are not staked or tied, and fungal disease is a minor problem. Harvest begins in late August and continues until the first rains of the new wet season, usually in late October or early November. Tomatoes harvested from this system have a reputation for more concentrated flavor (Figure 6.5).

The sustainability of dry farming systems must be weighed against the potential loss of soil organic matter from the upper soil levels with the dust mulch system, the danger of soil erosion from wind and rain because of the low level of soil cover, and the unpredictability of soil moisture availability as a result of variable rainfall during the fallow period. But as a way of farming in areas with low and upredictable rainfall, dry farming can be a low external-input alternative.

## WATER HARVESTING SYSTEMS IN ARID REGIONS

In warm regions of the world with arid climates (less than 250 mm annual precipitation), lack of rainfall is a severe limiting factor for agriculture. In many such places, however, rainfall does occur with some regularity in the form of short, torrential showers, and it is possible to "harvest" this water by collecting and concentrating rainfall runoff.

**FIGURE 6.5 Dry-farmed tomatoes, Santa Cruz, CA.** A cultivated soil mulch keeps moisture close to the surface and controls weeds during the rainless summer growing season.

In the Negev desert of Israel, once-abandoned systems of small catchment runoff farms have been reconstructed and made to produce crop yields equivalent to those of irrigated farms in the same region (Evenari, et al., 1961). The farm unit consists of catchment areas for rainfall on the slopes of the watershed surrounding flattened drainage channels where runoff is collected. Low rock walls channel rain runoff down into the small flood plain of the channels. This system can collect 20 to 40% of the rainfall that occurs, and removing loose rock from the soil surface on the hillsides can increase runoff collection to as much as 60%. Small rock checkdams in the larger channels at the bottom of the slopes concentrate runoff to a depth sufficient to allow water to penetrate to approximately 2 m into the soil, after which the soil dries and leaves a crust relatively impervious to evaporative water loss. As each check dam fills, it spills over into others below, watering a complex system of floodplain farm plots. Crop yields of grains such as barley and wheat, and fruits such as almonds, apricots, and grapes, are quite respectable for such an arid region. Rather than attempt to create large reservoirs of water that would mostly evaporate in such a climate (and accumulate nutrient-rich sediments), both water and nutrient-rich sediments are stored on-site in the water harvest system (Figure 6.6).

A similar system still is used in the arid American Southwest, where native American groups such as the Hopi and Papago have been practicing a form of water harvesting for many centuries (see the accompanying case study). The flow from heavy convective rainfall in the mountains during the summer is diverted over alluvial fans as a shallow sheet of runoff, rather than being allowed to concentrate in a stream channel. This sheet of water then "irrigates" annual crops of corn, beans, squash, and other local crops. The upper watershed is not manipulated as in the Negev system, but similar manipulation of runoff on the floodplain below takes place. The goals of both agroecosystems are to work within the constraints and limits of the natural rainfall regime.

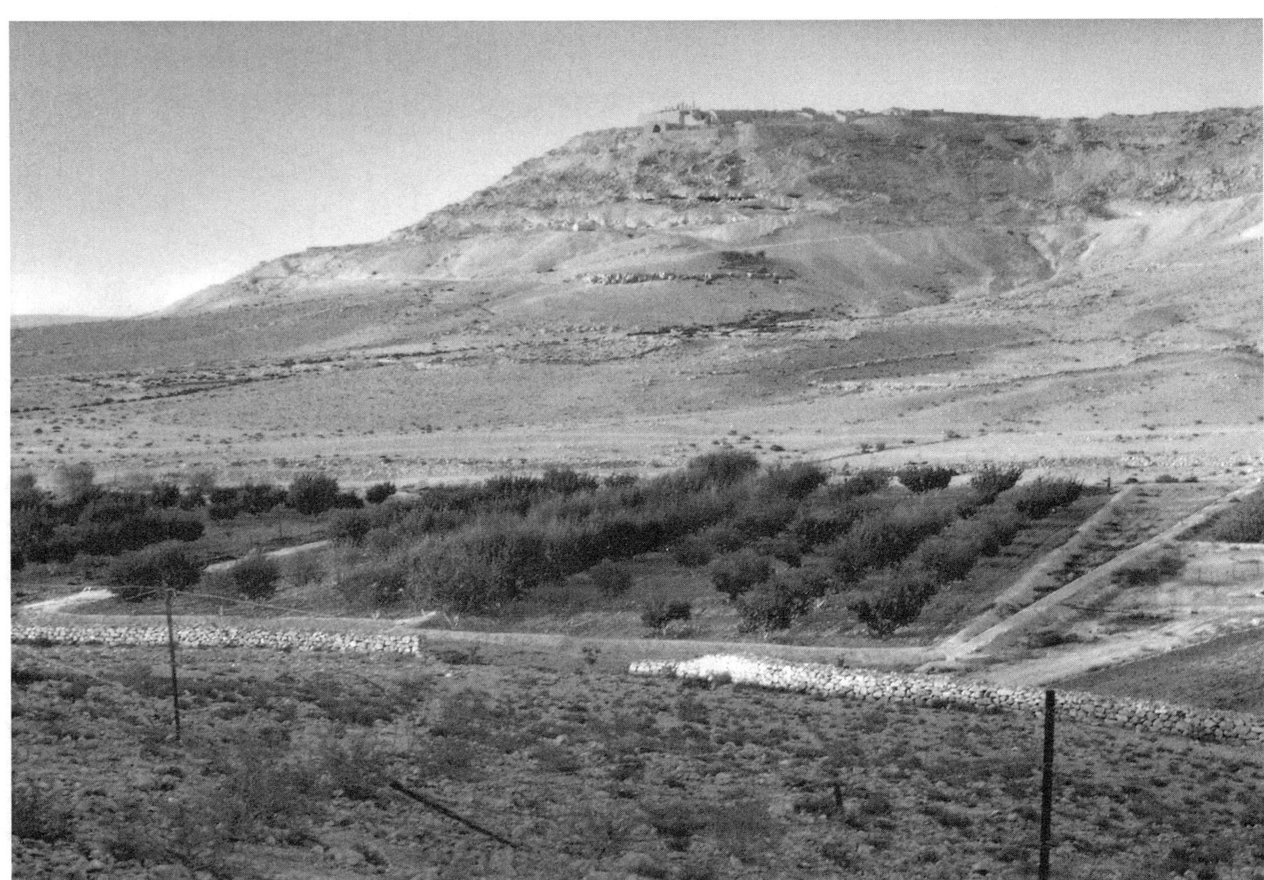

**FIGURE 6.6 Fruit and olive trees in the Negev Desert near Avdat in Israel.** Rainwater is harvested from the surrounding hillsides to provide soil moisture for the orchard.

## HOPI AGRICULTURE

In the southwestern U.S., the Hopi have been practicing agriculture for more than 500 years in an arid landscape vegetated mostly by desert-adapted plants. Their success is based on a multi-faceted strategy: they take advantage of natural water concentration and storage, build structures to harvest water, and plant crop varieties adapted to the local conditions.

Total annual precipitation in the areas where the Hopi grow crops, averages only 24 to 34 cm. Nearly all this precipitation is concentrated into two short periods of the year. In the winter, it falls mostly as snow in the mountains; in the late summer, it comes in short downpours during convective thunderstorms. This precipitation regime poses several challenges to agriculture. Crops can't be grown during the cold winter when precipitation is most plentiful, and the summer rainfall is usually too intense and short-lived to percolate down into the soil; most of it is lost as runoff.

The Hopi have learned that the local topography and soil allow them to turn these challenges to their advantage. They plant their main crop, corn, in arroyos. The soil in the arroyos is a sandy loam deposited by many years of flash floods, and it is covered by pure sand, blown up over the soil surface by summer winds. It sits on top of a layer of shale, which forms an impermeable barrier. When the winter snows melt, the runoff flows down the arroyos. It easily penetrates the sandy soil and is trapped there by the layer of shale. When the dry spring begins, the top layer of sand is quickly dried by sun and wind into a crust, which serves to protect the lower layers of soil from desiccation.

The Hopi plant their corn crop in late spring, placing each seed in a six- to ten-inch deep hole to give it access to the belowground moisture. This moisture is sufficient to allow germination and development of the corn until the summer rains come in July or August.

To get the most from the later summer rains, the Hopi build a system of dams and ditches each year. These structures serve a dual purpose: they protect the corn crop from potential flash floods and they spread the flow of water into a sheet, slowing it down and allowing it to percolate into the soil. Managed in this way, the summer rains provide enough additional moisture to allow the corn to mature, and deposit alluvial soil that renews the fertility of the fields. As an added benefit, the alluvial deposit dries out to produce a cracked and hardened surface that protects the water stored in the soil below, much like the sand crust had done previously.

The final component of the Hopi management strategy is to plant a local strain of corn. The plants stay relatively short, allowing them to withstand the gusty winds that frequently occur during the season, and they produce a long taproot that can access moisture deep in the soil.

## GRAZING SYSTEMS

In regions where rainfall is both limited and highly unpredictable, natural vegetation is made up of a mixture of water-seeking, drought-resistant shrubs and perennial grasses and annual species that can germinate and complete their life cycles in the short period that water is available. The drought tolerance of the perennials is combined with the drought avoidance of the annuals to form a system that can produce biomass during most of the year. In many parts of the world, this type of ecosystem is associated with extensive populations of native grazing animals. When we consider the ability of grazing animals to move in search of adequate forage, such ecosystems reflect considerable adaptability and diversity.

Interestingly, some of the earliest domesticates of grazing animals are thought to have come about in areas of more extreme, semiarid environments. Animals that were preadapted in the wild state to subsisting on sparse vegetative cover, such as wild relatives of sheep and goats, provided an important means for humans to survive in an otherwise hostile environment. Nomadic herding is thought to be an important form of early agriculture.

Today, many managed grazing systems take advantage of the ability of pasture or range ecosystems to maintain production of biomass in the face of low and highly variable rainfall. In most cases, natural range is managed with specific stocking rates and timing to adjust to the natural dynamics of plant growth in response to rainfall. Animals are moved from one part of a range to another during the year as forage availability shifts. In other cases, such range is improved with the introduction of drought-tolerant forage species that are very successful under drier conditions.

In a world in which increasing consumption of animal products and ecologically inefficient and degrading methods of raising livestock represent some of the most serious threats to the integrity and long-term productivity of our food systems, many traditional and managed grazing systems in low-rainfall regions stand as good examples of sustainable animal-based food production. We will discuss grazing systems in this context in Chapter 19, Animals in Agroecosystems.

## LESSONS FROM SUSTAINABLE SYSTEMS

Irrigation technologies have been developed all over the world to compensate for the vagaries of the rainfall factor, but the ecological consequences of such technologies have begun to manifest themselves in many ways. Soil erosion, sedimentation, salinization, and loss of natural wetlands and watershed systems are just some of the problems. It is hoped that by examining the nature of humidity and rainfall as we have done in this chapter, as well as the examples of agroecosystems that work with local rainfall conditions rather than against them, we can get a glimpse of an important aspect of sustainability.

For a factor such as rainfall, nature can serve as a useful model for developing a sustainable agriculture. Much of present-day agricultural development has approached the lack or excess of rainfall, intent upon eliminating or altering conditions to fit the needs of the cropping systems being introduced. This usually involves high levels of external inputs of energy or materials. There are many well-known examples of massive irrigation, drainage, or desalination projects that attempted to alter existing ecological conditions, but achieved only limited success when evaluated in terms of crop productivity, economic viability, and social welfare. We need to intensify the search for ways to accommodate agriculture to the natural variability and unpredictability of rainfall.

## FOOD FOR THOUGHT

1. What are some of the benefits and detrimental effects of irrigation as a means of overcoming limiting rainfall, from the point of view of sustainable agriculture?
2. How are rainfall patterns affected by topography? How has agriculture been adapted to the variation in rainfall patterns caused by topographic variation?
3. What are some of the possible ecological roles of a dry season for ecosystems?
4. What is the best way to prepare an agroecosystem for the unpredictable nature of precipitation?
5. What are some ways by which farming systems of the future might adjust to the probable changes in rainfall patterns caused by global climate change?

## INTERNET RESOURCES

Climate Rainfall Data Center (CRDC) at Colorado State University
rain.atmos.colostate.edu/CRDC

Global Change Data and Information System (GCDIS)
globalchange.gov
Comprehensive data sets on all aspects of global climate change, including precipitation.

Global Water Partnership
www.gwpforum.org/servlet/PSP

National Acid Precipitation Assessment Program
www.oar.noaa.gov/organization/napap.html

USDA Agricultural Research Service: National Program on Water Quality and Management
www.ars.usda.gov/research/programs/programs.htm?NP_CODE = 201

United States Geological Survey
bqs.usgs.gov/acidrain/
On-line data and reports on acid rain, atmospheric deposition, and precipitation chemistry.

The World's Water: Information on the World's Freshwater Resources
www.worldwater.org

## RECOMMENDED READING

Barry, R.C. and Chorley, J. 2003. *Atmosphere, Weather, and Climate.* 8th ed. Routledge: London and New York. A recently revised and updated edition, which emphasizes the ways in which the complex interactions between atmosphere and weather create world climates.

Bonan, G.G. 2002. *Ecological Climatology: Concepts and Applications.* Cambridge University Press: UK. This book integrates the perspectives of atmospheric science and ecology to describe and analyze climatic impacts on natural and managed ecosystems. In turn, it discusses the feedback loop whereby the use and management of land by people affects climate. The book includes detailed information on the science of climatology as well as chapters on the interactions between climate and terrestrial ecosystems, including agroecosystems and urban ecosystems.

Glieck, P. 2005. *World's Water, 2004–2005: The Biennial Report on Freshwater Resources.* Island Press: Washington, D.C. The latest of a biennial series starting in 1998, this comprehensive volume discusses global freshwater resources and the political, economic, scientific, and technological issues associated with them.

Nabham, G.P. 1982. *The Desert Smells Like Rain: Naturalist in Papago Indian Country.* North Point Press: San Francisco, CA. A sensitive look at how water is the lifeblood of desert ecosystems and the humans who live there.

Oliver, J.E. and J.D. Hidore. 2002. *Climatology: an Atmospheric Science*. 2nd ed. Prentice Hall. A systematic coverage of climate and climatology, as well as a thorough examination of the impact climate has on life and the basic processes of the atmosphere.

Postel, S. and B. Richter. 2003. *Rivers for Life: Managing Water for People and Nature*. Island Press: Washington, D.C. A realistic and positive focus on how to develop ways to ensure the sustainability of the world's vital water resources.

Reisner, M. 1986. *Cadillac Desert: The American West and its Disappearing Water*. Island Press: Covelo, California. A perceptive political history of the capture and control of water for human development in the western U.S.

Shiva, V. 2002. *Water Wars: Privatization, Pollution and Profit*. South End Press: Cambridge, MA. A critical analysis of the historical erosion of communal water rights, this book examines the international conflicts related to water, including trade, damming, mining, and aquafarming.

Whiteford, L. and S. Whiteford (eds.) 2005. *Globalization, Water, & Health: Resource Management in Times of Scarcity*. James Currey: UK. Addresses global disparities in health and access to water as the two major threats to world stability, from a medical and ecological anthropology approach. Focused on deepening our understanding of the management, sale, and conceptualization of water as it affects human health.

Wilken, G.C. 1988. *Good Farmers: Traditional Agricultural Resource Management in Mexico and Central America*. University of California Press, Berkeley. An excellent study of the sustainability of traditional farming systems, with water management practices providing some of the best examples.

# 7 Wind

Wind is not always present as a factor of the environment, but it is nevertheless capable of having very significant impacts on agroecosystems. These impacts are a result of the wind's ability to (1) exert a physical force on the plant body, (2) transport particles and materials — such as salt, pollen, soil, seeds, and fungal spores — into and out of agroecosystems, and (3) mix the atmosphere immediately surrounding the plants, thus changing its composition, heat-dispersal properties, and effect on plant physiology.

When all these types of effects are taken into consideration, what may seem a relatively simple environmental factor becomes quite complex. Wind can simultaneously have both positive and negative impacts, or be desirable in some instances and undesirable in others. Wind is, therefore, a challenging factor to manage.

## ATMOSPHERIC MOVEMENT

The earth's atmosphere is constantly in motion, circulating in ever-changing, complex, and locally variable patterns. This circulation is responsible for moving air masses and driving the changes in weather. It is also responsible for creating the surface air movement we experience as wind.

The most basic process driving the atmosphere's movement is the differential heating and cooling of the earth's surface. In the equatorial regions, intense heating of the surface and the atmosphere just above it causes the air to expand and rise high into the atmosphere, creating a zone of low pressure. Cooler surface air, further away from the equator, moves in to take the place of the rising air mass, while high in the atmosphere the heated air moves poleward. In the polar regions, the opposite occurs. Air at the colder poles cools much more rapidly higher in the atmosphere, and descends to the surface, creating a high-pressure zone and the movement of surface air toward the equator.

As a result of the equatorial low-pressure zone and the polar high-pressure zones, large cells of circulation are created in each hemisphere, as shown in Figure 7.1. The flow of air in the equatorial cells and the polar cells creates an additional cell in the temperate region of each hemisphere. As a result, there is a zone of low pressure (rising air) at about 60°N latitude and 60°S latitude, and a zone of high pressure (descending air) at about 30°N and 30°S.

The rotation of the earth alters the flow of these large-scale circulation cells. Air currents are deflected to the right of the pressure gradient, north of the equator, and to the left in the south. This deflection is known as the *Coriolis effect*. At the surface, the end result is winds that tend to blow from the northeast and southwest in the Northern Hemisphere, and from the southeast and northwest in the Southern Hemisphere. These winds, typical of certain latitudinal bands, are known as the *prevailing winds*. They are shown in Figure 7.2.

Although they describe overall, macro patterns of atmospheric circulation at the surface, the prevailing winds are subject to a great deal of local and seasonal modification. This modification is the result of a number of factors, including the presence of mountain masses on the continents and the temperature gradients created by the differential heating and cooling rates of land and water.

All these factors together result in the formation of large, mobile high-pressure and low-pressure air masses that greatly influence the local wind patterns as they move. In the Northern Hemisphere, air circulates around high-pressure cells in a clockwise direction and around low-pressure cells in a counter-clockwise direction. In the Southern Hemisphere, the directions are reversed. In both hemispheres, air flows outward from areas of high pressure toward areas of low pressure.

## LOCAL WINDS

Winds are also generated by local conditions that have to do with such factors as local topography and proximity to bodies of water. In certain areas these winds are relatively predictable.

In coastal areas in the summer, as well as around large bodies of water such as lakes or reservoirs, daytime winds (called sea or lake breezes) typically blow toward the land because the nearby land mass heats up faster than the body of water. The air above the land heats up, expands, and rises, and then the cooler air over the ocean flows inland to take the place of the rising air. At night the process can reverse as the land mass cools more rapidly than the water, and winds begin to move toward the water.

*Slope winds* are another form of local wind. In areas of mountainous topography, as the land radiates heat back to the atmosphere at night, the air close to the surface cools as well. Since cooler air is heavier, it begins to flow

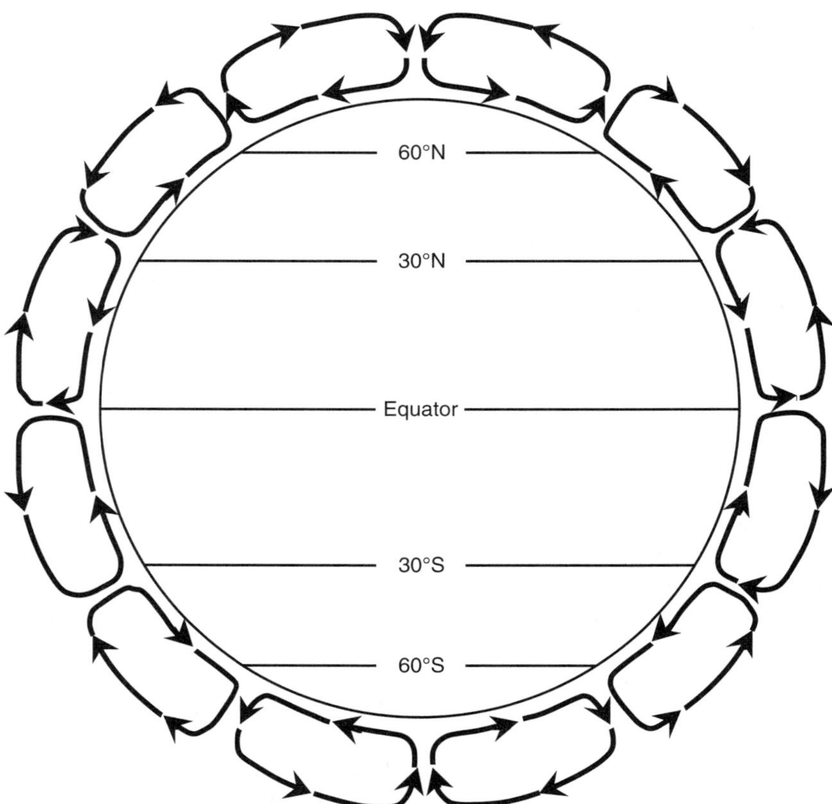


**FIGURE 7.1 Latitudinal arrangement of atmospheric circulation cells.**

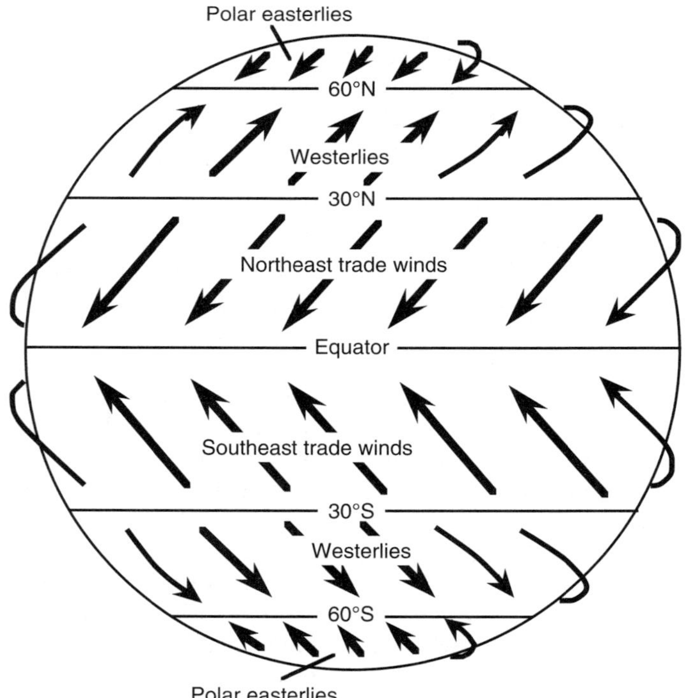


**FIGURE 7.2 Pattern of prevailing winds.**

downslope. Such movement is very localized at first, but eventually winds moving down single canyons can join in an entire valley system to create a *mountain wind.* During the day, the opposite effect can occur, and a *valley wind* forms as heating of the valley floor causes warm air to rise upslope.

When large air masses are forced over a mountain range and down onto a plain or valley below, the falling air mass expands. As a result, it heats up and its relative humidity falls. This heating and drying process is called *catabatic warming* and is responsible for the familiar rain shadow effect. Winds caused by catabatic warming occur commonly in the winter along east-facing slopes of the Sierra Nevada and Rocky Mountain systems when a cyclonic storm system moves inland and pushes air ahead of itself, forcing the air over these mountain ranges. As the air descends down the eastern or lee side of the mountains, it creates warm winds known as Chinooks that can be very gusty and cause rapid melting of snow on the surface. Since the ground usually stays frozen during these rather short-duration winds, plants can suffer considerable damage from desiccation.

A similar kind of wind occurs occasionally during the summer on the coastal slopes of southern California and central Chile. When high-pressure cells form inland, the falling air associated with these cells is pushed over the coastal range mountains and down to the coastal plains below. Called sundowners or Santa Anas, these warm winds can come up quickly at the end of the day, forcing temperatures to rise 10 to 15°C and relative humidity to plummet from near dew point to less than 20%, all in just a few minutes. This is a time of high fire danger, and crops can be damaged by the dry, gusting winds. A similar phenomenon can occur on the Isthmus of Tehuantepec in southern Mexico, where during the dry-season months high-pressure systems on the western side of the country create hot and dry downslope winds on the eastern side. Called southers or sures, these winds accentuate the dryness of the dry season months.

## DIRECT EFFECTS OF WIND ON PLANTS

The physical effects of wind on organisms can be of considerable ecological importance. This is especially true in areas prone to more constant wind, such as flat plains, near the edge of the ocean, or in high mountain areas. In general, as with all factors of the environment, the magnitude of the wind's effect is dependent on its intensity, duration, and timing.

### Desiccation

Each stomatal opening in the leaf of a plant leads to an air space in which gas exchange occurs at the surrounding cell wall membranes. This air space is saturated with humidity, and as long as the stomata are open, water vapor from inside the leaf flows out. When there is no air movement, the movement of saturated air outward from the stomata creates a *boundary layer* of saturated air around the leaf's surface. Air movement removes this boundary layer, increases transpiration, and increases overall water loss from the plant. The rate of desiccation increases proportionately with wind speed until a wind speed of about 10 km/h, where a maximum rate of loss is reached.

Normal water loss from the plant can be readily replaced by uptake from the roots and subsequent transport to the leaves. But if the rate of desiccation exceeds replacement, wilting can occur. Excessive wilting can seriously affect the normal leaf function, especially photosynthesis, leading to slower growth of the entire plant and even death.

### Dwarfing

There is a direct correlation between the wind and shortening of plant stature. The plants in alpine and coastal dune ecosystems are often short because of relatively constant high wind velocities. Crop plants that grow in areas with constant wind normally have shorter stature than the same crops planted in areas free of wind. Short stature is the result of constant desiccation causing smaller cells and a more compact plant. Where winds are more variable, and extensive periods of calm alternate with periods of high wind, plants tend not to be dwarfed.

### Deformation

When winds are both relatively constant and mostly from the same direction, they can permanently alter the growth form of plants. Windbreaks that show bent or deformed plant development are good indicators of a constant prevailing wind. Deformation can take many forms, from a permanent lean away from the wind, to a flag shape or a prostrate habit. Windborne ice is especially effective in contributing to the deformation of vegetation.

### Plant Damage and Uprooting

If excessive winds are relatively unusual events, and especially if they occur during heavy rain or snowfall, wind can cause damage to standing plants. Leaves can be shredded or removed, leaf surfaces can be abraded, branches can be broken off the trunk, tops can be removed, and whole plants can be uprooted. In areas where hurricanes, cyclones, or tornadoes occur, even mature plants that have been growing many years can suffer severe damage. Single tall trees left, following selective logging, are very prone to wind fall once they lose the protective environment of surrounding trees in a forest. This kind of damage demonstrates the importance of windbreaks (discussed later in this chapter).

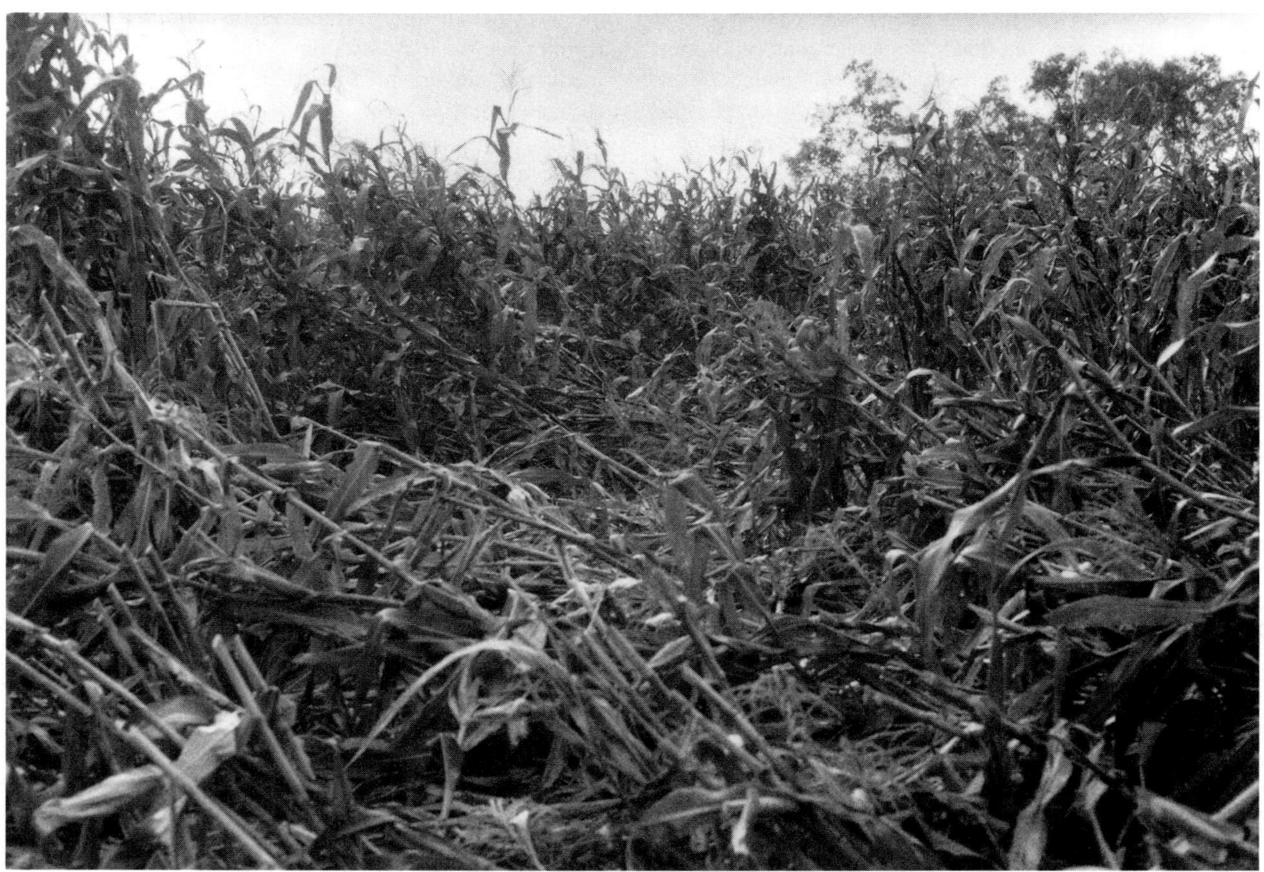

**FIGURE 7.3 Lodged corn knocked over by gusty, rain-laden winds near Cárdenas, Tabasco, Mexico.**

In agroecosystems, wind damage occurs most frequently in annual crops nearing maturity, when the plants are top heavy with grain or fruit. This type of damage, where the crop stand is flattened to the ground, is called *lodging*. In fruit crops, such as apples or plums, wind can both diminish pollination at the flowering stage and knock fruits off the tree before picking can occur (Figure 7.3).

### Changes in the Composition of Air Surrounding Plants

Apart from desiccation and the physical alteration of plant form, wind can also change the quality of air surrounding the plants. The air immediately around an organism is important since it is through the atmospheric medium that gas exchange and heat exchange can take place. The atmosphere directly affects plants by providing the $CO_2$ used in photosynthesis and the oxygen used for respiration.

Normal air is composed of 78% nitrogen, 21% oxygen, and 0.03% $CO_2$. (The remaining less-than-1% is a mixture of water vapor, dust, smoke, pollutants, and other gases.) In the immediate atmosphere surrounding plants, however, levels of oxygen and $CO_2$ vary considerably since plants produce oxygen and take in $CO_2$. During the day, oxygen levels close to plants can rise dramatically, accompanied by a drop in $CO_2$ as a result of photosynthetic uptake. Plant growth can be slowed if the concentration of $CO_2$ goes too low, because photosynthesis is limited. Air movement, however, acts to mix the air around plants, disturbing the oxygen-rich boundary layer around leaves and accelerating the diffusion of $CO_2$ toward the stomata. In this way, wind can actually be beneficial to plants.

## OTHER EFFECTS OF WIND

Wind impacts individual plants directly, as detailed above. But wind has agroecosystem-level effects as well because of its ability to transport materials.

### Wind Erosion

In any region with low and variable rainfall (or the potential for drought), occasional or frequent high-velocity winds, and high evaporation losses from the soil surface, wind erosion of soil can be a problem. Under such conditions, a loose, dry, smooth, and finely granulated soil surface lacking or partially lacking vegetative cover is easily eroded by wind.

Loss of soil by wind erosion involves two processes: detachment of particles and transport of particles. Wind agitates loose soil particles and eventually lifts and detaches them from the soil aggregates they may have been part of. These particles are then transported in different ways depending on their size and the velocity of the wind. Small soil particles that bounce across the surface, staying within 30 cm of the surface, are transported by a process called *saltation*. Under most conditions, saltation accounts for 50 to 70% of the wind movement of soil. The impact of saltating particles makes larger particles roll and slide along the surface, creating *soil creep*, which accounts for 5 to 25% of soil movement. The most visible form of transport is when particles — the size of fine sand or smaller, are moved parallel to the surface and become airborne. Wind turbulence can carry clouds of these airborne particles several kilometers upward into the atmosphere and hundreds of kilometers away to eventually settle or be washed out of the air. Generally, such erosion is about 15% of the total, but in some cases has been known to surpass 40%.

When agriculture is practiced in regions of the world where unprotected soil is subject to wind erosion, great amounts of topsoil can be lost (Nordstrom and Hotta, 2004). Desertification in the Sahel of Africa was greatly intensified in the 1970s by wind erosion of the soil caused by drought, over-grazing, and intensive cultivation of soils on marginal lands. The giant clouds of wind-blown soil and dust generated during the great "dust bowl" of the 1930s in the U.S. are still one of the most graphic examples of the physical impact of wind on farming systems through soil loss.

Soil removal from one place and its deposition in others are dual sides to the wind erosion problem when it occurs. Reduced soil productivity and crop performance are the ultimate results unless appropriate precautions are taken when agriculture is practiced in locations subject to wind erosion.

## Transport of Ocean Salt

At locations along seacoasts, the physical effect of wind can be combined with the injurious chemical effect of salt deposition. When waves break, bubbles and tiny droplets of salt water are formed and lifted into the air; in the presence of wind, they can be carried inland and the salt they contain deposited on leaf surfaces. Wind-blown salt and salt spray can burn the edges of leaves and even cause leaf drop.

Damage from wind-transported salt can occur many kilometers inland from the coast, but the most damaging effects of salt are seen close to the coastline. Windstorms without rain cause the most salt damage (Figure 7.4).

**FIGURE 7.4 A coastal shrub showing leaf burn and leaf drop caused by wind-deposited ocean salt near Paraiso, Tabasco, Mexico.** Note the accumulated pruning effect at the left on the part of the plant that is directly exposed to the wind.

The transport and deposition of salt by wind can have major impact on the zonation of vegetation along the coast, and requires that only salt-tolerant crops be planted in areas subject to deposition. In some locations, natural topographic features along the coast, such as sand dunes, block wind-blown salt, allowing salt-sensitive crops to be planted on their leeward side. Avocado trees, for example, were once planted in such protected locations along the coast of California from Santa Barbara to San Diego (but more recently such protected areas have become much sought after locations for residential home construction). Windbreaks may also be used to achieve the same effect.

### Transport of Disease and Pest Organisms

Wind serves as a means of transport for a range of organisms that are pests or diseases in agroecosystems. Bacteria and fungi depend on wind to transport spores from infected plants to new hosts, and many insect pest species take advantage of the wind to move long distances in the environment. Several aphids, for example, have a winged stage for dispersal and a wingless stage for development of sedentary pest populations on host plants. The wings of these aphids do not serve for much more than holding the insects aloft while the wind carries them where it may. Of course, if the landing site is an uninfested host plant, a pest problem can develop.

The females of many insect pests, such as the apple codling moth, release a sex pheromone and then depend on wind dispersal of the chemical in order to attract males for mating. The seeds of a large number of unwanted plants or weeds in agroecosystems are dispersed by wind as well. Since small propagules and even small organisms can be lifted hundreds of meters into the air on wind currents and then transported several hundred kilometers away, it is very difficult for farmers to escape the constant "rain" of potential problems. We will deal with the agroecological management of such dispersal problems in Chapter 15.

### Beneficial Effects of Wind

Some of the most important beneficial effects of wind take place at the microclimatic level. Internal to the agroecosystem, especially in the canopies of cropping systems, air movement is essential for mixing the atmosphere. Good air circulation maintains optimal gradients of $CO_2$, disperses excess humidity, and can even increase active gas exchange. Adequately mixed air lowers humidity levels at the leaf surface, thereby reducing the potential for many diseases. In warm climates, wind also has the important effect of enhancing convective and evaporative cooling in the direct sun.

Wind is also required for the production of grain crops such as corn, oats, and wheat. These crop plants are wind-pollinated, and depend on wind to distribute pollen from the male structures of plants to the seed-producing female structures of other plants.

## MODIFYING AND HARNESSING WIND IN AGROECOSYSTEMS

An understanding of the impacts that wind can have on agroecosystems, as well as the mechanisms of those impacts, gives farmers the opportunity to develop means of both mitigating the negative effects, as well as taking advantage of positive effects. In addition, the energy of wind can be harnessed for an array of uses in agriculture.

### Measuring Wind

Wind is usually measured with a device known as anemometer. Cup anemometers consist of three or four horizontally rotating arms with small cups on the ends fixed to a vertical shaft that activates a dial or recorder as it turns. Such a device will record wind from any horizontal direction, and based on the total revolutions measured, average wind velocity over time can be determined. A fan anemometer can record lower wind speeds more accurately, but has to be pointed in the direction of the wind. Thermal anemometers, which operate on the basis of the relation between ventilation and heat transfer, are used for very low wind speeds that are not recorded well with fan or cup systems. Other types of equipment exist to record wind gusts and wind direction.

Measuring average wind speed and direction is only one part of gaining an understanding of patterns of air movement in an agroecosystem. It is also important to know how local wind patterns are reduced to microclimatic patterns as wind encounters barriers. The barriers can be individual plants, natural topographic variation, or intentionally placed barriers of some kind. Use of such barriers will depend on how they effect the wind we are trying to modify or take advantage of.

### Techniques for Modifying Wind Patterns and Mitigating Wind Effects

There are many ways to manage the wind environment in cropping systems. Some are as simple as orienting the planting of rows of a crop in such a way as to funnel a prevailing wind through the crop; others are more dramatic, such as planting windbreaks or shelterbelts, or using intercropping systems that combine wind-sensitive crops with more tolerant ones.

#### Windbreaks

Windbreaks (also known as shelterbelts and hedgerows) are structures — usually made up of trees — that modify wind flow for the purpose of reducing soil erosion by wind, increasing crop yields, protecting the farmstead and other structures, or realizing any combination of these goals. Windbreaks are not meant to stop the wind, but rather to change its course and rate of flow. They are usually oriented perpendicularly to the wind (if their goal is modification of flow rate) or along the flow angle of the wind (if their goal is redirection). When trees are used to create permanent windbreaks in agroecosystems, the result is a form of agroforestry (Figure 7.5).

Extensive research has been carried out on windbreak technology and the role of such structures in cropping systems all over the world (Brandle and Hintz, 1988; Brandle et al., 2004). Windbreaks have been shown to dramatically alter wind flow patterns and velocity, and as a result, to reduce many of the negative impacts of wind described above, while taking advantage of some of the positive effects. Ultimately, crop plant and animal yields benefit (Figure 7.6).

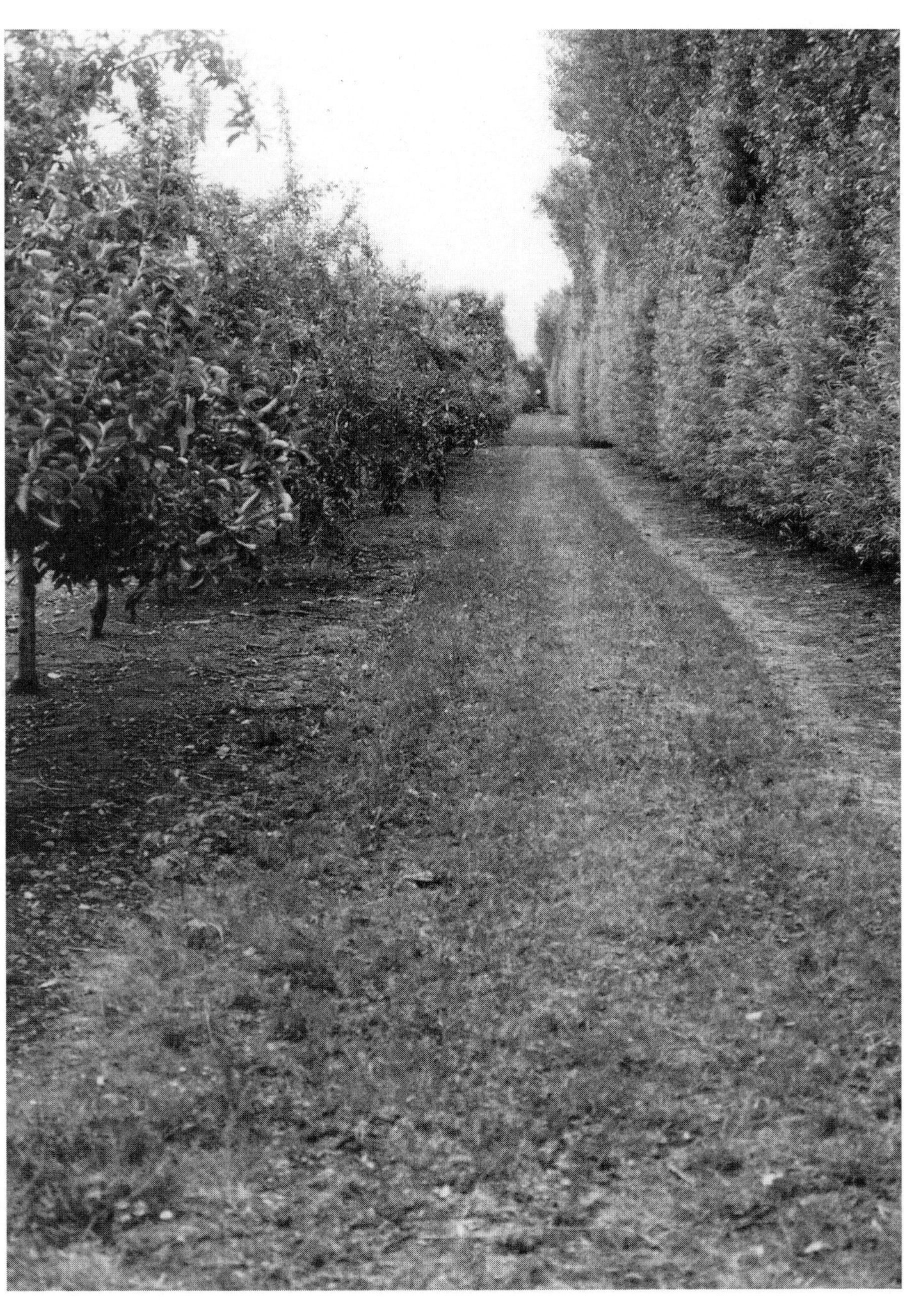

**FIGURE 7.5 Windbreak for improving the microclimate of an adjacent apple orchard near Lincoln, New Zealand.** This windbreak is made up of willow trees (*Salix* sp.).

**FIGURE 7.6 Windbreaks in the arid region near Eilat, Israel.** These windbreaks reduce evapotranspirational water loss for the irrigated annual crops grown between them.

The primary effect of a windbreak is reduction of wind velocity. A good windbreak can reduce wind velocity as much as 80% for a distance of up to ten tree heights downwind from the windbreak, and often for a distance as long as two tree heights to the windward side. The area in the lee of the barrier is known as the "quiet zone," a wedge-shaped area of greatly reduced wind speed with moderate turbulence and small eddies. Above the quiet zone and for a distance of several tree heights more downwind, there is a "wake zone" of large eddies, more turbulence, and less reduction in wind speed (Figure 7.7).

Since a windbreak creates an obstacle to the wind, flow is actually deflected upward as it approaches the barrier. Near the top of the windbreak, flow is compressed and accelerated. Just downwind and behind the barrier, flow is reduced to close to zero with a solid windbreak, and to intermediate speeds with a porous barrier. There is a zone of strong velocity shear just above the top of the windbreak that widens and follows the flowline as the air moves downwind, eventually mixing with the air in the zone of turbulence until it returns once again to its normal speed at as much as 20 to 30 heights to the leeward.

The density and porosity of a windbreak have a significant effect on the distance over which the windbreak can alter wind flow. Denser barriers produce the largest velocity reductions directly to the leeward, but the largest wind shear between the retarded air behind the windbreak, and the accelerated zone above. Denser barriers also create more turbulence, since kinetic energy loss from the original flow must be balanced by an increase in kinetic energy in the eddies. This leads to a quicker recovery of wind speed behind the barrier, and therefore, a reduced protected area. A barrier with a porosity of 40% has been shown to reduce wind speed effectively for a distance of 30 heights downwind (Tibke, 1988).

Besides reduction of soil erosion, the most tangible effect of windbreaks is enhancement of the final yield of the crop. Higher yield volume is the most obvious gain, but earlier harvest time and better harvest quality are important benefits as well. Less stress in the lee of the barrier allows crops to allocate more energy to vegetative or reproductive growth and less to maintenance. Less physical damage occurs, transpirational losses are minimized, and higher temperatures and humidity contribute to better quantity and quality of production.

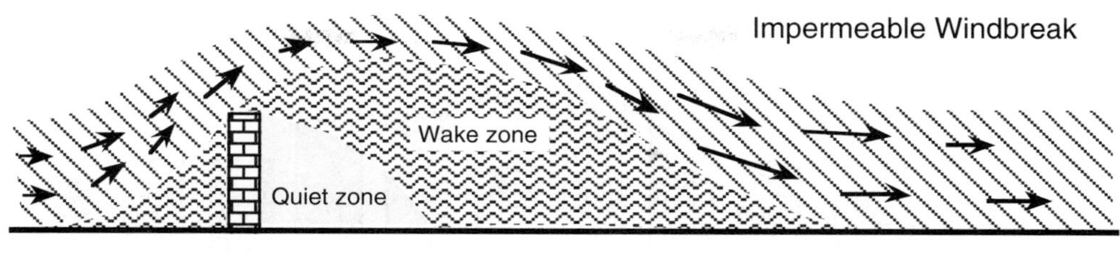


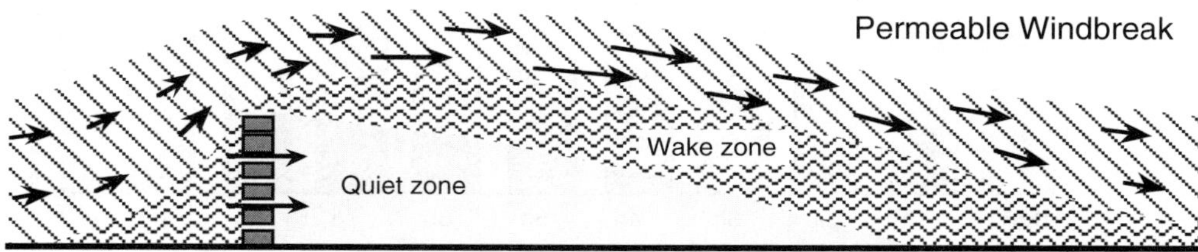


**FIGURE 7.7 Wind profiles of a barrier windbreak and filter windbreak.** A filter (permeable) windbreak reduces windspeed more effectively than a barrier (impermeable) windbreak and does so over a greater distance. Adapted from McNaughton 1988 and Guyot 1989.

In an extensive review of research on the benefits of windbreaks to field and forage crops around the world, Kort (1988) found that most of these crops show better yields when grown in fields with windbreaks, but that some benefit more than others. A broad-leafed forage crop such as alfalfa, with a high rate of transpirational water loss in the wind, appears to benefit most from a windbreak, and short-cycle grains such as spring wheat and oats benefit the least. Kort's findings are presented in Table 7.1

In a review of the influence of windbreaks on vegetable and specialty crops, Baldwin (1988) reports that there is overwhelming evidence to support and illustrate the positive benefits of wind shelter. Yield increases range from 5 to 50% for a variety of crops including beans, sugar beets, tomatoes, potatoes, melons, tobacco, berries, cacao, coffee, cotton, rubber, and okra. Most benefits occur within ten heights on the leeward side, with maximum benefits seen between three and six heights. Benefits are also seen within zero to three heights to windward. An example of how the improved yield caused by a windbreak varies with distance from the windbreak is shown for soybeans in Figure 7.8. With this crop, peak benefit was seen at four heights to the leeward; interestingly, however, yields were reduced within a distance of one height, presumably from either shading, root competition, or allelopathy.

**TABLE 7.1**
**Relative Impacts of Windbreaks on Yields of Various Grains and Forage Crops**

| Crop | Yield Increase, in Percentage, Relative to Fields without Barriers |
|---|---|
| Alfalfa | 99 |
| Millet | 44 |
| Clover | 25 |
| Barley | 25 |
| Rice | 24 |
| Winter wheat | 23 |
| Rye | 19 |
| Mustard | 13 |
| Corn | 12 |
| Flax | 11 |
| Spring wheat | 8 |
| Oats | 3 |

*Source*: Kort J. 1988. Agriculture, Ecosystems and Environment 22/23: 165–190.

With vegetable and specialty crops, crop quality improvement may be as important a benefit as increased yield. Crop quality can be improved in a variety of ways, including an increase in sugar content in crops such as sugar beets and strawberries, reduced abrasion by windblown sand on crops such as melons, and earlier ripening for most crops. Since vegetable and specialty crops are usually highly susceptible to wind damage and wind abrasion, improvements in crop quality are easily converted into better economic return, which adds to the gains from yield increases.

Windbreaks have also been shown to provide substantial benefits in the production of orchard and vineyard crops (Norton, 1988). Year-round protection is critical to the survival and proper development of trees and vines. Orchard microclimate modification in the form of a windbreak can improve pollination and fruit set, in turn leading to greater yields. Mechanical damage is also reduced, improving fruit quality and economic gain. Proper windbreak design and management can also reduce evaporation, increase the flexibility of the application of pest management materials, and even assist in frost management. Wind-protected temperate fruits such as plums,

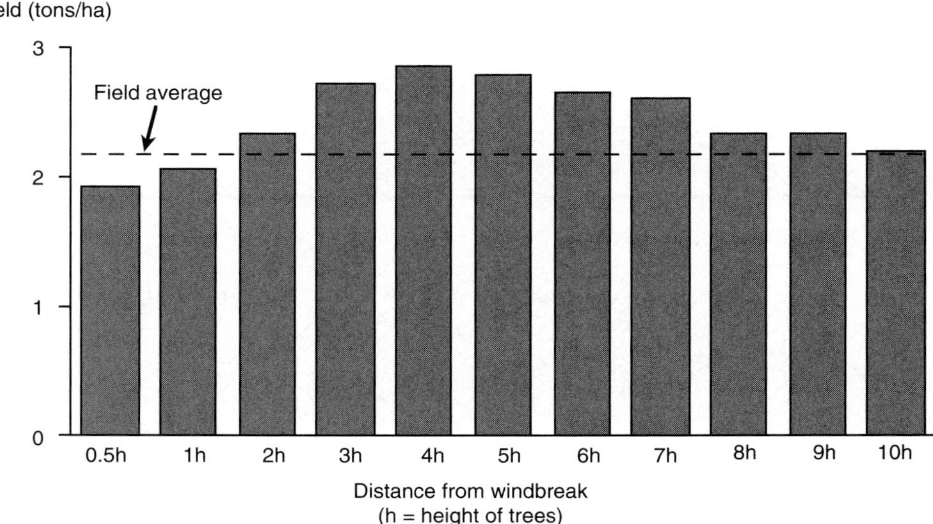


**FIGURE 7.8 Influence of windbreak protection on soybean yield at varying distances from the windbreak.** *Source:* Data from Baldwin and Johnston 1984.

pears, and grapes show yield increases from 10 to 37%, subtropical fruits such as kiwi, oranges, and lemons show yield increases up to 30% (as well as important gains in fruit quality), and tropical fruits such as bananas show yield gains of at least 15%, primarily due to a reduction in lodging of the mature stems.

## Planting Techniques

An alternative to permanent windbreaks made up of trees or shrubs is the planting of annuals within the field that work to protect the main crop from wind. Corn (*Zea mays*), sunflowers (*Helianthus annus*), and a range of grain crops such as sorghum (*Sorghum bicolor*) and pearl millet (*Pennisetum americanum*) are examples of annual plants used for this purpose. Such annual barriers have certain advantages over perennial woody shelterbelts in that they are easier, faster, and cheaper to establish, and may allow more flexibility in the farming operations. Like windbreaks, annual barrier plants reduce wind speed, thus improving moisture and temperature conditions for adjacent plants. They are usually planted at the same time as the main crop, often as individual rows interspersed among the main crop. Another technique is to plant the barrier plants (often rye) as a fall cover crop and then to reduce this crop to alternating strips in the spring by tilling when the main crop is planted. Research has shown that barrier porosity of 40 to 50% has the best impact on crop yields, and that plants need to be resistant to lodging, spaced according to the needs of the associated crop and the local wind conditions, and established early enough to give the necessary protection. Because the planting of annual windbreaks is incorporated into the process of planting the primary crop, this technique offers considerable flexibility to the farmer. Minimal time is lost and minimal space is occupied by the barrier.

Sunflowers are frequently used as annual wind barriers to improve crop conditions for tomatoes, broccoli, lettuce, and other annual crops in windy areas of the Salinas Valley of California, and corn is often used to protect strawberry crops from abrasion of the leaves, fruit damage, and reduction of the dispersal of pest mites in coastal areas of central California. Yields of annual crops such as snap beans and fresh market tomatoes have been shown to be improved by as much as 30% with the use of such barriers (Bilbro and Fryrear, 1988).

Crop plants themselves can also be planted to make them more resistant to lodging and other forms of wind damage. For crops that are able to produce adventitious roots on the lower stem, deeper planting can help anchor the plant more firmly in the ground. Cruciferous crops such as Brussels sprouts, cabbage, and broccoli benefit greatly when transplanted seedlings are buried deeply enough to cover most of the stem below the cotyledons, allowing the plant to form more roots as it develops. Otherwise, the small seedling with a few leaves can be whipped around like a kite on a string if it is too windy, eventually breaking off at ground level. In windy areas of Mexico, corn seed is often planted deeply in the base of a furrow, so that as the plant develops, soil can be built up around the base of the stem as a part of cultivation for weed control. By the time the crop is almost fully developed, the corn plants appear to be planted on the top of the rows, and as a result of their stronger anchoring in the soil are much more resistant to the lodging that can occur when convective thunderstorms create high-velocity winds (Figure 7.9).

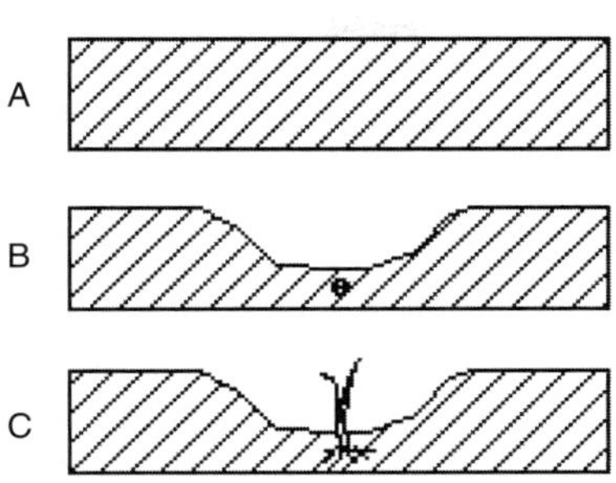


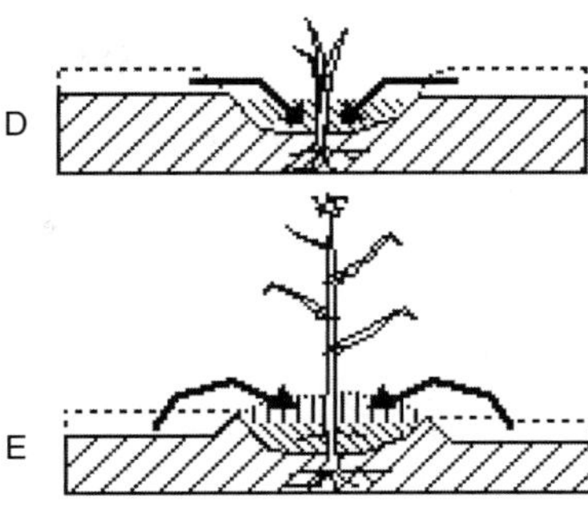


**FIGURE 7.9 Soil mounding to reduce lodging in corn.** Seeds are planted at the bottoms of furrows (B). After a period of growth (C), the furrows are filled in with soil from between the rows (D). Soil continues to be mounded around the corn plants as they grow (E), creating raised rows in which the corn is firmly anchored. The technique also has the advantages of collecting scarce rainfall for the seed (B) and allowing removal and burial of weeds when the soil between the rows is moved (D and E).

### Timing of Planting

Crop rotations can be used to adjust cropping systems to wind patterns. Crops prone to wind damage can be planted during less-windy seasons (assuming that other conditions are adequate) and followed by wind-tolerant crops. If wind erosion is more of an issue than wind damage to the crop, it might be advisable not to open up an entire field to the wind. Instead, a portion of the field can be planted earlier to one crop, which can then serve as a barrier for strips of crops planted at a later time. Another option for prevention of wind erosion is to grow low-residue crops in protected areas and high-residue crops in more exposed areas of the farm.

### Genetic Varieties Resistant to Wind Effects

A useful way to prevent lodging in grain crops is to plant a genetic stock that is shorter in stature than usual. Local farmers on the Isthmus of Tehuantepec in southern Mexico, for example, where wind occurs throughout the growing season, have selected for corn with a short stature, thicker stem, and well-developed root system. These local varieties are highly resistant to lodging. One of these varieties, called *tuxpeño*, was used as the genetic stock for breeding with improved Green Revolution varieties to develop shorter, lodging-resistant corn with a higher seed load, as well as to develop varieties more appropriate for harvesting by mechanized combines.

## Harnessing Wind

We have primarily discussed ways by which a farmer can manipulate wind in order to take advantage of its positive effects or to mitigate the negative impacts. But wind has other uses in farming systems that help contribute to the larger goal of sustainability. Harnessing the energy of wind can help reduce external-input and nonrenewable energy use, especially the burning of fossil fuels. This is becoming especially important for small farm systems and farmers in the developing world.

Many methods of harnessing or using the wind are quite simple. For example, the wind can be used to clean seeds of chaff and leaves (winnowing). The wind can also be used for drying. Harvested bean plants can be hung in preparation for thrashing, or fruits such as raisins or apricots can be laid out to be dried by the wind. A light breeze aids considerably in removing the boundary layer of moisture that can form around the plant or plant product.

Finally, windmills have been used to harness wind power for a large range of farming activities, from pumping water to generating electricity for use in farming operations or the farm homestead. Farms in isolated areas, especially in developing countries, where wind is a constant factor, are especially appropriate candidates for the use of wind power.

## Wind and Sustainability

Wind is an important component of climate and weather all over the world. It is also a factor that often has disruptive or damaging impacts on agroecosystems. By learning how to design agroecosystems so they are capable of withstanding and even mitigating the negative aspects of wind, we take steps toward sustainability. But the most important steps will come with the development of design and management strategies that accentuate the very positive role that air in motion can play in agriculture. In some ways, these steps may involve a return to the use of old technologies, such as windbreaks and hedgerows. Nevertheless, there is a critical need to understand the ecological basis for using such practices or strategies. Only then can we develop another measurable component of sustainability, and as a result, help establish a more active role for windbreaks, wind turbines, and the management of daily wind patterns in sustainable farming systems.

## FOOD FOR THOUGHT

1. In certain cases, an ecological factor may be limiting in the absence of wind but not limiting when wind is present. What are some examples?
2. The most common argument for not using (or even removing) windbreaks and shelterbelts is that they take up valuable crop production land. What are the primary counter-arguments for this "fencerow to fencerow" farming mentality?
3. Wind is one of those factors that can simultaneously have negative and positive effects. What are some possible examples of this situation? How would you manage the wind in these examples?
4. What are some of the primary barriers to the broader use of the free and renewable source of energy contained in wind?

## INTERNET RESOURCES

Wind Erosion Research Unit of the U.S. Department of Agriculture
www.weru.ksu.edu

Union of Concerned Scientists: Wind Power and Agriculture
www.ucusa.org/clean_energy/renewable_energy_basics/farming-the-wind-wind-power-and-agriculture.html

## RECOMMENDED READING


Brandle, J.R., L. Hodges, and X.H. Zhou. 2004. *Windbreaks in North American agricultural systems. Agrofor Syst*, 61(1):65–78.

Brandle, J.R. and D.L. Hintz (eds.) 1988. Special issue: *Windbreak Technology. Agric Ecosyst Environ*, 22/23:1–598. Proceedings of a symposium that brought together experts from all over the world on the design and use of windbreaks in agriculture. It continues to be a primary windbreak reference.

Burke, S. 2001. *Windbreaks*. Elsevier Science: New York. This comprehensive book includes both theoretical and practical considerations for establishing and utilizing windbreaks. Written in a way that will be useful to a wide audience, including students, researchers, and farmers.

Cleugh, H.A. 1998. *Effects of Windbreaks on Airflow, Microclimates and Crop Yields. Agrofor Syst*, 41(1):55–84. The mechanisms by which a porous windbreak modifies airflow, microclimates and hence crop yields are addressed, based upon recent wind tunnel experiments, field observations, and numerical modeling. This paper is thus an update to the excellent reviews in Brandle and Hintz (1988).

Coutts, M.P. and J. Grace. 1995. *Wind and Trees*. Cambridge University Press: New York. A full review of the ecological and physiological impacts of wind on trees, and the adaptations trees have developed to withstand these impacts.

Geiger, R. 1965. *The Climate near the Ground*. Harvard University Press: Cambridge. The definitive source of information on the formation of microclimates and how they impact living organisms.

Morgan, R.P.C. 2005. *Soil Erosion and Conservation*. 3rd ed. Blackwell Publishing: Willingston VT. A complete review of the processes, control methods, and conservation programs related to soil erosion, including updated information on the mechanics of and responses to wind erosion.

Moss, A.E. 1940. *Effect of Wind-driven Salt Water. J For*, 38:421–425. A key research review on how wind and salt combine to form an important factor in the environment.

Nordstrom, K.F. and S. Hotta. 2004. *Wind Erosion from Cropland in the USA: A Review of Problems, Solutions, and Prospects. Geoderma*, 121(3–4):157–167. An excellent review of the multiple strategies that can be employed to reduce or eliminate soil erosion caused by the wind.

Reifsnyner, W.S. and T.O. Darnhofer. 1989. *Meteorology and Agroforestry*. International Council for Research in Agroforestry: Nairobi, Kenya. A general reference on wind energy, and an excellent review of how trees in agriculture can play important roles in the modification of microclimatological factors and conditions.

Shao, Y. 2001. *Physics and Modeling of Wind Erosion*. Springer: New York. A summary of the recent developments in wind-erosion research, providing a key resource for researchers and postgraduate students engaged in wind-erosion studies. Topics range from global climate change to air quality and land conservation.

# 8 Soil

Soil is a complex, living, changing, and dynamic component of the agroecosystem. It is subject to alteration, and can either be degraded or wisely managed. In much of present-day agriculture, with the availability of an array of mechanical and chemical technologies for rapid soil modification, soil is all too often viewed primarily as a growth medium, something from which to extract a harvest. Farmers often take the soil for granted, and pay little attention to the complex ecological processes that take place below the surface. The premise of this chapter, in contrast, is that a thorough understanding of the ecology of the soil system is a key part of designing and managing sustainable agroecosystems in which the long-term fertility of the soil is maintained.

The word *soil*, in its broadest sense, refers to that portion of the earth's crust where plants are anchored; this includes everything from the deep soils of a river bottomland to a crevice in a rock with a bit of dust and plant debris. More specifically, the soil is that weathered superficial layer of the earth that is mixed with living organisms and the products of their metabolic activities and decay (Odum and Barrett, 2004). Soil includes material derived from rocks and organic and inorganic substances derived from living organisms, and the air and water occupying the spaces between soil particles.

From an agricultural perspective, an "ideal" soil is made up of 45% minerals, 5% organic matter, and 50% space, with the "space" filled half with water and half with air. It is hard to find anything that we can call a "typical" soil, since each site or location has unique properties that ultimately determine the final outcome of the soil formation process.

## PROCESSES OF SOIL FORMATION AND DEVELOPMENT

Biological processes combine with physical and chemical processes in each particular climatic region and location to form soil. Once formed, soil changes and develops due to these and other biological, physical, and chemical processes. With variations in slope, climate, and type of vegetative cover, many different soils can form in close juxtaposition with one another, even though the parent material may be fairly similar.

Natural processes of soil formation and development take considerable time. For example, it is estimated that only about 0.5 to 1.5 t of topsoil per acre is formed annually in areas of corn and wheat production in the central Midwest region of the U.S. (Daily, 1995). In contrast, about 5 to 6 t of soil per acre are estimated to erode from conventionally farmed land in these areas, with soil losses often exceeding 15 to 20 t/acre in some years (NRCS, 2005).

### Formation of Regolith

As a whole, the layer of unconsolidated material between the soil surface and the solid bedrock of the earth below is called the *regolith*. The most basic element of the regolith is its mineral component, made up of soil particles formed from the breakdown of the bedrock or parent material. At any particular location, these soil particles may have been derived from the bedrock below, or they may have been transported from elsewhere. Where a soil's mineral particles have been formed in place from the bedrock below, the soil is a residual soil. Where the mineral particles have been carried from some other location by wind, water, gravity, or ice, the soil is a *transported soil*.

#### Physical Weathering

The weathering of rock and rock minerals is the original source of mineral soil particles, whether the particles remain in a location or are moved elsewhere. The combined forces of water, wind, temperature, and gravity slowly peel and flake rock away, accompanied by the gradual decomposition of the minerals themselves. Water can seep into cracks and crevices in rock, and with heating and cooling causing alternating swelling and contracting, rock begins to fragment. In addition, the carbon dioxide contained in the water that seeps into cracks can form carbonic acid, pulling elements such as calcium and magnesium from the minerals of the rock and forming carbonates, and in the process weakening the crystalline structure of the rock and making it more susceptible to further physical weathering. Finer particles mix with larger particles, promoted by the physical movement created by the combined forces of gravity, temperature change, and alternating wetting and drying. Even the abrasive forces of rocks against each other during this movement can form smaller particles. Eventually the unconsolidated regolith takes form.

Depending on local conditions and geological history, the regolith can be recently formed, lightly weathered, and made up of mostly primary minerals, or it may have been subjected to intensive weathering and be made up of more resistant materials such as quartz.

### Transport

As rock is broken down into smaller and looser materials, it can remain in place and eventually form residual soils, but a more likely fate is for it to be carried some distance and deposited. The forces of wind, water movement, gravity, and glacial ice movement can all transport weathered soil particles. Transported soils have different classifications depending on the manner in which their particles were transported. Soil is called

- Colluvium, where it has been transported by gravity
- Alluvium, where it has been transported by the movement of water
- Glacial soil, where it has been transported by the movement of glaciers
- Eolian soil, where it has been transported by wind

## Biotic Processes

Sooner or later, depending on the consistency of the regolith, plants establish themselves on the weathered material. They send roots down that draw nutrients from mineral matter, store them for a while in plant matter, but eventually return them to the soil surface. Deep roots further break down the regolith, capture nutrients that have leached from the upper surface, and add them to the soil surface in an organic form. Plant residue then serves as an important source of energy for the bacteria, fungi, earthworms, and other soil organisms that establish in the area.

Organic matter is broken down into simpler forms through decomposition and mineralization. Macroorganisms in the soil — centipedes, millipedes, earthworms, mites, grasshoppers, and others — consume freshly deposited plant debris and convert it to partially decomposed material, either in the form of excreta or their own dead bodies. This material is then further decomposed by microorganisms, mostly bacteria and fungi, into an array of compounds such as carbohydrates, lignins, fats, resins, waxes, and proteins. Mineralization eventually breaks these complex compounds down into simple products such as carbon dioxide, water, salts, and minerals.

The fraction of organic matter left in the soil as a result of decomposition and mineralization is called *humus*. It has a certain lifetime in the soil, after which it is broken down itself. New humus, however, is constantly replacing old humus, and the equilibrium point between the two is an important factor in soil management.

## Chemical Weathering

While the regolith is forming and living organisms begin to have their impacts on it, chemical weathering is occurring as well. Chemical weathering includes natural chemical processes that aid in the breakdown of parent materials, the conversion of materials from one form to another in the soil, and the movement of materials within the soil. Four different chemical processes are of primary importance in soil formation and development: hydration, hydrolysis, solution, and oxidation.

*Hydration* is the addition of water molecules to a mineral's chemical structure. It is an important cause of crystal swelling and fracturing. *Hydrolysis* occurs when various cations of the original crystalline structure of silicate minerals are replaced by hydrogen ions, causing decomposition. In regolith with low pH, the greater concentration of $H^+$ accelerates hydrolysis. The release of organic acids as a byproduct of the metabolic activities of living organisms, or from decomposition of dead organic matter, can add to this process as well. *Solution* occurs when parent materials with a high concentration of easily soluble minerals (such as nitrates or chlorides) go into solution in water. Limestone is particularly susceptible to solution in the presence of water, high in carbonic acid; in extreme cases the solution of limestone leads to the formation of limestone caves in areas of underground water flow. Finally, *oxidation* is the conversion of elements such as iron from their original reduced form into an oxidized form in the presence of water or air. Softening of the crystalline structure usually accompanies this process.

Once minerals are released from the consolidated parent material, another chemical process that is of great importance is the formation of secondary minerals — the most important being clay minerals. Clay mineralogy is a very complex field of study, but it is important to understand some basic aspects of clay formation, since they have such dramatic impacts on plant growth and development.

Clay minerals are very small particles in the soil, but they affect everything from water retention to nutrient availability, as will be discussed elsewhere. They are formed by complex processes in which silicate minerals are chemically modified and reorganized. Depending on the combination of climatic conditions and parent material, the secondary minerals that are formed are of two basic types: *silicate clays* that are predominantly made up of microscopic aluminum silicate plates with different arrangements and the presence or absence of other elements such as iron and magnesium; and *hydroxide clays* that lack a definite crystalline structure and are made up of hydrated iron and aluminum oxides in which many of the silicon ions have been replaced.

Eventually, the clays found in any soil will be a mixture of many subtypes of these two basic types of secondary clay minerals, although one or a few subtypes may predominate. When silicate clays dominate, there are abundant sites for absorbing cations, giving the soil a relatively high productive potential. When hydroxide clays dominate — as in many humid tropical regions — fewer cation sites are available, making the soil more difficult to farm because of its poor ability to exchange nutrient cations.

Organic matter, either from plant residues or the activities of living organisms, has important impacts on all of these chemical weathering processes of parent material and greatly accelerates the formation of the regolith. However, it is the array of biological and chemical interactions that go on once rock and minerals have been converted to small soil particles that are of the greatest agricultural interest.

## SOIL HORIZONS

Over time, the localized chemical, physical, and biological processes in the regolith lead to the development of observable layers in the soil, called *horizons*. Together, the horizons in a particular location give each soil a distinctive *soil profile*. Each horizon of the soil profile has a distinct combination of characteristics.

### The Soil Profile

In general terms, a soil profile is made up of four major horizons: the organic, or O horizon, and three mineral horizons. The O horizon lies at the soil surface; immediately below it is the A horizon, where organic matter accumulates and where soil particle structure can be granular, crumb like, or platy. Under the A horizon is the B horizon, where materials leached from the A horizon can accumulate in the form of silicates, clay, iron, aluminum, or humus, and soil structure can be blocky, prismatic, or columnar. Finally there is the C horizon, made up of weathered parent material, either derived from the local parent material below or from material transported at some earlier time to that location. Some material leached or deposited from the A and B horizons can be found here, such as carbonates of calcium and magnesium, especially in areas of low rainfall. Depending on the depth of the upper four horizons, an R horizon made up of consolidated bedrock may also be included as part of the soil profile.

Since the separations between each horizon are rarely distinct, the horizons described above actually form a continuum in the soil profile. A typical soil profile is presented in schematic form in Figure 8.1. The depth, characteristics, and differentiation of each horizon of each soil profile is the result of the combined impacts of the properties of the soil material (its color, organic matter content, and chemical and physical traits), the type of vegetative cover, and the climate.

The processes that differentiate soil horizons function in different ways depending on regional and local conditions. These differences result in four basic types of soil development, which are summarized in Table 8.1. The process of calcification is most characteristic of areas of grassland vegetation in subhumid-to-arid and temperate-to-tropical climates of the world. Podzolization is most characteristic of humid, temperate areas of the world

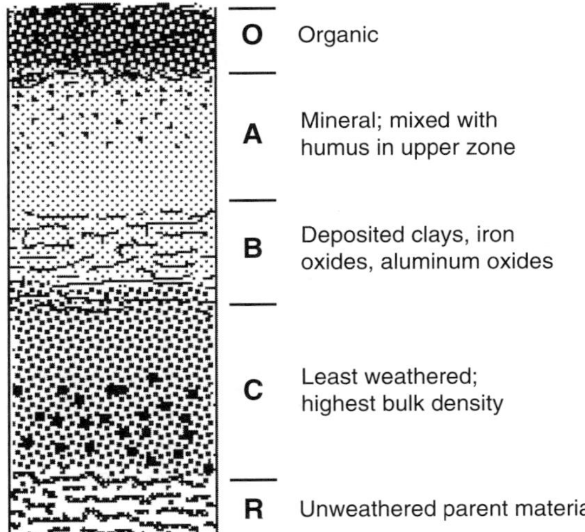


**FIGURE 8.1 Generalized soil profile.**

**TABLE 8.1**
**Four Types of Soil Development**

| Development Process | Moisture | Temperature | Typical Vegetation | Resulting Characteristics |
|---|---|---|---|---|
| Gleization | High | Cold | Tundra | Compact horizons, little biological activity |
| Podzolization | High | Cool to warm | Needle-leaf forest, deciduous forest | Light-colored A horizon; yellow-brown B horizon high in iron and aluminum |
| Laterization | High | Warm to hot | Rainforest | Weathered to great depth; indistinct horizons; low in plant nutrients |
| Calcification | Low | Cool to hot | Praire, steppe, desert | Thick A horizon rich in calcium, nitrogen, and organic matter (except in deserts) |

where forests have been the dominant vegetative cover for a long time. Laterization takes place on older and heavily weathered soils of the humid subtropical and tropical forested regions of the world, and gleization is most common on soils where water stays at or near the surface for a good part of the year. But depending on localized conditions of slope, drainage, vegetation, depth to bedrock, etc., combinations of these processes can be found. On the whole, soil formation and development is an ecological process, where soil affects the vegetation, and the vegetation affects the soil.

### Importance of the Organic Horizon

In natural ecosystems, the O horizon is the most biologically active part of the profile and the most important ecologically. It plays a significant role in the life and distribution of plants and animals, the maintenance of soil fertility, and in many soil-development processes. Macro- and microorganisms responsible for decomposition are most active in this layer and in the upper part of the A horizon. Significantly, the O horizon is usually greatly reduced or even absent from cultivated soils.

The combination of local climate and vegetation type contributes to the conditions that promote activity in this layer; yet at the same time, the quality of the layer has profound influence on what kinds of organisms prosper. Bacteria, for example, favor nearly neutral or slightly alkaline conditions, whereas fungi favor more acid conditions. Soil-dwelling mites and collembola are more important under acid conditions, whereas earthworms and termites tend to predominate at or above neutrality.

The complex process of soil particle aggregation, which creates what is called the crumb structure of the soil, is greatly influenced by humus formed during decomposition in the organic soil layer. In addition, many valuable soil fertility processes, discussed later in this chapter, are related closely to the ecological characteristics of this important layer.

## SOIL CHARACTERISTICS

In order to develop and maintain a healthy soil system, as well as make sound judgments about particular soil management strategies, it is important to understand some of the most essential properties of soils as they affect crop response.

### Texture

Soil texture is defined as the percentage, by weight, of the total mineral soil that falls into various particle size classes. These size classes are gravel, sand, silt, and clay (Table 8.2). Particles greater than 2.0 mm in diameter are classified as gravel. Sand is easily visible by the naked eye, and feels gritty when rubbed between the fingers. Its low surface-to-volume ratio makes it porous to water and less able to adsorb and hold nutrient cations. Silt, although finer than sand, still is grainy in appearance and feel, but more actively holds water and nutrient ions. Clay particles are difficult to see separately with the naked eye, and look and feel like flour. Clay particles are colloidal in that they can form a suspension in water and are active sites for the adhesion of nutrient ions or water molecules. As a result, clay controls the most important soil properties, including plasticity and ion exchange between soil particles and water in the soil. A soil very high in clay content, however, can have problems with water drainage, and when dry can exhibit cracking.

**TABLE 8.2**
**Soil Texture Classifications**

| Category | Diameter Range[a] (mm) |
|---|---|
| Very coarse sand | 2.00–1.00 |
| Coarse sand | 1.00–0.50 |
| Medium sand | 0.50–0.25 |
| Fine sand | 0.25–0.10 |
| Very fine sand | 0.10–0.05 |
| Silt | 0.05–0.002 |
| Clay | <0.002 |

[a] According to the US Dept. of Agriculture system.

Most soils are a mixture of texture classes, and based on the percentage of each class, soils are named as shown in Figure 8.2. From an agricultural perspective, sand gives a soil good drainage and contributes to ease of cultivation, but a sandy soil also dries easily and looses nutrients to leaching. Clay, at the other extreme, tends not to drain well and can become easily compacted and difficult to work, yet is good at holding soil moisture and nutrients.

What soil texture is best, depends on the crops grown in it. Potatoes, for example, do best in a sandy, well-drained soil, which helps prevent rotting of the tubers and makes harvest easier. Paddy rice does best on heavy soils high in clay content due to this crop's particular adaptations to the wet environment. A clay loam soil may be best overall in a drier environment, whereas a sandy loam might be better in a wet one. The addition of organic matter changes the relationships of the particles in mixtures, as we will see below.

### Structure

In addition to the aspects of texture described above, soils possess a macrostructure formed by the ways individual particles are held together in clusters of different shapes and sizes called aggregates (Figure 8.3). Soil aggregates tend to become larger with increasing depth in the soil. Soil texture is one important determinant of

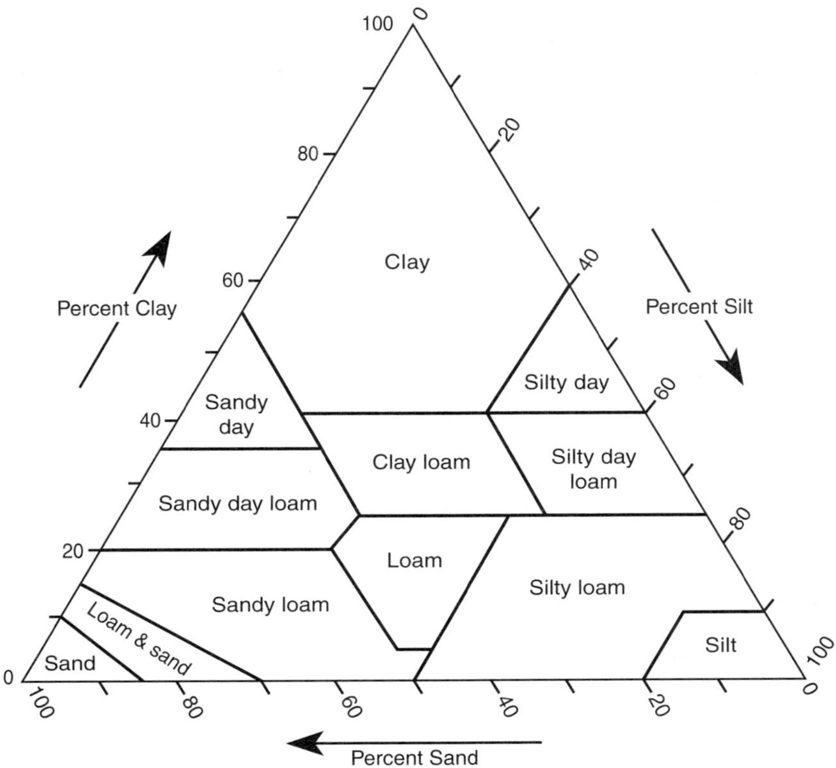


**FIGURE 8.2 Soil textural names.** The best type of soil is determined by the crop and local conditions; generally, however, soils containing relatively equal amounts of clay, sand, and silt — called loams — are best for agricultural purposes (From USDA diagram).

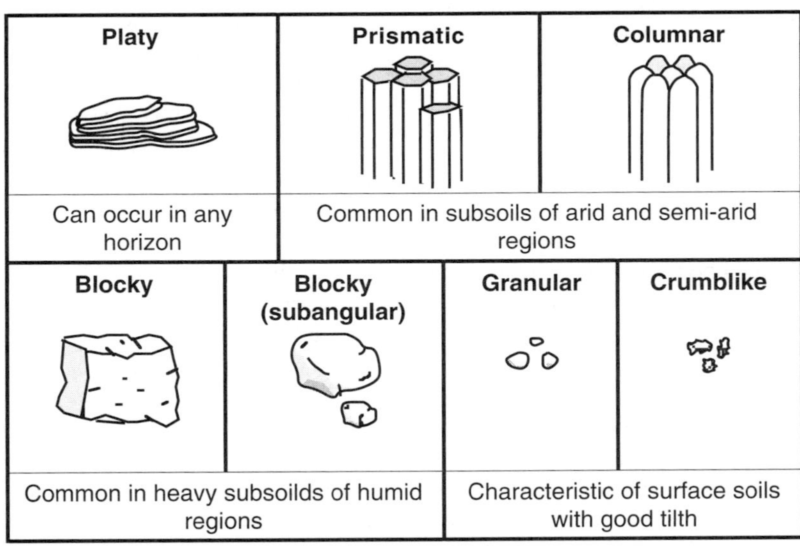


**FIGURE 8.3 Patterns of soil aggregation** (Modified from Brady, N. C., and R. R. Weil. 1996. 11th ed. Prentice Hall: Upper Saddle River, New Jersey).

structure, but structure is usually more dependent on soil organic matter content, the plants growing in the soil, the presence of soil organisms, and the soil's chemical status. Structure of the crumb or granular type is of the most benefit for agriculture, since good "crumb structure" improves soil porosity and ease of tillage, which together are known as tilth. When a lump of soil is crushed in the hand, and easily breaks into the crumb or granular structure noted in Figure 8.3, good crumb structure is present.

From an agroecological perspective, good crumb structure is of considerable significance. Soil particles that are bound together resist wind and water erosion, especially during any time of the year when vegetative cover is minimal. Good structure also helps maintain low *bulk density*, defined as the weight of solids per unit volume of soil. Soil with a low bulk density has a higher percentage of pore space (higher porosity), more aeration, better water percolation (permeability), and more water storage capacity. Obviously, such a soil is easier to till and allows plant roots to penetrate more easily. Excessive cultivation accelerates breakdown of soil organic matter and increases the potential for compaction, causing bulk density to go up and many of the advantages of good crumb structure to be lost.

The formation of soil aggregates has essentially two components: the attraction between individual soil particles, the degree of which is very dependent on soil texture, and the cementing of these attracted groups of particles by organic matter. The first component cannot be very easily manipulated by the farmer, at least in any practical manner, but the second can be very much impacted by farming practices. Thus good crumb structure can be maintained, degraded, or improved.

For example, excessive tillage with heavy equipment while the soil is too wet can lead to the formation of large blocky clods of soil that can dry on the surface and later be broken apart only with great difficulty. Compaction, or the loss of pore spaces and a rise in bulk density, is an indication of the loss of crumb structure, and can be caused by the weight of farm machinery, by the loss of organic matter from excessive tillage, or by a combination of the two.

### Color

Soil color plays its most important role in the identification of soil types, but at the same time it can tell us much about the history of a soil's development and management. Dark-colored soils are generally an indication of high organic matter content, especially in temperate regions. Red and yellow soils generally indicate high levels of iron oxides, formed under conditions of good aeration and drainage, but these colors can also be derived directly from the parent material. Gray or yellow-brown colors can be indicators of poor drainage; these colors form when iron is reduced to a ferrous form rather than oxidized to the ferric form in the presence of abundant oxygen. Whitish light-colored soils often indicate the presence of quartz, carbonates, or gypsum. Standardized color charts are used to determine a soil's color.

Hence, a soil's color can be an indicator of certain kinds of soil conditions that a farmer might want to look for or avoid, depending on the kinds of crops or cropping systems that might be used. More specific analysis of soil structure and chemistry is necessary to complete the picture, but color is a good beginning. In addition, soil color can influence the interaction of the soil with other factors of the environment. For example, it may be an advantage to have a lighter-colored, sandy soil on the surface in some tropical farming systems in order to reflect the sun's rays and keep the soil cooler; conversely, a darker soil surface in areas with cold winters will help the soil temperature rise earlier in the spring, dry the soil sooner, and permit soil preparation for planting at an earlier date.

### Cation-Exchange Capacity (CEC)

Plants obtain the mineral nutrients described in Chapter 2 and Chapter 3 from the soil in the form of dissolved ions, whose solubility is determined by their electrostatic attraction to molecules of water. Some important mineral nutrients, such as potassium and calcium, are in the form of positively charged ions; others, such as nitrate and phosphate, are in the form of negatively charged ions. If these dissolved ions are not taken up immediately through plant roots or fungi, they risk being leached out of the soil solution.

Clay and humus particles, separately or in aggregates that form platelike structures known as micelles, have negatively charged surfaces that hold the smaller, more mobile positively charged ions in the soil. The number of sites on the micelles available for binding positively charged ions (cations) determines what is called soil CEC, which is measured in milliequivalents of cations per 100 g of dry soil. The higher the CEC the better the soil's ability to hold and exchange cations, prevent leaching of nutrients, and provide plants with adequate nutrition.

CEC varies from soil to soil, depending on the structure of the clay/humus complex, the type of micelle present, and the amount of organic matter incorporated into the soil. Multisided polyhedrons form lattices that vary in their sites of attraction and flexibility in relation to moisture content. Cations cling to the negatively charged outer surfaces of the micelles and humates with differing degrees of attraction. The most tenacious cations — such as hydrogen ions added by rain, positively charged acids from decomposing organic matter, and acids given off by root metabolism — can displace other important nutrient cations such as $K^+$ or $Ca^{2+}$. Organic matter in the form of humus is many times more effective than clay in increasing CEC since it has a much more extensive surface area-to-volume ratio (hence more adsorption sites) and because it is colloidal in nature. Farming practices that reduce soil organic matter content can also reduce this important component of soil fertility maintenance.

Negative ions that are important for plant growth and development, such as nitrate, phosphate, and sulfate, are more commonly adsorbed to clay micelles by means of ion "bridges." Under acid conditions these bridges form by association of additional hydrogen ions with functional groups such as the hydroxyl group (OH). An important example is the binding of nitrate ($NO_3^-$) with $OH_2^+$ formed following the dissociation of water molecules under acid conditions. Because soil acidity influences electrical charge on micelle surfaces and controls whether other ions are displaced from soil micelles, it greatly affects the retention of ions in the soil and the short-term availability of nutrients, both of which are key components of soil fertility.

### Soil Acidity and pH

Any experienced gardener or farmer is aware of the importance of a soil's pH, or acid–base balance. The typical pH range of soils is between very acid (a pH of 3) and strongly alkaline (a pH of 8). Any soil over a pH of 7 (neutral) is considered basic, and those less than pH 6.6 are considered acid. Few plants, especially agricultural crops, grow well outside the pH range of 5 to 8. Legumes are particularly sensitive to low pH due to the impacts acid soils have on the microbial symbiont in nitrogen fixation. Bacteria in general are negatively impacted by low pH. Soil acidity is well known for its effects on nutrient availability as well, but the effects are less due to direct toxicity on the plant than they are to the plant's impaired ability to absorb specific nutrients at either very low or very high pH. It becomes important, then, to find ways to maintain soil pH in the optimal range.

Many soils increase in acidity through natural processes. Soil acidification is a result of the loss of bases by leaching of water moving downward through the soil profile, the uptake of nutrient ions by plants and their removal through harvest or grazing, and the production of organic acids by plant roots and microorganisms. Soils that are poorly buffered against these input or removal processes will tend to increase in acidity.

### Salinity and Alkalinity

It is common for the soils of arid and semiarid regions of the world to accumulate salts, in either a soluble or insoluble form. Salts released by the weathering of parent material, combined with those added in limited rainfall, are not removed by leaching. In areas of low rainfall and high evaporation rates, dissolved salts such as $Na^+$ and $Cl^-$ are common, combined with others such as $Ca^{2+}$, $Mg^{2+}$, $K^+$, $HCO_3^-$, and $NO_3^-$. Irrigation can add even more salts to the soil, especially in areas with a high evaporation potential (Chapter 9), where added salts migrate to the surface of the soil by capillary movement during evaporation. In addition, many inorganic fertilizers, such as ammonium nitrate, can increase salinity as well because they are in the form of salts.

Soils with a high concentration of neutral salt (e.g., NaCl or $Na_2SO_4$) are called saline. In cases where sodium is combined with weak anions (such as $HCO_3^-$), alkaline soils develop, which have a pH generally greater than 8.5. Soils with high levels of neutral salts are a problem for plants due to osmotic imbalances. Alkaline soils are a problem because of excess $OH^-$ ions and difficulty in nutrient uptake and plant development. In some regions, saline-alkaline conditions occur when both forms of salt are present. Proper irrigation and soil water management becomes a key part of dealing with these conditions.

## SOIL NUTRIENTS

Since plants obtain their nutrients from the soil, the supply of nutrients in the soil becomes a major determinant of an agroecosystem's productivity. Many nutrient analysis methodologies have been developed for determining the levels of various nutrients in the soil. When a particular nutrient is not present in sufficient quantity, it is called a *limiting nutrient* and must be added. Fertilization technologies have grown and evolved to meet this need. It must be kept in mind, however, that the presence of a nutrient does not necessarily mean it is *available* to plants. A variety of factors — including pH, CEC, and soil texture — determine the actual availability of nutrients.

Because of the loss or export of nutrients out of the soil due to harvest, leaching, or volatilization, fertilizers must continually be added in large amounts to most agroecosystems. But the cost of fertilizers as an input is increasing, and leached fertilizer pollutes ground and surface water supplies; therefore, an understanding of how nutrients can be cycled more efficiently in agroecosystems becomes essential for long-term sustainability.

As described in Chapter 2, the major plant nutrients are carbon, nitrogen, oxygen, phosphorus, potassium, and sulfur. Each of these nutrients is part of a different biogeochemical cycle and relates to management of soil in a unique way. The management of carbon will be discussed below in terms of organic matter; nitrogen in the soil will be included in a discussion of mutualisms and the ecological role of nitrogen-fixing bacteria and legumes in Chapter 16. Here, as an example of an important soil nutrient, we will examine the nutrient phosphorus. Because the efficient recycling of phosphorus depends principally on what happens in the soil, it can teach us a lot about sustainable nutrient management (Figure 8.4).

Unlike carbon and nitrogen, whose principal reservoirs are in the atmosphere, the principal reservoir of phosphorus is in the soil. It occurs naturally in the environment as a form of phosphate. Phosphates can occur in the soil solution as inorganic phosphate ions (especially as $PO_4^{3-}$) or as part of dissolved organic compounds. But the primary source of phosphate is the weathering of

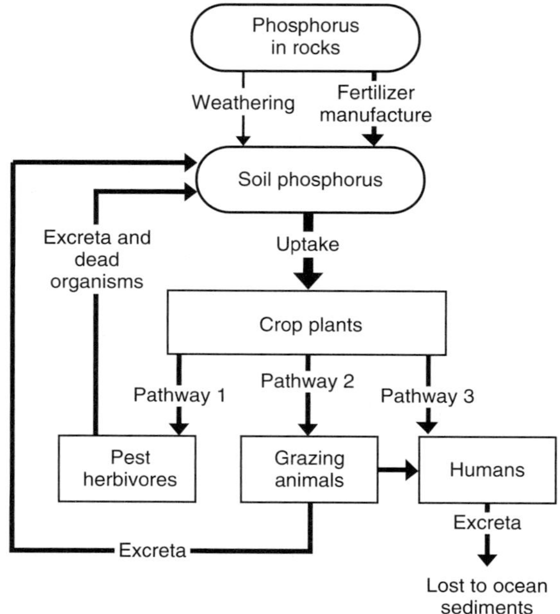


**FIGURE 8.4** Pathways of phosphorus cycling in agroecosystems.

parent material; therefore, the input of phosphorus into the soil and the phosphorus cycle in agroecosystems is limited by the relatively slow rate of this geologic process.

Inorganic soluble phosphate ions are absorbed by plant roots and incorporated into plant biomass. The phosphorus in this biomass can be sent along one of three different pathways, depending on how the biomass is consumed. As shown in Figure 8.4, consumption of plant biomass by pest herbivores, by grazing animals, or by humans who harvest the biomass comprises the three pathways. Phosphorus in the first pathway is returned to the soil as excreta, where it decomposes and enters the soil solution. Phosphorus in the second pathway can be recycled in the same way, but if the grazing animal goes to market, some phosphorus goes with it. In the third pathway, there is little chance of the phosphorus returning to the soil from which it was extracted (except in much of China, where human excreta is used as fertilizer).

Much of the phosphorus consumed by humans in the form of plant biomass or the flesh of grazing animals is essentially lost from the system. An example of what may happen to phosphorus in the third (human consumption) pathway may serve to illustrate the problem: phosphate is mined from phosphate-rich marine deposits that have been geologically uplifted and exposed in Florida, processed into soluble fertilizer or crushed into rock powder, and shipped to farms in Iowa where it is applied to the soil for the production of soybeans. A part of the phosphorus, in the form of phosphates, is taken up by the plant and sequestered in the beans that are harvested and sent to California, where they are turned into tofu. Following consumption of the tofu, most of the liberated phosphate finds its way into local sewer systems, and eventually ends up returning to the sea 3000 mi from where it originated. Since the time necessary to build up sufficient sediments of phosphate-rich rock and to go through the geological process of uplifting is very much beyond the realm of the human time frame, and since the known easily-available phosphate reserves are quite limited, current practices of phosphate fertilizer management in many modern agroecosystems can be said to be unsustainable.

For sustainable management of phosphorus to occur, phosphate needs to pass quickly through the soil component of the cycle and back to plants for it not to be fixed in sediments or washed to sea. Ways must be found to better keep phosphorus in an organic form, either in standing biomass or in soil organic matter, and to ensure that as soon as phosphorus is liberated from this organic form, it is quickly reabsorbed by soil microorganisms or plant roots.

An additional component of sustainable management of soil phosphorus has to do with the formation of insoluble phosphorus compounds in the soil. Phosphates in the soil solution often react chemically (especially with iron and aluminum) to form insoluble compounds, or become trapped in clay micelles out of reach of most biological recovery. Low pH in the soil exacerbates the problem of phosphate fixation in an insoluble form. At the same time, however, these processes provide a strong mechanism for retaining phosphorus in the soils of the agroecosystem; phosphate fertilizers added to the soil are retained almost completely. Some agricultural soils in California show very high levels of total (through not easily available) phosphorus after several decades of farming. So leakage of phosphorus from agroecosystems can be quite small, but the unavailability of phosphorus from the soil component of the system once it is fixed requires further addition of available phosphorus in the form of fertilizer. Of course, biological means of liberating this "stored" phosphorus might contribute better to sustainability. These means have a lot to do with the management of soil organic matter.

## SOIL ORGANIC MATTER

In natural ecosystems, the organic matter content of the A horizon can range up to 15 or 20% or more, but in most soils it averages 1 to 5%. In the absence of human intervention, organic matter content of the soil depends mostly on climate and vegetative cover; generally, more organic matter is found under the conditions of cool and moist climates. We also know that there is a very close correlation between the amount of organic matter in the soil and both carbon and nitrogen content. A close estimate of soil organic matter content can be obtained by either multiplying total carbon content by 2 or total nitrogen content by 20.

Soil organic matter is comprised of diverse, heterogeneous components. Its living material includes roots, microorganisms, and soil fauna; its nonliving material includes surface litter, dead roots, microbial metabolites, and humic substances (humus). The nonliving component is by far the greater proportion.

Interaction between the living and nonliving organic matter is constant. The complex carbon compounds in fresh plant litter are rapidly metabolized or decomposed, undergoing a process known as humification that eventually imparts a darker color to the soil as it produces humic residues, or humus. Humic residues consist of condensed aromatic polymers that are normally relatively resistant to further breakdown, and normally are capable of becoming stabilized in soil. The organic matter fraction that becomes stabilized, though, eventually undergoes mineralization, releasing mineral nutrients that can be taken up by plant roots. An equilibrium is reached between humification and mineralization, but this equilibrium is subject to shifts depending on farming practices.

During its life in the soil, organic matter plays many very important roles, all of which are of importance to sustainable agriculture (Magdoff and Weil, 2004). Apart from providing the more obvious source of nutrients for plant growth, organic matter builds, promotes, protects, and maintains the soil ecosystem. As we have already discussed, soil organic matter is a key component of good soil structure, increases water and nutrient retention, is the food source for soil microorganisms, and provides important mechanical protection of the soil surface. Depending on the cropping practices used, however, these traits can be rapidly altered — for the better as well as for the worse. Of all of the characteristics of soil, the factor that we can manipulate the most is soil organic matter.

Once a soil is put under cultivation, the original organic matter levels begin to decline unless specific steps are taken to maintain them. After an initial rapid decline, the drop slows. Several kinds of changes occur in the soil as a consequence of the loss of organic matter. Crumb structure is lost, bulk density begins to rise, soil porosity suffers, and biological activity declines. Soil compaction and the development of a hardened soil layer at the average depth of cultivation, called a plow pan, can become problems as well.

The extent to which organic content declines in soil under cultivation is dependent on the crop and cropping practices. Some examples follow.

In one study, the organic matter content of the upper 25 cm of soil in two agroecosystems used for intensive vegetable production in coastal central California were compared with each other and to an unfarmed grassland control. One system had been farmed for 25 years using organic farming practices; the other for 40 years under conventional practices. The study showed that the organic matter content had been reduced from 9.869 to 8.705 kg/m$^3$ in the organic system and to 9.088 kg/m$^3$ in the conventional system (Waldon, 1994). Even with the higher inputs of organic matter in the form of composts and winter cover crops in the organic system, cultivation and cropping significantly reduced soil organic matter even more than in the conventional system.

In another case, after 15 years of continual production of grains such as corn and rice, organic matter in the initial 15 cm of a heavy alluvial clay in the humid lowlands of tropical Tabasco, Mexico, had lowered to less than 2%, as compared to an organic matter content of more than 4% in an adjacent area of uncut tropical forest (Gliessman and Amador, 1980). A tree-covered cacao plantation on the same soil type was able to maintain the soil organic matter in the same layer at 3.5%, demonstrating the negative impact of soil disturbance on organic matter in cropping systems and the role of vegetative cover in retaining it.

A study comparing soils after 75 years of organic and conventional wheat production in eastern Washington found that organic matter was not only maintained in the organic system, but also actually increased over time, while production levels for the organic farmer were near equal to the conventional (Reganold et al., 1987). We can see from these three examples that crop type, input management, local environment, and cultivation practices all help determine the long-term impacts of farming on soil organic matter.

## SOIL MANAGEMENT

In present-day farming systems, soil is treated as if it were mainly a medium for holding the plant up. When soil is managed for sustainable production and emphasis is placed on the role of soil organic matter, however, the role of soil is greatly expanded.

Many farmers feel that if a high yield is obtained from the land, then this is evidence of a productive soil. However, if the perspective is agroecological and the goal is to maintain and promote all of the soil-forming and soil-protecting processes involving organic matter, then a productive soil is not necessarily a fertile soil. The processes in the soil that enable us to produce a crop take on greater importance in sustainable agriculture. Fertilizers can be added to raise production, but only through an understanding of nutrient cycles and soil ecological processes — especially soil organic matter dynamics — can soil fertility be maintained or restored.

### Management of Soil Organic Matter

The first step in developing soil organic matter is to maintain constant inputs of new organic matter to replace that, which is lost through harvest and decomposition. If the agroecosystem were more similar to a natural ecosystem, a diversity of plant species would be present in addition to the crop or crops being grown for harvest. Many agroforestry systems (Chapter 17), especially in tropical agriculture, have a large number of plants, many of them noncrop species, whose primary role is biomass production and the return of organic matter to the soil. But present-day agriculture, with its focus on the market, has reduced plant diversity so greatly that very little organic matter is returned to the soil.

## Crop Residue

An important source of organic matter is crop residue. Many farmers are experimenting with better ways of returning to the soil the parts of the crop that are not destined for human or animal use. A major concern has been how to deal with potential pest or disease organisms that residue may harbor and pass on to a subsequent crop. Proper timing of incorporation of the residue into the soil, rotating crops, and composting the residue away from the field and then returning the finished compost are possible ways of overcoming this problem. Research on these and other management strategies are helping transform crop residue from a problematic by-product into a valuable part of soil organic matter management (Franzluebbers, 2004) (Figure 8.5).

## Cover Crops

Cover cropping, where a plant cover is grown specifically to produce plant matter for incorporation, as a "green manure" into the soil, is another important source of organic matter. Cover crop plants are usually grown in rotation with a crop or during a time of the year that the crop can't be grown. When legumes are used as cover crops, either alone or in combination with nonlegume species, the quality of the biomass can be greatly improved. The resultant biomass can be incorporated into the soil, or left on the surface as a protective mulch until it decomposes.

In research done at UC Santa Cruz (Gliessman, 1987), a local variety of fava bean called bellbean (*Vicia faba*) was grown as a cover crop in combination with either cerealrye or barley during the winter wet season fallow period. It was shown that the total dry matter produced in the grass/legume mixtures was almost double that of the legume alone. After 3 years of cover crop use, organic matter levels in soils under mixed covers improved as much as 8.8%. Interestingly, soils under the legume-only cover actually dropped slightly in organic matter content

**FIGURE 8.5 Burning of crop residue in Taiwan.** Burning is a common method of removing crop residue. Although it returns some nutrients to the soil and helps control pests and diseases, burning can cause significant air pollution and prevents crop residue from being incorporated into the soil as organic matter. When crop residue is seen as a valuable and useful resource for maintaining soil organic matter, techniques for incorporating it into the soil can be developed as alternatives to burning.

after three years, probably because the lower C/N ratio of the incorporated organic matter caused more rapid microbial breakdown.

A more recent innovation in the cover-cropping approach is the use of a living mulch, where a noncrop species is planted between the rows of the crop during the cropping cycle. Living mulches have become especially popular in vineyard, orchard, and tree crop systems. Research has focused on ways of minimizing negative interactions between cover crop and crop species, especially living mulches in annual crops. Studies are also finding that living mulches can provide and conserve nitrogen for grain crops, reduce soil erosion, reduce weed pressure, and increase soil organic matter content (Hartwig and Ammon, 2002).

## Manure

It is a long-standing practice, both in conventional and in alternative farming systems, to add animal manures to the soil to improve organic matter content. The application of animal manure is an important tool for an integrated nutrient management strategy because applications can simultaneously increase soil organic matter and supply nutrients for crop growth (Seiter and Horwath, 2004). Dairies and feedlot operations produce large amounts of animal wastes that are converted to a useful resource when returned to fields, but as we have already noted in Chapter 1, there are many problems involved in containing, storing, transporting, and applying such large quantities of animal manures. Small, integrated farm operations can more easily use animal manures that accumulate in stables or pens for intensive vegetable production or use on other crops (Chapter 19). The use of silkworm droppings in Chinese agriculture is yet another example of the use of animal manures.

At any scale, the direct application of animal manures can have many drawbacks, however. Smell and flies are often associated with direct manure application. Nitrogen loss through ammonification can be quite high. Runoff of nitrates and other soluble materials can be a problem. And once fresh manures are incorporated into the soil, there often is a waiting period for decomposition and stabilization before planting can take place. To avoid these problems, current organic certification standards in the U.S. require that fresh or raw animal manures be composted under specific conditions before they are applied (Figure 8.6).

**FIGURE 8.6 Manure spreader used on a dairy farm near Cody, Wyoming.** Aged manure is returned to fields in which feed is grown for the farm's dairy cows.

## Composts

Compost amendment of soil is an attractive way to add organic matter for a variety of reasons. The particle-size distribution of compost favors uniform field application; the ratio of carbon to nitrogen is optimal; compost is usually free of weed seeds; and soil diseases are often suppressed by compost addition (Chen et al., 2004). Many different sources of organic materials, from manures to agricultural by-products to lawn clippings, are being converted into useful soil amendments through the composting process (Figure 8.7). Under controlled conditions, raw organic matter goes through the first stages of decomposition and humification, so that when it is added to the soil, it has stabilized considerably and can contribute more effectively to the soil fertility-building process. In this way, wastes — including materials that would otherwise go to already bulging landfills — are being converted into resources (Table 8.3).

**Table 8.3**
**Organic Waste Materials Employed in the Production of Compost**

| Agricultural By-Products | Manures |
|---|---|
| Alfalfa leaf meal | Feedlot beef cattle manure |
| Apple and grape pomace | Dairy cattle manure |
| Blood meal | Broiler chicken litter |
| Bone meal | Laying chicken litter |
| Cottonseed meal | Turkey litter |
| Feather meal | Swine manure |
| Almond and walnut hulls | Horse manure |
| Coffee pulp | Sheep manure |
| Cocao pulp | Goat manure |
| Soybean cakes | |
| Rice hulls | |
| Green garden and yard wastes | |

Vermicompost, or compost produced through the action of worms, is also becoming a popular source of soil organic matter, especially for smaller-scale farm and garden systems. Fresh, wet organic matter, especially food waste, is consumed by worms specifically known for their composting ability (redworms such as *Eisenia foetida* are especially good), and systems have been developed, where a small household vermicomposting chamber can produce up to 25 kg of worm castings a month. These castings are known for their high levels of phosphate, nitrogen, and other nutrients, and also contain polysaccharides that glue soil particles together and aid in soil organic matter development. Cuban researchers have recently developed farm-scale vermicomposting systems that are designed to replace difficult-to-obtain imported fertilizers. Further development of larger-scale systems could aid greatly in improved soil management.

**FIGURE 8.7 Farm wastes being turned into compost on a farm on the central coast of California.** The breakdown of vegetative matter by microorganisms releases significant amounts of energy in the form of heat.

### Other Soil Amendments

A range of other types of organic soil amendments can be used as well. Humates, kelp, fishmeal, animal by-products, mined guano, and others are on the market. Each one has specific applications, advantages and disadvantages, and optimal scales of use. Each organic matter source needs to be examined for short-term crop response, but more importantly for possible long-term contributions to soil organic matter development and maintenance.

### Sewage

A final source of organic matter — underutilized except in a few parts of the world — is sewage. To complete nutrient cycles, nutrients that leave the farm should ultimately come back to the farm. If they can come back in an organic form, then they will also add to the soil-building process.

Solid material removed from wastewater during treatment, known as sewage sludge, has been spread on the land for decades. As a percentage of dry weight, sewage sludge can contain 6 to 9% nitrogen, 3 to 7% phosphorus, and up to 1% potassium. It can be applied as dried cake or granules, with water content of 40 to 70%, or as a liquid slurry that is 80–90% water. Sewage sludge is widely used on turf grass, degraded rangeland, and even on the ground below fruit trees. The liquid portion of treated sewage, known as effluent, has been applied to land for a long time in Europe and selected sites in the U.S. Some cities operate, what are called, sewage farms where effluent is used to produce crops, usually animal feeds and forages, that partially offset the cost of disposal, where in other cases, it is used for irrigating golf courses, highway landscaping, and even forests.

There is much to learn, however, about how to treat sewage so that pathogens are dealt with properly. Collection, treatment, and transport all need to be examined with an eye toward the goal of linking waste management with sustainable agriculture. The fact that many sewage systems around the world do not separate human from industrial wastes, contaminating the resultant sludge with toxic amounts of heavy metals, complicates the process immensely.

Nevertheless, sewage will undoubtedly become a more important resource in the future as a source of organic matter, nutrients, and water for crop production. Many small-scale and traditional practices for turning sewage into a useful resource can serve as an important basis for future research on this important link to sustainability.

## Tillage Systems

The conventional wisdom in agriculture is that soil must be cultivated to control weeds, incorporate organic matter, and allow root growth. Despite its potential benefits, however, cultivation can degrade soil structure and organic matter content, and cause the soil to lose some of the elements of productivity. For this reason, paying attention to how the soil is cultivated must be an integral part of soil organic matter management.

Many different patterns of soil tillage exist, but the main pattern employed in conventional agriculture is a three-stage process involving a deep plowing that turns the soil, a secondary tilling for preparation of a seed bed, and finally postplanting cultivations (often combined with herbicide use) for controlling weeds. Soil erosion, loss of good soil structure, and nutrient leaching are well-known problems associated with this pattern of tillage. Despite these problems, most conventional farming systems, especially those producing annual grains and vegetables, are dependent on extensive and repeated tillage.

At the other extreme, there are many traditional farming systems in which no tillage is used at all. In swidden agriculture, traditional farmers clear land using slash and burn techniques and then poke the soil with a planting stick to sow seeds. Such systems, which have the longest history of sustained management, respect the need for a fallow period to control weedy vegetation and to allow natural soil building processes to replace removed nutrients. Many agroforestry systems, such as coffee or cacao under shade, depend on the tree component of the system to provide soil cover and nutrient cycling, and only receive occasional surface weeding. Permanent pasture is rarely cultivated either.

Alternative tillage techniques, many of them borrowed from traditional farming practices, have been developed for and tested in conventional annual crop systems. These have demonstrated that annual crop systems do not have to remain dependent on extensive and repeated tillage and that reduced tillage can help improve soil quality and fertility (Franzluebbers, 2004).

Using the technique of *zero tillage*, soil cultivation is limited to the actual seedbed and is done at the time of seed planting. In some cases, special equipment is used that allows planting directly into the crop residue left from the previous crop. Other steps, such as fertilization and weed control, can be completed at the same time as planting. Unfortunately, many zero tillage systems have developed a great dependence on herbicides, which may create other ecological problems.

In order to reduce herbicide use, a number of *reduced tillage* systems have been developed. One in particular that has been quite successful for corn and soybean production is *ridge tillage*. After an initial plowing and formation of planting beds or ridges, the only cultivation that occurs is seed planting and weed management with specially designed tillers that cultivate the surface of the soil only. Some ridge till systems can go through many years of repeated planting without deep tillage, and the reduced soil disturbance helps preserve soil organic matter and structure. The Thompson Farm in Boone County, Iowa,

## REDUCED TILLAGE ON THE THOMPSON FARM

Farmers who want to reduce the frequency and intensity of cultivation and soil disturbance face a major quandary: how to cultivate less without simultaneously increasing herbicide use? Decades ago, Dick Thompson, a farmer in Boone, Iowa, realized this was a false choice — it should be possible to reduce both tillage and herbicide use and yet still control weeds effectively. Dick reasoned that while cultivation killed weeds, it also created the ideal disturbed environment for their regrowth. The solution must involve cultivating less, and doing so in a way that aided crops but deterred weeds.

Through extensive experimentation and tinkering, Dick has developed a modified ridge tillage system that, combined with other farming practices, produces the desired result: the Thompson Farm uses virtually no herbicides, the soil is tilled much less frequently and intensively than conventional wisdom would dictate, and weeds are not a problem. In addition, the Thompsons generate about $147 more profit per acre than their conventional neighbors, and their soil contains about twice as much organic matter.

The ridge tillage works like this: in the spring, special planters shave a little soil — and the weeds and cover crop growing in it — off the top of each ridge, burying plant residue, weed seeds, and previously applied compost in the inter-row zone. There, the allelopathic effect of the residue suppresses weed germination. Later, a rotary hoe lifts soil up around the crops planted in the weed-free area in the middle of the ridges. Weeds are controlled by this process until the crop is developed enough to suppress weeds itself.

The Thompson's modified ridge tillage system is the centerpiece of their farming operation — or at least the part that differs most from conventional practice — but it is only one part of a broader management system with many components.

- *Rotations*. Whereas many conventional farms in Iowa grow corn continuously, the Thompsons always rotate their crops. One rotation they've used with success is the five-year sequence of corn–soybeans–corn–oats–hay. Rotating crops avoids problems with soil pests and helps maintain soil fertility. The rotation has a direct connection with the ridge-till system as well: after the four years of row crops planted on minimally tilled ridgetops, the ridges are flattened for the final crop of hay, which is a mixture of grasses and legumes. This crop is cut multiple times to help control weeds. During this ridgeless period, the field is cultivated conventionally, with incorporation of manure deep into the soil where its fertility is most needed by crops. Then the ridges are re-formed for the corn that begins the cycle again.
- *Cover crops*. During winter, the soil in the fields is always covered by a cover crop. At first, the Thompsons used winter rye and hairy vetch, but after much experimentation, they found that grains worked best as cover crops. They now plant oats ahead of corn, and rye ahead of soybeans. At the point in the rotation when hay will be followed by corn, however, they plant a cover crop of rye. In addition to reducing erosion and nitrate leaching, these cover crops suppress weed growth allelopathically. The weeds that do grow with the cover crop are treated as part of the cover vegetation — they help provide cover, suppress later-germinating weeds, and contribute organic matter to the soil after cultivation.
- *Livestock integration*. The Thompsons raise both beef cattle and hogs, and all their manure — composted aerobically in special containers — goes into the fields as organic fertilizer. To complete the nutrient cycling, the livestock are fed hay and crop residue from the crop fields.

Combined, these principles result in both profitability and sustainability, making the Thompson Farm an excellent model. Every year, hundreds of visitors — ranging from college professors to organic farmers — tour the farm to learn alternative practices for soil management, erosion control, weed management, and reduced tillage.

featured in the Case Study, is a well-known example of a successful diversified farm operation that has as its centerpiece the use of a modified ridge tillage program (Practical Farmers of Iowa, 2002).

The challenge for research on reduced tillage systems is how to find ways to reduce tillage without increasing input costs elsewhere in the system, especially those involving the use of chemicals or fossil fuels.

### Sustainable Soil Management

When soil is understood to be a living, dynamic system — an ecosystem — management for sustainability becomes an integrated, whole-system process. Focusing on the processes that promote the maintenance of a healthy, dynamic, and productive system becomes paramount. Fertility management is based on our understanding of nutrient cycles, organic matter development, and the balance between the living and nonliving components of the soil. The application of our understanding of the ecological processes that maintain the structure and function of the soil ecosystem over time takes on the greatest importance. And since the soil ecosystem is a complex, dynamic, and ever-changing set of components and processes, our understanding of this complexity must increase.

Good soil management is an important part of attaining overall sustainability of agroecosystems. Many of the indicators of sustainability discussed in Chapter 21 relate directly to soil.

## FOOD FOR THOUGHT

1. Organic matter is considered to be one of the most important components of a healthy soil ecosystem, but most agricultural activities (i.e., plowing, burning, cultivation, harvest) remove, reduce, or degrade organic matter. What are some of the most practical ways of maintaining this valuable resource in the soil?
2. What are the key factors that determine how long a degraded soil will take to be restored to a condition similar to its previous healthy condition?
3. What is the difference between dirt and soil?
4. It has recently been proposed that we develop some indicators of "soil health" in order to determine the sustainability of different farming practices. What indicators do you think should be used to evaluate the health of the soil?
5. Why is it important for farmers to learn how to use the concept of the soil ecosystem?

## INTERNET RESOURCES

www.pedosphere.com
An on-line soil science magazine.

Natural Resources Conservation Service: Soil Quality
soils.usda.gov/sqi
The Soil Quality portion of the NRCS Soils website, with information about soil management practices, soil biology, and soil quality assessment.

National Sustainable Agriculture Information Service: Sustainable Soil Management
www.attra.org/attra-pub/soilmgmt.html#Resources
Contains a wealth of information about the biology and chemistry of soil and the relationships between soil management and sustainability.

Rodale Research Foundation
www.newfarm.org
A wealth of examples of alternative farm management practices, especially of soils.

US Department of Agriculture, Natural Resources Conservation Service
soils.usda.gov
The NRCS soils website, providing extensive science-based soil information, including soil surveys from across the nation.

## RECOMMENDED READING

Brady, N.C. and R.R. Weil. 2001. *The Nature and Properties of Soils.* 13th ed. Prentice Hall: Upper Saddle River, New Jersey. One of the most complete reference books on soil as a natural resource; highlights the many interactions between soil and other components of the ecosystem. The recognized primer of soil science.

Frissel, M.J. (ed.) 1978. *Cycling of Mineral Nutrients in Agricultural Ecosystems.* Elsevier: Amsterdam. A pioneering work on the need for an ecological approach to the study of nutrient use in agriculture.

Jenny, H. 1994. *Factors of Soil Formation.* Reprint edition of the 1941 original. Dover Publications: Toronto, Canada. The classic textbook on soil and the soil formation process; emphasizes the soil as a complex system that changes through time.

Juo, A.S.R. and K. Franzluebbers. 2003. *Tropical Soils: Properties and Management for Sustainable Agriculture.* Oxford University Press USA: Cary, NC. A text that uses an agroecological approach to describe the tropical soil environments of sub-Saharan Africa, Southeast Asia, and South and Central America, focusing on production and management systems unique to each region.

Magdoff, F. and H. van Es. 2000. *Building Soils for Better Crops.* 2nd ed. Sustainable Agriculture Network Handbook Series. Sustainable Agriculture Publications: Burlington, Vermont. Very farmer-friendly and practical information that explains how ecological soil management boosts soil fertility and yields, while reducing pest pressures and environmental impacts.

Magdoff, F. and R.R. Weil. 2004. *Soil Organic Matter in Sustainable Agriculture.* Advances in Agroecology Series. CRC Press: Boca Raton, Florida. This book provides the essential scientific background to understand soil organic matter and develop improved soil and crop management systems.

Paddock, J., N. Paddock, and C. Bly. 1986. *Soil and Survival: Land Stewardship and the Future of American Agriculture.* Sierra Club Books: San Francisco. An important work on the need for linking sound soil management with sustainability in American agriculture.

Stevenson, F.J. and M.A. Cole. 1999. *Cycles of Soil Carbon, Nitrogen, Phosphorus, Sulfur, and Micronutrients*. 2nd ed. John Wiley and Sons: New York. An examination of the processes and mechanisms of cycling of both macro- and micronutrients in the soil.

Sylvia, D.M., J.J. Fuhrmann, P.G. Hartel, and D.A. Zuberer. 2004. *Principles and Applications of Soil Microbiology*. 2nd ed. Prentice Hall: Upper Saddle River, NJ. A comprehensive and balanced introduction to soil microbiology, including habitats and organisms, microbially mediated transformation, and applied environmental topics.

Vitousek, P.M. 2004. *Nutrient Cycling and Limitation: Hawai'i as a Model System*. Princeton University Press: Princeton, NJ. A volume that makes use of Hawaiian ecosystems to explore the mechanisms that shape productivity and diversity in ecosystems throughout the world.

Wardle, D.A. 2002. *Communities and Ecosystems: Linking the Aboveground and Belowground Components*. Princeton University Press: Princeton, NJ. A synthetic volume that analyzes the interactions between biotic communities aboveground and belowground, focusing on their important roles in defining community structure and ecosystem functioning.

Woomer, P.L. and M.J. Swift (eds.) 1994. *The Biological Management of Tropical Soil Fertility*. John Wiley and Sons: New York. An ecosystem approach to managing soil fertility in natural and agricultural ecosystems, with case studies from a diverse array of tropical regions whose emphasis is on long-term sustainability.

# 9 Water in the Soil

Water is continually flowing through the body of a plant: leaving the stomata via transpiration (T) and entering through the roots. For this reason, plants depend on having a certain amount of water available to their roots in the soil. Without adequate soil moisture, they quickly wilt and die. Thus, maintaining sufficient moisture in the soil is a crucial part of agroecosystem management.

Yet, soil moisture management is not simply a matter of there being adequate inputs of water into the soil from precipitation or irrigation. Soil moisture is part of the ecology of the soil and of the whole agroecosystem. Water availability and retention is affected by a myriad of factors and water itself plays many roles. It carries soluble nutrients, affects soil aeration and temperature, and impacts soil biotic processes. A farmer, therefore, must be aware of how water acts in the soil, how water levels in the soil are affected by weather conditions and cropping practices, how inputs of water affect soil moisture, and what the water needs of the crop are.

Rarely is the moisture availability of a soil exactly optimum for a crop for a very long period of time. Water supply varies between deficiency and surplus from day to day and throughout the season. The actual optimum is hard to determine, since it is affected by a range of other factors, and conditions are constantly changing. But we do know a lot about the range of moisture conditions that promote highest yields for most crops. The challenge is to manage water in the soil in ways that keep conditions within this range.

## MOVEMENT OF WATER INTO AND OUT OF THE SOIL

In natural ecosystems, water enters the system as rainfall or snowmelt at the surface of the soil. In agroecosystems, water enters from the same sources, as described in Chapter 6, or is added as irrigation. Sustainable management of soil moisture depends greatly on understanding the fate of this applied water, with a goal of maximizing efficiency of water use by the system.

### Infiltration

For the water falling on or applied to the soil surface to become available to plants, it must infiltrate into the soil. Infiltration is by no means a given: water can be lost to surface runoff or even evaporation if it cannot penetrate the soil surface easily. Infiltration is affected by soil type, slope, vegetative cover, and characteristics of the precipitation itself. Soils with greater porosity such as sandy soils or those with high organic matter content are more open to easy infiltration of water. Flat terrain is more apt to allow better infiltration than sloping ground, and a smooth slope loses more water to runoff than one that is broken by microtopographic variation caused by rocks, soil clumps, slight depressions, or other obstructions on the surface. Vegetative cover, both alive and as litter on the surface, greatly aids initial water entry. In general, assuming optimal conditions, the greater the intensity of rainfall, the greater the infiltration rate until saturation is achieved. However, with excessively intense rainfall, increased runoff will occur.

### Percolation

Once saturation of the upper layers of the soil occurs, gravitational forces begin to pull the excess water more deeply into the soil profile. This process, known as *percolation*, is shown in Figure 9.1. The rate of percolation is determined by soil structure, texture, and porosity. A soil with good crumb structure and aggregate stability will allow water to move freely between soil particles. Sandy-textured soils have larger pore spaces and less soil-particle surface area to hold water than more finely textured soils, and will therefore allow the most rapid movement of water. A soil that is very high in clay content may allow rapid percolation initially, but once the clay micelles swell with water, they may close the pore spaces and impede movement. Root channels and animal burrows, especially those of earthworms, are important pathways for percolation, but soil texture and structure are probably of greater importance, especially in frequently cultivated agroecosystems.

### Evaporation

Once moisture enters the soil, it can be lost to the atmosphere through evaporation. The rate of evaporation from the soil surface depends on the moisture content and temperature of the atmosphere above the surface, as well as the temperature of the soil surface itself. Wind greatly accelerates the evaporation process, especially at higher temperatures.

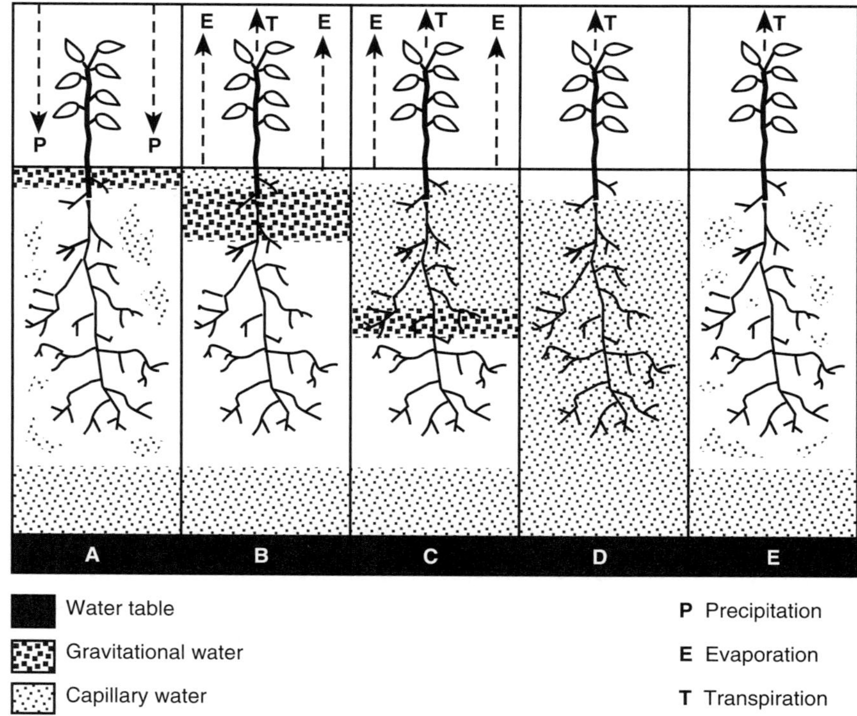


**FIGURE 9.1 Movement of water in the soil of a cropping system.** A. Water infiltrates the surface after falling as precipitation, B. gravitational water percolates downward, leaving the soil above moistened to field capacity with capillary water. At the same time, evaporation and T begin to remove water from the soil C as gravitational water continues to percolate downward, the soil near the surface begins to dry out. D. When the gravitational water reaches the water table, most of the soil profile is moistened close to field capacity. The exception is the upper layer of soil, which has dried out from evaporation. E. Most of the soil above the capillary fringe, the region kept moist by the water table, has dried out, and the soil once more nears the wilting point. (Adapted from Daubenmire, R. F. 1974. *Plants and Environment.* 3rd ed. John Wiley and Sons: New York.)

Even though evaporation occurs at the surface, it can affect soil moisture deep into the soil profile. As evaporation creates a water deficit at the soil surface, the attractive forces between water molecules draw water from below through capillary action. This process continues until the saturated zone reaches too deep or the upper soil layer becomes so dry that capillarity is broken. Any kind of mulch or soil surface cover that slows the heat gain of the soil surface and presents a barrier between the soil and the atmosphere will slow the rate of evaporation.

### Transpiration

As described in Chapter 3, plants lose water through stomata in the leaves as transpired moisture, creating a water deficit in the plant that is balanced by uptake of water by the plant roots. This biotic removal of water from the soil, especially by roots that penetrate the soil layers below those affected by evaporation, constitutes a major avenue of water movement out of the soil ecosystem. If water is not added to replace this loss, plants either have to go dormant or are eliminated from the ecosystem.

## SOIL MOISTURE AVAILABILITY

The attractive forces operating between water and individual soil particles play a key role in determining how soil moisture is retained, lost, and used by plants. Understanding these forces means looking at the physical and chemical properties of the *soil solution*, the liquid phase of the soil, and its dissolved solutes that are separate from the soil particles themselves.

The percentage of moisture available for plant use in a soil has traditionally been determined by collecting a soil sample, measuring its weight, drying the soil at 105°C for 24 h, and then measuring its dry weight. The amount of moisture lost during drying is divided by the sample dry weight, giving a figure that is expressed as a percentage.

This procedure, however, is not adequate for measuring the amount of water actually available to plants in the soil because it does not take into account the important variable of water adhesion to soil particles. As both clay and organic matter content increase in a soil, water is attracted more tightly to soil particles and becomes more difficult for roots to take up. Lettuce may wilt, for example, in a clay soil with 15% moisture, whereas in a sandy soil, moisture may drop as low as 6% before the crop will wilt.

Because water is held more tightly in some kinds of soil compared to others, another measure besides just percent moisture content is needed, which better reflects the attractive force between soil particles and moisture. This measure is achieved by expressing soil moisture in energy terms. The force of attraction of water molecules to soil particles, the soil water potential, is expressed as bars of suction, where one bar is equivalent to standard atmospheric pressure at sea level (760 mm Hg or 1020 cm of water). This method provides a means of measuring the availability of water in the soil solution and takes into account the varying forces of attraction determined by soil particle size and organic matter content.

A number of special terms are used to describe water moisture content and availability in terms of attractive forces. These are defined below and illustrated in Figure 9.2.

- *Gravitational water* is water that moves into, through, and out of the soil under the influence of gravity alone. Immediately following rain or irrigation, this water begins to move downward into the soil, occupying all macropore spaces.
- *Capillary water* is the water that fills the micropores of the soil and is held to particles with a force between 0.3 and 31 bars of suction.
- *Hygroscopic water* is the water held most tightly to soil particles, usually with more than 31 bars of suction. After soil has been oven dried, the remaining nonchemically bound water is hygroscopic water.
- *Water of hydration* is the water that is chemically bound with the soil particles.
- *Easily available water* is the portion of the water in the soil that is readily absorbed by plant roots — usually capillary water between 0.3 and 15 bars of suction.
- *Field capacity* is the moisture left in the soil after the downward pull of gravity has drained the macropores of gravitational water, leaving the micropores filled with capillary water held with at least 0.3 bars of suction to soil particles.
- *Permanent wilting point* is the moisture content of the soil at which a plant wilts and does not recover even when placed in a dark, humid environment. Permanent wilting point usually occurs when all the capillary water held at less than 15 bars of suction has been removed from the soil.

Since every soil is a different mixture of particle sizes and is variable in organic matter content, and because these characteristics determine water retention ability, it is important to determine the soil type as a part of developing a water management plan. In most soils, optimum growth takes place when soil moisture content is kept just below field capacity. It is clear that the moisture needed for optimum growth does not extend over the complete range of soil moisture content.

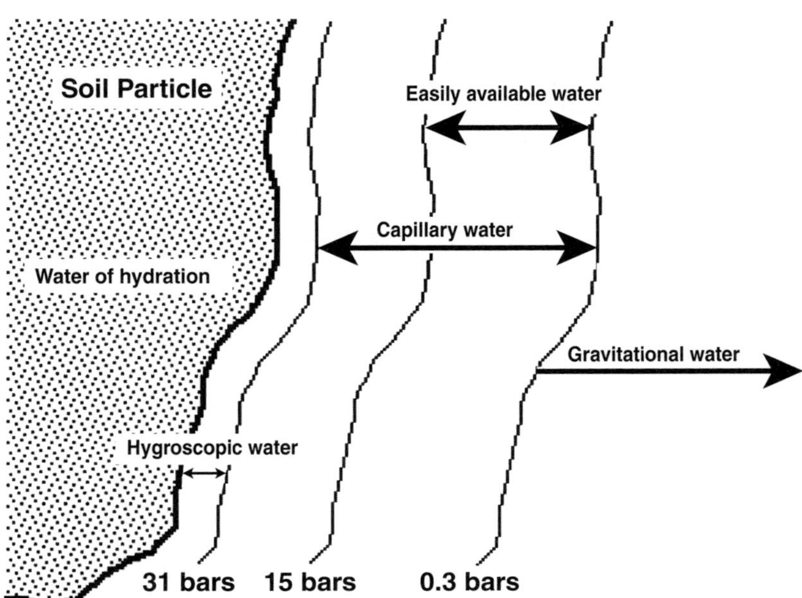


**FIGURE 9.2 Soil moisture in relation to force of attraction to soil particles.** Permanent wilting point is reached when easily available water has been depleted. Field capacity is the amount of water remaining after gravitational water has drained away.

## PLANTS' UPTAKE OF SOIL MOISTURE

While they are transpiring, plants must continually replace the significant amount of water they lose through their stomata. At any one time, however, only a small proportion of available soil water is close enough to the root surfaces that actually absorb the water. Two processes compensate for this limitation. First, water is drawn passively through the soil to root surfaces through capillary movement of water, and second, plant roots actively grow into the soil toward areas with sufficient moisture for uptake.

### Capillary Movement of Water

As a plant takes in water through its roots to replace that which it loses through T, the soil moisture content of the area immediately surrounding the root is reduced. This increases the energy of suction in that region, creating a gradient of lower water potential that tends to draw moisture in all directions from the surrounding soil. Most water is probably drawn from deeper in the soil profile, especially when the water table is close to the surface. Capillary movement is due partly to the attraction of water molecules to soil particle surfaces, and partly to the attraction of water molecules to each other. The speed at which capillary movement occurs depends on the intensity of the water deficit and the type of soil. In most sandy soils, movement is fairly rapid because the larger-sized particles hold water less tightly. In soils with more clay, especially those with poor crumb structure, movement is much slower.

It has been shown that water can move only a few centimeters a day through capillary action. But due to the extensive volume of soil occupied by most root systems, movement of any greater distance is probably not needed. Plants can obtain a large proportion of their water needs through capillary movement even when T rates are very high. The increased suction pressure created in the immediate root zone during the day is replaced by water movement from areas of lower suction during the night. It is at times when soil moisture content has been severely depleted and plant growth has slowed that such movement is of greatest significance. Otherwise, the plant reaches permanent wilting point.

### Extension of Roots Into the Soil

Plants are continually extending roots into the soil, ensuring that new sites of root contact with the soil are being established. Roots, rootlets, and root hairs all combine to produce an extensive network of soil–root interface. Despite continued root penetration and the large volume of the root network, the total amount of any particular soil volume that is in contact with a plant's roots at any one time is very small. According to most estimates, less than 1% of the total soil-particle surface area within the volume of soil occupied by a plant's roots is actually in contact with root surfaces. This fact underlines the importance of capillary movement of water and the complementarity of water movement and root extension.

Most annual plants distribute most of their roots in the upper 25 to 30 cm of the soil, and as a result, absorb most of their water from that horizon. Many perennial plants such as grapes and fruit trees have roots that extend much more deeply and are able to pull moisture from deeper in the soil profile. But even these plants probably rely heavily on water that is absorbed by roots in the upper horizons when it is available — the usual situation during the cropping cycle. When water is not sufficient, even annual plants such as squash and corn will rely on their deeper roots in an attempt to replace transpirational losses.

The relationship between soil moisture and plants' water needs is the result of a complex interaction between soil conditions, rainfall, or irrigation regimes, and the needs of the crop. Farmers try to maintain a balance between these components during the cropping season, but many times events or conditions occur, which shift the balance toward an excess of soil moisture or a deficiency.

## EXCESS WATER IN THE SOIL

When excess water is present in an agroecosystem for an extended period of time, or movement of excess water out of the system is impeded, the condition known as waterlogging can occur. High rainfall, poor irrigation management, unfavorable topography, and poor surface drainage can bring about waterlogging and associated changes in the soil ecosystem. Waterlogged soils occur throughout the world, ranging from riverbank sediments to marshes, swamps, and peat bogs. Even well drained soils can experience periods of waterlogging if they are subject to seasonal flooding.

Waterlogging occurs frequently and broadly enough that agricultural systems around the world have developed ways of dealing with excess water. More recently, this has involved the construction of costly draining and damming infrastructures. Simpler and traditional techniques, in contrast, have the goal of working with the condition of excess water rather than getting rid of it. In many wet areas of the world, for example, rice is cultivated as a crop ideally suited to wetland agriculture.

### Negative Effects of Excess Water

In a soil where air fills the pore spaces between soil particles, oxygen diffusion is rapid and there is rarely a deficiency of $O_2$ for ecological process (i.e., root metabolism, decomposer activity, etc.). But when the pores are filled or saturated with water, the diffusion rate of $O_2$ is greatly reduced. Oxygen movement in saturated soil can

**FIGURE 9.3 Corn damaged by waterlogging in Tabasco, Mexico.** Excess soil moisture creates conditions that can stunt or even kill a crop.

be one-thousandth or less of what it is in well-aerated soil. Lack of $O_2$ can severely limit the respiration of root cells, allow populations of anaerobic microorganisms to build up, and establish chemically reducing conditions (Figure 9.3).

The depressed rates of gas exchange in waterlogged soils also allow the buildup of $CO_2$ and other gases. $CO_2$ accumulates wherever respiration occurs such as in the area of the roots, displacing needed oxygen and limiting many metabolic processes. Other gases begin to accumulate under the same conditions; for example, methane and ethylene can increase to toxic levels as a result of anaerobic breakdown of organic matter. Phytotoxic water-soluble breakdown products of anaerobic organic matter decomposition also accumulate, a problem that has been noted even for rice production systems (Chou, 1990).

Under conditions of limited O2 supply, many soil microorganisms make use of electron acceptors other than oxygen for their respiratory oxidations. As a result, numerous compounds are converted into a state of chemical reduction, where oxygen is lost and hydrogen is gained. This in turn leads to imbalance in the oxidation-reduction (redox) potential of the soil, measured as the electrical potential of the soil to receive or supply electrons. Ferrous and manganous ions (rather than ferric or manganic) build up to toxic levels under reducing conditions.

Some anaerobic-tolerant microorganisms that can use nitrate as an oxygen source for respiration cause denitrification by liberating $N_2$ gas or toxic levels of nitrous oxide ($N_2O$). Ammonia, too, can build up after flooding, but this is due more to the anaerobic breakdown of organic matter. In addition, anaerobic activity reduces sulfates to phytotoxic soluble sulfides, producing the familiar rotten-egg hydrogen sulfide ($H_2S$) smell.

Each of the conditions described above can become limiting for plant development, either alone or in some combination. When a plant is weakened by these conditions, it becomes more susceptible to diseases, especially in the root zone. The timing of flooding is also important. The susceptibility of a crop to negative effects from excess soil water conditions may depend on what stage of development the crop is in when the waterlogging occurs. The data in Figure 9.4 illustrate how waterlogging can affect crop growth, development, and yields in different ways, depending on the timing of the waterlogging.

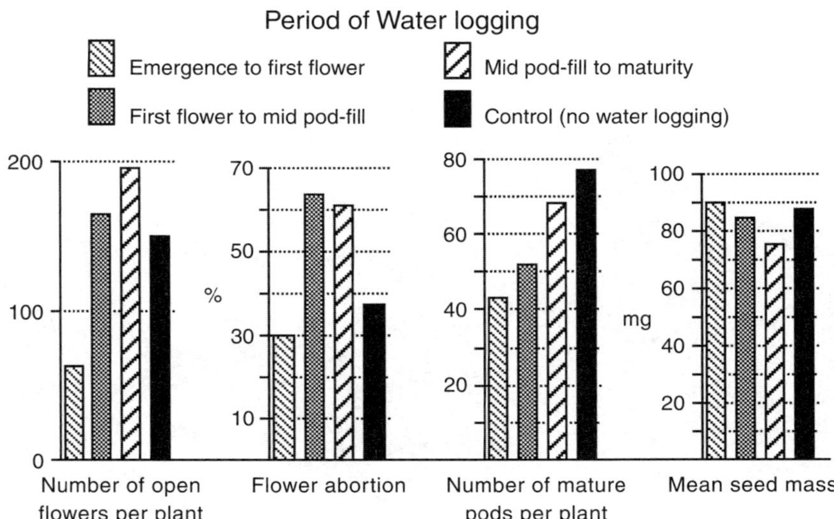


**FIGURE 9.4 Effects of the timing of waterlogging on components of cowpea (*Vigna unguiculata*) yield.** (Data from Minchin, F. R., R. J. Summerfield, A. R. J. Eaglesham, and K. A. Stewart. 1978. *Journal of Agricultural Science* 90: 355–366.)

### Drainage Systems

Drainage systems have long been employed to remove excess water from the root zone of crops and to prevent flooding of farmland. By lowering water levels or preventing flooding, the soil ecosystem is kept aerobic, promoting healthy root systems and leading to increased yields. Drainage systems are known to have been used by Roman and Chinese farmers over 2000 years ago. Much of the Yangtze River Valley of China, the lowlands of The Netherlands, and the Delta region of California would not be farmable without complex drainage systems.

Drainage systems involve the construction of levees, canals, and ditch systems that either keep low-lying areas from being flooded or permit the water table to be lowered so that cropping can take place. In some locations with saturated soils, mounding or raised beds are used. More recently, stricter control of soil moisture has become possible with the development of subsurface drainage systems employing perforated plastic pipe that can be laid with special trenching machines.

But drainage systems are not without costs. Apart from the economic costs of installation and maintenance, drainage systems have ecological costs. The removed water carries with it nutrients and sediments that are lost to the system and must be replaced. In areas of variable rainfall, excess drainage can cause increased drought damage during a dry year. In some regions with high evapotranspiration (ET) during the growing season and where drains are used extensively, the disposal of the drainage water itself can be a problem, especially, when it carries pesticide residues and high salt loads that can damage nearby natural ecosystems.

### Wetland-Adapted Crops

Instead of treating flooding as a problem to be solved with drainage systems or other infrastructures, it can be viewed as an opportunity for growing crops with adaptations that allow them to tolerate waterlogging. Rice (*Oryza sativa*) is probably the most well-known example of such a crop. Originally, an aquatic or swampland plant, rice has been cultivated as a crop that flourishes in wet habitats. Its adaptations include special air-space tissue in the stems that allow air to diffuse to the roots, roots that can grow under conditions of low oxygen concentration, the ability to oxidize ferrous ions to reddish brown ferric hydroxide in the rhizosphere and thus tolerate soils with high redox potential, and seeds that will germinate underwater due to their low oxygen requirement. Other crops are not completely wetland-adapted, yet have adaptations that allow them to tolerate periodic flooding. Taro (*Colocasia esculenta*), for example, may be able to tolerate flooding because of its ability to store oxygen in the swollen cormlike base of the leaves.

### Agroecosystem-level Adaptation to Excess Soil Water

When an agroecological focus is applied to coping with excess water, an intermediate approach is taken. Rather than trying to eliminate the water or restricting production to waterlogging-adapted crops, various forms of traditional raised-field agriculture can be used. In areas with a high water table or periods of inundation, topographic variation in soil levels is created. Soil is dug to build up raised beds, and in the process, canals or ditches are formed, which can also serve for drainage, if too much water enters the system at some time of the year. But the main purpose of the ditches is to serve as a catchment for erosional sediments and organic matter, and, in some cases, to make possible fish production. Instead of changing the level of the water table, cropping areas are raised above the water table. If the system is installed in an area with an extended dry season, capillary movement of water upward from the water table can be sufficient to maintain crops. In some cases, irrigation water can be drawn from the nearby canal. (Figure 9.5). The most well-known examples of such raised-bed systems are the Chinampas of central Mexico (described in more detail in Chapter 6), the pond-dike systems of the Pearl River Delta of southern China, and the canal-field systems of The Netherlands. Many of these agroecosystems have a very long history of successful management.

**FIGURE 9.5 Constructing a raised-field farming system in a wetland in Tabasco, Mexico.** Soil dug from lateral ditches is being layered with waste sugar cane fiber to create a raised planting surface.

## PRE-HISPANIC RAISED FIELD SYSTEMS OF QUINTANA ROO, MEXICO

Many areas of the world rely on food grown in wetlands or periodically flooded areas. Farmers in these areas have adapted to conditions of excessive water in a variety of different ways, and some of the resulting systems are highly productive. Some are also very old. Studies indicate that the Maya of the Yucatan peninsula developed, more than 2000 years ago, a productive wetland agroecosystem in the lowlands of what is now the state of Quintana Roo, Mexico (Gliessman, 1991).

The remnants of huge systems of canals and raised beds encompassing more than 40,000 ha have been discovered and studied (Gomez-Pompa et al., 2003). Dating studies show that the canal and bed systems were formed as early as 800 BC and were used continuously for more than 1000 years. It is unclear why the fields were eventually abandoned.

By analyzing the remains of the system, scientists have been able to deduce the technique used to prepare the canals and platforms. The Mayans dug the canals down to the bottom of the topsoil, and formed the platforms by mounding up the soil from the canals onto the surface between the canals. The canals would have been periodically cleaned, allowing the Mayans to move soil and organic matter that collected in the canals back up to the mounds.

Because plant material decomposes so quickly in the tropics, very little evidence remains to indicate what types of crops were cultivated on the raised beds. Also impossible to deduce are the planting patterns of the Mayans or the frequency of cleaning the canals.

However, many traditional farmers in lowland regions of Mexico farm wetlands and their margins today, mainly in areas with populations of indigenous ancestry, and it is highly possible that the practices they use were handed down from Mayan ancestors. By looking at these current systems, it may be possible to infer how Mayan agriculture may have developed. Corn and beans form the basis of the current cropping system, and it is important to note that when corn is grown in these wetland-adapted agroecosystems, farmers achieve yields over four times higher than those of nearby fields that have been cleared and drained using modern technology. Researchers are attempting to use the archeological information from pre-Hispanic systems, combined with knowledge of present-day wetland use, to reconstruct more sustainable farming systems for present day farmers of the region (Jimenez-Osornio & Rorive, 1999).

## SOIL WATER DEFICIENCY

When the rate of moisture loss from a soil through ET is greater than the input from rainfall or irrigation, plants begin to suffer. Evaporation depletes the water supply in the upper 15 to 25 cm of the soil, and depending on the rooting characteristics and T rates of the plants in the soil, depletion can extend to a greater depth as plants lose water to the atmosphere through T. As moisture is depleted from the soil, soil temperatures near the surface begin to rise, increasing the rate of evaporation even more. When the easily available water held to soil particles is depleted through these processes, levels of soil moisture may reach the permanent wilting point for plants.

If temporary wilting consistently occurs, leaves begin to yellow, and growth and development are generally retarded. Leaves expand more slowly, are smaller, and age sooner. Photosynthetic rates drop in a stressed leaf, and a larger amount of assimilated photosynthate is stored in the plant roots. From a crop production point of view, such responses are usually negative since they result in a reduction in harvestable product. But from an ecological perspective, such responses may provide some adaptive advantage to the plant. For example, the allocation of more carbon to the roots of a water-stressed plant may promote more root growth, allowing the plant to draw moisture from a broader area. Water stress may force earlier flowering, fruiting, and seed formation, helping to ensure the survival of the species. In some cases, farmers can actually take advantage of such drought responses, as when water is withheld from cotton plants in late summer to force defoliation and avoid the need for chemical defoliants before harvest.

Many plants have specific structures or metabolic pathways that aid in survival under water-stressed conditions. Farmers in an area subject to periodic water stress would do well to look for crop species and varieties that demonstrate some of these adaptive traits. Some examples of drought-tolerant crops are certain cacti species, garbanzo beans, sesame, nut crops such as pistachio, and certain deep-rooted perennials such as olives and dates (Figure 9.6).

## THE ECOLOGY OF IRRIGATION

In natural ecosystems, vegetation is adapted to the soil moisture regime set by climate and soil type. Agroecosystems, on the other hand, often introduce plants with water needs that exceed the ability of the natural ecosystem to supply those needs. When this is the case, irrigation is used to provide adequate soil moisture for crops.

Irrigation represents a major change in ecosystem function and generates its own particular ecological problems. At the same time, water supply systems are costly in terms of both money and energy. Their use must balance ecological and economic costs if long-term sustainability is to be achieved.

Water harvesting, storage, and delivery systems can have major impacts on surface and subterranean water flow. Aquifers can be overdrafted, and the ecology of riverine, riparian, and wetland ecosystems can be severely damaged. Since maintaining healthy waterways and water supplies is as important as maintaining profitable crop production, the impacts of water-supply systems on local and regional hydrology must be taken into account (Postel and Richter, 2003).

### SALT BUILDUP

Nearly all irrigation waters contain salts that can damage crops if allowed to accumulate. Since irrigation is used primarily in areas with high ET potential, the deposition of salts at the soil surface over time is inevitable. If uncontrolled, this buildup, called salinization, can reach levels unfavorable for crop production, especially when the salts contain toxic trace elements such as boron and selenium (Figure 9.7). Total salt content is measured as electrical conductivity in mhos. For each 1.0 millimhos per centimeter of applied irrigation water, the salt content of the water increases by about 640 ppm. Careful monitoring of salt levels in irrigated soils along with analysis of the salt content of incoming irrigation water can help avoid excessive buildup.

**FIGURE 9.6 Dry-farmed olives in Andalucia, Spain.** This deep-rooted perennial crop is well suited to regions with limited rainfall and difficult access to irrigation.

**FIGURE 9.7 Land damaged by salt buildup near Kesterson in central California.** Irrigation water draining from surrounding farmland and then evaporating has left toxic salts in the soil. (Photo courtesy of Roberta Jaffe.)

Because of the inevitability of salt buildup in most irrigated systems, long-term sustainability is not possible without adequate natural or artificial drainage that removes the accumulated salts from the upper layers of the soil. Rainfall is the primary natural leaching agent. In the absence of sufficient rainfall, it is necessary to construct systems of drains, ditches, and canals as described above. Excess irrigation water is applied periodically to dissolve salts, and the salt-laden water either leaches below the productive root zone or is removed through surface drainage from the crop fields.

A natural consequence of farming in dry areas where ET is high and irrigation water carries appreciable salt loads is that the water leaving the agroecosystem will have a higher salt concentration than the water applied. Care needs to be taken, therefore, not to salinize the areas receiving the outflow, be they soils, the groundwater, or surface water systems.

### Ecological Changes

The introduction of irrigation water into a farming region during a normally dry part of the year may have profound effects on natural ecological cycles and the life cycles of both beneficial and pest organisms. Under natural conditions, seasonal drought may have been a very important means of reducing the buildup of pests and diseases, acting much as frost or flooding does in other regions to disrupt the life cycles of these organisms. Loss of this natural control mechanism can have serious consequences in terms of outbreaks and increased resistance to artificial control strategies.

Another type of change that may result from introducing irrigation into naturally dry areas is local or regional climate change caused by the increased evaporation from surface water storage areas or from farm fields where water is applied. Elevated humidity in the atmosphere can be connected to increased pest and disease problems, and might also be associated with shifts in the distribution and quantity of precipitation. The off-farm effects of irrigation must be considered along with its on-farm effects when the larger context of sustainability is applied.

## OPTIMIZING USE OF THE WATER RESOURCE

Soil moisture is managed optimally in agroecosystems designed to ensure that the primary route for water out of the soil is through the crop. The focus for management, therefore, is to reduce evaporation and increase the flow through T. Farming practices that encourage this differential water movement are important components of sustainability.

### Efficiency of Water Use

The biomass produced by a plant with a given amount of water can be used as a measure of the efficiency of the use of water applied to an agroecosystem. When this efficiency is expressed as dry matter produced per unit of water transpired, it is called transpiration efficiency, and when it is calculated on the basis of dry matter produced per unit of water lost through both evaporation from the soil surface and T, it is called ET efficiency.

#### T Efficiency

Plants vary in their relative T efficiencies, although actual T efficiency depends on the conditions that exist where the crop is growing. Data suggest that crops such as corn, sorghum, and millet have relatively high T efficiencies, since they use less water to produce 1 kg of dry matter. In contrast, legumes such as alfalfa have low T efficiencies and depend on high moisture inputs for each kg of dry matter produced. Most cereal and vegetable crops are intermediate. Average T efficiencies for a number of important crop plants are shown in Figure 9.8.

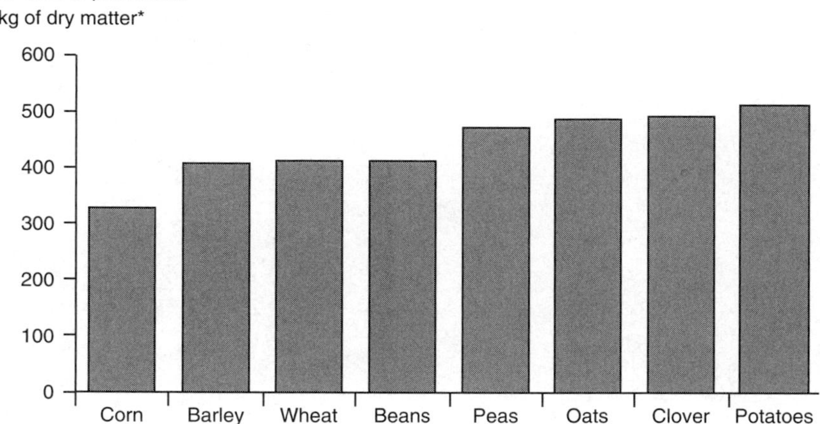


**FIGURE 9.8 Average T efficiencies of various crop plants.** The averages were computed from data compiled by Lyon et al. from various locations around the world. (From Lyon, T. L., H. O. Buckman, and N. C. Brady. 1952. *The Nature and Properties of Soils.* 5th ed. Macmillan: New York.)

It takes a large amount of water to bring a crop plant to maturity. For example, a representative crop of corn containing 10,000 kg/ha of dry matter and having a T ratio of 350 would draw the equivalent of 35 cm of water per ha from the soil. This moisture must be in the soil at the time the plants need it, or growth will suffer. Add evaporation losses to this figure, and it can be seen how moisture is often the most critical factor in production in moisture-limited regions.

Research focusing on increasing the T efficiency of crops has begun to show some promise (Davies, 2003; Shan and Deng, 2003), but even these efforts show little success in significantly altering the T efficiency ratio. Without other conditions being limiting, the amount of water needed to produce a unit of dry matter of a crop species or variety in a given climate is relatively constant. This suggests that we need to continue focusing on control of evaporation from the soil surface.

### ET Efficiency

Since soil itself is quite variable, ET efficiency is also extremely variable. However, by changing soil and crop management practices that affect evaporation from the soil as described below, desirable changes in ET efficiency can be readily obtained. Ideally, the ratio of transpirational water loss to evaporative water loss should be as high as possible. A higher T to E ratio indicates more movement of water through the plant, and, hence, a higher potential for production of plant biomass per unit of water used. Sustainable water management places greatest emphasis, then, on reducing E so as to have more moisture for T and related plant growth and development processes.

## Managing ET

Since T is a plant process that is subject to only minor control if a plant is otherwise growing normally, it is best to focus on reducing evaporative loss by managing the way the plants are grown.

### Crop Choice and Agroecosystem Design

The choice of plant species and the timing of cropping can influence both T and ET efficiency. Choosing a crop with less-intensive water needs, such as corn or sorghum, in an area with very high ET and limited water for irrigation is one good strategy for soil moisture management. It may also be useful to shift the growing of more water-intensive crops to a cooler time of the year when moisture loss potential is less.

Greater vegetative cover can reduce evaporation dramatically. One way of gaining more cover is to use intercropping techniques. A forest plantation, for example, shades the soil surface, whereas an apple orchard with widely separated rows of trees has much more evaporative soil surface exposed. But an increase in plant cover (higher LAI) can also be a liability in drier regions, since lower evaporation rates can be offset by much higher T rates, depleting soil moisture reserves more rapidly.

### Fallow Cropping

In moisture-limited parts of the world such as the Great Plains of the U.S. and the southeastern wheat belt of Australia, farmers sometimes alternate between cropping one year and fallow the next to conserve soil moisture (Figure 9.9). The elimination of transpirational losses from a crop during the fallow year allows soil moisture to be stored for the planting year. Stubble from the previous crop is usually left on the soil surface during the fallow year to limit evaporative losses, and then some kind of soil cultivation or herbicide treatment is used during the fallow season to minimize T losses from weeds. Alternatively, a pasture crop is sown toward the end of the cropping year and left as a grazed cover during the fallow year. Although low rainfall during the fallow year can cause lower crop yields during the cropping year, a crop planted following a year of fallow will generally have a higher yield than if planted without fallow. In fact, as long as sufficient rainfall for recharge is received during the fallow year, there is much less risk of crop failure if the crop season turns out to be a drought year.

### Managing Surface Evaporation

Evaporation directly from the soil surface normally returns to the atmosphere more than half the moisture gained from precipitation. This degree of evaporative loss occurs not only in dry land regions, but in irrigated arid and rain-fed humid regions as well. Plant growth suffers as a result of the loss of moisture through surface evaporation. Any practice that covers the soil will aid in the reduction of evaporative losses.

#### *Organic Mulches*

A wide range of plant and animal materials can be used to cover the surface of the soil as mulch in order to reduce evaporation (and to reduce weed growth and transpirational losses from the weeds). Commonly used materials include sawdust, leaves, straw, composted agricultural wastes, manure, and crop residues (Figure 9.10). Mulches provide a very effective barrier to moisture loss, and have special application in intensive garden and small-farm systems, or with high-value crops such as strawberries, blackberries, and some other fruit crops (Figure 9.11). Mulches work best when the cropping system requires only infrequent cultivation or relies mostly on hand weeding.

Mulching provides a viable option for soil water management, but at the same time has many other beneficial effects. It protects the soil from erosion, returns organic matter and nutrients to the soil, alters the surface reflectivity (albedo), increases the boundary layer for gaseous diffusion, and allows better infiltration of incoming rainfall. All of these factors interact.

**FIGURE 9.9 Sheep-grazed fallow on an Australian wheat farm.** The sheep control moisture-using herbs and serve as a cash crop during the fallow year. Soil moisture gained during the fallow year combines with rainfall during the following cropping year to permit a successful wheat harvest. Successive years of wheat production with no fallow are impossible, except when there is unusually high rainfall. (Photo courtesy of David Dumaresq.)

**FIGURE 9.10 Water hyacinth mulch between rows of chilis in Tabasco, Mexico.**

**FIGURE 9.11** **Redwood bark mulch on the tops of strawberry beds near Aromas, California.**

### *Artificial Mulches*

A range of specially manufactured papers and plastics are now available for use as mulches. Such materials can be easily spread out and firmly secured to the soil surface. When these "mulches" are spread directly over planted beds, slits or holes can be made for the crop plants. Moisture loss is greatly reduced and crop yields, very often, are increased. Some plastics provide a concentrated greenhouse effect as well, raising soil temperatures several degrees. This is a very important benefit for crops that are planted during the colder time of the year, such as strawberries in coastal California (Figure 9.12).

### *Crop Residues and Reduced Tillage*

By leaving a high percentage of the residue from the cropping season on the surface of the soil, a protective barrier that lowers evaporation is created. The residue mulch protects the boundary layer at the surface of the soil and provides a barrier against the capillary flow of water to the surface. The lower temperatures created by the mulch barrier probably help reduce evaporation as well.

Reduced tillage and no-till techniques are often combined by using crop residues as mulch. A major goal of most reduced tillage systems is to develop greater soil cover to reduce evaporative losses from the surface. In no-till systems, seeds are sown directly into the sod or under residues of the previous crop with no plowing or disking, allowing the plant material to remain as a barrier to evaporative loss. Stubble mulching is a common practice in subhumid and semiarid areas where enough biomass is produced by the previous crop to provide sufficient soil cover. The residue is chopped or mown, spread evenly over the surface, and then special tillage implements that can penetrate the mulch are used to plant the following crop. Despite their positive impact on soil moisture, reduced tillage systems have potential drawbacks. These include increased dependence on herbicides for weed management, buildup of soil pathogens from crop residues, and the need for more complex and costly farming equipment.

### *Soil Mulch*

Natural soil mulch made from a cultivated dry soil layer on the surface of the soil can conserve moisture in regions with a distinct alternation between the wet and dry season. This dry layer breaks the capillary flow of water to the surface, and the process of its creation eliminates weeds that might tap moisture below the dry layer and increase transpirational losses. These benefits, however, must be

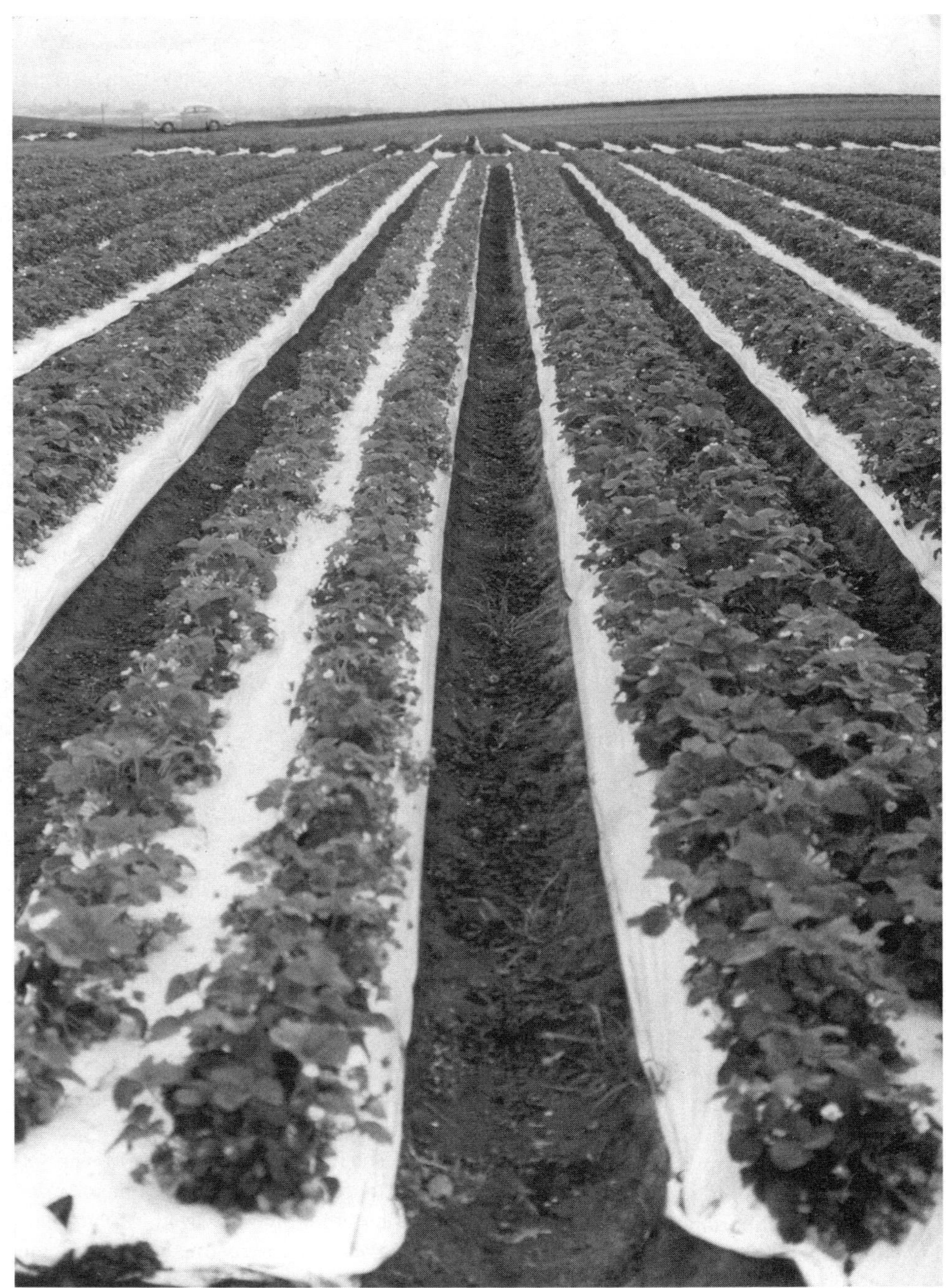

**FIGURE 9.12 Plastic mulch on strawberry beds in coastal California.** The plastic is applied after the small strawberry plants are transplanted, and then slits are cut in the plastic for the plants to grow through.

weighed against potential negative impacts such as increased costs for cultivation, a greater threat of soil erosion from rain and wind, and the loss of organic matter from the dry layer.

## FUTURE RESEARCH

When sustainability is the primary goal, moisture in the soil is managed so that it remains as close as possible to the optimum required to maintain the best growth and development of the crop. This means going beyond simply removing water when it is in excess and adding it when it is deficient. Sustainability requires an in-depth understanding of how water functions in the soil and at the plant–soil interface. Efficiency of uptake of water and its conversion to plant biomass can be one indicator of agroecosystem sustainability. Further development and testing of water management strategies is needed, especially those that view water in the context of the larger cycles and patterns that link the farm with the surrounding environments from which water comes and ultimately returns after passing through the farm.

## FOOD FOR THOUGHT

1. In rainfall-deficient regions, the lack of soil moisture for crop production can be dealt with in two ways: (1) developing crops or cropping systems that are adapted to the low levels of moisture or (2) introducing irrigation to overcome the water deficit. What are the advantages and disadvantages of each approach?
2. What are some of the reasons that farmers must be aware of the "downstream" effects of their use of irrigation?
3. A period without rainfall long enough to create moisture stress in the soil, or a period of waterlogging long enough to create limiting conditions of anerobiosis in the soil ecosystem, can help control pest populations and diseases in the soil that might otherwise cause crop loss. When these natural events are removed from a particular soil system, what alternative pest and disease management strategies could be employed?

## INTERNET RESOURCES

International Water Management Institute
www.iwmi.cgiar.org

National Program for Sustainable Irrigation
www.lwa.gov.au/irrigation
An information-packed site pertaining to Australia, where much agriculture requires irrigation.

Sustainable Waters Program of the Nature Conservancy
www.freshwaters.org/eswm/
All about ecologically sustainable water management.

Water Resources, Development and Management Service of the FAO
www.fao.org/landandwater/aglw/index.stm

## RECOMMENDED READING


Brady, N.C. and R.R. Weil. 2001. *The Nature and Properties of Soils.* Thirteenth Edition. Prentice Hall: Upper Saddle River, New Jersey. The most recent edition of this comprehensive soils textbook, with an extensive section on how water functions in the soil ecosystem.

Ehlers, W. and M. Goss. 2003. *Water Dynamics in Plant Production.* CABI: Oxon and Cambridge, MA. Explains the basic principles of water transport, taking into account soil–plant–atmosphere interactions, and their use in soil and agricultural management.

Essington, M.E. 2003. *Soil and Water Chemistry: An Integrative Approach*. CRC Press, Boca Raton FL. This book balances agricultural and environmental perspectives in its analysis of the chemical properties and processes that affect organic and inorganic substances in soil and soil water.

Hargreaves, G.H. and G.P. Merkley. 1998. *Irrigation Fundamentals.* Water Resources Publications LLC: Highlands Ranch, CO. A very comprehensive text on the principles and practices of irrigated agriculture.

Kirkham, M.B. 2004. *Principles of Soil and Plant Water Relations.* Academic Press: London. Explores the methods used to measure the status of water in soil and plants, including details on instruments and basic sampling methods.

Lal, R. and M.K. Shuk (eds.). 2004. *Principles of Soil Physics.* Taylor and Francis: London and New York. This book analyzes the impact of the physical and hydrological properties and processes of soil on agricultural production, the environment, and sustainable use of natural resources.

National Research Council. 1993. *Soil and Water Quality: An Agenda for Agriculture.* National Academies Press: Washington, DC. Discusses strategies for national policy to protect soil and water quality, without adversely affecting agriculture.

Singer, M. J. and D. J. Munns. 2006. *Soils: An Introduction.* Sixth Edition. Prentice-Hall: New Jersey. A very useful introductory text on soils, with a very good treatment of the management of the soil–water interface.

Sparks, D.L. 2002. *Environmental Soil Chemistry.* Academic Press: San Diego, CA. This book illustrates fundamental principles of soil chemistry, and the interactions of soil with other important environmental factors and materials.

# 10 Fire

Fire is a major form of environmental change or disturbance. It removes dominant plant species, displaces animals, returns nutrients to the soil, and burns accumulated litter on the forest floor. Nearly all the vegetation of the earth has been influenced in some way by fire. Periodic fires of varying frequencies and intensities are thought to occur in most ecosystems, especially in regions with pronounced dry seasons.

The most common fires are natural in origin, but anthropogenic (human-induced) fires have a considerable history as well. There are reports in the literature of charcoal deposits in tropical rain forest areas dating back as far as 6000 B.P., many of which appear to be associated with human activity. Before the development of early agricultural tools, fire may have been the most important "tool" early humans had for vegetation management.

Some natural vegetation types that have evolved in areas where fire is relatively frequent are actually dependent on fire for their long-term stability; these include certain prairie, savanna, shrub, and forest types. Chaparral is probably the best-known fire-dependent vegetation, often being described as a "fire climax" community (Figure 10.1).

In early ecological research, fire was not studied much, because it was seen only as a destructive force, and because it was hard to observe its actual effects. More recently, however, detailed studies of fire in ecosystems such as California chaparral have helped make fire an important topic of ecological investigation. Today, fire is seen as an integral part of many ecosystems, as witnessed by the rising use of controlled or prescribed burns in the management of parks and nature reserves. Fire plays very important roles in agroecosystems as well: it is an important part of the practice of shifting cultivation, and is used to manage crop residue, kill weeds, and clear slash following logging.

## FIRE IN NATURAL ECOSYSTEMS

A fire can occur in an ecosystem when three conditions are met: an accumulation of sufficient fuel or organic matter, dry weather, and a source of ignition. For millions of years, lightning was the primary source of ignition. It is still important today, with over 70% of the wildfires in the western U.S. being caused by lightning strikes. In very recent geologic time, humans have become another important "source of ignition." Humans have used fire since the Paleolithic, as long as 500,000 years ago. Fire was probably used first for the hunting or herding of animals, and then evolved into a vegetation management tool. Burning may have been used to provide better feed for animals, or even to promote the presence of certain plants that served as food or materials sources. Eventually, fire became a tool to prepare the ground for planting, with evidence thus far showing that early slash and burn agriculture began about 10,000 years ago.

From an ecological perspective there are primarily three types of fires (Figure 10.2):

**Surface Fire:** This is the most common type of fire. Fire temperatures are not too hot, with flames burning the trash, grass, or litter that has accumulated on the surface of the soil. Such a fire can move along under a forest canopy and not burn the trees. Changes that occur in soil conditions during a surface fire are usually short lived, although the understory vegetation can be greatly altered. Surface fires can be used to either control or promote the growth of weedy or invasive vegetation, depending on the circumstances.

**Crown Fire:** This type of fire can be very damaging for some types of vegetation, whereas it may be an integral part of rejuvenating other types. During crown fires, the canopy of the vegetation is consumed, and usually the mature plant species are killed. Crown fires are usually very fast moving and often combine with a surface fire to burn everything above the soil surface.

**Ground or Subsoil Fire:** This type of fire is not very frequent, but when it does occur, it can be very destructive. It is characteristic of soils that are high in organic matter, especially peat or muck soils. Organic matter in the soil can be burned down to the mineral soil layer. These are usually slow fires, with more smoke than flame, that dry the soil as they burn. Roots and seeds in the soil are killed, and animal habitats are severely altered.

Any individual fire can combine aspects of all three fire types. In general, the intensity of a fire is very closely related to the frequency of fires in the area.

**FIGURE 10.1 Chaparral fire in the Santa Ynez mountains near Santa Barbara, CA.** Periodic fires are part of the evolutionary history of chaparral; humans have only recently disrupted the natural pattern of burning.

## EFFECTS OF FIRE ON SOIL

Much of the ecological significance of fire revolves around its effects on the soil. Fire has very noticeable impacts on a range of abiotic and biotic components of the soil ecosystem, and knowledge of these impacts is important in employing fire as a tool for agroecosystem management. It must be pointed out, however, that the effects of fire will vary widely depending on the type and stage of development of the vegetation, the type of soil, the season of burning, the prevailing weather conditions, the amount of time since the last fire, and other conditions.

### Abiotic Factors

When a fire occurs, the temperature of the surface layers of the soil is raised. The actual heating rate and depth depends on the amount of moisture in the soil and the type of fire. Temperatures during a burn at the surface of the soil almost always exceed 100°C and can reach as high as 720°C for brief periods of time. Increases in temperature below the surface are usually restricted to the upper 3 to 4 cm of soil, where they rise 50 to 80°C above the temperature present before the fire, usually for only a few minutes (Raison, 1979). These temperatures are high enough to modify the soil environment in ways that can be useful for agroecosystem management.

The complete burning of above-ground organic matter combusts most nitrogen and organic acid components, returning inorganic cations to the soil (mainly $K^+$ and $Ca^{2+}$) which then have an alkalizing effect. The strength of this effect depends on the intensity of the fire and the thoroughness of the combustion of plant biomass, but increases in soil pH during the first several days following fire, especially once the soil is moistened by precipitation, are commonly 3 or more pH units.

Following the fire, the blackened soil surface will tend to have more solar gain; however, if the standing biomass was considerable before the fire and burn temperatures were very high, enough white ash may be present at the surface to actually have the opposite effect for a short period of time. The higher albedo of the white surface will reflect solar energy and limit soil heating.

The hot temperatures caused by fire can greatly reduce the amount of organic matter in the upper layers of the soil. At a temperature of 200 to 300°C for 20 to 30 min there is an 85% reduction in organic matter, with an

**FIGURE 10.2 The three types of fires.** A slow-moving, cool surface fire (top) burns litter in the understory of summer deciduous forest in northwestern Costa Rica. A fast-moving crown fire (center) in chaparral burned everything from the surface to the plant crowns near Santa Barbara, CA. A subsurface fire (bottom), visible in the distance, burns in a swamp near Coatzacoalos, Veracruz, Mexico.

accompanying release of $CO_2$, a loss of nitrogen and sulfur in volatilized forms, and the deposition of minerals.

After fire there is usually a reduction in soil moisture-holding capacity, although with the removal of vegetative cover, actual moisture availability in the soil can increase because of reduced demand. Soil aggregate size is reduced, bulk density goes up, and permeability and water infiltration rates are reduced. Often there is also an increase in rainfall runoff and nutrient leaching, and the possibility of greater soil erosion until the soil is covered once again with vegetation. It is not uncommon just after a fire for the immediate surface of the soil to actually be water repellent, but this condition is usually overcome after some exposure to moisture.

Generally speaking, most of the abiotic effects listed above are of a rather short-term nature. Regeneration of the vegetation, coupled with replacement of soil organic matter, leaching rainfall, and plant modification of the burned conditions, rapidly begins the process of recovery. In the case of severe fire intensity following excessive fire suppression and abnormal fuel buildup, or in the case of a fire burning thick organic layers of peat or muck that reaccumulate at a very slow rate, abiotic conditions can be altered for longer periods of time. Unnaturally frequent fires, usually human induced, can also lead to more lasting change.

### Biotic Factors

Obviously, any living plants or animals caught in the path of a fire are in peril. Plants that are not adapted to fire are easily killed, especially if the bark type does not protect the living cambium. If the fire is hot enough and other conditions are right, living plant matter can be killed, dried out, and ignited very rapidly, reducing all above-ground material to ash. Then, if the plants do not sprout from below-ground structures, recovery will only begin with the germination of seeds. Seeds of some species of plants are killed by fire, whereas others are either stimulated by the breaking of specific dormancy factors or by the creation of soil conditions that favor germination and establishment (Figure 10.3).

Repeated fire can retard the vegetation recovery process to the point that another vegetation type, more tolerant of fire, can establish dominance. The conversion of shrubland to grassland is a good example of this process.

**FIGURE 10.3 Fire response by pines.** Young lodgepole pines reestablish following devastating crown fires that killed the parent trees in Yellowstone, Wyoming.

On the other hand, some vegetation types are in a sense kept healthy by periodic fire, because the fire removes old and dying individuals, returns stored nutrients to the soil, and stimulates renovation by new or younger individuals.

Many larger animals can avoid fire by moving away from it, but even when they are killed by fire, their populations in the burned area can recover through recolonization from nearby unburned areas. Some animals actually seek out recently burned areas because of the concentration of new growth and forage for feed, or because the ash can aid in removal of parasites such as ticks and fleas.

Following a fire there is an immediate reduction in the populations of nearly all soil-dwelling organisms, including fungus, nitrifying bacteria, spiders, millipedes, and earthworms. Many die as a result of the high temperatures, but some organisms are impacted by the changes in pH that follow the fire or by the flush of certain nutrients into the soil that comes from burned organic matter. After a fire, however, there is fairly rapid recolonization, especially by bacteria that are stimulated by the increase in pH.

On the whole, fire can have both negative and positive impacts on the environment, but regardless, it must be remembered that the intensity, duration, and frequency of fires in natural ecosystems are incredibly variable. From one year to the next, conditions that favor fire are going to vary tremendously. And when a fire does occur, it effects will not be uniform. Some areas will be burned very thoroughly, whereas a short distance away the same type of ecosystem may be spared the impacts of fire completely.

## PLANT ADAPTATIONS TO FIRE

In any location where fire has a long evolutionary history, most plants and at least a few of the animals have developed adaptations to fire. It is interesting that the adaptations that provide resistance to fire in plants are in many cases also traits that enable the plants to deal with excess light or drought stress.

Plants can be adapted to fire in three different ways.

**Fire resistance:** Plants with fire resistance have traits that help prevent the living parts from being burned in a fire. These traits include such characteristics as thick bark, fire-resistant foliage, or a litter mat that will support frequent but less damaging fires.

**Fire tolerance:** Fire-tolerant plants have traits that allow the plant to survive being burned in a fire. A common fire-tolerant trait is the ability to resprout from the crown following a fire.

**Fire dependence:** Fire-dependent plants actually require fire for reproduction or long-term survival. Some fire-dependent plants have seeds that need fire before they will germinate, or cones that will not open unless exposed to fire. Other fire-dependent plants will not flower until after a fire, or will become senescent unless exposed to periodic fires.

## FIRE IN AGROECOSYSTEMS

Fire has a long history of use in agriculture. But from an agroecological perspective, there can be good fires and bad fires, overuse or underuse of fire, and careful or careless use of fire. The challenge is the appropriate application of the knowledge of the ecological impacts of fire.

### Shifting Cultivation

The agroecosystem with the longest history of fire use is shifting cultivation, or slash and burn agriculture. Shifting cultivation with the use of fire continues today to be the most important form of subsistence agriculture in many parts of the world. Although thought to be practised primarily in the tropics, fire-based shifting cultivation was used in early agriculture even in Europe, where wheat and barley were grown on a 10 to 25 years fallow cycle (Russell, 1968). Although it might seem quite simple to clear, burn, and plant, good shifting cultivators have learned through experience that the timing of all activities, especially the fire, make the difference between a sustainable system and a degrading system. Shifting cultivation works when the system is allowed enough time for natural successional processes to restore the soil fertility lost through disturbance and crop harvest (Figure 10.4).

Immediately following a fire, nutrient mobility in the system is quite high, often resulting in high leaching losses. This accentuates the need for a fallow period in order to recover the lost fertility. Crops in slash and burn systems need to quickly pick up the nutrients added to the soil from ash, or else leaching will remove them or invading noncrop plant species will begin to capture them. Depending on soil types, climatic regimes, and cropping practices, the rate of nutrient loss varies considerably. But studies have shown that the loss can be rapid and high, especially for nutrients such as calcium, potassium, and magnesium (Ewel et al., 1981; Jordan, 1985; Nye and Greenland, 1960). Repeated fires in short succession, as well as soil cultivation, can accelerate nutrient loss even more (Sanchez, 1976).

Shifting cultivation systems are generally thought to be able to sustain relatively low human population levels. In well-managed shifting cultivation systems, most of the soil carbon and nitrogen remains following a fire, the root mat stays intact and alive, the soil surface is protected by some form of biomass cover, and even soil mycorrhizae survive. As a result, nutrient loss and soil erosion are minimized, and the system is sustainable.

**FIGURE 10.4 Managing fire in a slash and burn agroecosystem in Tabasco, Mexico.** A small firebreak separates the fire from future slash and nearby crops.

**FIGURE 10.5 Pattern of shifting cultivation in the mountains of Chiapas, Mexico.** Fallow plots of various ages are clearly seen next to plots being farmed. Farmers say that a 15 to 20 years fallow period is required for the system to be sustainable over the long term. Pressures to shorten this fallow period are many.

But many of these systems have recently begun to move in an unsustainable direction, because an array of social, economic, and cultural factors create pressures that shorten the fallow period, remove fallen timber for firewood, introduce inappropriate crops, or overgraze animals, eventually promoting the invasion of noxious weedy species or leading to a breakdown of the processes that enhance the recovery of native species ground cover. Overuse of fire is often one cause of the breakdown in sustainability (Figure 10.5).

### Modern Agricultural Systems

In modern agricultural systems, fire plays many diverse roles. The examples presented below represent different levels of technology and have different levels of use depending on the agroecosystem type, part of the world, and cultures involved. They can be used at any time during the cropping cycle, from preplant to harvest, depending on the system and the purpose. The biggest challenge in the use of fire overall is to understand how to take advantage of the beneficial effects of fire while avoiding or minimizing the negative ones. Skill, experience, and knowledge are all required.

#### Land Clearing

In many parts of the world today, fire continues to be the most accessible and affordable tool for clearing vegetation and plant biomass from the soil surface prior to preparing the land for planting, especially in present-day versions of shifting cultivation. The use of fire for land clearing is particularly important in many forestry systems, where the large slash load left after logging is burned to make replanting easier, as well as to reduce the chance of a wildfire moving through the dry slash and suppressing the establishment of seeded or transplanted tree seedlings.

The amount of dry matter that needs to be cleared will obviously have a great impact on the type and intensity of the fire. As shown in Table 10.1, these amounts, called slash loads, vary considerably depending on the system. Slash left on the soil in tropical shifting cultivation systems can easily exceed 4 kg/m$^2$, and if adequately dried and burned at an appropriate time, will carry a hot, uniform fire that will consume most all of the plant material except large diameter branches and trunks (Ewel et al., 1981). Even young second-growth produces 1 to 2 kg/m$^2$ of dry matter and can easily carry a fire (Gliessman, 1982).

Logging of older forest systems invariably leaves the forest littered with logs, tops, and branches, which can become a fire hazard as they dry out. Such slash can also harbor pests and be detrimental to the recovery of tree seedlings. On the other hand, as the debris decomposes, it improves soil structure and nutrient status while protecting the soil against erosion. All of these factors need to be taken into account in deciding if slash should be burned uniformly over the surface, piled so that impacts of burning can be localized, or left unburned as a mulch. In some traditional systems, when slash is limited in supply (usually less that 0.5 kg/m$^2$), it is piled, burned, and the ash scattered uniformly over the cleared fields as a fertilizer (Figure 10.6).

**TABLE 10.1**
**Slash Loads Available for Burning as a Part of Land Clearing in a Range of Ecosystems**

| System | Location | Slash Load (kg/m$^2$) | Ref. |
|---|---|---|---|
| Napier grassland | Tabasco, Mexico | 1.63 | Gliessman (1982) |
| 2 years second growth | Tabasco, Mexico | 1.18 | Gliessman (1982) |
| 8 years second growth | Turrialba, Costa Rica | 3.85 | Ewel, et al. (1981) |
| Mature tropical dry forest | Jalisco, Mexico | 1.18–1.35 | Ellingson, et al. (2000) |
| Upland rice and barley | Central Japan | 0.34 | Koizumi, et al. (1992) |
| Upland rice | Tabasco, Mexico | 0.51 | Gliessman (1982) |
| Paddy rice | Central Valley, California | 0.7–0.9 | Blank, et al. (1993) |
| Douglas fir with red alder (9 years old) | Oregon, U.S. | 0.986 | Cromack, el al. (1999) |
| Conifer forest | Pacific Northwest, U.S. | 0.5–3.0 | Dell and Ward (1971) |
| Annual pasture | Central Coast, California | 0.2–0.3 | Gliessman (1992b) |

A unique example of the use of fire for land clearing is a system for renovating old cacao plantations in Tabasco, Mexico that are no longer profitable (Figure 10.7). First, bananas are planted in the understory. The next year, all overstory shade trees and old cacao trees are cut, leaving a heavy slash load of more than 5 kg/m$^2$ that covers the corms of the bananas. Once adequately dried, the slash is burned. Immediately after the fire, a traditional corn/bean/squash intercrop is planted in the same way as in local shifting cultivation systems, allowing for a harvest within 6 months after cutting of the trees. While the annual crops are being planted and cared for, sprouting bananas and new shoots from the trunks of the leguminous shade trees are protected and allowed to develop. After the annual crop has completed its cycle, short-lived perennial crops such as yuca (cassava) or papaya are planted. By the time these crops are harvested, the bananas have formed a fairly continuous canopy, producing bananas (or plantain) for local use or sale. By the third year, the resprouted shade trees have also begun to become part of the shade-producing canopy. At this point, shade conditions at the soil surface have returned to the reduced levels appropriate for

**FIGURE 10.6 Burned slash piles in Chiapas, Mexico.** When biomass production is limited by climate or short fallow, slash can be piled for burning and the ash spread.

the replanting of new cacao seedlings. Bananas are harvested up to the time the new cacao plants come into production (5 to 7 years after planting), at which point the renovation cycle is complete. Local farmers claim that without the use of fire, it would be at least 10 years before cacao could begin to be replanted on such a site — a long time to wait for this valuable cash crop. Research is needed to tell us exactly how fire benefits this agroecosystem.

## Nutrient Additions to the Soil

In many cropping systems in the world, the ash left after burning crop residues, noncrop slash, and even wood for cooking or heating is seen as a valuable nutrient source that should be returned to the soil. Ash is quickly carried into the soil with rainfall and the nutrients it contains are readily available as part of the soil solution. The loss of nitrogen and sulfur to volatilization during burning is more than offset by a gain in all other nutrients and by an increase in their availability to plants. Ash has been shown to contain as much as 2.6% potassium, and appreciable amounts of phosphorus, calcium, magnesium, and other mineral elements. Since ash can amount to between 0.4 and 0.67 kg/m$^2$, it has significant potential as a nutrient input to agroecosystems (Ewel et al., 1981; Seubert et al., 1977).

Of course, being so soluble, these nutrients can easily be washed out of the system, so effective plant cover and good root development should accompany the addition of nutrients from ash. Timing of ash application is very important. There must be active plant roots in the soil to rapidly take up the highly soluble nutrients. And knowledge of rainfall patterns is needed to avoid having heavy rains follow burning or ash application, so that nutrients are not leached below the root zone or washed off the surface. Research is needed that determines which crop systems or combinations can best take advantage of fire-released plant nutrients.

## Crop Residue Management

Fire is often used as a tool for crop residue management. One of its main benefits is to make nitrogen from the residue more easily available to the following crop. When the residue is very high in carbon as compared to nitrogen (C/N 25 to 100), the nitrogen in the residue can be immobilized by incorporation into microbial biomass (and then

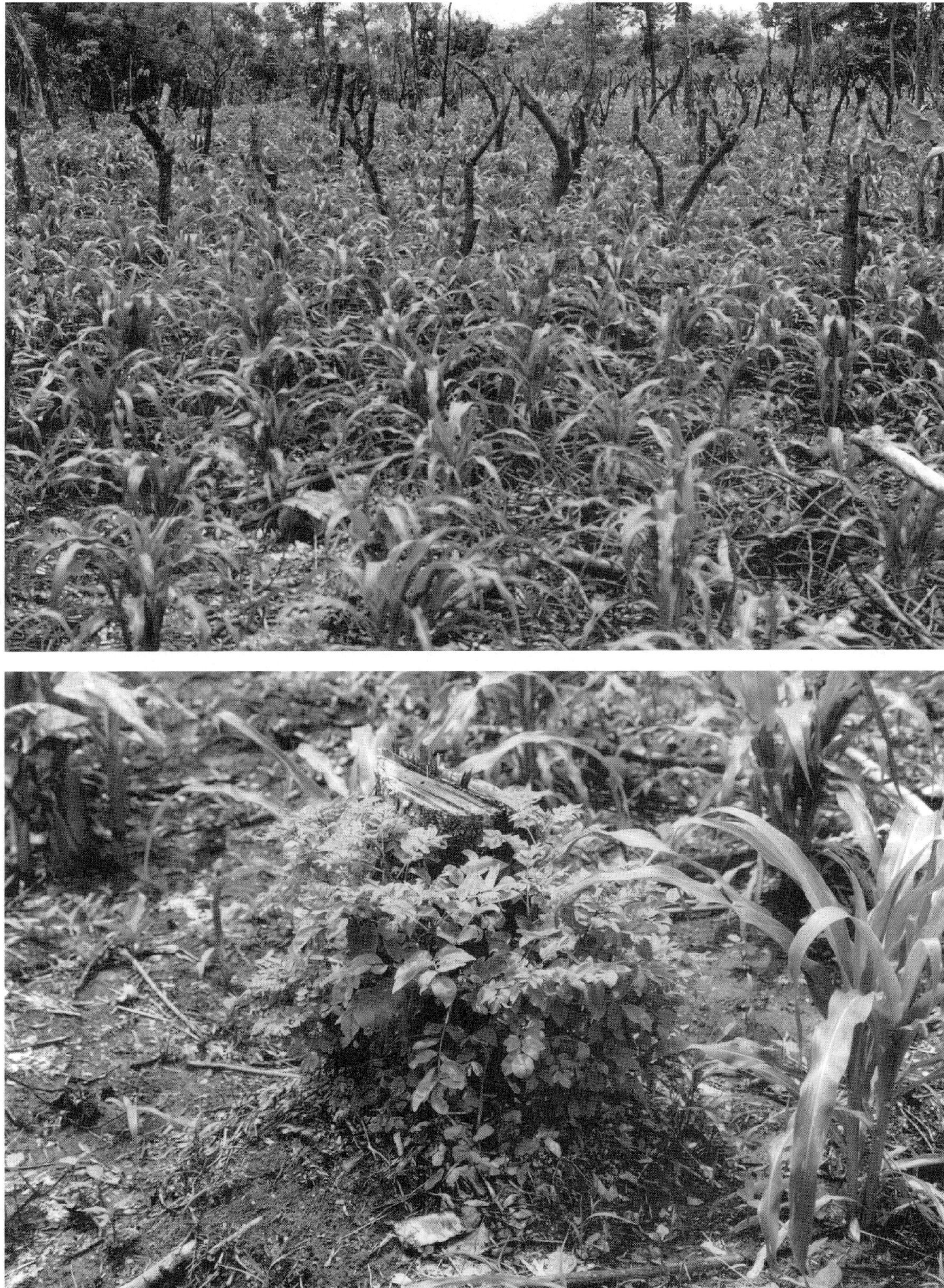

**FIGURE 10.7 Using fire to renovate old cacao plantations in Tabasco, Mexico.** An annual crop of corn, beans, and squash (top) grows through the ash left from burning old cacao plants (standing) and associated shade trees. A leguminous shade tree (bottom; *Pithecelobium saman*) begins to recover following the fire. It will be pruned to one or two stems and eventually provide shade for new cacao plants.

more permanently into soil humus). Burning, however, makes the nitrogen readily available for uptake by plants. Even though most nitrogen is lost through volatilization during burning, the C-to-N ratio of the ash is lower relative to that of unburned residue, making the nitrogen that remains more readily available and reducing the need for external nitrogen amendments.

Another benefit of residue burning is reduction in the amount of tillage needed. Also, in many parts of the developing world, residue is burned not to eliminate the residue, but as fuel for home heating or cooking. Sometimes the ash is collected and returned to fields as a soil amendment.

Rice production is often associated with fire. In any part of the world where rice is grown, the straw and stubble left following harvest can amount to as much as 0.95 to 1.0 kg/m$^2$. Traditionally, this straw has been used as animal feed, fuel, or construction material, or as raw material for compost. In many present-day rice systems, however, the increasing need to get another crop into the ground as soon as possible following the rice harvest has led to the use of fire to quickly reduce the straw to ash. Burning does reduce stubble-borne diseases and insects, and also reduces the potential of methane being produced during decay under flooded conditions in amounts that might become toxic to some following crops. But due to the perceived impact of the smoke on atmospheric quality, regulations increasingly limit burning and force farmers to deal with the reincorporation of the straw into the soil, or to find alternative uses for harvested straw (Blank et al., 1993). In Amazonian slash and burn systems, including rice, researchers found that nutrient losses from burning exceeded inputs for N, K, Ca and Mg, but not for P. Thus, this group is currently searching for alternatives that do not use fire (Denich et al., 2005).

From the standpoint of sustainability, the many advantages of residue burning must be weighed against disadvantages that include loss of nutrients through volatilization or leaching, air pollution, exposure of soil surface, and loss of organic matter inputs to the soil.

## Weed Management

Fire is used for weed management most effectively and practically when the weeds are either in the litter or soil as seed, or shortly after the seeds have germinated. Seeds or seedlings in the litter are most likely to be killed by fire, since litter at the surface burns at high temperatures and down to the soil surface. For this reason, it is necessary to have some kind of mulch cover or crop residue to carry the fire. Slash and burn systems are very effective at destroying seed in the litter and on the immediate soil surface.

A more recently developed practice for weed control has been used in Europe for many years. A propane tank is connected to a hose and a nozzle so that a flame can be moved rapidly over the soil surface to destroy weed seedlings. Both backpack- and tractor-mounted flame weeders are available. Specially shaped nozzles and an assortment of deflectors and shields protect any crop seedlings while desiccating the weeds. Weed seedlings must be very small to be effectively controlled with this technology, or the seedlings of the crop must be at a stage of development that gives them greater resistance than the weeds to the heat. Under some field conditions, a crop such as corn in its first and second leaf stage has a structure and moisture content that will keep it from suffering damage while most surrounding weed seedlings are killed. The necessary equipment can be expensive to purchase and use, and depends greatly on the use of fossil fuel, but in some very weed-prone crops like carrots and onions, flame weeders are a very cost-effective means of weed control.

But fire must be used on weeds with care. Perennial weeds and those with fire-resistant roots, rhizomes, crowns, or other structures that resist burning may actually be stimulated by fire. Bracken (*Pteridium aquilinum*), for example, is a very aggressive plant that can act as a weed in deforested or pasture areas, and is favored by fire in two ways (Gliessman, 1978d). Its deep underground rhizomes permit it to survive fire, and there is some evidence that removal of above-ground litter of bracken actually promotes more vigorous regrowth of the fern. At the same time, spores of the fern are favored by the soil conditions created by fire and ash, allowing for initial establishment of the fern where it did not occur before and the potential for its aggressive vegetative growth from then on. In shifting cultivation systems, where fire is used to help clear the fallow, fire can begin to have negative effects if the fallow period is too short. These effects can include leaching of nutrients and invasion of fire-resistant weeds. In general, use of fire for weed control requires careful consideration of its potential impacts, based on the unique characteristics of the system.

## Management of Arthropods

Fire is a very effective means of eliminating damaging arthropods, such as insects and mites, from an agroecosystem. Heat, smoke, and loss of habitat all combine to either kill these organisms (as well as their eggs or larva) or drive them from the system. In some natural ecosystems, fire is probably as much a factor in the natural fluctuations of arthropod populations as climatic factors or trophic interactions. Fire suppression in forests may actually be upsetting the natural equilibrium, allowing outbreaks of such common pests as bark beetles, leaf miners, and lepidopterous leaf eaters such as tent caterpillars. In some ecosystems, however, fire may not impact arthropod populations. Joern (2005), for example, found that different burn frequencies had no effect on grasshopper species diversity or density in North American tall grass prairies. In general, studies

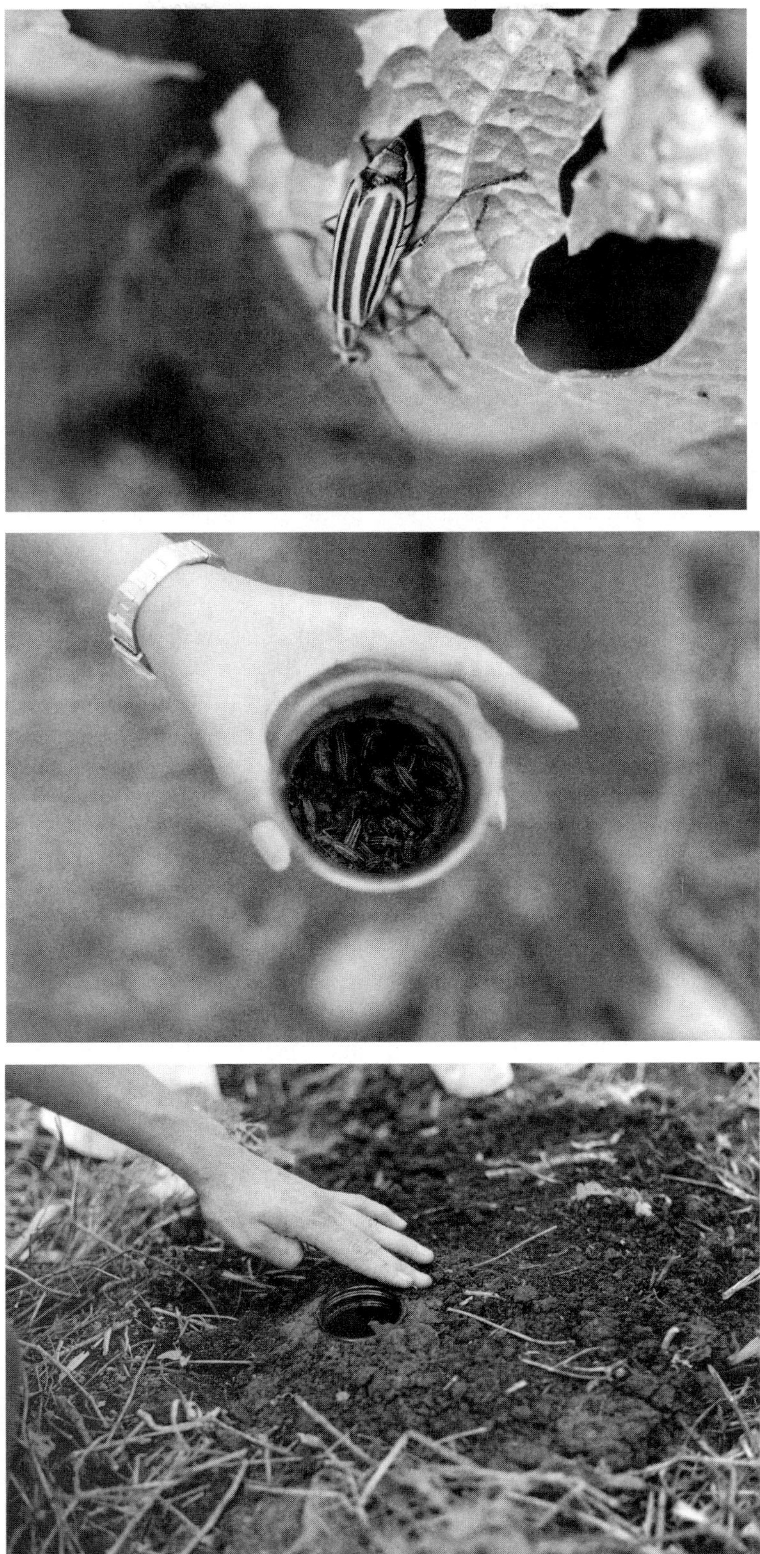

**FIGURE 10.8 Using burned beetles to repel other beetles in Tabasco, Mexico.** The beetle pest *botijón* feeds on a bean plant (top). The beetles are put in jars and heated just enough to kill them (center). The open jars are then placed in the soil around the bean planting (bottom).

specifically focusing on the effects of fire frequency on arthropod populations are still rare.

In agroecosystems, even less is known about pest management in relation to fire. Many insect pests can pass the time between cropping seasons in some part of the plant, either living or dead, left over from the previous season. Bollworm problems in cotton are dramatically reduced if all plant residue is destroyed, and fire is one tool for achieving this end. Stemborers in grain crops overwinter in straw remaining in the field after harvest, and appropriate use of fire might aid in their management.

For ground-dwelling arthropod pests, fire that penetrates the soil surface can be a useful method of pest management. Burning mulch or crop residues, and artificial flaming of the soil surface, are ways of introducing fire for this purpose.

A traditional practice that uses fire to protect a crop from insect damage is known from Tabasco, Mexico (Figure 10.8). A large Coleopteran beetle has a reputation for being able to invade a bean planting and defoliate the crop in a very short period of time. The beetles invade in large numbers and can be seen consuming the plant leaves in the early morning hours. Farmers report that an old practice was to come into the infested field in the morning, collect enough of the live beetles to place 25 to 50 of them in each of several fire-resistant containers. At the end of the day, each container was placed over a fire long enough to kill the insects but not to burn them. Shortly thereafter, the open containers were partly buried in the soil in the bean field, about one to every 400 m$^2$. By the next morning, farmers report, there were no signs of living or actively feeding beetles in the field. An alarm pheromone released by the dying beetles is suspected of alerting living beetles to danger so they leave the field, but further research is needed. Farmers have stopped using this practice since synthetic chemical pesticides have been introduced.

## Pathogen Management

Because of fire's ability to elevate temperatures in the soil, especially close to the surface, fire should be expected to have a significant impact on plant pathogens living in the soil, such as fungi, bacteria, and nematodes. Relatively little research has been done on the effect of fire in relation to plant disease management, but in a review done some years ago (Hardison, 1976), it was found that fire could effectively reduce inoculum of diseases of various forest crops, fruits, ornamentals, cotton, potatoes, small grains, and grasses and forages. In a recent study investigating the relationship between fire and Sudden Oak Death Disease (*Phytophthora ramorum*), researchers found that infections were extremely rare within areas that had been burned in the last 50 years. However, the mechanisms underlying these results were not completely understood and the authors pointed to the need for more research in order to better understand the use of fire as tool to fight this disease (Moritz and Odion, 2005). It is interesting to note that the burning of grass fields, a practice that has become very important in fields used to produce commercial grass seed in the Pacific Northwest region of the U.S., was started originally for the purpose of disease control in the late 1940s.

Heat and desiccation probably have the greatest direct impact on pathogenic organisms. The high temperatures registered at the soil surface during a fire, and the penetration of heat down to several centimeters below the surface, can kill large numbers of living pathogens and their inoculum. In addition, the sudden increase in pH caused by the wetting of ash deposited on the soil after a fire can have an inhibitory affect on fungi, since fungi prefer neutral to acid conditions for optimal development. Many bacteria, on the other hand, are actually stimulated by the higher pH, and might become more of a problem if they are pathogenic.

The effect of burning above-ground plant material, especially crop residues, on potential plant pathogens is well documented. Since a well-managed fire can consume as much as 95% of the above-ground biomass and generate extreme heat, it can kill most pathogens present in the biomass. This effect of fire is the most common reason for burning crop residues, as described above.

## Preparing a Crop for Harvest

Fire can be used to prepare a crop for harvest. A common example is the burning of sugar cane fields a few days ahead of harvest of the canes. Cane cutters claim that fire is important for removing the leaves from the stems, facilitating the cutting process when done by hand, making access to the canes easier, and displacing bothersome animals such as rats and snakes. But ease of harvest in such a system has to be weighed against ecological impacts such as loss of organic matter, volatilization of certain nutrients, and nutrient leaching with heavy rainfall. For sugar cane in particular, another possible negative impact of fire may be to degrade the quality of the sugar extracted from overheated canes.

Another simple role for fire at harvest time is in the collection of pine nuts. Cones of several pinyon pine species are collected from trees before they open and disperse their seeds (called nuts). Usually the cones are coated by dense pitch. Fire is used to heat rocks that are then placed with the cones, melting away the pitch and opening the cones to release the seed. Fire can also be used to heat an oven into which the pitch-covered cones can be placed.

## Pasture and Range Management

Despite the fact that in most grassland areas of the world, natural fire is frequent and an important aspect of the environment, the effective use of fire as a tool for managing grazing systems is not really that common. When fire is used in grazing systems, it is employed in the form of a controlled fire known as a **prescribed burn.** A prescribed burn in a grazing agroecosystem can play many roles. It can:

- burn off unpalatable growth from previous seasons that is not eaten by most animals and that would otherwise compete with more desirable species
- stimulate growth (in the form of fire-response sprouting of perennial plants) during times of the year when very little green growth would normally be available
- destroy parasites such as ticks and fleas that can carry stock disease
- control the spread of undesirable plants in pasture or range
- remove the fire hazard of accumulated old browse or grass
- establish fire breaks as a system of protection from wildfire
- prepare a seedbed for natural or artificial seeding of desired plant species
- stimulate some plants to produce seed
- encourage growth of native legumes for forage and soil improvement
- promote more rapid nutrient cycling and uptake

All of these potential effects of fire can play important roles in determining the most appropriate regime of management using fire.

The relative importance of each of the impacts of burning varies with the type and intensity of grazing system, time since the last fire, season of the year, and the stage of development of the edible plants. In open grassland, for example, there is little tendency for woody species to invade; therefore fire is employed to remove the accumulation of inedible growth. In savanna regions, or areas where natural succession would favor shrub or tree vegetation, burning is of much greater importance for suppressing some plants while establishing or maintaining the pasture components (Figure 10.9).

**FIGURE 10.9 Chaparral species invading grassland, Santa Barbara County, CA.** Fire is needed to periodically repress the shrubs and promote grass for grazing.

When fire is withheld from a grazing area that normally burns with some regularity, grasses lose their dominance and can be replaced by nonedible or poorly consumed shrubs or tree species. For example, the rangeland in the Great Basin of the western U.S. converts to sagebrush (*Artemesia tridentata*) with lack of fire, especially when combined with excessive grazing pressure. The open savanna areas of the parts of the southwest U.S. or northern Mexico, where grasses grow between mesquite and juniper, become virtual forests of the tree species when fire is not incorporated into the management of the rangelands. In other areas, where grassland borders shrub or tree vegetation, lack of periodic fires can allow the gradual invasion of the grassland by the more aggressive woody species. Annual grasslands in the foothills of the coastal mountains of central and southern California are encroached upon by allelopathic chaparral shrubs when fire is withheld for more than a few years (Muller, 1974).

## FUTURE RESEARCH

Probably one of the oldest tools used in agriculture, fire is still of considerable value in the present-day search for sustainable farming practices. But being able to use fire to benefit the system depends on having knowledge of the long-term impacts that fire will have on different components of agroecosystem structure and function. Research is needed that goes beyond thinking of fire as a destructive factor in the environment and helps us make use of its ability to release nutrients from organic matter, quickly alter agroecosystem structure, kill undesirable organisms, and emulate the disturbance regimes of natural systems.

## FOOD FOR THOUGHT

1. What kind of knowledge and information is needed to convince farmers to use fire as a tool for contributing to sustainability?
2. What are some of the ways that the different types of fires that occur in natural ecosystems might be combined in order to find useful ways of applying them to agroecosystem management?
3. Smoke in the atmosphere is often considered wholly undesirable, with new restrictions being placed on smoke-generating activities every day. How would we justify the use of fire in agriculture even though smoke may be one of the byproducts?
4. Which do you consider to be of greater agroecological significance in management — the abiotic effects of fire or its biotic effects? Explain why.
5. Under what conditions would it be possible to effectively use fire in diverse, mixed-crop, perennial species cropping systems?

## INTERNET RESOURCES

Western Fire Ecology Center
www.fire-ecology.org/
The mission of the Center is to reform federal fire management policies and restore fire ecology processes through research, education, and advocacy.

The Association for Fire Ecology
www.fireecology.net/
The AFE is an organization of professionals dedicated to improving the knowledge and use of fire in land management. They publish a peer-reviewed journal, *Fire Ecology* (www.fireecology.net/fe/).

The Fire Ecology Center
www.rw.ttu.edu/fec/
The Fire Ecology Center is a research and outreach organization at Texas Tech University.

The Association for Fire Ecology of the Tropics
www.tropicalfire.org/

## RECOMMENDED READING

Biswell, H. 1999. *Prescribed Burning in California Wildlands Vegetation Management*. University of California Press, Berkeley, Los Angeles, and London. An introduction to the principles and practice of prescribed burning in California.

Bond, W.J. and B. van Wilgen. 1995. *Fire and Plants*. Chapman & Hall, New York. A unique text on the many and varied responses and adaptations of plants to the fire factor.

Debano, L., D.G. Neary, and P.F. Folliott. 1998. *Fire Effects on Ecosystems*. John Wiley and Sons, Hoboken, NJ. A broad exploration of the effects of fires on ecosystems, including forests and other landscape components, such as watersheds, plants, and air.

Hecht, S. and A. Cockburn. 1990. *The Fate of the Forest: Developers, Destroyers, and Defenders of the Amazon*. Harper Perennial, New York. An enthralling examination of the complex human drama in the Amazon basin, where fire has played such an important role in the development of agriculture.

Spencer, J.E. 1966. *Shifting Cultivation in Southeast Asia.* University of California Press, Berkeley. One of the best authorities on an agricultural system that uses fire and has existed for many centuries.

Watters, R.F. 1971. *Shifting Cultivation in Latin America.* FAO, Rome. An exhaustive review of how fire is employed in shifting cultivation throughout the entire Latin American region.

West, O. 1965. Fire in Vegetation and its Use in Pasture Management. Publication 1/1965. Commonwealth Agricultural Bureau. Hurley, Berkshire. An excellent review that examines both the ecology and management of fire in grazing ecosystems.

Whelan, R.J. 1995. *The Ecology of Fire.* Cambridge Studies in Ecology. Cambridge University Press, New York. An analysis of fire as an ecological factor in the environment.

# 11 Biotic Factors

Chapter 4 through Chapter 10 have focused on how individual plants are impacted by abiotic factors of the environment, such as light, temperature, and mineral nutrients. In this chapter, we will complete the picture of how the environment impacts plants by exploring how biotic factors of the environment — that is, conditions created and modified by living organisms — affect individual plants.

In agroecosystems, the farmer is, in a sense, the organism with the greatest impact on the environment in which crops are grown. The farmer alters and adjusts conditions of the physical as well as the biological environment to meet the needs of the crop or crops. To do so sustainably, the farmer must have an understanding of the biotic interactions of the agroecosystem — how each member of the community impacts the agricultural environment and alters conditions for its neighbors.

To conceptualize biotic factors in ecological terms, we must enter an area of overlap between autecology and synecology. Even though we begin from the perspective of the individual organism confronting an environment made up of various factors, we must deal with interactions between organisms when the factors we are concerned with are biotic. Despite their synecological origin, however, the concepts developed in this chapter to describe these interactions can be applied in an autecological way by considering interactions in terms of their impact on each individual organism in the agroecosystem.

There are two basic frameworks for conceptualizing the interactions between organisms in a community or ecosystem; each has its respective advantages. Traditionally in ecology, interactions have been understood in terms of the effects that two interacting organisms have on each other. This framework is the basis for such foundational concepts as competition and mutualism. In agroecology, however, it is often more helpful to view interactions as deriving from the impact that organisms have on their shared environment. Organisms remove substances from, alter, and even add substances to the areas they occupy, in the process, changing the environmental conditions for themselves and other organisms. Thus each biotic factor that an individual organism faces can be understood as a modification of the environment created by another organism. Both of these frameworks, or perspectives, are explained in more detail below.

## THE ORGANISM–ORGANISM PERSPECTIVE

A broadly accepted system for classifying interactions between organisms was developed by Odum (1971). This system has many useful applications and has served ecologists well in understanding the biotic environment. Interactions between two organisms of different species are seen as having a negative effect (–), a positive effect (+), or a neutral effect (0) for each member in the interaction. For example, in the interaction classified as mutualism, both organisms are impacted positively (+ +). The degree to which the interaction is positive or negative for each organism depends on the level of interdependence and the level of intensity of the interaction.

In this scheme, there is an important distinction between situations in which both members of the mixture are present together and the interaction is actually taking place, and situations in which the two are separate, or together and not interacting. In Table 11.1, the "not interacting" column shows the results in this latter situation and gives an indication of the degree of dependence or need for interaction that each member may have developed over evolutionary time.

The interaction that has probably received the greatest attention, especially in the design of conventional agroecosystems, is *competition* (– –). Competition occurs in an environment where resources are in limited supply for both members of the relationship, and even though one member of the mixture may end up dominating the other, both do worse when they are interacting in this way than if there had been no interaction at all. The organisms interact by removing something from the environment that they both need. Two crop varieties of the same species are highly likely to compete in a resource-limited environment — for example, a crop field with low nitrogen levels in the soil.

When two organisms have become so dependent on each other that they suffer when not in interaction, then it can be said that the interaction is a *mutualism* (+ +). Both organisms depend upon the way in which the other modifies the environment for both. Some interactions between legumes and *Rhizobium* bacteria, for example, are thought to be mutualistic: neither organism does as well alone as they do together.

When an interaction benefits both members, but neither is negatively impacted in the absence of interaction,

**TABLE 11.1**
**Types of Two-Species Interactions as Defined by Odum**

| | Interacting | | Not Interacting | | |
|---|---|---|---|---|---|
| **Interaction** | **A** | **B** | **A** | **B** | **Nature of Interaction** |
| Neutralism | 0 | 0 | 0 | 0 | Neither organism affects the other |
| Competition | – | – | 0 | 0 | Both A and B affected negatively |
| Mutualism | + | + | – | – | Obligate interaction |
| Protocooperation | + | + | 0 | 0 | Not obligate |
| Commensalism | + | 0 | – | 0 | A obligate commensal; B host |
| Amensalism | – | 0 | 0 | 0 | A harmed by presence of B |
| Parasitism | + | – | – | 0 | A parasite, B host |
| Predation | + | – | – | 0 | A predator, B prey |

*Note:* + organism growth increased; – organism growth decreased; 0 organism growth not affected.

the interaction is termed *protocooperation* (+ +). Pollination can be an example of such an interaction: when there are several species of pollinating insects available and many species of nectar-producing plants, one species of pollinator and one species of plant benefit each other if they interact, but neither are harmed if they do not interact. Both mutualism and protocooperation are considered examples of *symbiosis*, a term formed from the Greek words for "living together."

When one organism maintains or provides a condition necessary for the welfare of another but does not affect its own well-being by doing so, the interaction (+ 0) is termed *commensalism*. The assisted organism suffers, though, when the organism creating the needed conditions is not present. A shade tree species in a cacao agroforestry system, for example, creates the reduction in light intensity needed by the obligate shade-loving cacao plants below, but the shade tree does equally well with the cacao present or not.

When one species negatively affects another, but is not directly affected itself, then the interaction is termed an *amensalism* (– 0). An example of an amensal interaction is when a plant releases a chemical from its leaves in raindrip that can negatively impact other plants around it, but which does not impact the producer of the chemical. Such a process is a form of allelopathy, which will be discussed in more detail below. An example of this kind of amensalism is the relationship between the black walnut (*Juglans nigra*) and almost any plant that attempts to grow under the canopy of a black walnut. Chemicals leached from the husks, leaves, and root exudates of black walnut are toxic to most plants.

In the two remaining types of interactions, one organism is negatively impacted by the actions of the other (+ –). The perpetrator of the actions generally has an obligate relationship with the other, whereas the organism receiving the brunt of the negative impacts does better if left alone (i.e., the relationship becomes – 0). In *parasitism*, one organism (the parasite) feeds on another (the host), but the host is rarely killed outright. The parasite may live together with the host for a long period, with the host eventually surviving, but its fitness is reduced. Some parasites, known as parasitoids, cause the death of the host (e.g., parasitic wasps in the genus *Trichogramma*); we take advantage of such interactions for biological control in agroecosystems. *Predation* is a much more direct interaction, where one organism actually kills and consumes its prey. We depend greatly on predation by certain beneficial organisms for the management of pests in farming systems.

This classification scheme is very useful for distinguishing the types of interactions that are observed in most natural environments. But it focuses on the end result of each type of interaction, rather than on the mechanisms involved as the interaction takes place.

## THE ORGANISM–ENVIRONMENT–ORGANISM PERSPECTIVE

Each of the interactions described above can be understood alternatively as the result of one organism modifying the environment in a way that impacts the other organism in the interaction. By focusing on how the environment mediates the effects that organisms have on each other, it is possible to understand the *mechanisms* through which the effects occur. With knowledge of the mechanisms, the agroecosystem manager is in a much better position to manipulate or take advantage of the interactions.

When an organism modifies the environment in some way that impacts another organism, that modification is termed as *interference*. Interferences can be divided into two types:

- In *removal interference*, one organism removes something from the environment, reducing the availability of that resource for other organisms.
- In *addition interference*, one organism adds something to the environment that can have a positive, negative, or neutral impact on other organisms.

Usually only one or the other of these interferences takes place in a particular interaction, but they can occur together in some interactions, as discussed below. When conceptualized with this framework, an interaction between two or more organisms is comprised of an impact on the environment (an addition or a removal) perpetrated by one organism (and in some cases an additional impact created by the other organism), followed by a response on the part of both organisms to the resulting changes in the environment. Note that the "environment" is not necessarily external to the interacting organisms — it can include the tissues or body of either or both organisms. Types of removal and addition interferences are described in greater detail below and then summarized in Table 11.2.

## Removal Interferences

When one organism removes something from the environment as part of its life activities or interaction with other organisms, it can affect other organisms. This type of interference is generally negative for one or more members of the interaction, but it can have positive effects as well. There are several types of removal interferences in agroecosystems.

### Competition

Only a shift of emphasis is needed to understand competition as removal interference. Competition occurs when two organisms are removing a resource from the environment — light, nitrogen, or water — that is not abundant enough to meet the needs of both. Many of the earlier chapters in this book have described the conditions under which resources may become limiting and thus set the stage for competition.

Viewing competition as removal interference provides an alternative way of understanding what is commonly thought of as competition for space. Under this framework, "space" is seen as a complex mixture of resources that is impacted by the removal effects of the organisms that occupy that space; thus, organisms are in competition over the resources of the space, not the space itself.

Competition between individuals of the same species — *intraspecific competition* — can be quite intense since the needs of the interacting individuals are so similar. Monoculture agriculture has invested much energy in determining how densely crops can be planted without competition between individual plants negatively affecting production.

Competition between individuals of different species, called *interspecific competition*, can also be important when levels of resources are not sufficient to meet the needs of both. The mechanisms of the interaction involve either removal of a resource or its direct protection or sequestration by an organism (e.g., when an animal defends a territory and its resources). In either case, the resource is the primary focus of the interaction.

Competition is a very important concept in ecology, but it also has a history of controversy and discussion. On the one hand, interspecific competition is a cornerstone of evolutionary ecology. Competition is considered the engine of natural selection and a force, which all organisms must contend within their struggle to survive and

**TABLE 11.2**
**Summary of Interference Interactions**

| | Creator of Interference (A) | Receiver(s) of Interference (B) | Type and Identity of Interference | Location of Interference | Effect on A[a] | Effect on B[a] |
|---|---|---|---|---|---|---|
| Competition | Roles interchangeable | Roles interchangeable | Removal of resources | Shared habitat | – | – |
| Parasitism | Parasite | Host | Removal of nutrients | Body of host | + | – |
| Herbivory | Herbivore | Consumee | Removal of biomass | Body of consumee; shared habitat | + | – or + |
| Epiphytism | Host | Epiphyte | Addition of habitat surface | Body of host | 0 | + |
| Proto-cooperation | Roles interchangeable | Roles interchangeable | Addition of material or structure | Shared habitat or body of A/B | + (0) | + (0) |
| Mutualism | Roles interchangeable | Roles interchangeable | Addition of material or structure | Shared habitat or body of A/B | + (–) | + (–) |
| Allelopathy | Allelopathic plant | Potential habitat associates | Addition of active compound | Habitat of organism A | + or 0 | +, –, or 0 |

[a] Symbols in parenthesis refer to the effect when the organisms are not interacting.

leave offspring. Interestingly, however, ecologists also see that avoiding competition can actually be advantageous for a species, and that doing so has probably played a key role in the development of species diversity.

Without actually studying the mechanisms of interference that are involved in competition, and identifying the removal process from the environment that leads to it, we can only assume that competition occurs. Agroecosystem management requires a more detailed determination of competitive interactions; otherwise, the farmer is left with no other option but to overload the system with excess resources.

## Parasitism

As described above, parasitism is an interaction in which two organisms live together, with one (the parasite) deriving its nourishment from the tissues of the other (the host) without killing it. In interference terms, the environment from which removal takes place is the body of the host. Parasites are physiologically dependent on their hosts, live much shorter lives, and have a high reproductive potential.

The relationship between mistletoe and various species of trees is an example of this kind of removal interference (Figure 11.1). The mistletoe plant actually penetrates and taps into the vascular system of the host tree, drawing its water and nutrients from the host. If the parasite becomes too abundant on the host tree, the tree is stunted and often deformed, and can become subject to debilitating attacks from other pests. Farm and range animals are especially susceptible to parasites; these include ticks that attach externally to the host, screwworm flies that lay eggs in the flesh of the animal, and stomach parasites ranging from bacteria to worms.

Under natural conditions, parasitism probably represents something of a compromise between the host and the parasite. They have evolved together over time, with the host being tolerant of a constant low-grade infection, and the parasite depending on the continuity of the host's life for its own reproductive success. In agricultural situations, however — especially the human-maintained conditions of concentrated monocultures — heavy parasite loads become a serious form of disease that puts the entire crop or herd at risk of developing secondary diseases and dying.

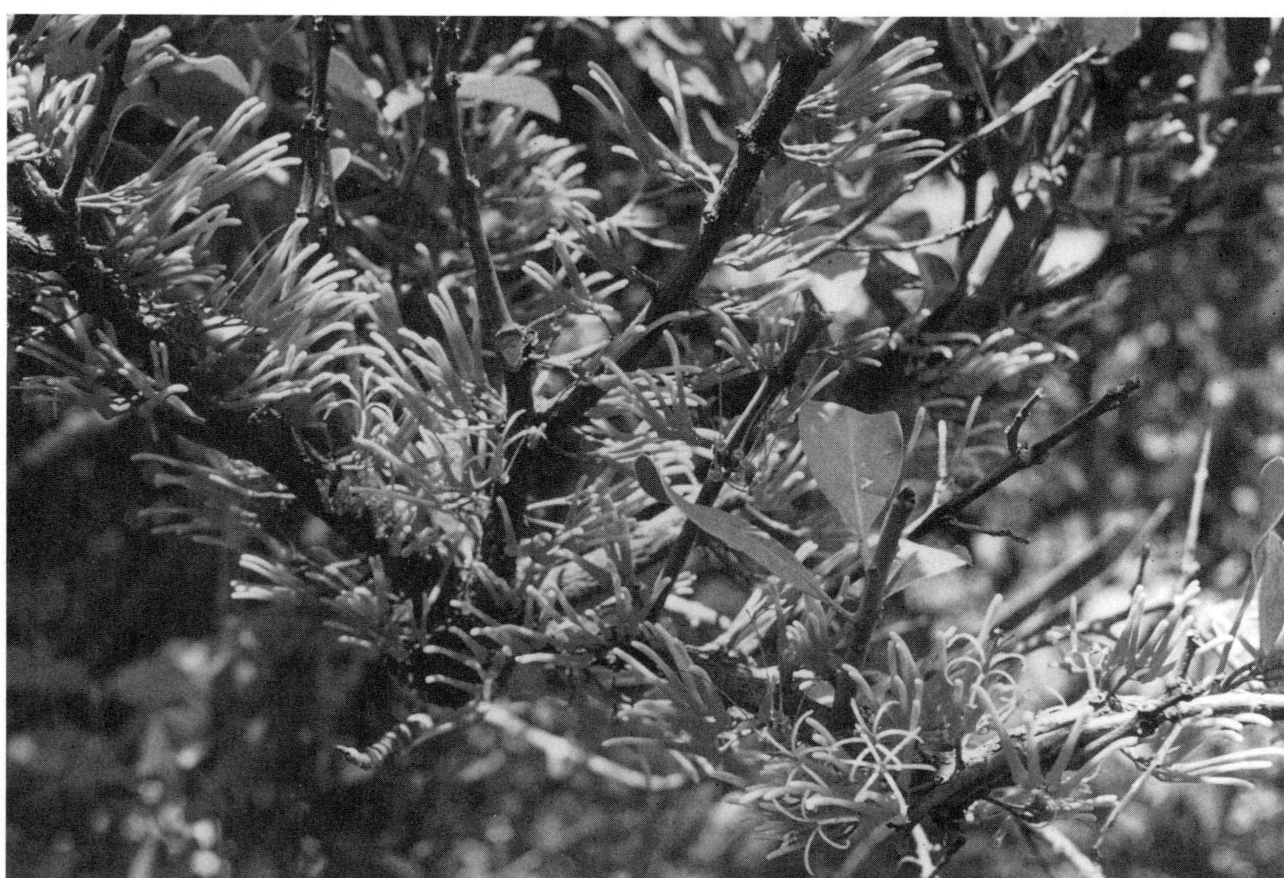

**FIGURE 11.1 Parasitic mistletoe on a guava tree, Monteverde, Costa Rica.** The guava branch is so heavily infested by the parasite that only the red-orange flowers of the mistletoe are visible.

### Herbivory

The interference relationship between a herbivore and the plant it consumes — like that of parasite and host — is a very direct one, with plant tissue being the part of the environment that is removed. Beyond the scope of the individual plant, however, herbivory is removal interference in an even broader sense in that biomass and its associated nutrients are removed from the environment. The consumption of plant material reduces the return of biomass to the soil, and if the removal is too intense and takes place over an extended time frame, it can lead to depletion of nutrients in the system.

From an agricultural perspective, herbivory can have three types of negative impacts. First, herbivory removes photosynthetic surface area that may be of importance in the development of the crop plant. Second, if the plant part that is consumed were going to return to the soil as crop residue, herbivory is reducing this input to the system. Third, if the herbivory damages a part of the crop that is intended to be harvested and sent to market, the product's sale value may be reduced.

The effects of herbivory, however, are not always negative. In some pasture or range situations, for example, grazing can be beneficial to the productivity of the forage species. Removal of excessive plant material can stimulate the production of new biomass, or even allow certain plant species that are suppressed by old or excessive plant cover to germinate or become more predominant in the pasture mixture. The evolutionary role of such removal interference has been well documented for the Serengeti Plains in Africa (McNaughton, 1985), where it has been shown that the highest productivity and species diversity of both plants and animals has developed under cyclical patterns of multispecies grazing. Good range managers know that periodic rotational grazing promotes the most production in pasture systems.

In natural systems as well, herbivory plays an important role in removing excess biomass, directing energy flow, and recycling nutrients. These processes have the potential for playing important and positive roles in agroecosystems, but humans have tended to view herbivory as wholly negative, a constant challenge to be overcome. Further research needs to be focused on how the pressure of such removal interference can be directed away from the economically valuable parts of the agroecosystem and concentrated in parts that stimulate other components of the system in ways that contribute to sustainability.

## Addition Interferences

Many organisms in the course of their daily life processes add something to the environment that impacts associated organisms. These impacts can be negative, such as when the addition causes a reduction in growth or development for the associated organisms, or when it excludes them from the area entirely. In other cases, the impact of the addition interference can be positive for the associated organisms, as when they use the added substance or material to improve their own standing in the community, or when the exclusion of intolerant organisms from the habitat allows them to occupy it. Ultimately, associated organisms benefiting from the addition may develop a dependence on the organism making the addition, creating a relationship of coexistence or even of symbiosis.

### Epiphytism

When one organism lives on the body of another without drawing any nutrition from it, addition interference is occurring because the host is adding a physical structure to the environment that is providing another organism with a habitat. When the two organisms are plants and the habitat is a trunk or stem, the perched plant is called an *epiphyte;* when the habitat is a leaf, it is called an *epiphyll.* In Odum's terms, epiphytism is a form of commensalism.

Epiphytes and epiphylls do not obtain water or food from the supporting plant, nor do they have connections to the soil. Water is derived from precipitation, and nutrients from wind-borne particles, the decay of the supporting plant's bark, and minerals and organic compounds dissolved in raindrip. Most epiphytic plants face frequent drought conditions in their aerial environment, even in the moist habitats where they are most common. Algae, lichens, mosses, and a few ferns are the most common epiphytes in cold and wet environments; a wide variety of vascular plants have evolved the epiphytic lifestyle in warm and wet climates, especially ferns and species belonging to the families *Bromeliaceae* and *Orchidaceae*. A large number of species in these two families have taken on considerable economic importance in horticulture and floriculture, and are raised on artificial perches in greenhouses and lathhouses for commercial sale.

An epiphytic plant of considerable economic importance in agriculture in several tropical countries is vanilla (*Vanilla fragrans*). Vanilla produces long whitish aerial adventitious roots at each leaf that adhere firmly appressed to the trunk or branches of the host plant. Sometimes roots climb down the trunk to the ground, but only ramify in the humus or mulch layer. Capsule-like fruits up to 25 cm long (called beans in the trade) form on the aerial stems, and are dependent on hand pollination for successful formation in many parts of the world into which the crop has been introduced from its native Mesoamerica (Figure 11.2).

### Symbioses

When two organisms make additions to the environment they share so as to benefit each other, they form a symbiotic relationship. If the relationship is nonobligatory and

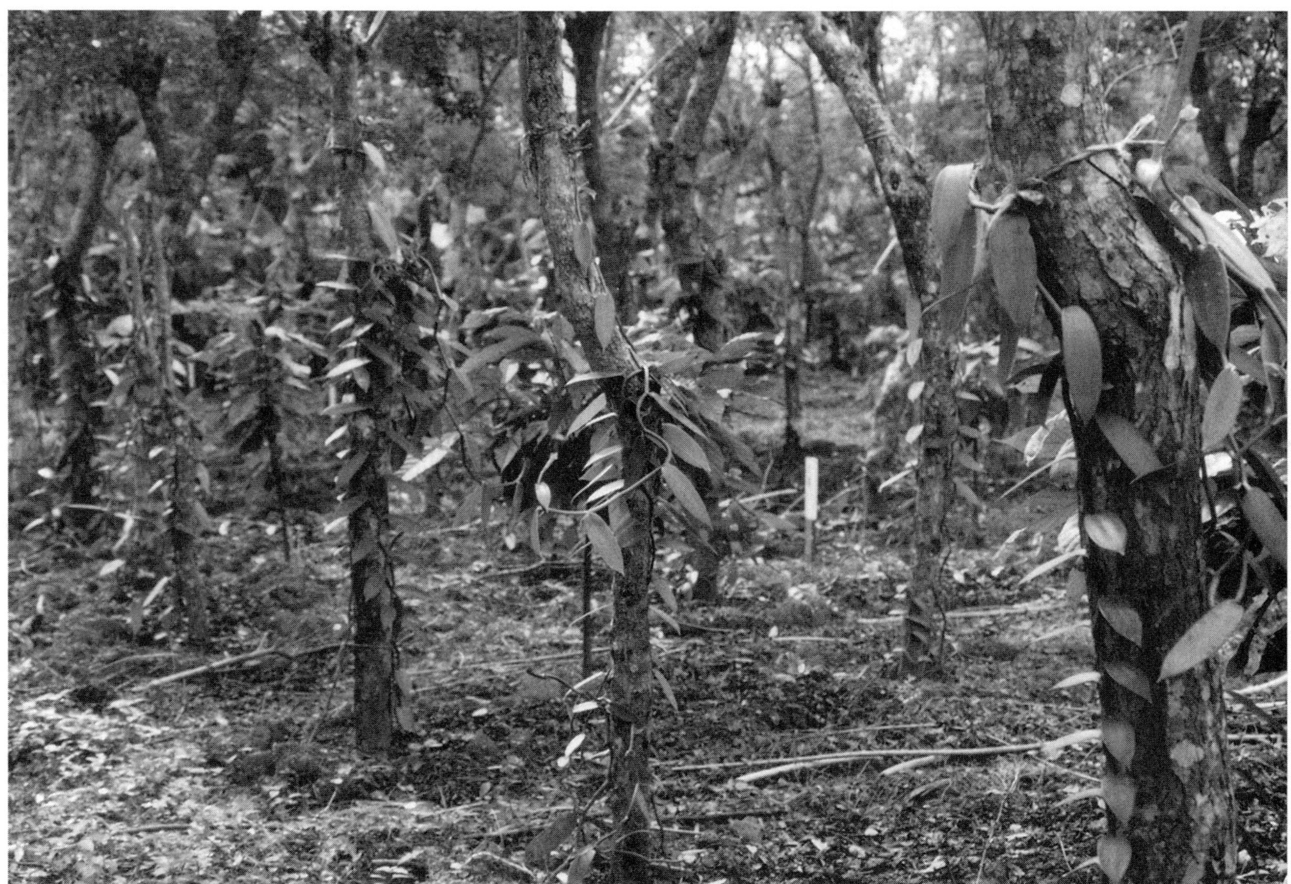

**FIGURE 11.2 A plantation of the epiphytic vanilla orchid in Tabasco, Mexico.** The vanilla plants (*Vanilla fragrans*) grow on the shade tree *Glyricidia sepium*.

nonessential for the survival of either organism, the resulting relationship is called *protocooperation*. An example of protocooperation is the relationship between the European honeybee (*Apis mellifera*) and the plants it pollinates. The plant a bee visits is adding pollen and nectar to the environment, serving to attract the pollinator. The actual gathering of the nectar or honey by the bee is removal interference, but then the pollen is added back into the environment when the bee deposits it onto the stigma of another flower — this is the point at which the positive effects of the interaction are realized. Honeybees visit a wide range of plant species, most of which are visited by other pollinators as well, making the relationship between the honeybee and any particular plant species nonobligatory. In many agricultural landscapes, however, the dramatic reduction in biotic diversity that has accompanied the expansion of monocultures, heavy use of pesticides, and fencerow-to-fencerow farming has created an artificial dependence on honey bees that are raised by beekeepers and transported in hives to the crop fields during pollination time.

When the organisms benefiting each other through addition interferences become dependent on each other for optimal performance and even survival, then the relationship is *mutualism*. A good example of mutualism is the relationship between certain soil-dwelling fungi and their vascular-plant associates. The fungi are made up of *mycorrhizae*, special compound structures that can form connections with plant roots. The mycorrhizae allow the root to provide sugars for the fungus, and the fungus in return to provide water and minerals to the plant. There are two types of mycorrhizae: (1) *ectotrophic*, in which the fungal mycelium forms a dense mantle covering the surface of the root, with many hyphae that extent outward into the soil, and others that extend inward and force themselves between the cells of the epidermis and cortex of the root (very common in the *Pinaceae*); and (2) *endotrophic*, the most common type, in which there is no surface mantle but instead some of the hyphae actually inhabit the protoplasts of parenchymatous tissues and extend outward into the soil (common in most flowering plant families, especially important crop species such as corn, beans, apples, and strawberries).

Another important example of a mutualism is the relationship between legumes (plants in the *Fabaceae* family) and Rhizobium bacteria (Figure 11.3). The bacteria enter

**FIGURE 11.3 Nodules on the roots of fava beans.** The nodules are inhabited by nitrogen-fixing *Rhizobium* bacteria in mutualistic association with the legume.

the root tissue of a legume plant, causing the tissue to form nodules in which the bacteria live and reproduce. The nodules, formed from root tissue, represent addition interference on the part of the legume plant. The legume also provides the bacteria with sugars. The bacteria's addition interference comes in the form of fixed (useable) nitrogen, which the bacteria produce from atmospheric nitrogen. The legume would be greatly handicapped in its growth without the fixed nitrogen provided by the bacteria, and the bacteria require the root nodules for optimal growth and reproduction. The fixing of nitrogen by Rhizobia is one the most important means by which nitrogen is moved from the vast atmospheric reservoir into soil and biomass.

As we will see in later chapters, such beneficial mutualisms, where two or more members of the relationship interact through addition interference, is of major importance in the design and management of many intercropping agroecosystems.

## Allelopathy

A form of interference that has received considerable attention recently, especially in agriculture, is *allelopathy* (Gliessman, 1989, 2002a). Allelopathy is the production of a compound by a plant that when released into the environment has an inhibitory or stimulatory impact on other organisms. Allelopathic interactions have been shown to occur in a wide variety of natural ecosystems and agroecosystems.

Allelopathic compounds are natural products that may be direct metabolites, byproducts of other metabolic pathways, or breakdown products of compounds or biomass. The compounds are often toxic to the plant that produces them if they are not stored in some nontoxic form or released before they build up internally to toxic levels. In some cases, even when the toxins are released from the plant, they may build up in the immediate environment and become toxic to the plant that produced them. Allelopathic compounds take many forms, from water soluble to volatile, simple to complex, and persistent to very short-lived. The most common allelopathic compounds fall into such chemical groups as tannins, phenolic acids, terpenes, and alkaloids.

Allelopathic products are released from the plant in a variety of ways. They can be washed off of green leaves, leached out of dry leaves, volatilized from the leaves, exuded from roots, or released from shed plant material during decomposition. Even flowers, fruits, and seeds can be sources of allelopathic toxins. There are also cases in which products do not become toxic until they have been altered once they are in the environment, either by normal chemical degradation or by conversion to toxic compounds by microorganisms.

In natural ecosystems, allelopathy may help explain some important phenomena:

- the dominance of a single species or group of species over others;
- successional change and species replacement, or the maintenance of a deflected stage in the successional process;
- reduced ecosystem productivity; and
- unique patterning or distribution of plant species in the environment.

In agroecosystems, allelopathy may play important roles in biological control, the design of intercropping systems, and crop rotation management. Examples are presented below and in more detail in later chapters.

### Comparison of Types of Interference

Table 11.2 provides a brief summary of the most salient characteristics of each type of interference. Study of this table may reveal that the grouping of interferences into addition interferences and removal interferences does not exhaust the ways in which interferences can be classified. Mutualism, for example, shares with competition the property of involving symmetrical roles; that is, the organism creating the interference is simultaneously the organism receiving the interference created by the other interacting organism (note that this symmetry does not necessarily extend to the results of the interaction). As another example, parasitism and epiphytism both involve interferences that act directly on one organism's body rather than on the external, physical environment. These observations suggest that interferences may be grouped as either direct or indirect, and as either symmetrical or asymmetrical. Allelopathy, for example, is asymmetrical and indirect. Table 11.3 shows the typology resulting from such a classification. Most forms of interference occupy only one cell in the matrix, but protocooperation and mutualism can be either direct or indirect.

**TABLE 11.3**
**Types of Interference**

| | Direct (occurs in or on the body of one or both organisms) | Indirect (occurs in the shared habitat of the organisms) |
|---|---|---|
| Symmetrical (both organisms create interference) | Protocooperation<br>Mutualism | Competition<br>Protocooperation<br>Mutualism |
| Asymmetrical (interference created by one organism) | Herbivory<br>Parasitism<br>Epiphytism | Allelopathy |

### Interferences at Work in Agroecosystems

In most multiple-species interactions, plants are removing and adding things to the environment simultaneously. It is very difficult to separate removal and addition interactions, much less show how they may interact in ways that determine which species and how many individuals of each are able to coexist in a specific habitat. Ultimately, the combination of interference types is going to play an important role in determining the structure and function of the ecosystem.

It is easy to imagine how allelopathy and competition, for example, can both play a part in a polyculture cropping system. The members of the mixture are simultaneously adding materials to and removing resources from the environment, modifying the microclimatic conditions of that environment at the same time, and interacting with each other in ways that permit coexistence or favor mutualistic interdependence. It is important, though, to understand the mechanisms of each interaction, beginning with the impacts of each species on the environment in which they all occur. The ability of farmers to successfully manage complex crop mixtures and rotations depends on the development of this understanding.

## ALLELOPATHIC MODIFICATION OF THE ENVIRONMENT

Ecological research has placed the greatest emphasis on competitive interactions. This has been especially true in agronomy, where great efforts have been made to understand what the conditions of the environment are that limit optimal crop development, and what kinds of inputs or technologies are needed to correct the situation when something that the crop needs is missing or is in short supply. Crop arrangements and densities have been researched and developed to avoid the effects of competition.

More recently, however, the role of allelopathy in agroecosystems has begun to receive considerable attention. The growing desire to replace synthetic chemical inputs to agroecosystems with naturally produced materials has spurred a burst in applied research on allelopathy, especially in Europe and India (Kohli et al.,

2001). Allelopathy thus serves as an excellent example of how a research focus on the mechanisms of interference can have important applications in agroecology. Because allelopathy has such potential importance in agroecological research and for sustainability, the remainder of this chapter will be devoted to exploring it in greater detail.

There are many possible allelopathic effects of weed and crop species that need to be taken into account in agroecosystem management. The production and release of phytotoxic chemicals can originate from crops or weeds, and they can play very important roles in crop selection, weed management, crop rotations, the use of cover crops, and intercropping design. Many examples of such interactions have appeared in the international publication *Allelopathy Journal.*

Our purpose in this section is to gain more insight into the actual mechanisms of allelopathic interactions. The implications and applications of these interactions will be more fully explored in Chapter 13.

## THE HISTORY OF THE STUDY OF ALLELOPATHY

The effects of allelopathy have been observed since the times of the Greeks and Romans, when Theophrastus suggested that the "odors" of cabbage caused vine plants to "wilt and retreat" (Willis, 1985). Japanese sources dating back to at least the 1600s independently document what we now know to be allelopathic interactions, and such knowledge may have developed earlier and independently in other areas.

In Europe, scientific observations of allelopathic plant interactions were not made until the 17th century, when De Candolle published an influential work describing his observations of the excretion of droplets of some sort from the roots of *Lolium temulentum*. De Candolle believed that plants used their roots as excretory organs and that these excretions contained chemicals that stayed in the soil and affected subsequent plant growth. His theory fell out of favor, however, when Justus Von Liebig developed the theory of mineral nutrition, and the focus on plant interactions shifted to nutrient depletion and competition.

It was not until the late 19th century that careful experiments in the U.S. and U.K. scientifically demonstrated that allelopathy was an important plant interaction. In England, certain grasses were found to negatively impact the growth of nearby trees, and the research indicated that the effects could not have been due to soil nutrient depletion. In fact, leachates of soil from pots planted with the grasses impacted the trees as much as the grass itself. In the U.S., Schreiner and his associates published a series of papers between 1907 and 1911 documenting the "exhaustion" of soils planted continually in one crop and the extraction of the chemicals responsible for the exhaustion. This was the first time researchers demonstrated the ability of plant chemicals to inhibit germination and seedling growth of a plant species.

During the 1920s, some important work focused on the black walnut. Cook documented the tree's ability to inhibit nearby plants, and Massey found that an extract of walnut bark in water caused tomato plants to wilt.

In 1937, the term *allelopathy* was coined by Molisch to describe any biochemical interaction between plants and microorganisms, positive or negative. Soon afterward, studies by Benedict, Bonner and Galston, Evenari, and McCalla and Duley again documented chemotrophic plant effects, and the term *allelopathy* came into common usage for the first time (Willis, 1985).

Muller introduced the concept of interference in 1969 as a way of explaining both competition and allelopathy in a single theory. Ecologists began to realize that competitive or allelopathic effects may work in tandem in any given system, and that allelopathic interactions can be particularly important in multiple cropping systems (Rice, 1984). More recently, recognition of the importance of allelopathy in agriculture has lead to research on ways phytotoxins can be involved in such practices as weed control, cover-cropping, soil biofumigation, and even pest management (Gliessman, 2002a).

The difficulty of demonstrating how allelopathy actually works in the field has kept ecologists from attributing a significant role to chemical interference in overall vegetation process. But recent work by Bais et al. (2003) has firmly placed allelopathy back on center stage. They meticulously documented the displacement of native plant species by the Eurasian spotted knapweed (*Centaurea maculosa*) in the western U.S., and the role that allelopathy plays in the process. They identified the phytotoxin that this economically destructive plant invader produces, showed how it is released from the roots, and characterized the mechanisms that trigger the death of susceptible native plant neighbors. Such research clearly demonstrates how allelopathy must be reckoned with in plant species interactions (Fitter, 2003), both in natural ecosystems and agroecosystems.

## Demonstrating Allelopathy

In order for allelopathy to be fully implicated in an interference interaction, the following steps must be followed:

1. Determine the presence of a potential allelopathic compound in the suspected plant and plant part. A screening system that employs some type of bioassay is a common procedure for doing this test (Leather and Einhellig, 1986). A positive bioassay can only be used to imply that there is a potentially allelopathically active chemical present in the plant.
2. Show that the compounds are released from the donor plant.
3. Determine that the compounds accumulate or concentrate to toxic levels in the environment.
4. Show that uptake or absorption of the compounds by the target organism takes place.
5. Demonstrate that inhibition (or stimulation) of the target species takes place in the field.
6. Identify the chemical compounds and determine the actual physiological basis for the response.
7. Finally, determine how the allelopathic compound interacts with other factors in the environment so as to either reduce or enhance its effect. (Rarely does an allelopathic compound kill another organism outright).

Under ideal situations, all of these steps could be carried out before attempting to manage allelopathy in an agroecosystem setting. But most of the time, such intensity of research is not possible, and farmers are faced with the need to make decisions on their farms every day. Astute observation coupled with research results can make allelopathy one more tool for managing the farm environment for the benefit of the crop.

## Allelopathic Effects of Weeds

Weeds are responsible for the loss of crop production all over the world. The literature abounds with reports on the "competitive effects" of weeds, but seldom is allelopathy considered or even mentioned as one of the mechanisms by which weeds impact crops. Whenever weeds and crops are in the same planting together, many possible forms of interference are going to be working together or in sequence. Allelopathic potential has been suggested for a large number of weed species (Putnam and Weston, 1986), but we are just beginning to understand the mechanisms of release of the potentially phytotoxic compounds into the environment, how they are taken up by crops, how they inhibit crops, and how the negative impacts of the compounds can be ameliorated.

Allelopathic chemicals released by weeds can directly influence crop seed germination and emergence, crop growth and development, and the health of associated crop symbionts in the soil. Recent research on weed allelopathy has shown that many weeds use multiple mechanisms to inhibit crop growth and development, and such knowledge is an important component in developing alternative weed management strategies (Qasem and Foy, 2001).

An example of an allelopathic weed is bitter grass (*Paspalum conjugatum*), an aggressive weed in annual cropping systems in Tabasco, Mexico. Figure 11.4 illustrates the inhibitory effect of bitter grass when it is present in a corn crop. As the dominance of the grass increases, the stunting of the corn becomes more noticeable, reaching a point where the corn is not even able to establish.

Water extracts made from the dry grass that has not yet been leached by rains showed the ability to affect both germination and early growth of corn seed. Local farmers recognize the negative impacts of the grass on the soil, referring to a heating effect that can cause the stunting or yellowing of the crop. When researchers could find no temperature differences in the field with thermometers, allelopathy became suspect. Although the evidence is not sufficient to rule out competitive interference from the grass, the inhibitory effect exists even when farmers add recommended levels of chemical fertilizers to the crop and when rainfall is more than sufficient.

In a study in California, two common weeds — lambsquarters (*Chenopodium album*) and red root pigweed (*Amaranthus retroflexus*) — were tested for allelopathic potential against green beans (*Phaseolus vulgaris*). Both weed species showed allelopathic potential in laboratory bioassays; in the field it was found that bean plants grown with pigweed were stunted but had normal numbers of nodules of symbiotic Rhizobium bacteria, and that beans grown with lambsquarters were both stunted and had greatly reduced numbers of nodules (Espinosa, 1984). These results indicate that the chemicals released by the two different weeds were impacting the crop plants in different ways, with one affecting the growth of the beans directly and the other inhibiting the activity of N-fixing bacteria. Since the farm field was irrigated, had recently been fertilized, and crop spacing ensured that adequate light reached the beans, removal interference was probably minimal.

A weed species that has been studied in great detail in order to demonstrate its allelopathic mechanisms is quackgrass (*Agropyron repens*). The following findings are described in a review by Putnam and Weston (1986):

- Quackgrass inhibited several crop types (e.g., clover, alfalfa, and barley), and this inhibition could not be explained by removal interference (that is, competition).
- Laboratory and greenhouse bioassays demonstrated the inhibitory potential of both quackgrass

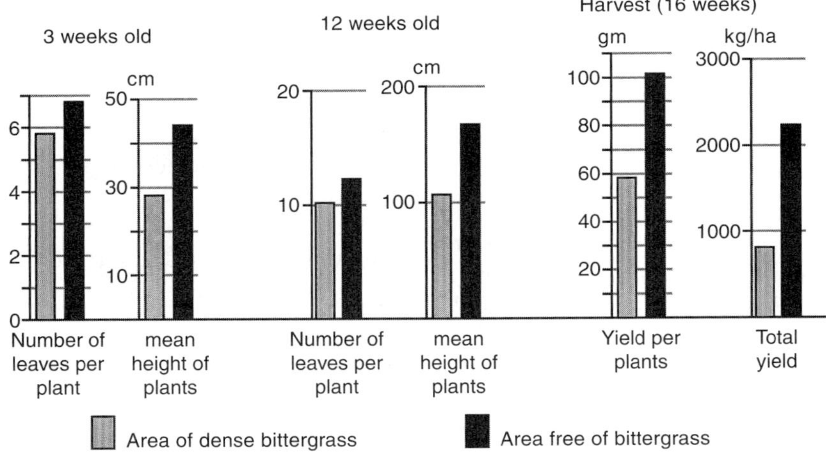


**FIGURE 11.4 Allelopathic inhibition of corn by bitter grass (Paspalum conjugatum), Tabasco, Mexico.** Rain washes phytotoxins off of dead and living parts of the grass, and additional compounds are exuded from the roots. (Data from Gliessman S.R. In A. Amador ed., *Memoirs of Seminar Series of Ecology.* pp. 1–8. Colegio Superior de Agricultura Tropical: Cardenas, Tabasco, Mexico. 1979.)

foliage and quackgrass rhizomes, although foliage residue was twice as toxic as rhizomatous material. Water extracts and incorporated residues were both phytotoxic.

- There is some evidence that greater inhibition is observed in the presence of soil fungi.
- Decaying quackgrass residues were shown to produce water-soluble inhibitors, explaining the inhibition that has been observed when quackgrass residues are a significant part of no-till systems.
- Inhibition of nodulation in legumes and reduction of root hair formation in other plants are suspected as being possible mechanisms of inhibition.
- Several compounds have been isolated and identified from water extracts and decomposition products, and include several phenolic acids, a glycoside, a compound known as agropyrene, and a flavone tricin and related compounds.
- Even when quackgrass is killed with herbicides, the plant residues and toxins in the soil must be allowed to degrade prior to successful establishment of the succeeding crop.

The case of quackgrass demonstrates that allelopathic interference can be very important, but it also suggests that different plant parts may play different roles, and that phytotoxic compounds can enter the environment through different mechanisms and have varying impacts on crops.

## Allelopathic Effects of Crops

Although much research has focused on the allelopathic potential of weeds in agroecosystems, many crop plants have been shown to release phytotoxins as well. Such mechanisms of interaction have important possibilities for farmers looking for alternative management practices.

### Cover Crops

Cover crops are usually grown during a fallow period in a crop field in order to protect the soil from erosion, contribute organic matter to the soil, improve soil conditions for water penetration and retention, and "smother" weeds. Cover crops of wheat, barley, oats, rye, grain sorghum, and sudangrass (*Sorghum sudanense*) have been used effectively to suppress weeds, primarily annual broadleaf species. The weed suppression ability of many of these and other cover crops is due, at least in part, to allelopathy (Overland, 1966).

Because the phytotoxic compounds released by cover crops typically break down relatively quickly in the environment, they generally have little effect on the subsequent crop. The compounds inhibit weeds during the time they are actively produced by the cover crop plants, but after the plants die or are killed through tillage, the compounds quickly degrade (Mamolos and Kalburtji, 2001).

The allelopathic potential of winter rye (*Secale cereale*) has been particularly well studied (Barnes et al., 1986). Rye produces considerable biomass early in the growing season, and has found much success as a green manure crop in poor soils. But it is most notable for its ability to suppress weed growth while it is actively growing, as well as after rye residues are incorporated into the soil with tillage or left on the soil surface after cutting. Allelopathic effects are even seen from residues left on the soil after herbicide spraying has killed the cover. Extensive chemical analysis has identified two benzoxazolinones and associated breakdown products as the probable phytotoxic agents.

The cover crop called velvet bean (*Mucuna puriens*), used extensively in rural Tabasco, Mexico, has been shown to inhibit weeds through allelopathy. This annual vining legume is planted into a corn crop near the end of the cropping cycle. It covers the open space between the corn plants, effectively suppressing weed growth, both before and after harvest. The weed suppression is due, in part, to shading, but release of allelopathic compounds is also at work. After the velvet bean plants complete their life cycle, they are left dead on the ground, covering the soil with nitrogen-rich mulch into which the next corn crop will be planted. Large areas are managed in this manner without the use of fertilizers or herbicides (Gliessman and Garcia, 1982).

As more information is generated on the mechanisms of phytotoxin release in cover crops, farmers will be better able to optimize the use of cover crops for weed control by maximizing the addition of the chemicals into the soil and improving the timing of incorporation. Since cover crop species will vary from region to region, an understanding is also needed of how local climates affect the mechanism of release of the toxins into the environment where they can impact weeds. Proper species selection and management will vary accordingly.

### Organic Mulches Derived from Crops

Plant materials and crop residues can be brought to the field and spread over the soil, serving as organic mulch. Waste plant material from farm fields or the processing of farm products is particularly useful for this purpose. Such materials were already discussed for their value as soil amendments (Chapter 8), but an important benefit of many types of mulch that often get overlooked is their potential for allelopathic weed control.

Like the phytotoxins produced by covercrops, the biologically active compounds found in organic mulches

**FIGURE 11.5 Cacao pod hulls used as allelopathic mulch, Tabasco, Mexico.** The dark cacao hulls, seen between rows of zucchini, suppress weed growth.

degrade relatively quickly, as a rule. However, breakdown rates do differ. For this reason, the timing of mulch application as well as the amount and age of the mulch must be carefully considered so as to maximize weed inhibition and limit the effect on crops.

An excellent example of allelopathically active mulch is dried and crushed cacao pods, obtained in the cocoa production process after the seeds and pulp have been removed from the pods. Spread over the surface of the soil or between established crop plants, the crushed pods leach tannic substances that can inhibit the germination and establishment of weeds (Figure 11.5). Laboratory bioassays of water extracts of the pod material show considerable allelopathic potential. Other types of crop and processing residue with allelopathic potential include coffee chaff from the dried beans, almond hulls, rice hulls, apple pomace, and grape skins and seeds.

Walnut hulls were one of the plant parts initially studied in detail for allelopathic potential, since it was noted long ago that very few other plants (especially weeds) would grow under walnut trees where the outer covering over the walnut nut fell during fruit maturation.

## Crop Inhibition of Weeds

When a crop plant itself is able to inhibit weeds through allelopathy, farmers have a very important tool to add to their toolbox. Several crops are known to be effective in suppressing weeds that grow near them (Batish et al., 2001). The list includes beets (*Beta vulgaris*), lupine (*Lupinus* sp.), corn, wheat, oats, peas, buckwheat (*Fagopyrum esculentum*), millet (*Panicum* sp.), barley, rye, and cucumber (*Cucumis sativa*). Allelopathy can be implicated in all cases, but research needs to thoroughly determine the role phytotoxins play in relation to other forms of interference. In some cases, the inhibition appears to occur from substances released by the living crop plants, but in others it appears that the effect is left over from decomposition products of crop residues incorporated into the soil at the end of the crop cycle. Care has to be taken to keep these inhibitory effects on weeds from affecting the crops that follow. Mixtures of these crops might express even greater allelopathic activity through complementary combining of phytotoxins.

Squash has been implicated as being an especially effective allelopathic crop (Fujiyoshi et al., 2002). Rain leaches inhibitors out of the large, horizontally arranged leaves, and once in the soil, these compounds can suppress weeds. The shade that the leaves cast probably enhances the effect, combining a removal interference with addition interference. Bioassays show the allelopathic potential of water extracts of intact leaves on a range of species, with weeds often being inhibited to a greater extent than crop plants (Table 11.4). When squash is added to an intercropped agroecosystem such as corn and beans, it takes on the important role of weed suppressor for the entire mixture.

Other research has shown that older varieties of some crops, especially the varieties most closely related to wild stock, show the greatest allelopathic potential (Putnam and Duke, 1974; Batish et al., 2001). Crop breeding may have selected against allelopathic potential in exchange for higher crop yields. Screening for allelopathic types in germplasm collections of crops could lead to incorporation of greater allelopathic potential in current crop types through conventional crop breeding or the use of more recently developed genetic engineering technologies.

Considering the problems associated with currently used weed control strategies — possible environmental pollution, groundwater contamination, increased cost of developing and testing new herbicides, increased herbicide resistance by weeds, and the difficulties of registering new herbicides — allelopathic potential in crops will become a more attractive alternative. Connecting the plant's allelopathic potential with an understanding of the fate and activity of the phytotoxic compounds once they leave the plant will make these alternatives most useful.

**TABLE 11.4**
**Initial Root Elongation of the Germinating Seeds of Two Weeds and Two Crops in Laboratory Bioassays of Squash Leaf Extracts**

| Target Species | Distilled Water Control[a] (%) | 2.5% Squash Leaf Extract[b] (%) | 5.0% Squash Leaf Extract[b] (%) |
|---|---|---|---|
| *Avena fatua* | 100 | 61.0 | 40.1 |
| *Brassica kaber* | 100 | 48.2 | 30.7 |
| *Raphanus sativa* | 100 | 112.1 | 57.1 |
| *Hordeum secale* | 100 | 122.0 | 57.8 |

[a] Root elongation after 72 h at 25°C in distilled water defined as 100% growth.

[b] Air-dried intact squash leaves were soaked in distilled water for 2 h and the resulting solution filtered and used to irrigate seeds. Concentration based on ratio of grams of squash leaf to grams of water.

*Source:* Data from Gliessman 1988a.

### Growth Stimulation

The emphasis in the foregoing discussion has been primarily on the inhibitory or negative impacts of chemicals added to the environment by plants. There are, however, limited reports of plants releasing compounds into the environment that have stimulatory effects on other plants around them. Such stimulatory addition interferences can be classified as allelopathy as well, since the term was originally coined to include them along with inhibitory effects.

In some cases, low concentrations of otherwise inhibitory chemicals may actually have a stimulatory effect. Bioassays for allelopathic potential often show increased root elongation in newly germinated seeds when plant extracts are at low concentrations. In other cases, plants produce compounds with wholly stimulatory effects. For example, a study reported in a review by Rice (1986) found that a weed known as corn cockle (*Agrostemma githago*) had an appreciable stimulatory effect on wheat yields when grown in mixed stands as compared to wheat grown alone. A stimulatory substance isolated from corn cockle was named agrostemmin, and when applied separately to wheat fields, was shown to increase wheat yields in both fertilized and unfertilized areas. Rice also reports on work where chopped alfalfa added to soil stimulated the growth of tobacco, cucumber, and lettuce, and a substance known as triacontanol was identified as the stimulant. Even some substances isolated from weeds have stimulatory effects at certain concentrations. Researchers are challenged to demonstrate ways that some of these effects can be practically incorporated into cropping system management, but the potential certainly exists once the full mechanisms of the interference are worked out.

## THE IMPORTANCE OF INTERACTIONS AMONG ORGANISMS

Organisms can have positive and negative influence on each other depending on the nature of their interactions. These interactions have dynamic and potentially important impacts on the environment of agroecosystems. This chapter proposes a model for the study and understanding of such interactions that focuses on the mechanisms through which one organism adds to or removes from its immediate environment some resource or material that can have consequences for the other organisms living there.

As we will see in Section III, finding effective ways of harnessing and managing the interactions among organisms is at the very heart of developing more sustainable practices in agriculture. The autecological perspectives on these interactions developed in this chapter will be a necessary basis for exploring their action and management at the level of crop populations, crop communities, whole agroecosystems, and the landscape in Chapter 13 through Chapter 19.

## FOOD FOR THOUGHT

1. Describe a situation where an organism appears to be competing for a specific space in the environment, but actually is competing for limited or potentially limiting resources in that space.
2. Why is the organism–environment–organism model for understanding the mechanisms of biotic interactions of such great potential importance for designing sustainable agroecosystems?
3. Describe a situation that you have seen in which allelopathy plays an important role in the development of an alternative strategy for weed management in an agroecosystem.
4. How do you differentiate between the influence of an abiotic factor on an organism and the influence of another organism on that organism?
5. What are some of the ways of avoiding competition in a crop ecosystem?

## INTERNET RESOURCES

International Allelopathy Society
www-ias.uca.es
Information on allelopathy research, publications, and meetings.

## RECOMMENDED READING

Booth, B.D., S.D. Murphy, and C.J. Swanton. 2003. *Weed Ecology in Natural and Agricultural Systems*. CABI Publishing: Oxfordshire, UK and Cambridge, MA. A textbook discussing ecological principles within the context of weed ecology and management.

Chou, C.H. and G.R. Waller (eds.) 1989. *Phytochemical Ecology: Allelochemicals, Mycotoxins and Insect Pheromones and Allomones*. Institute of Botany, Academia Sinica Monograph Series No. 9, Taipei, Taiwan. An important collection of research reports and reviews of the ecological role of natural plant chemicals in a range of interactions in ecosystems.

Combes, C. 2001. *Parasitism: The Ecology and Evolution of Intimate Interactions*. University of Chicago Press: Chicago. An exploration of the adaptations and interactions that have developed between parasites and their hosts.

Darwin, C. 1979. *The Illustrated Origin of Species*. Abridged and Introduced by R.E. Leakey. Hill and Wang: New York. A classic of scientific literature, presented in a very readable and beautifully illustrated manner; relates Darwin's hypothesis to the scientific advances of recent years, with a focus on species interactions.

Daubenmire, R.F. 1974. *Plants and Environment*. Second edition. John Wiley and Sons: New York. The classic textbook of autecology, with several chapters that emphasize the role of biotic interactions as factors in the environment.

Grace, J.B. and D. Tilman (eds.) 2003. *Perspectives on Plant Competition*. The Blackburn Press: Caldwell, New Jersey. A compilation of research reports and reviews on the concept of competition in ecosystems.

Herrera, C.M. and O. Pellmyr, (eds.) 2002. *Plant–Animal Interactions: An Evolutionary Approach*. Blackwell Science: Oxford, UK. A text covering the role of plant–animal interactions in the evolution and conservation of biodiversity.

Inderjit and A.U. Mallik, 2002. *Chemical Ecology of Plants: Allelopathy in Aquatic and Terrestrial Ecosystems*. Birkhauser Verlag: Switzerland. A current review of research methods and approaches for the study of allelopathy in plants, with sections devoted specifically to the role of allelopathy in agriculture.

Kohli, R.K., H.P. Singh, and D.R. Batish. 2001. *Allelopathy in Agroecosystems*. Food Products Press (imprint of The Howarth Press, Inc.): New York. A collection of research reports and reviews on how plant-produced chemicals can play important ecological roles in agroecosystems, with many suggestions for future research and application.

Lars, C. and J.D. Thomson (eds.) 2001. *Cognitive Ecology of Pollination: Animal Behaviour and Floral Evolution*. Cambridge University Press: Cambridge, UK. A compilation of contributions from scholars in various disciplines working on pollination biology.

Radosevich, S., J. Holt, and C. Ghersa, 1997. *Weed Ecology: Implications for Management*. Second Edition. John Wiley & Sons: New York.

Rice, E.L. 1984. *Allelopathy*. 2nd ed. Academic Press: Orlando, Florida. The key reference on the ecological significance of allelopathy in both natural and managed ecosystems.

Tow, P.G., and A. Lazenby. 2001. *Competition and Succession in Pastures*. CABI Publishing: Oxfordshire, UK. A volume describing competition and succession of plants in grasslands and grazed pastures of several continents.

van der Heijden, M.G.A. and Ian R. Sanders (eds.) 2003. *Mycorrhizal Ecology*. Ecological Studies 157. Springer-Verlag: New York. An overview of research on mycorrhizal ecology, including mycorrhizal types, multitrophic interactions, biodiversity, and ecosystem functioning.

# 12 The Environmental Complex

Previous chapters have considered the separate influences of individual environmental factors — light, temperature, precipitation, wind, soil, soil moisture, fire, and other organisms — on the crop plant. Although it is important to understand the impact that each of these factors has by itself, rarely does any factor operate alone or in a consistent manner on the organism. Moreover, all the factors that have been discussed as separate components of the environment also interact with and affect each other. Therefore, the environment in which an individual organism occurs needs to be understood as a dynamic, ever-changing composite of all the interacting environmental factors — that is, as an *environmental complex.*

When all the factors that confront a crop plant are considered together, it is possible to examine characteristics of the environment that emerge only from the interaction of these factors. These characteristics — which include complexity, heterogeneity, and dynamic change — are the main topics of this chapter. Their examination in terms of their impact on the crop plant, represents the final step in analyzing agroecosystems autecologically, and prepares us for the synecological level of analysis that begins in the following chapter.

## THE ENVIRONMENT AS A COMPLEX OF FACTORS

The environment of an organism can be defined as the sum of all external forces and factors, both biotic and abiotic, which affect the growth, structure, and reproduction of that organism. In agroecosystems, it is vital to understand which factors in this environment — due to their condition or level at the time — might be limiting an organism, and to know what levels of certain factors are necessary for optimum performance. Agroecosystem design and management are based largely on such information. The foundations of this understanding have been presented in the earlier chapters of this book. Individual factors have been explored, and many agricultural options for their management have been reviewed. Since the environment is a complex of all of these factors, it becomes just as important to understand how each factor affects or is affected by others, singly or in complex combinations that vary in time and place. It is the complex interactions of factors that make up the total environment of the organism.

### Factoring the Environment

The concept of an environmental complex is presented schematically in Figure 12.1. Although lines representing connections have not been drawn, the figure is intended to show that interactions occur between factors themselves, as well as between each factor and the crop organism. The component factors of the environment discussed in the previous chapters are all included, as well as several others. Since it is impossible to divide the entire environment neatly into components, or to include every possible factor, the factors shown in Figure 12.1 involve some simplification and overlap. Furthermore, each of the factors is not of equal importance at any particular time. For this reason, time is not listed as an independent factor, but should instead be considered as the background context within which the entire complex of factors is changing.

Because of the complexity of the environment, it is clear that its factors can combine to affect organisms in the environment in addition to doing so independently. Factors can work together simultaneously and synergistically to affect an organism, or they can make their effects felt through a cascade of changes in other factors. An example of such factor interaction is the lush weed growth on the north-facing side of the furrow illustrated in Figure 4.4. In this particular microclimatic site, lower temperatures, higher moisture, higher biological activity, and possibly higher nutrient availability were simultaneously associated with the small amount of shading that occurred, and this combination of factors effectively altered the conditions for plant growth. As another example, an allelopathic compound released from the roots of a crop can interact with shading, moisture stress, herbivory, susceptibility to disease, and other factors to either enhance or reduce the effectiveness of the phytotoxic compound in limiting weed growth in a cropping system. Because of such interactions, it is often a challenge to predict the consequences of any single modification of the agroecosystem.

One of the weaknesses of the conventional agronomic approach to managing agroecosystems is that it ignores factor interactions and environmental complexity. The needs of the crop are considered in terms of isolated, individual factors, and then each factor is managed separately to achieve maximum yield. Agroecological

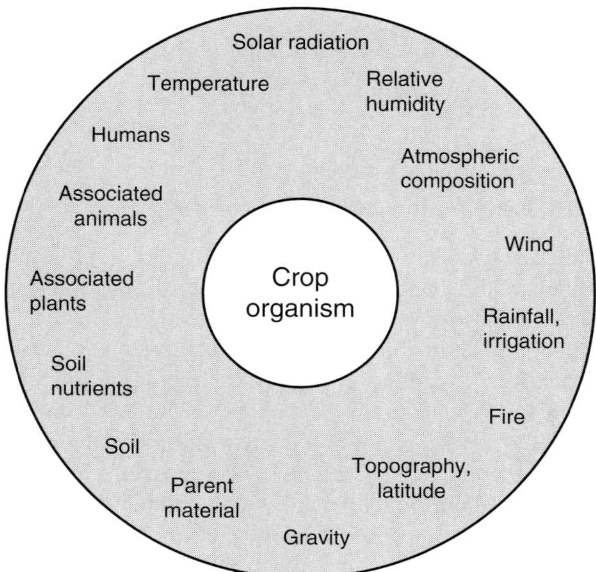


**FIGURE 12.1 Representation of the environmental complex.** The environment of an individual crop plant is made up of many interacting factors. Although the environment's level of complexity is high, most of the factors that make it up can be managed. Recognizing factor interactions and the overall complexity of the environment is the first step toward sustainable management. (Adapted from Billings, W. D. 1952. *Quarterly Review of Biology* 27: 251–265.)

management, in contrast, begins with the farm system as a whole and designs interventions according to how they will impact the whole system, not just crop yield. Interventions may be intended to modify single factors, but the potential impact on other factors is always considered as well.

### Complexity of Interaction

The way in which a complex of factors interacts to impact a plant can be illustrated by seed germination and the "safe site" concept of Harper (1977). We know from ecophysiological studies that an individual seed germinates in response to a precise set of conditions it encounters in its immediate environment (Naylor, 1984). The locality at the scale of the seed that provides these conditions has been termed the *safe site*. A safe site provides the exact requirements of an individual seed for the breaking of dormancy, and for the processes of germination to take place. In addition, there must be freedom from hazards such as diseases, predators, or toxic substances. The conditions of the safe site must endure until the seedling becomes independent of the original seed reserves. The requirements of the seed during this time change, and so the limits of what constitutes a safe site must also change.

Figure 12.2 describes some of the environmental factors that influence the germination of a seed and make up the "safe site." Factors immediately surrounding the seed are what influence the seed most directly. Factors around the outside perimeter of the diagram are factors and variables that influence the effect, degree, or presence of the direct factors.

## HETEROGENEITY OF THE ENVIRONMENT

The environment of any individual organism varies not only in space but also in time. The intensity of each factor in Figure 12.1 shows variation from place to place through time, with an average for each factor setting the parameters of the habitat within which each organism is adapted. When variation in a factor exceeds the limits of tolerance of an organism, the effects can be very damaging. Farming systems that take this variation into account are much more likely to have a positive outcome for the farmer.

### Spatial Heterogeneity

The habitat in which an organism occurs is the space characterized by particular combinations of factor intensities that vary both horizontally and vertically. Even in a field planted to a single variety of grain crop, for example, each plant will encounter slightly different conditions because of spatial variation in factors such as soil, moisture, temperature, and nutrient levels. The amount of variation in these factors will depend upon the extent to which the farmer tries to create uniformity in that field with equipment, irrigation, fertilizers, or other inputs. Regardless of these attempts, however, there will be slight variation in topography, exposure, soil cover, and so on that will create microenvironmental differences across the space of the field. Very small variations in microhabitat, in turn, can bring about shifts in crop response.

In a wet tropical lowland environment, for example, where soils are poorly drained and rainfall is high, slight topographic variation can make a big difference in soil moisture and drainage. In such an area, the lower lying areas of a field may be subject to much more waterlogging than the rest of the field, and crop plants growing there may experience arrested root development and poorer performance, as illustrated in Figure 9.3. Some farmers in the region of Tabasco, Mexico, where the photograph in Figure 9.3 was taken, plant waterlogging-tolerant crops, such as rice or local varieties of taro (*Colocasia* spp. or *Xanthosoma* spp.), in the lower lying areas of their farms as a way of making a better match between crop requirements and field conditions. Finding ways to take advantage of the spatial heterogeneity of conditions by adjusting crop types and arrangements is often more ecologically efficient than trying to enforce homogeneity or ignore heterogeneity.

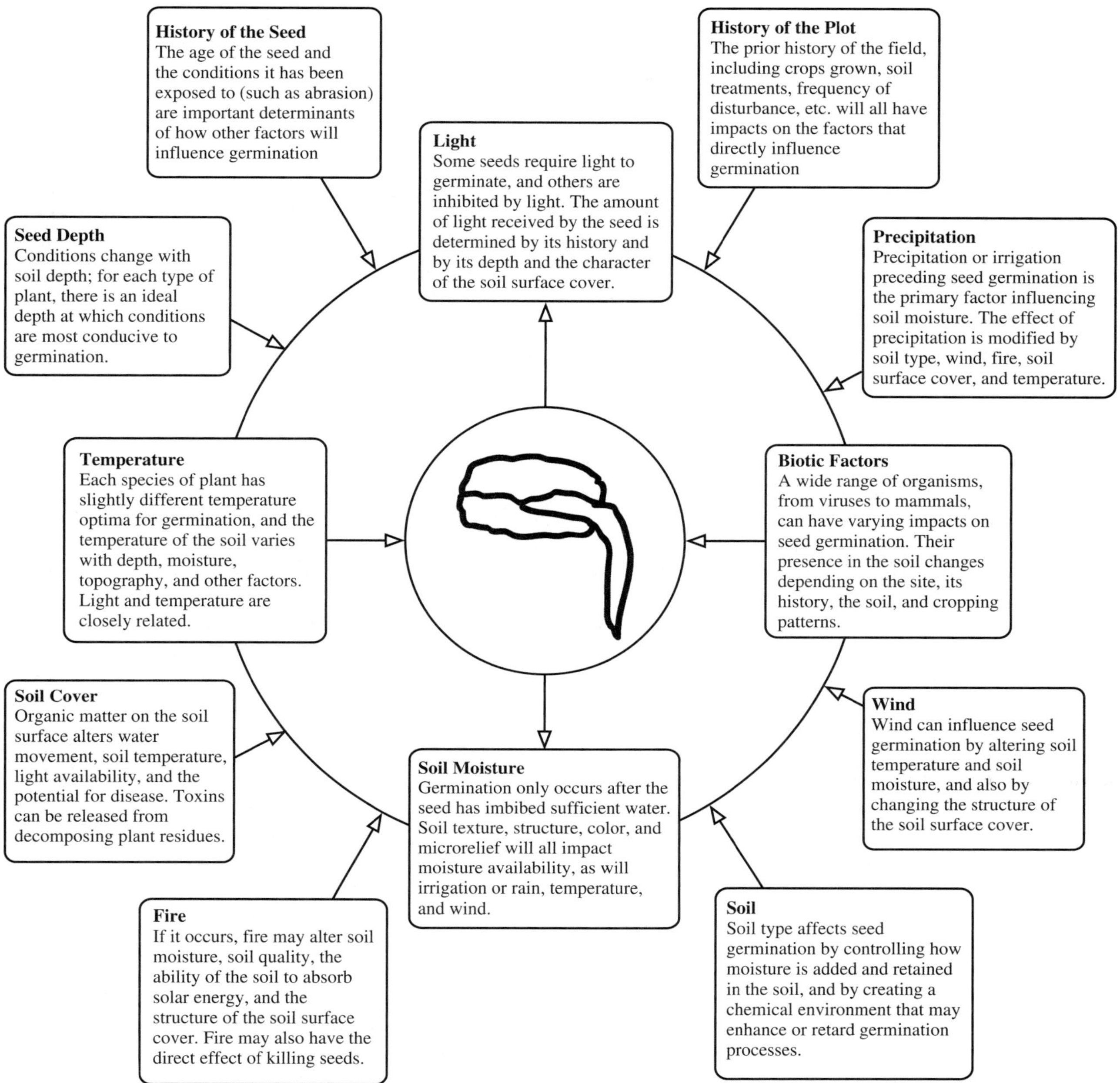

History of the Seed
The age of the seed and the conditions it has been exposed to (such as abrasion) are important determinants of how other factors will influence germination
Light
Some seeds require light to germinate, and others are inhibited by light. The amount of light received by the seed is determined by its history and by its depth and the character of the soil surface cover.
History of the Plot
The prior history of the field, including crops grown, soil treatments, frequency of disturbance, etc. will all have impacts on the factors that directly influence germination
Seed Depth
Conditions change with soil depth; for each type of plant, there is an ideal depth at which conditions are most conducive to germination.
Precipitation
Precipitation or irrigation preceding seed germination is the primary factor influencing soil moisture. The effect of precipitation is modified by soil type, wind, fire, soil surface cover, and temperature.
Temperature
Each species of plant has slightly different temperature optima for germination, and the temperature of the soil varies with depth, moisture, topography, and other factors. Light and temperature are closely related.
Biotic Factors
A wide range of organisms, from viruses to mammals, can have varying impacts on seed germination. Their presence in the soil changes depending on the site, its history, the soil, and cropping patterns.
Soil Cover
Organic matter on the soil surface alters water movement, soil temperature, light availability, and the potential for disease. Toxins can be released from decomposing plant residues.
Wind
Wind can influence seed germination by altering soil temperature and soil moisture, and also by changing the structure of the soil surface cover.
Soil Moisture
Germination only occurs after the seed has imbibed sufficient water. Soil texture, structure, color, and microrelief will all impact moisture availability, as will irrigation or rain, temperature, and wind.
Fire
If it occurs, fire may alter soil moisture, soil quality, the ability of the soil to absorb solar energy, and the structure of the soil surface cover. Fire may also have the direct effect of killing seeds.
Soil
Soil type affects seed germination by controlling how moisture is added and retained in the soil, and by creating a chemical environment that may enhance or retard germination processes.


**FIGURE 12.2 Environmental factors affecting seed germination.** Factors immediately surrounding the seed affect it most directly; factors in the outer perimeter mostly affect the intensity, level, and presence of the direct factors. The importance of each factor will vary depending on the species of the seed.

In multiple cropping systems, variations in the vertical dimension must also be taken into account because one crop or canopy layer will generally create strata of varying conditions for other crops or canopy layers. This is especially true if a new crop is being planted into an already established canopy, such as into an agroforestry or tree-dominated home garden agroecosystem. To complicate matters even more, a large mature plant member of such a system is occupying a range of microhabitats simultaneously. Which portion of the habitat and combination of microenvironmental conditions are affecting the organism the most?

Because of the difficulty involved in creating absolutely uniform conditions in farm fields, especially in resource-limited or small-scale traditional agroecosystems, farmers often plant multiple species or a variety of crop mixtures, with the idea that a diverse combination of crops with a range of adaptations will do better in a variable environment (Smith and Francis, 1986). It is a real challenge in experimental agronomic studies to adequately take such variability into account. High standard deviations do not necessarily mean that something was wrong with the research methodology. It may just mean that the sample area was extremely variable!

### Dynamic Change

Since the combination of factors in any environment is constantly changing through time, a farmer must also take into account temporal heterogeneity. Changes take place hourly, daily, seasonally, yearly, and even as part of longer-term climatic shifts. Some of this change is cumulative and some of it is cyclic. For any particular factor, there is a need to be aware of how rapidly its intensity can change over time, and how the changes can affect a particular organism, based on its length of exposure and its limits of tolerance. At the same time, each organism, as it goes through its life cycle, will undergo shifts both in the way it responds to different factor intensities and in its tolerance for those intensities.

A crop plant, for example, experiences a continually changing environment as it progresses through its life cycle. If a factor or combination of factors reach some critical level at the same time the plant reaches some particularly sensitive stage in its life cycle, suppression of further development can occur and result in crop failure. Germination, initial seedling growth, flowering, and fruiting are the stages during which extreme or unusual variation in environmental factors is most likely to impact crop performance. As was seen in Figure 9.4, for example, a period of waterlogging during the growth of cowpeas had a negative effect on yield, but the nature and extent of this effect depended on when the waterlogging occurred.

Because of dynamic change, interventions in the field often need to be carefully timed. For example, a farmer wanting to use a propane-fired burner (described in Chapter 10) to kill weed seedlings is limited to a small window of time in the early stages of development of the crop. If the crop is too small and delicate, flaming can kill the crop seedlings along with the weed seedlings. If the crop is too tall, it might be difficult to avoid damaging the plants with the flaming apparatus itself. The effective window for using flame weeders might be as short as 4 or 5 d in delicate crops such as carrots or onions, both of which have little ability to deal with interference from weeds on their own.

## INTERACTION OF ENVIRONMENTAL FACTORS

Each of the many factors that make up the environmental complex has the potential to interact with other factors and thereby modify, accentuate, or mitigate their affects on organisms. The interaction of factors can have both positive and negative consequences in agroecosystems.

### Compensating Factors

When one factor overcomes or eliminates the impact of another, then it is referred to as a *compensating factor.* When a crop is growing under conditions that would otherwise be limiting for its successful growth or development, one or more factors may be compensating for the limiting factor.

The effect of a compensating factor is commonly seen in fertilization trials, when a particular soil nutrient (e.g., nitrogen) is limiting as determined by the plant response. Reduced growth and lower yields are signs of the deficiency. But rather than simply adding more of the deficient nutrient, it is sometimes possible to alter some other factor of the environment that renders more of the "limited" nutrient available to plants. In the case of nitrogen deficiency, it may be that poor soil drainage is restricting nitrogen uptake by roots, so that once soil drainage is improved, the lack of nitrogen uptake is compensated for.

Another case of compensation for a limiting factor occurs when a farmer counters the negative impact of a leaf-eating herbivore by stimulating more luxurious or rapid growth of the affected crop through an intervention such as adding compost to the soil or applying a foliar fertilizer. The added biomass can allow the crop to carry the herbivore load and still produce a successful harvest. The added plant growth compensates for herbivory.

In coastal regions where fog is common during the dry summer season (e.g., the Mediterranean maritime region of coastal California), the fog can compensate for the lack of rainfall. This occurs through the reduction in transpirational water loss, and the lower evaporative stress due to less direct sunlight and lower temperatures. The leafy vegetable crops common in the lower Salinas and Pajaro Valleys of California could probably not be grown profitably during the middle of the summer without such compensation, because these crops are subject to considerable water loss through transpiration on hot days.

### 12.3.2 Multiplicity of Factors

When several factors are closely related, it may be particularly difficult to separate the effect of one factor from another. The factors can act as a functional unit, either simultaneously or in a chainlike manner. One factor influences or accentuates another, which then affects a third; but in terms of crop response, where one factor stops and another takes over is impossible to determine. The factors of temperature, light, and soil moisture often function in such a closely interrelated manner. For a corn crop in an open field, for example, increasing light levels during the morning increase temperature, and the higher temperature increases evaporation of water from the soil while transpiration also increases. Thus the intensity of each factor varies simultaneously with every change in the intensity of solar radiation, and the relative effect of each factor on the crop is practically inseparable from the multiplicity of effects they have together.

### 12.3.3 Factor Predisposition

A particular environmental factor may cause a crop response that renders the crop more susceptible to damage by another factor. In such cases, the first factor is said to predispose the plant to the effects of the second factor. Low light levels caused by shading, for example, can predispose a plant to fungal attack. The lower light levels usually mean higher relative humidity for the plant and cause it to develop thinner, larger leaves that then may be more susceptible to attack by a pathogenic fungus that occurs more commonly when excess moisture is present in the environment. Similarly, research has shown that some crop plants are more susceptible to herbivore damage when they have been given large amounts of nitrogenous fertilizer. The plant tissue is predisposed to the herbivory due to its higher nitrogen content — apparently the nitrogen serves as an attractant for the pest (Scriber, 1984).

## 12.4 MANAGING COMPLEXITY

Sustainable agroecosystem management will require an understanding not only of how individual factors affect crop organisms but also of how all factors interact to form the environmental complex. Part of this understanding comes from knowing how factors interact with, compensate for, enhance, and even counteract with each other. Another part comes from knowing the extent of variability present on the farm, from field to field and within each field. Conditions vary from one season to another as well as from one year to the next. From climate to soils, from abiotic to biotic factors, and from plants to animals, factors interact and vary in dynamic and ever-changing patterns. Perhaps an important component of sustainability is knowing not only the extent and form of factor interaction, but also the range of variability in interactions that can occur over time. Adapting the agroecosystem as much as possible to take advantage of complexity and variability where appropriate, and to compensate for both when not, is in many ways the challenge that will be addressed in the following chapters.

## FOOD FOR THOUGHT

1. What factors may have impacted seed before a farmer buys it for planting? How may these influences affect the performance of the seed once it is planted?
2. What are some ways that a farmer can manage an agroecosystem in a highly variable environment other than trying to control or homogenize the conditions that create the heterogeneity?
3. What are some of the disadvantages for a farmer who chooses to deal with or adapt to (rather than overcome) spatial and temporal heterogeneity in the agroecosystem?
4. What are some ways in which a farmer can successfully compensate for a limiting factor by altering or managing one or several other factors, and thus contribute to the sustainability of a farming system?

## INTERNET RESOURCES

Plant ecophysiology research group at the University of Wales
www.bangor.ac.uk/safs-new/research/plecophys.php
A good example of a research group doing work on the factors that impact plants in the environment, a field known as ecophysiology.

## RECOMMENDED READING


Daubenmire, R.F. 1974. *Plants and Environment.* 3rd ed. John Wiley & Sons: New York. The book that established the foundation for an agroecological approach to plant–environment relationships.

Forman, R.T.T. and M. Gordon. 1986. *Landscape Ecology.* John Wiley & Sons: New York. Essential reading in understanding the relationships between plant distribution and the temporal and spatial complexity of the physical landscape.

Harper, J.L. 1977. *Population Biology of Plants.* Academic Press: London. The key reference for understanding the foundations for modern plant population biology, with many references to agricultural systems.

Larcher, W. 2003. *Physiological Plant Ecology.* 4th ed.. Springer: New York. A very complete text of ecophysiology, covering plant adaptation to the factors of the environmental complex.

Schmidt-Nielsen, K. 1997. *Animal Physiology: Adaptations and Environment.* 5th ed. Cambridge University Press: New York. An important review of the physiological ecology of animals in the environment.

# Section III

## System-Level Interactions

With a grounding in the autecological knowledge developed in Section II, we can now expand our perspective to the *synecological* level — the study of how groups of organisms interact in the cropping environment. This whole-system perspective stresses the need for understanding the emergent qualities of populations, communities, and ecosystems, and how these qualities are put to use in designing and managing sustainable agroecosystems.

Chapters 13 and 14 begin at the population level, exploring the population ecology of mixtures of species in the crop environment and the management of genetic resources. Chapter 15 examines species interactions at the community level, explaining the benefits of complexity and the role of cooperation and mutualisms in sustainable agriculture. Chapters 16 through 18 cover a range of important ecological concepts — including diversity, stability, disturbance, succession, and energy flow — that function at the ecosystem level, showing how these emergent qualities of whole systems are key aspects of agroecosystem design and management. To conclude Section III, Chapter 19 explores the role of livestock animals in agroecosystems, with a focus on creating integrated food production systems.

# 13 The Population Ecology of Agroecosystems

In agronomy and conventional agriculture, the crop plant or animal population is the center of attention. A farmer attempts to maximize the performance of this population by managing the various factors of the environmental complex. When sustainability of the entire agroecosystem becomes the primary concern, however, this narrow focus on the needs of one genetically homogenous population becomes wholly inadequate. The agroecosystem must be viewed as a collection of interacting populations of many kinds of organisms, including crop and noncrop species, plants and animals, and microorganisms.

Consideration of the agroecosystem as a collection of interacting populations involves several levels of study. First, we require the conceptual tools necessary to understand and compare how each population goes about surviving and reproducing itself in the environment of the agroecosystem. These tools and their application are the subject of this chapter. Second, we need to look at the genetic basis of crop populations and how the manipulation of this genetic potential by humans has affected crop plants' adaptability and range of tolerance. We will turn our attention to this topic in Chapter 14. Finally, we need to consider the community and ecosystem-level processes of interacting populations, which will be explored in Chapter 15, Chapter 16, and Chapter 17.

## PRINCIPLES OF POPULATION ECOLOGY AND PLANT DEMOGRAPHY

The single species population has long been the main subject of agronomic research. Crop breeders adjust the genetic potential of crop populations, and production specialists develop management technologies that get the most out of that potential. This has led to a type of crop ecologist skilled at adjusting one factor of the system at a time or developing technologies that solve single problems such as controlling a particular pest with a specific pesticide. But since the agroecosystem is made up of complex interactions between many populations of organisms, an agroecological approach requires a broader analysis. Studies of interactions between populations at the same trophic level must be carried out at the same time studies are going on that focus on the interactions between populations at different trophic levels. Integrated pest management, for example, requires a simultaneous analysis of the population ecology of each member of the specific crop/pest/natural enemy complex, as well as other populations of organisms with which the entire complex interacts. Ultimately, we must consider this complex of populations as the entire crop community, a level of ecological analysis we will turn to in Chapter 15. But first, several basic principles of population ecology that help us understand the dynamics of each population will be discussed.

### POPULATION GROWTH

Ecologists view population growth as the net result of birth rates, death rates, and the movement of individuals into and out of a particular population. Population growth is thus described by the formula

$$r = (N + I) - (M + E)$$

where $r$ is the intrinsic rate of population increase in a population over time, taking into account natality ($N$), immigration ($I$), mortality ($M$), and emigration ($E$). Any population changes over time are described by

$$\frac{dP}{dt} = rP$$

where $P$ is the population under study over a specific time ($t$) period. If resources do not become limiting, and negative interactions between members of the population do not reach some critical level as the population increases, a population would increase exponentially. Since this very simple equation does not take into account the effect of abiotic and biotic factors of the environment on a population, nor the limits to growth that an environment can impose on a population, the following equation was developed:

$$\frac{dP}{dt} = rP(\frac{K-P}{K}) = rP(1 - \frac{P}{K})$$

The rate of growth of the population is unaffected by interference when $P$ approaches 0, and slows when $P$ approaches $K$ (the population size at the carrying capacity of the environment). This equation describes a logistic, sigmoidal, or S-shaped growth curve, as shown in Figure 13.1. The leveling-off of the curve indicates that problems are eventually encountered in allocating

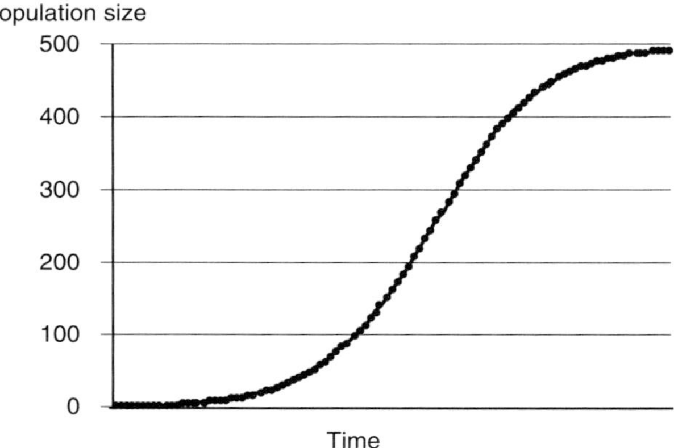


**FIGURE 13.1 The population growth curve.** This graph shows the theoretical rate of increase of a population over time. In this case, carrying capacity ($K$) is reached at a population size of 500.

resources to an expanding population. This curve could apply to a weed species in a crop field or a particular pest organism on the crop. Population increase is slow at first, begins to accelerate until it reaches a maximum rate of increase, and then slows as density increases. When the carrying capacity of the environment is reached, the curve levels off, and in many cases, will begin to drop if impact on the environment has created conditions that affect the entire population.

In natural ecosystems, complex feedback mechanisms can slow population increase before carrying capacity is reached, buffering the species against population crashes. Sometimes these mechanisms are directly determined by the number of individuals already present — in which case they are *density-dependent*. An example is competition for a limited resource. In other cases, the mechanism is due more to some external factor of the environment such as a frost or flood, and is therefore *density-independent*. In cropping systems, humans have devised different interventions and technologies that allow a crop population to increase in number or develop beyond the normal carrying capacity of that environment. Usually these interventions are associated with intensive habitat modification or inputs, and can include the control or elimination of other species (both plant and animal) and the use of fertilizers and irrigation.

## Colonization of New Areas

The study of population growth is concerned mainly with the potential of a population to increase in size over time. It is incomplete without attention being paid to the potential of a population to increase in area — that is, to colonize new habitats. The process of colonizing new areas is especially important to the agroecologist, who is concerned with how organisms besides crop plants — both beneficial and not — invade a field and establish populations there.

### Stages of Colonization

The manner in which a weed or animal pest colonizes a field is related to its life cycle. The initial invasion is accomplished as part of the species' reproduction and dispersal process; the establishment of the population is dependent on the requirements of its seeds and seedlings or eggs and juveniles; whether the population remains in the area over time is a function of how it grows, matures, and reproduces. Each of the stages in a species' life history offers specific opportunities for intervention on the part of the farmer — either to encourage the colonization of a desired species or to restrict that of an unwanted one. Below, the colonization process is divided into four stages, based on the life stages of the colonizing organisms: dispersal, establishment, growth, and reproduction. For the sake of clarity, these stages are discussed mostly in terms of plants.

#### *Dispersal*

The dispersion of organisms is an important phenomenon in natural ecosystems, and has some interesting applications to agroecology. Dispersal allows progeny to "escape" the vicinity of the mother organism, lessening the potential for intraspecific interference from an overpopulation of ecologically very similar siblings. It also allows a species to reach new habitats.

In agriculture, dispersal is important because of the continual disturbance of fields. This disturbance — whether wholesale in the case of conventional tillage or piecemeal in the case of perennial/annual polycultures such as tropical home gardens — continually creates new habitats available for colonization. Although many organisms maintain resident populations in a field despite their disturbance and manipulation, many noncrop organisms — including beneficial and detrimental weeds, insects, other animals, diseases, and microorganisms — all arrive in the field through dispersal. In this context, ecological barriers to dispersal take on important significance. Barriers may be as simple as a weedy border around a field, or a border made up of a different crop plant. In general, a more in-depth understanding of the mechanisms of the dispersal of noncrop organisms, and how they are affected by barriers, can become important in the design and management of the agroecosystem.

How plants and animals get from one place to another during the dispersal stages of their life cycles depends on the mechanisms they each have for dispersing themselves. These mechanisms are quite variable, but most often involve wind, animals, water, or gravity. Research on the long-distance dispersal of plants and animals has given us much insight into what these mechanisms are and how they work.

One of the best works on dispersal is Sherwin Carlquist's (1965) *Island Life*. He reviews the natural history of islands of the world, discussing how animals and plants reach islands that either have had a physical connection to an adjacent mainland colonizing source or that have never had such a link. The work of Van der Pijl (1972) on the *Principles of Dispersal in Higher Plants* goes into great detail on the incredible diversity of mechanisms that aid seeds in moving from one place to another. These mechanisms can move an organism only a short distance, or great distances across amazing barriers of ocean or desert. They can also get a weed seed to a new field.

An important aspect of dispersal mechanisms is how many of them seem to provide a selective advantage for "getting away" from the source of reproduction. This is illustrated by field studies done on the distribution of seedlings around "mother trees" in the forests of Costa Rica. As shown in Figure 13.2, most of the newly germinated seeds and very young seedlings were concentrated close to the tree, but the older saplings (with potential for becoming adult, reproductive individuals) were found at a greater distance. Some intraspecific mechanism (e.g., competition, and allelopathy) seems to eliminate seedlings from near the tree, and does not function at a greater distance. It is interesting to consider why there is advantage in establishing at some distance from the parent, especially in relation to resource availability, potential competition, and susceptibility to predation or disease.

Plant seeds are incorporated into the soil soon after they fall onto the soil surface, with the largest numbers found in the upper layers of soil. The population of each species of seed combines with others to form the *seed bank*. In cropping systems, the analysis of the weed seed bank can tell us a great deal about the prior history of management of a site and the potential problems that weeds may pose; this information can be important for designing appropriate management.

Since most crop organisms are dependent on humans for dispersal, their adaptations for dispersal have become irrelevant for the most part. Indeed, most crop species have lost the dispersal mechanisms they had as wild species. Their seeds have become too large, the seeds have lost appendages that once facilitated dispersal, or their inflorescences no longer scatter seed. The loss of dispersal adaptations is seen particularly in annual crops, whose seed or grain is the portion of the crop that is harvested.

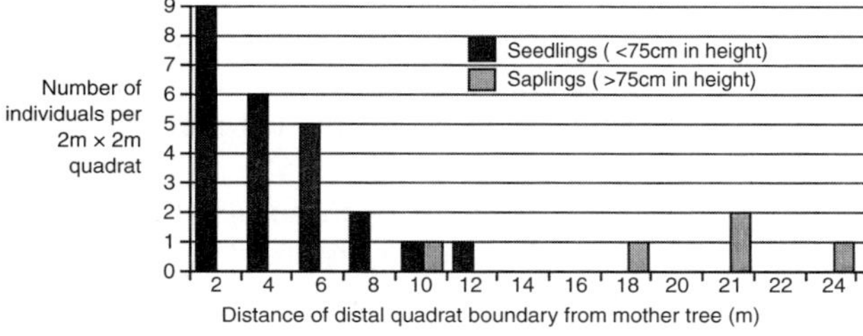


**FIGURE 13.2 Distribution of seedlings and saplings of *Gavilan schizolobium* on a westerly strip transect away from the mother tree, Rincon de Osa, Costa Rica.** Data from Ewert and Gliessman, 1972.

### Establishment

There really is no bare area on the earth that propagules of plants and animals cannot get to. The incredible diversity of dispersal mechanisms mentioned above makes sure of that. But once a propagule arrives at a new location, it most certainly can have problems getting established. Restricting our attention to plants, a dispersing seed cannot determine where it will land; so it is the condition of the site that determines if the propagule can establish. Seeds fall into a very heterogeneous environment, and only a fraction of the sites encountered will meet the needs of the seed. Only those microsites that fulfil the needs of the seed — the "safe sites" — can support germination and establishment. The greater the number of a species' seeds that land in safe sites, the greater the chance of that species establishing a viable population in the new habitat.

The seedling stage is generally known to be the most sensitive period in the life cycle of the plant, and is therefore a critical stage in the establishment of a new population. This is true for crop species, weeds, and plants in natural ecosystems. A dormant seed can tolerate very difficult environmental conditions, but once it germinates, the newly emerged seedling must grow or die. Any one of the many extremes of environmental conditions the seedling might face can eliminate it, including drought, frost, herbivory, and cultivation. Human intervention can help ensure the successful and uniform establishment of crop seedlings, but the variability of the environmental complex still makes this the most sensitive phase for most crop plant populations. Early juvenile stages of most animals show the same sensitivity to environmental stress.

### Growth and Maturation

Once a seedling has successfully established, its main "goal" is continued growth. The environment in which seedlings are located, as well as the genetic potential the seeds contain, combine to determine just how quickly they will grow. In natural ecosystems, environmental factors such as drought or competition for light generally limit the growth process at some phase of the plants' development. If these factors become too extreme, individuals in the population will die.

Plants generally grow fastest, as measured by net biomass accumulated over time, in the early stages of growth. Their rate of growth slows as maturation begins — more energy is allocated to maintenance and the production of reproductive organs than to the production of new plant tissue. Growth may also slow if the resources available for each member of the population become limiting.

The time period from germination to maturity can range from a matter of days for some annuals to several decades for some perennials. A species that matures quickly will colonize a new area differently than a species that matures slowly, and each will present different challenges for management.

#### *Reproduction*

Once the original colonizing individuals have reached maturity, they can reproduce. The extent to which they are successful determines whether the new population will remain in the area, how it will grow, and how it will affect populations of other species over the long run. Reproduction can take place asexually through vegetative reproduction or sexually through the production of seeds. Some species depend on the rapid early growth of the colonizing seed supply and strong early control of the environment to inhibit later colonizers, followed by abundant reproduction. Other species may allocate more resources to developing fewer but larger and more dominant individuals in the population, sacrificing the production of new seeds in the process but ensuring the success of the individuals that reach maturity.

### Factors Affecting Success of Colonization

At any stage in the colonizing process described above, some event or condition can occur that may eliminate a certain percentage of the population. For an invading plant species, part of this elimination occurs when only a fraction of the seeds find an appropriate safe site for germination. Another large percentage of the population is lost shortly after germination, especially if weather conditions are not ideal. At any time during the development of the juvenile plants, more loss can occur. The final outcome is often a very reduced number of mature adults that begin to reproduce. The attrition can be even more pronounced in the presence of human management, which can, in some cases, threaten the survival of a whole population or species.

For some species, especially long-lived perennials, attrition of individuals at early stages of colonization may be so complete that environmental conditions may all come together in a sequence that permits survival of seedlings only one or a few years out of many. Several oak species (*Quercus* spp.) in California, for example, show clusters of equal-aged individuals in populations that are separated by 40–200 yr, indicating that opportunities for establishment of new population clusters occur very infrequently.

## Life-history Strategies

Each species that is successful in a particular environment has a unique set of adaptations that allow it to maintain a population in that environment over time. These adaptations can be thought of as comprising a "strategy" for organizing the life cycle to insure reproduction and the continuation of a viable population. Across species, life history strategies can be classified into general types.

Two important ways of classifying life history strategies are discussed below. They help provide an understanding of how the populations of specific organisms are able to grow in number or colonize new areas. They can also help explain the ecological role of each species in the agroecosystem, aiding greatly in the management of both crop and noncrop species.

### r-and K-Selection Theory

Plants and animals have a limited amount of energy to "spend" on maintenance, growth, and reproduction. Allocation of more energy to reproduction reduces the amount available for growth, and vice-versa. Ecologists have used observed differences in the allocation of energy to growth or to reproduction to develop a classification system that defines two basic types of life-history strategies at opposite ends of a continuum: *r*-selection and *K*-selection. This system is known as *r*- and *K*-selection theory (MacArthur, 1962; Pianka, 1970, 1978).

At one extreme, we find species that live in harsh or variable environments in which mortality is mostly determined by limiting environmental factors rather than the density of the population, and where natural selection favors genotypes with a high intrinsic growth value. Members of the populations of these species allocate more energy to reproduction and less to growth and maintenance once they are established. Members of such species are called *r-strategists* because environmental factors keep the growth of such populations on the most rapidly increasing point of the logistic curve (Figure 13.1). Their population sizes are limited more by physical factors than by biotic factors.

At the other extreme, we find species that live in stable or predictable environments where mortality is more a function of density-dependent factors such as interference with individuals of other populations, and where natural selection favors genotypes with the ability to avoid or tolerate interference. These organisms allocate more resources to vegetative or nonreproductive activities. Members of such species are called *K-strategists* because they maintain the densest populations when the population size is close to the carrying capacity (*K*) of the environment. Their population sizes are limited more by biotic factors than by physical factors.

In general, *r*-strategists are opportunists; they have the ability to colonize temporary or disturbed habitats where interference is minimal, can rapidly take advantage of resources when they are available, are usually short-lived, allocate a large proportion of their biomass to reproduction, and occupy open habitats or early successional systems. In the animal kingdom, *r*-strategists require minimal parental care while young; in the plant kingdom, *r*-strategists usually produce large numbers of easily dispersible seeds. In contrast, *K*-strategists are tolerators; they are usually long-lived, have a prolonged vegetative or growth stage, allocate relatively small amounts of total biomass to reproduction, and occur in natural ecosystems in the later stages of succession. Animal *K*-strategists care for their young, whereas plant *K*-strategists produce relatively few large seeds that contain significant reserves of stored food.

The categories of *r*-selection and *K*-selection, however, are not clearly delineated. Most organisms are not purely *r*-selected or *K*-selected, but display a life history strategy making use of traits from both strategies. Therefore, *r*- and *K*-selection theory has to be applied with caution in the understanding of population dynamics and development.

Even so, the concepts of *r*- and *K*-selection can be very useful in understanding population dynamics in agroecosystems. Most invasive and weedy organisms, especially weeds, pathogens, and pest insects, are *r*-selected. They are opportunistic, easily dispersed, reproductively active organisms that can very rapidly find, occupy, and dominate habitats in the disturbed agricultural landscape. Interestingly, most of the crop plants that we depend upon today in the world for the production of most of our basic food materials can also be classified as *r*-selected species. The largest proportion of their biomass is in the reproductive portion of the plant. This is especially true of all of the annual grains we consume. It is thought that these crop plants were derived mainly from species that evolved in open, disturbed habitats; their *r*-selected ability to grow rapidly is what made them good candidates for domestication.

One reason that *r*-selected weeds are a problem in cropping systems is that the crop plants themselves are also *r*-selected, and the open, disturbed conditions under which the crop plants thrive are the same as those under

## DEVELOPING A PERENNIAL GRAIN CROP

The grain crops that form the cornerstone of the American diet — wheat, corn, and rice — can all be considered *r*-strategists. They are annuals that grow rapidly in the disturbed environment of a cultivated field and use a large portion of their energy producing reproductive structures. In the course of domestication, humans have, if anything, enhanced the *r*-selected nature of these plants, creating varieties of grains that are highly productive but dependent on extensive external inputs and human intervention.

Researchers at the Land Institute are concerned about the erosion and degradation of the soil that goes along with the frequent tilling and application of pesticides and inorganic fertilizers necessary in annual grain production. They are working on an interesting solution to the problem: breeding a perennial grain crop (Piper 1994; Cox et al. 2002; Glover 2005; DeHaan et al. 2005).

Unfortunately, developing a perennial grain productive enough for agriculture is not easy. Perennial plant species that produce edible carbohydrate-rich seeds do exist in nature; the problem is that they are *K*-selected and devote a relatively small proportion of their energy to seed production. For example, the natural perennial cousins of our annual grain crops — wild prairie grasses — have large rhizomes in which the plant stores substantial food reserves. The rhizomes help the plant survive harsh winters and occasional droughts, and enable it to reproduce asexually as well. For these plants, reproduction by seed is not a high priority, energetically speaking.

The Land Institute researchers are attempting to breed new grain crops that will maintain the rhizome and at the same time produce enough seed to make harvest worthwhile. There would be many ecological benefits from growing such plants extensively. In particular, they would help prevent soil erosion, a critical problem for annual grain crops. The soil would not have to be tilled each season, and the plants' larger root systems would effectively hold soil in place. Perennial plants would also be hardier, reducing the need for fertilizer and pesticide inputs each year.

The researchers originally surveyed more than 4000 perennial species for their potential to produce a grain crop, and have focused their research on the most promising candidates for domestication. These include eastern gamagrass (*Tripsacum dactyloides*), intermediate wheatgrass, wild rye, lymegrass, Indian ricegrass, and the non-grasses maximilian sunflower (*Helianthus maximilianii*) and Illinois bundlefower (*Desmanthus illinoensis*). Another avenue of research involves hybridizing annual grain crops such as rice, oats, sorghum, rye, wheat, and maize with closely or distantly related perennial relatives.

Even if the breeding program is successful, widespread use of the new crops would depend on changes in the ways farmers and consumers think. Consumers will need to be open about the possibility of cream of eastern gamagrass on the breakfast table, and grain farms will have to be redesigned to exploit the advantages of permanent cover.

Perennial grains, if they can be developed, would likely be grown in relatively diverse agroecosystems very different from fields of monocropped annuals. These systems would more closely resemble natural prairies. Such "natural systems agriculture," generally applied, is another topic being explored by Land Institute researchers (e.g., Jackson & Jackson 1999; Piper 1999; Jackson 2002).

which the weeds grow best. Annual cropping systems, or perennial cropping systems with frequent disturbance, are in a sense selecting for the very problems farmers are constantly using an array of technologies to stop or eliminate. From this perspective, it can be seen that *K*-strategists might be able to play important roles in agroecosystems as crop species. Perennial crop systems place a premium on the health and development of the vegetative part of the plant, even in cases where it is the fruit that is harvested. Lesser disturbance is created in the process of farming, and fewer opportunities are created for weedy *r*-strategists.

An interesting proposal is to combine the strengths and advantages of both strategies in a single crop population. The fast-growing, opportunistic, high reproductive effort of the *r*-strategist might be combined with the resistance, biomass accumulation, and stress tolerance of the *K*-strategist. An example of such an effort — the attempt to develop a perennial grain crop — is discussed in *Developing a Perennial Grain Crop*. In later chapters, when the ecosystem concepts of diversity and succession are presented, additional attention will be given to the use of *K*-strategists in agroecosystems.

In the paragraphs above, *r*- and *K*-selection theory has been discussed in the context of crop plants and their herbivorous pests, but it also has relevance for livestock animals. As a general rule, what have proved most valuable to humans in livestock are *K*-selected traits, and this is reflected both in the animal species humans chose to domesticate and the traits selected for in the domestication process. The *K*-selected trait of large size was of obvious value to humans seeking both a food supply and animals that could do work and transport goods. In the case of cattle, goats, and sheep, the *K*-selected trait of milk production (a clear example of parental investment) was also valuable. Once species such as horses, oxen, cattle, sheep, goats, and hogs were domesticated, their *K*-selected characters became the basis for further human-directed selection in the "*K*" direction (large size and more milk production, for example). This was not so much the case with avian livestock, such as chickens, where higher offspring numbers, more rapid growth rates, and greater mobility indicate some *r*-selected traits. However, even in poultry, human breeding has often introduced characteristics of *K*-selection, such as greatly increased body size. In nature, this might be considered to be a negative adaptive trait, but in an agroecosystem context, humans can step in to compensate for such disadvantages.

## Stress/Disturbance-Intensity Theory

As an alternative to *r*- and *K*-selection theory, ecologists have developed a life-history classification system for plants with three categories instead of two. It is based on the premise that there are two basic factors — stress and disturbance — that limit the amount of biomass a plant can produce in a given environment. Stress occurs through external conditions that limit production such as shading, drought, nutrient deficiency, or low temperature. Disturbance occurs when there is partial or total disruption of plant biomass due to natural events such as grazing or fire or to human activities such as mowing or tillage. When habitats are described using both dimensions — as high stress or low stress and low disturbance or high disturbance — four types of habitats are defined. Each of these habitats is then associated with a particular life-history strategy, as shown in Table 13.1. This scheme may have more direct application to agricultural environments than *r*- and *K*-selection theory, and may be of particular use in weed management.

**TABLE 13.1**
**Life History Strategies Based on Stress and Disturbance Levels in the Environment**

| | High Stress | Low Stress |
|---|---|---|
| **High disturbance** | [Plant mortality] | Ruderals (R) |
| **Low disturbance** | Stress tolerators (S) | Competitors (C) |

*Source:* Grime, J. P. 1977. *American Naturalist.* 111: 1169–1194.

Since an environment characterized by both high stress and high disturbance cannot support much plant growth, there are three useful classifications in this system:

- *Ruderals* (R), which are adapted to conditions of high disturbance and low stress
- *Stress tolerators* (S), which live in high-stress, low-disturbance environments
- *Competitors* (C), which live under conditions of low stress and low disturbance and have good competitive abilities

Most annual cropping systems present conditions of high disturbance because of frequent cultivation and harvest, but have relatively low stress since conditions have been optimized through agricultural inputs and crop system design. Ruderals are highly favored under these conditions, where the characteristics of short life span, high seed production, and ability to colonize open environments has such advantage. Most plants that fall into the ruderal category — annual weeds for example — can also be categorized as *r*-selected.

Degraded agroecosystems such as eroded hillsides in wet environments, or heavily cropped grain systems in dry-farmed areas that suffer periodic drought stress and wind erosion, favor the growth of stress tolerators. Noncrop species that are tolerant of these conditions may become the dominant feature of the landscape; examples are *Imperata* grasses in the wet tropics of Southeast Asia and cheat grass (*Bromus tectorum*) in the Great Basin rangelands of the western U.S. Since stress tolerators have been selected to endure the environmental stress characteristic of highly degraded and altered environments, they can establish and maintain dominance even though the environment in which they occur is relatively unproductive.

Many natural ecosystems, as well as perennial cropping systems, support the competitor category of plants. These plants have developed characteristics that maximize the capture of resources under relatively undisturbed conditions, but are not tolerant of heavy biomass removal. Excessive disturbance through harvest would open the system up to the invasion of weedy ruderals, whereas increased intensity of stress, such as that which would accompany over-extraction of soil nutrients or water, would open the system to invasion from stress-tolerating organisms. When a forest system is clear-cut and the soil ecosystem is left intact, recolonization by stress-tolerant early successional species is an initial problem, but tree species can usually reestablish and eventually recolonize the site and exclude them. But if fire periodically removes vegetative cover following tree harvest, the intensity of disturbance opens the system to invasion and dominance by shorter-lived and aggressive ruderals that greatly retard the recovery of the forest species.

Both *r*- and *K*-selection theory and stress/disturbance-intensity theory provide opportunities for combining our understanding of the environment with our understanding of the population dynamics of the organisms we are dealing with. By focusing this knowledge on both crop and noncrop species, we can plan our agricultural activities accordingly.

## ECOLOGICAL NICHE

The concept of life history strategy helps us understand how a population maintains a role and place in an ecosystem over time. An additional conceptual framework is required for understanding what that role and place are. This is the concept of *ecological niche*.

An organism's ecological niche is defined as its place and function in the environment. Niche comprises the organism's physical location in the environment, its trophic role, its limits and tolerances for environmental conditions, and its relationship to other organisms. The concept of ecological niche establishes an important foundation for determining the potential impact that a population can have on an environment and the other organisms that are there. It can be of great value in managing the complex interactions between populations in an agroecosystem.

### CONCEPTUALIZATIONS OF NICHE

The niche concept was first introduced in the pioneering work of Grinnell (1924, 1928) and Elton (1927) as the place of an animal in the environment. By "place," they meant a species' maximum possible distribution, controlled only by its structural limits and instincts. Today, this aspect of niche is part of what is termed *potential niche*. Potential niche is contrasted with *realized niche*, the actual area that a species is able to occupy, as determined by its interactions with other organisms in the environment (that is, by the impacts of interference, positive and negative).

Both potential niche and realized niche are built on a conceptualization of niche as having two distinct facets — one is the habitat in which the organism occurs, and the other is what the organism does in that habitat. The latter facet can be understood as the organism's "profession" — the way it "makes a living" in the habitat it lives in. An animal's profession, for example, can be flower feeder, leaf feeder, or insect feeder. A microorganism can be a decomposer or a parasite. Many levels of interaction are involved in defining this aspect of a specie's niche.

An important contribution to the niche concept was made by Gause (1934), who developed a theory now known as Gause's Law: two organisms cannot occupy the same ecological niche at the same time. If the niches of two organisms in the same habitat are too similar, and there are limited resources, one organism eventually excludes the other through "competitive exclusion." Competitive exclusion, however, is not always the cause of two populations with similar niches not occurring together. Other mechanisms may be at work.

The idea of the niche being an organism's profession is often not adequate. To develop a more complex way of understanding niche, ecologists have focused on defining the separate dimensions that make it up. A set of factor–response curves (discussed in Chapter 3) is determined for a particular organism. These are then layered over each other to form a matrix of factor responses. In a simple two-factor matrix, the area delineated by the overlapping regions of tolerance can be envisioned as the two-dimensional area of resource space occupied by the organism. With the addition of more factor–response curves, this spaces takes on multidimensional form. This procedure is the basis for a conceptualization of niche as the multidimensional hypervolume that an organism can potentially occupy (Hutchinson, 1957). By including biotic interactions in the factor matrix, the hypervolume formed by overlapping factor–response curves comes close to defining the actual niche that an organism occupies.

### NICHE AMPLITUDE

When niche is thought of as a multidimensional space, it becomes apparent that the size and shape of this space is different for each species. A measurement of one or more of its dimensions is termed "niche breadth" or "niche amplitude" (Levins, 1968; Colwell and Futuyma, 1971; Bazzaz, 1975), or niche width (Odum and Barrett, 2005). Organisms with a narrow niche and very specialized habitat adaptations and activities are called specialists. Those that have a broader niche are referred to as generalists. Generalists are more adaptable than specialists, can adjust more readily to change in the environment, and use a range of resources. Specialists are much more specific in their distribution and activities, but have the advantage of being able to make better use of an abundant resource when it is available. In some cases, since a generalist is not that thorough in its use of resources in a habitat, it leaves niche space within its niche for specialists. In other words, there can be several specialist niches inside a generalist niche.

### Niche Diversity and Overlap

Natural ecosystems are often characterized by a high degree of species diversity. In such systems, many different species occupy what appear at first glance to be similar ecological niches. If we accept Gause's Law — that two species cannot occupy the same niche at the same time without one excluding the other — then we must conclude that the niches of the similar organisms are in fact distinct in some way, or that some mechanism must be allowing coexistence to occur. Competitive exclusion appears to be a relatively uncommon phenomenon.

In cropping systems as well, ecologically similar organisms occupy simultaneously what appears to be the same niche. In fact, farmers have learned from accumulated experience and constant observation of their fields that there can often be advantages to managing a mixture of crop and noncrop organisms in a cropping system even when many of the constituents of the mixture have similar requirements. Competitive exclusion rarely occurs; therefore there must be some level of coexistence or avoidance of competition.

This coexistence of outwardly similar organisms in both natural ecosystems and agroecosystems is made possible by some kind of ecological divergence between the species involved. This divergence is referred to as *niche diversity* or diversification of the niche. Some examples include the following:

- Plants with different rooting depths. Variable crop architecture below ground permits different species to avoid direct interference for nutrients or water while occupying very similar components of the niche above ground.
- Plants with different photosynthetic pathways. When one crop uses the C4 pathway for photosynthesis, and another uses C3, the two crops can occur together. One species thrives in full sunlight and the other tolerates the reduced light environment created by the shade of the emergent species. The traditional corn/bean intercrop common in Mesoamerica is a well-known example.
- Insects with different prey preference. Two similar parasitic insects may cooccur in a cropping system, but they parasitize different hosts. Host–parasite specificity may be one way of diversifying the niche so as to allow for coexistence of adult insects elsewhere in the cropping system.
- Birds with different hunting or nesting behaviors. Several predatory birds may all feed on similar prey in an agroecosystem, but since they have different nesting habits and sites, or since they feed at different times of day, they can cooccur in the cropping system and help control pest organisms. Nocturnal owls and diurnal hawks are a good example.
- Plants with different nutritional needs. Mixed populations of weeds can cooccur in the same habitat due, in part, to the differential nutritional needs that may have evolved over time in each species as a result of the selective advantage of avoiding competitive exclusion. A crop population may suffer less negative interference from a mixed population of weeds than from a population of a single dominant weed with niche characteristics similar to that of the crop.

It appears that natural selection acts to create niche differentiation by separating some portion of the niche of one population from that of another. Niche differentiation allows partial overlap of niches to occur without exclusion.

The concept of niche, combined with knowledge of the niches of crop and noncrop species, can provide an important tool for agroecosystem management. A farmer can take advantage of niche overlap to exclude a species that is a detriment to the agroecosystem; similarly, he or she can use niche differentiation to allow the combination of species that are of benefit to the system (Figure 13.3).

## APPLICATIONS OF NICHE THEORY TO AGRICULTURE

Farmers are constantly managing aspects of the ecological niches of the organisms that occupy the farming system, even though most never refer directly to the concept. Once it is understood as a useful tool of ecosystem management, however, it can be applied in a variety of ways, from ensuring maximum yield through an understanding of a main crop's niche, to determining whether a noncrop species is likely to cause negative interference with the crop. Some specific examples follow.

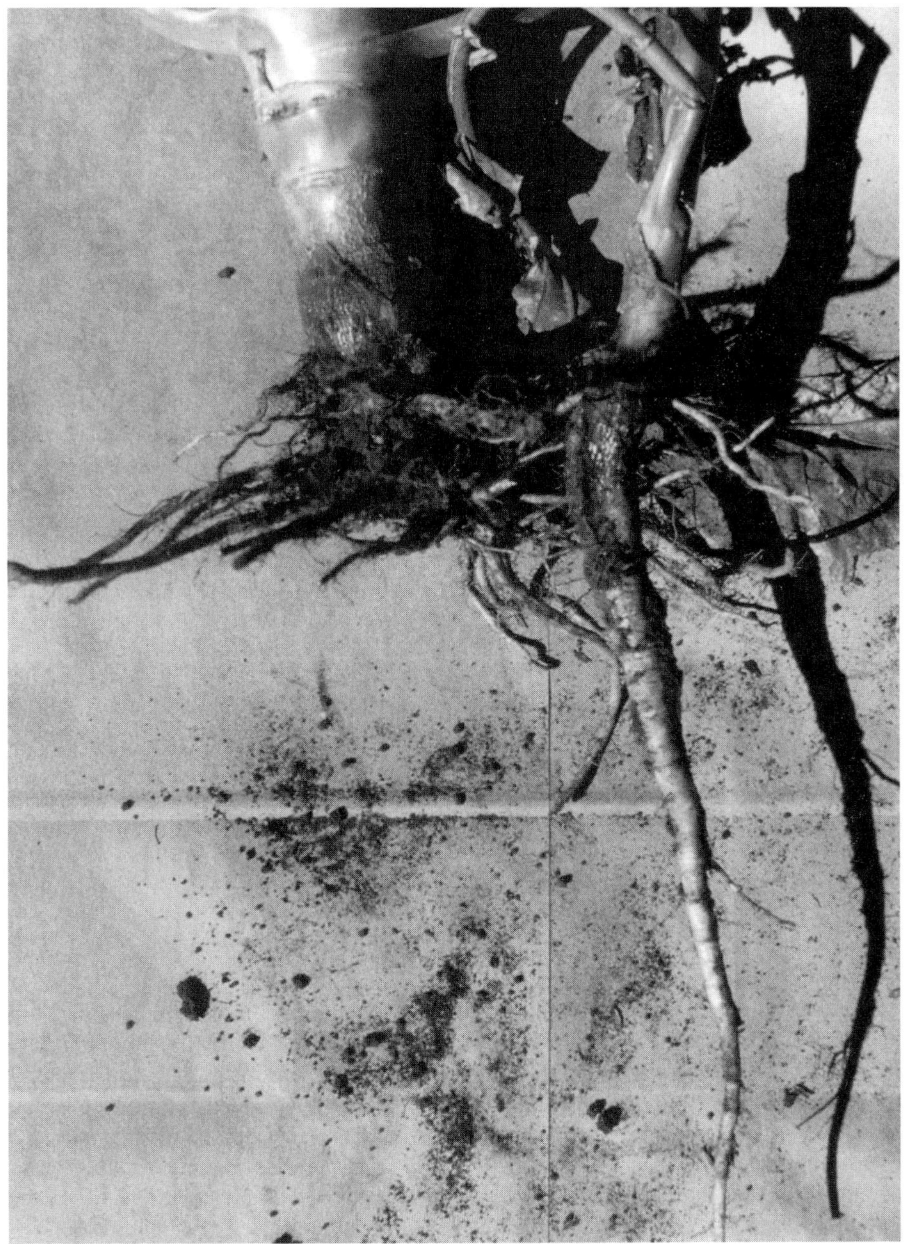

**FIGURE 13.3 Different root architectures permitting niche overlap.** The shallow root system of the transplanted broccoli (left) and the deeper tap root system of the direct-seeded wild mustard (right) take resources from different parts of the soil profile, allowing the plants to occupy the same habitat without negative interference.

### Promoting or Inhibiting Establishment of Weedy Species

Any part of the soil surface not occupied by the crop population is subject to invasion by weedy noncrop species. Specialized for being successful in what can be termed productive environments (i.e., farm fields), weeds occupy a niche that favors *r*-selected or ruderal populations of annual herbs. In cropping systems with lesser disturbance, where total plant biomass undergoes less disruption or removal, competitive (but still *r*-selected) biennial or perennial weeds become common. In a sense, weediness is a relatively specialized niche characteristic.

The habitat facet of the niche concept can be used to help guide how the environmental conditions of a farm field are manipulated in order to promote or inhibit the establishment of weedy species. The type of modification will depend on the niche specificity of each species in relation to the crop. With knowledge of the niche characteristics of a weed species, we can begin by controlling the conditions of the "safe sites" to the disadvantage of the weed. Additionally, we can look for some critical or susceptible phase in the life cycle of the weed population in which a particular management practice could eliminate or reduce the population. It may also be possible to promote the growth of a weed population that will inhibit other weeds. For example, wild mustard (*Brassica* spp.) has little negative effect on crop plants but has the ability to displace, through interference, other weeds that may have a negative influence on the crop. A more detailed discussion of this phenomenon is provided in the case study *Broccoli and Lettuce Intercrop*.

It is important to remember that most weeds are colonizers and invaders, and that crop fields that are disturbed annually are just the type of habitats they have been selected for. The challenge is to find a way to incorporate these ecological concepts into a management plan where planned activities, such as cultivation, are timed or controlled so that the weedy niche may be occupied by more desirable species.

### Biological Control of Insect Pests

Classical *biological control* is an excellent example of the use of the niche concept. A beneficial organism is introduced into an agroecosystem for the purpose of having it occupy an empty niche. Most commonly, a predatory or parasitic species is brought into a crop system from which it was absent in order to put negative pressure on the population of a particular prey or host that has been able to reach pest or disease levels due to the absence of the beneficial organism.

It is hoped that once the beneficial organism is introduced into the cropping system it will be able to complete its entire life cycle and reproduce in large enough numbers to become a permanent resident of the agroecosystem. But often the conditions of the niche into which the beneficial species is introduced may not meet its requirements for long-term survival and reproduction, so reintroductions become necessary. This can be especially true in a constantly changing agricultural environment with high disturbance and regular alteration of the characteristics of the niche needed to maintain permanent populations of both the pest and the beneficiary.

### BROCCOLI AND LETTUCE INTERCROP

An intercrop is successful when the potential competitive interferences between its component crop species are minimized. This is accomplished by mixing plants with complementary patterns of resource use or complementary life history strategies.

Two crops that have been shown to combine well in an intercrop are broccoli and lettuce. Studies at the University of California, Santa Cruz farm facility (Aoki, et al. 1989) have demonstrated that a mixture of these crops will produce a higher yield than a monoculture of lettuce and a monoculture of broccoli grown on the same area of land. (This result, called overyielding, is explained in greater detail in Chapter 16.)

In the study, broccoli and lettuce were planted together at three different densities and the yields from each compared to yields from monocrops of each crop. The lowest intercrop density was a substitution intercrop, in which the overall planting density was similar to that of a standard monocrop. The highest density intercrop was an addition intercrop, in which broccoli plants were added between lettuce plants planted at a standard density. The monocrops were planted at standard commercial densities, which are designed to avoid intraspecific competition.

All three densities of intercrop produced higher total yields than the monocultures. The yield advantages ranged from a 10% greater yield to a 36% greater yield (for the substitution intercrop). The addition intercrop produced lettuce heads of a slightly lower mean weight than the monoculture lettuce, but the combined production still exceeded the total that was produced by a combination of monocrops on the same amount of land. The intercrops also retained more soil moisture than the monocrops, indicating that the physical arrangement of the two species in the field helps to conserve this resource.

These results indicate that interspecific competitive interference did not negatively impact the plants in the intercrops, even when their density was approximately twice that of either of the monocrops. For this avoidance of competition to have occurred, the broccoli and lettuce must each have been able to utilize resources that were not accessible to the other species.

An examination of the two species' life histories and niches illuminates the complementarity of their resource use patterns and suggests mechanisms for the observed overyielding. Lettuce matures rapidly, completing nearly all its growth within 45 days of being transplanted into the field. It also has a relatively shallow root system. Broccoli matures much more slowly and its roots penetrate much deeper into the soil. Therefore, when the two are planted nearly simultaneously, lettuce receives all the resources it needs to complete its growth well before the broccoli grows very large; then after the lettuce is harvested, the broccoli can take full advantage of the available resources as it grows to maturity.

Another potential use of the niche in biological control is the introduction of another organism that has a niche very similar to that of the pest, but which has a less negative impact on the crop. The introduced herbivore, for example, may feed on a part of the plant that is not of economic significance. If the introduced herbivore has a niche similar enough to the target pest, it might be able to displace it. There might be similar applications for weeds.

### Design of Intercropping Systems

When two or more different crop populations are planted together to form an intercropped agroecosystem, and the resulting yields of the combined populations are greater than those of the crops planted separately, it is very likely that the yield increases were a result of complementarity of the niche characteristics of the member populations. For intercropping systems to be successful, each species must have a somewhat different niche. Therefore, full knowledge of the niche characteristics of each species is essential. In some intercrop cases, each species occupies a completely unrelated or otherwise unoccupied niche in the system, leading to niche complementarity. In most cases, however, the niches of the member species overlap, but interference at the interspecific level is less intense than interference at the intraspecific level.

Successful management of crop mixtures, then, depends on knowing each member's population dynamics, as well as its specific niche characteristics. Such knowledge then forms the basis for management of the intercrop as a community of populations, a level of agroecological management on which we will focus in Chapter 15.

## POPULATION ECOLOGY — A CROP PERSPECTIVE

In this chapter, the focus has been on populations in the context of their environment. Important similarities and differences between populations of crop, noncrop, and natural species have been discussed. Some of these characteristics, along with additional relevant ones, are summarized in Table 13.2.

Knowledge of these characteristics becomes especially important when we are trying to find ecologically-based management strategies for weedy noncrop species. Weedy species have maintained some of the characteristics of wild, natural ecosystem populations (e.g., dispersability, strong intra- and interspecific interference ability, and dormancy), but through a range of adaptations (e.g., high seed viability, even-aged population structure, high reproduction allocation, and narrower genetic diversity) have adapted to the conditions of disturbance and alteration of the environment common in agroecosystems, especially those systems that depend on annual crops. The ability of weeds to thrive in agroecosystems poses strong challenges for the agroecosystem manager.

Each species has certain strategies for ensuring that individuals of that species successfully complete their life cycles, thus enabling populations of that species to maintain a presence in a certain habitat over time. Principles of population ecology, applied agroecologically, help the farmer decide where and how to take advantage of each specie's particular life history strategy to either promote or limit the population growth of the species, depending on its role in the agroecosystem. Agroecosystem managers

**TABLE 13.2**
**Population Characteristics of Crop, Noncrop, and Related Natural Species Populations**

| | Crop Population | Noncrop Population | Natural Population |
|---|---|---|---|
| Dispersal | Little or none | Very important | Important |
| In-migration | Propagule input decoupled from output | Immigration very important | Most propagules from local population |
| Seed viability | High | High | Variable |
| Seed rain | Controlled | Relatively homogeneous | Patchy |
| Soil environment | Homogeneous | Homogeneous | Heterogeneous |
| Seed dormancy | None; seed not part of seed bank | Variable; seed bank present | Common; seed bank present |
| Age relationships | Often even-aged, synchronous | Mostly even-aged, synchronous | Age variable, mostly asynchronous |
| Intraspecific interference | Reduced | Can be intense | Can be intense |
| Seed density | Low and controlled | Usually quite high | Variable |
| Density-dependent mortality | Little or none | Significant | Significant |
| Interspecific interference | Reduced | Very important | Important |
| Reproductive allocation | Very high | Very high | Low |
| Genetic diversity | Usually very uniform | Relatively uniform | Usually diverse |
| Life-history strategies | Modified r-strategists | r, C, and R strategists | K and S strategists |

*Source*: Weiner, J. 1990. In C. R. Carroll, J. H. Vandermeer, and P. M. Rossett (eds.), *Agroecology*. pp. 235–262. McGraw Hill: New York.

and researchers need to build on population ecology concepts such as safe site, *r*- and *K*-strategies, and ecological niche to further develop techniques and principles for effective and sustainable management of crop and noncrop organisms.

## FOOD FOR THOUGHT

1. What might permit coexistence of two very similar crop species that would otherwise be thought to competitively exclude each other if allowed to grow in the same resource space?
2. How might the concept of niche diversity be used to design an alternative management strategy for a particular herbivorous pest in a cropping system?
3. Identify several particularly sensitive steps in the life cycle of a weed species, and describe how this knowledge might be of value in managing populations of the weed in a sustainable fashion.
4. What aspect of plant demographics have agronomists been able to use successfully in their quest for improved crop yields, but which has sacrificed overall agroecosystem sustainability? What changes would you make in the research agenda of agronomists in order to correct this problem?
5. What is your definition of a "good" weed?
6. Tropical environments seem to have more specialists, whereas temperate environments have more generalists. Where do agroecosystems fall in this spectrum?

## INTERNET RESOURCES

The Land Institute
www.landinstitute.org
The Land Institute is leading the effort to develop a perennial grain crop.

Plant Population Ecology section of the Ecological Society of America
pltecol.cas.usf.edu/index.pl
A site that facilitates interactions among researchers in plant population biology.

Centre for Population Biology
www.cpb.bio.ic.ac.uk/
This site has information on the centre's basic research in population biology and related disciplines, dedicated to understand and predict the functioning of ecological systems, from populations to ecosystems.

Aberdeen Population Ecology Research Unit
www.abdn.ac.uk/aperu/aperu.shtml
The APERU is a collaborative group of statisticians and population ecologists in England. This site contains information on their approaches and projects on population ecology.

## RECOMMENDED READING

Grime, J.R. 2002. *Plant Strategies, Vegetative Processes, and Ecosystem Properties.* 2nd ed. John Wiley and Sons: New York. A review of the relevance of the plant strategy concept in ecological and evolutionary theory.

Harper, J.L. 1977. *Population Biology of Plants.* Academic Press: London. Considered to be the key reference in modern plant population biology, this book thoroughly reviews plant demography and life history strategies.

Radosevich, S.R., J.S. Holt, and C. Ghersa. 1997. *Weed Ecology: Implications for Vegetation Management.* 2nd ed. John Wiley and Sons: New York. A thorough review of how ecological knowledge of weeds and weed populations forms an essential basis for successful weed management.

Silvertown, J.W. and D. Charlesworth. 2001. *Introduction to Plant Population Ecology.* 4th ed. Blackwell Science: Oxford, London, Vermont. An up-to-date introduction to the field of plant population ecology, with many references to studies of agricultural populations.

Van der Pijl, L. 1972. *Principles of Dispersal in Higher Plants.* 2nd ed. Springer-Verlag: Berlin. A review of the ecology of dispersal mechanisms in plants and their role in determining the success of different plant species in the environment.

# 14 Genetic Resources in Agroecosystems

Agriculture came about as early human cultures intensified their use and care of particular plants and animals that they found to be of value. During this process, humans inadvertently selected specific traits and qualities in these useful organisms, altering their genetic makeup over time. Their ability to produce edible or useful biomass was enhanced, but their ability to survive without human intervention was reduced. Humans came to depend on these domesticated species for food, feed, and fiber, and most of them became dependent on us.

Throughout most of human history, humans manipulated the genetic makeup of crops and livestock without explicit knowledge of genetics. Farmers simply made the choice to plant seed or breed animals from the individuals or populations that demonstrated the most desirable characteristics, and this was enough to direct the evolution of domesticated species (Figure 14.1). Gradually, plant and animal breeding developed into a science as we learned more about the genetic basis of selection and began to direct it more specifically to our advantage. Today, the fields of biotechnology and genetic engineering are rapidly changing the way humans manipulate the genes of domesticated species, making it possible to incorporate traits and characteristics into plants and animals in ways and at rates never before possible.

But from the viewpoint of sustainability, the direction of crop and livestock breeding efforts of the past several decades — and the directions proposed for the future — are causes for deep concern. The genetic base of agriculture has narrowed to a dangerous point as human societies have become increasingly dependent on a few species of food-producing organisms and on a smaller number of the genes and genetic combinations found in those species. Crop plants have lost much of the genetic basis of their pest and disease resistance and their ability to tolerate adverse environmental conditions, leading to crop failures and increased dependence on human-derived inputs and technologies for the maintenance of optimum growth conditions. In addition, genetic resources beyond the crops themselves — wild crop relatives, weedy derivatives, and traditional cultivated varieties, genetic lines, and breeding stocks — have been greatly reduced.

The relationship between genetics and agriculture is a vast topic. This chapter explores a small part of it, focusing on the foundations needed for understanding the role of genetic diversity in moving toward sustainability in agriculture. We examine genetic change in nature and how it results in genetic diversity, outline the processes humans use to direct and manipulate genetic change in domesticated species (with a focus on crop plants), and then discuss how genetic resources eventually need to be managed at the whole agroecosystem level. The important social and economic issues concerning crop and livestock breeding and control of access to genetic material are addressed in the recommended readings at the end of this chapter.

## GENETIC CHANGE IN NATURE AND THE PRODUCTION OF GENETIC DIVERSITY

From the perspective of geologic time, the earth's flora and fauna are constantly changing. The physical and behavioral characteristics of species change, new species appear, other species go extinct. This change, called evolution, is made possible by the manner in which traits are passed from parent to offspring and is driven by changes in environmental conditions. As ice ages come and go, continents move, and mountains emerge and erode, living things respond. Through natural selection, the changing and varied environment acts on species' genomes, causing them to change from generation to generation.

Natural selection has created the genetic diversity found in nature, the raw material that humans have worked with in domesticating plants and animals and creating agroecosystems. It is therefore important to understand how natural selection works and how it applies to human-directed genetic change and the maintenance of our agricultural genetic resources.

### ADAPTATION

The concept of *adaptation* is a basis for understanding natural selection because it relates the environment to a species' traits. The term refers both to a process and to a characteristic resulting from that process. In static terms, an adaptation is any aspect of an organism or its parts that is of value in allowing the organism to withstand conditions of the environment. An adaptation may be as follows:

- enable an organism to better use resources
- provide protection from environmental stresses and pressures

**FIGURE 14.1** *Diversity of beans for sale in an Oaxaca city market, Mexico.* Traditional varieties reflect local ecological and cultural diversity.

- modify local environments to the benefit of the organism; or
- facilitate reproduction

Any organism existing in nature must have a great many adaptations in order for it to survive; in theory, nearly all behaviors and physical characteristics of an organism are adaptations. Another way of saying this is that at any point in time a naturally existing organism as a whole is always adapted to its environment.

The adaptations possessed by a particular species, however, do not remain the same over long periods of time, because the environment is always changing and organisms are continuously adapting. The process by which adaptations change over time is also called adaptation, and is understood in terms of natural selection.

## Variation and Natural Selection

Individual members of sexually reproducing species are not identical to each other. The variation that exists among humans is mirrored in other species, even though we may not always be able to discern it. This natural variability exists both at the level of the *genotype* — the genetic information carried by an individual — and at the level of the *phenotype* — the physical and behavioral expression of the genotype.

An examination of a number of individuals of any population quickly demonstrates the existence of phenotypic variability. Any characteristic, from number of leaves on a plant to the length of the tail of an animal, shows a range of variability. An average value or mode for each characteristic occurs, and if variation in each trait were graphed as a frequency distribution, it would tend to follow a normal curve of probability (a bell-shaped curve). Some populations show a very narrow range of variation, while others show much more. Although phenotypic variation does not correlate directly with genotypic variation, it usually has a significant genotypic basis (Figure 14.2).

The genetic variability within a species is due mainly to the nature of DNA replication: DNA does not always replicate itself perfectly; errors of different types, called mutations, always occur at some frequency. Since DNA replication is a prerequisite to reproduction, new individuals are constantly coming into existence with mutations. Although some mutations are fatal, some detrimental,

**FIGURE 14.2 Squash fruit variability from a farmer's field in Tabasco, Mexico.** Seed from one fruit were used to plant the field.

some neutral, and only a few advantageous, all mutations represent genetic difference and thus genetic variability. Most mutations are simply single changes in the nucleotide sequence of DNA molecules; by themselves they may have no significant effect, but added together over time they can result in fundamental changes, such as bigger fruit, resistance to frost, or the addition of tendrils for climbing.

Variability is also produced by sexual reproduction. When two individuals reproduce sexually, the genes of each are distributed differently into different gametes (sex cells), and the genetic material carried in the gametes is mixed in novel ways when the gametes combine during fertilization. Variation is also introduced during meiosis (the formation of gametes) when chromosomes are deleted or translocated, or when homologous chromosomes fail to separate at the first meiotic division.

This latter kind of "error" creates gametes that have two copies of each chromosome (diploid) instead of the usual one (haploid). If one of these diploid gametes fuses with a normal haploid gamete, a zygote with three times the haploid number of chromosomes can result, and when one fuses with another unreduced diploid gamete, a zygote with four times the haploid number can be formed. Such increases in the number of chromosomes represent another source of genetic variety, particularly important in plants. Plants with more than the diploid number of chromosomes, called *polyploid*, typically have different characteristics than their diploid forebears, and occur relatively commonly in nature.

Because of natural genetic variation, some individuals of a population will have traits not possessed by others, or will express a certain trait to a greater or lesser degree than others. These traits may give the individuals who possess them certain advantages in living. These individuals may grow more rapidly, survive in greater numbers, or have some reproductive advantage. Due to such factors, they may leave more offspring than other individuals, thus increasing the representation of their genetic material in the population as a whole. It is through such a process of differential reproductive success that a species undergoes genetic change over time.

The direction and manner of this change is determined by *natural selection* — the process by which environmental conditions determine which traits confer an advantage and therefore increase in frequency in the population.

If the environment in which a population lived was totally optimal and never changed, genetic change would occur, but there would be no natural selection to direct it. However, since environmental conditions are always changing and never optimal for very long, natural selection is always occurring at some level. In addition to long-term changes in factors such as climate, natural selection is driven by such environmental changes as population growth of other species, the appearance of new species through migration, the evolution of predators and herbivores, and changes in microhabitats due to erosion, sedimentation, succession, and other processes.

Natural selection acts on populations, not whole species. If a population of a species becomes reproductively isolated from the rest of the species — that is, if physical barriers prevent its members from interbreeding with members of other populations — that population can undergo genetic change in a unique way. Because the environment is never homogenous over space and time, the isolated population will be subjected to somewhat different selective pressures than other populations of the species. The tendency, therefore, is for different populations to evolve somewhat differently. Biogeographically, the species becomes a mosaic of populations, each of which has unique genetically based physiological and morphological characteristics. Each distinct population is referred to as an *ecotype*. Through evolutionary time, an ecotype can become distinct enough from other ecotypes of the species that it becomes a distinct species in its own right.

The evolutionary processes that cause the development of ecotypes and drive speciation are constantly diversifying the genetic basis of earth's biota. Although species go extinct, new species are always evolving, and the genomes of many existing species are becoming more varied over time. One of our great fears today, however, is that human activity, including agriculture, is fundamentally altering this process. Our destruction, alteration, and simplification of natural habitats is greatly increasing rates of extinction and eliminating ecotypes, thus eroding natural genetic diversity and the potential for its renewal (Wilson, 1992).

## DIRECTED SELECTION AND DOMESTICATION

Genetic change in an agricultural context differs greatly from genetic change in naturally occurring populations. Humans construct and manipulate the environments in which agricultural species live, grow, and reproduce, thereby creating an entirely different set of selective pressures for them. Humans determine which traits are most desirable, and select these traits in the way they cultivate and propagate the species. Because humans "direct" genetic change in agricultural populations, the process by which this genetic change occurs is called *directed selection*.

Today's agricultural species — both plants and animals — were domesticated by gradually shifting their context from natural systems dominated by natural selection to human-controlled systems in which directed selection operated. Some 10,000 to 12,000 yr ago, humans did not create strictly controlled agricultural environments like farmers do today. In the case of plants, they cared for certain naturally-occurring species by modifying their habitats, facilitating their reproduction, controlling their competitors, and occasionally moving them to more convenient places. In the case of animals, they followed herds of herbivores more closely, began to protect them from predators, and often provided them with feed. Natural selection still had an important role in such systems, because the human intervention was not sufficient to overcome the fact that the useful species still had to survive the rigors of the natural environment.

The process of *domestication* began as humans became better able to alter and control the environment in which useful plants and animals occurred, and to manage the reproduction of these species to such an extent that they began to unintentionally select specific useful traits. As domestication progressed, selection became more intentional, with early agriculturalists choosing seed from the plants with higher and more predictable yields, and early pastoralists choosing, for example, to breed the goats that produced the most milk. Throughout the process of domestication, the screening effect of the natural environment became less important and directed selection took on a greater role. Eventually, crop and livestock species reached a point where their genetic makeup had been altered to such an extent that they could no longer survive outside of an agroecosystem.

A domesticated species is dependent on human intervention, and the human species is now dependent on domesticated plants and animals. In ecological terms, this interdependency can be considered an obligate mutualism. It has come about through a process of mutual change: human cultures have both caused changes in the genetic makeup of certain useful species and been transformed themselves as a result of those changes.

### Traits Selected in Crop Plants and Livestock Animals

Today's crop plants and stock animals have been subjected to many selection pressures over thousands of years. In plants, humans have selected optimized yield, appealing taste and appearance, and ease of harvest, and more recently, fast response to fertilizer and water application, ease of processing, resistance to shipping damage, longer shelf life, and genetic uniformity. In animals, we have selected docility, more easily manageable reproductive cycles, and rapid growth and maximal production of

## ORIGINS OF AGRICULTURE

Between 4000 and 10,000 years ago, agriculture arose independently in several different areas of the world, each with its own geography, climate, and indigenous flora and fauna. Six widely recognized centers of early agricultural development are shown on the map in Figure 14.3. The center in China is sometimes divided into two subcenters, the Yangtze River Valley in the south and the Yellow River Valley in the north. The Southeast Asia and South Pacific "center" is diffuse, spreading over a somewhat larger area than indicated. Some researchers add other centers to this list: one in the Ohio and Mississippi River valleys of North America, and one on the Indian subcontinent.

What these regions had in common is high natural biological diversity, variable topography, and climate, and human cultures ready to exploit the potential benefits of more intensive management of edible plant species. Since the local flora in each center was made up of a distinct assemblage of plant families and genera, the kinds of plants domesticated in each region varied greatly.

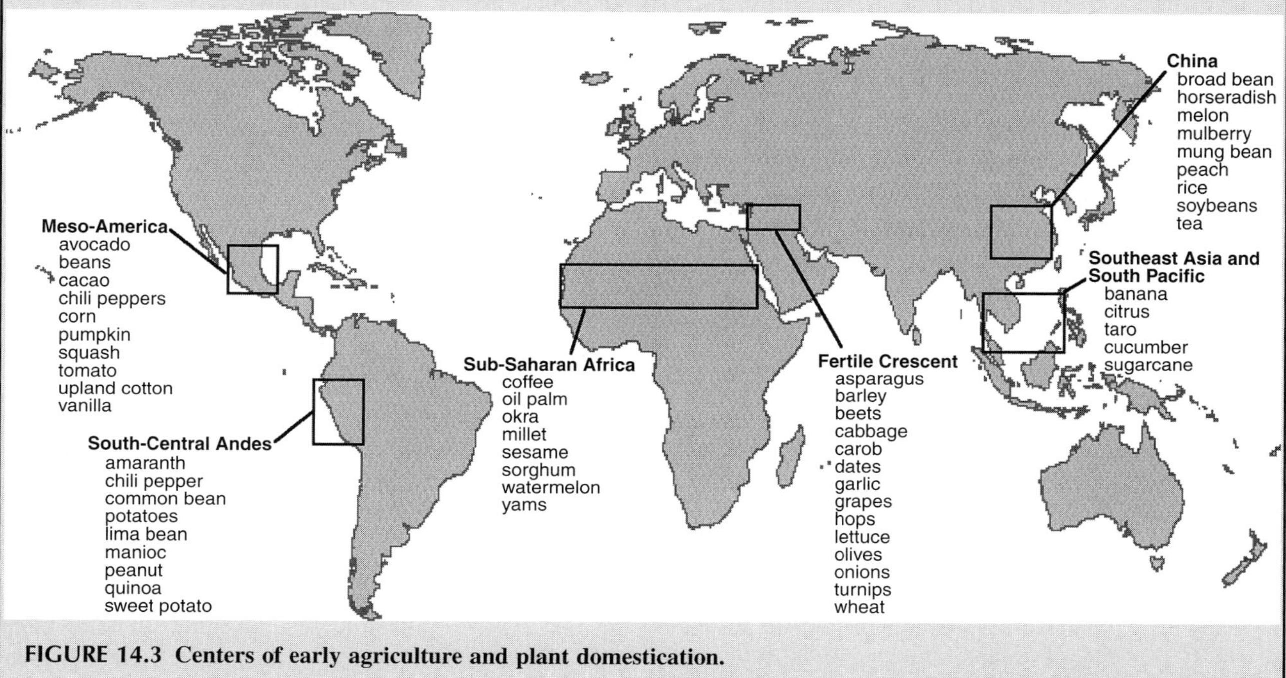


**FIGURE 14.3 Centers of early agriculture and plant domestication.**

desired parts and products — better wool, larger and more numerous eggs, more liters of milk, or more muscle tissue.

This selection process has, among other things, greatly altered the physiologies and morphologies of the domesticated species. In domesticated plants, carbon partitioning operates very differently than carbon partitioning in wild species. Crop plants store a much greater proportion of their biomass in their edible or harvestable parts than do the natural species from which they were derived. As a consequence, less energy is partitioned for use in traits or behaviors that confer *environmental resistance* — the ability to withstand stresses, threats, or limiting factors in the environment. In addition, many traits that once conferred environmental resistance have been lost from the genotype altogether. An analogous change has occurred in domesticated animal species. Because of these fundamental changes in the genetic basis of their physiology and morphology, many domesticated species and varieties require completely artificial and optimum conditions. For plants, this means ideal soil moisture, nutrient availability, temperature, and sunlight, as well as the absence of pests, in order to perform well and express the high-yield traits for which they were selected. For animals, it often means controlled climatic conditions, antibiotics, and artificial insemination (the turkey breed that accounts for nearly all turkey production in the U.S. cannot reproduce without human assistance because body structure and extra meat limit animal movement).

Directed selection in agriculture has therefore led us into a difficult situation. Our major crop varieties require external inputs of inorganic fertilizers, pesticides, herbicides, and irrigation water to perform as designed, and many of our domesticated animals require hormones and antibiotics, highly controlled conditions, and highly processed feed. But such external inputs are the major cause of agriculture's negative effect on the environment and on human health, and the degradation of the soil resource. If steps are taken to restrict the use of many of the practices and materials that humans have developed to protect and promote the growth of our crops and stock animals, yields and production can suffer.

This problem is particularly troubling with regard to pesticide use on crop plants. Plants' natural abilities to withstand herbivory — through morphological adaptations, mutualistic interactions, the production of obnoxious compounds, and other methods — have been largely lost at the expense of the development of other traits. Agroecosystems become dependent on pesticide use to prevent loss of the crop through herbivory, but pesticide use becomes a selective pressure on the herbivore populations, resulting in their evolution toward pesticide resistance and requiring the application of more pesticide or the continual development of new pesticide types.

A fundamental problem is that traits of environmental resistance have been lost not just from the genetic makeup of individual species and varieties, but from the structure and organization of the entire agroecosystem (Chapter 16). Attempts to reincorporate environmental resistance into domesticated species' genomes, therefore, must work at the agroecosystem level, not just at the level of individual species, breeds, and varieties.

## Methods of Directed Selection in Plants

Farmers and crop breeders change the genetic makeup of crop species and varieties in a number of ways, ranging from indirect means that resemble natural selection to high technology means that work directly on the plant genome. These latter methods are not selection per se, but they are discussed here since they have the same results as directed selection methods.

The methods that can be used on a particular species depend on its manner of reproduction. Some plant species (more annuals than perennials) reproduce primarily by *self-pollination* — the female parts of a plant's flowers are fertilized by pollen from the same plant, and often from the same flower. Other plant species (more perennials than annuals) reproduce mainly by *cross-pollination*. Such plants typically have some kind of morphological, chemical, or behavioral adaptation to assure that an individual's female flower parts are fertilized only by pollen from other plants.

### Mass Selection

Until relatively recently, the only method of directed selection was to collect seed from those individuals in a population that showed one or more desirable traits, such as high yielding ability or disease resistance, and to use that seed for planting the next crop. This method, called *mass selection*, can produce a gradual shift in the relative frequency of a trait or traits in the population (Figure 14.4). Through mass selection methods, farmers all over the world have developed varieties called *landraces*. Landraces are adapted to local conditions, and although a landrace, as a whole, is genetically distinct, its members are genetically diverse (Figure 14.5).

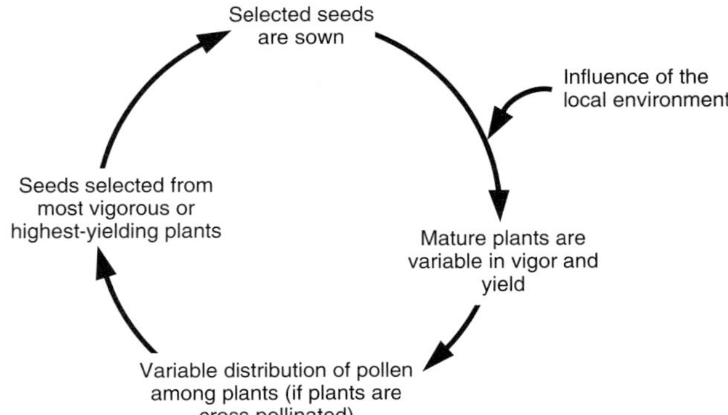


**FIGURE 14.4 The mass selection process.** This method of selecting desirable characteristics maintains adaptations to local conditions and allows for maximum genetic variability.

Mass selection works similarly for both self-pollinated and cross-pollinated plants. When cross-pollinated plants are involved, mass selection allows *open pollination* to occur. Also known as out-crossing, this natural mixing of pollen among the members of a population results in high genotypic variability. With self-pollinated plants, mass selection also allows the maintenance of relatively high variability.

This older, more traditional method of directed selection involves the whole organism and field-based selection; despite being a relatively slow process and more variable in its results, it has the advantage of being more like natural selection in natural ecosystems. Traits involving adaptation to local conditions are retained along with the more directly desirable aspects of yield or performance, and genotypic variability is maintained as well. Such characteristics are very important especially for small farm systems with limited resources and more variability in production conditions. All other methods of directed selection tend to increase genetic uniformity, and most greatly reduce or eliminate the role of local environmental conditions in the selection process.

### Pure Line Selection

In self-pollinating plants, a common method of selection is to choose several superior-appearing plants from a variable population and then subject the progeny of each to extensive testing over many generations. At the end of the testing period, any line sufficiently distinct from and superior to existing varieties is released as a new variety. Because the plants are self-pollinating, the selected genotype stays relatively stable over time.

**FIGURE 14.5 Four mass-selected, local landraces of corn from the lowlands of Tabasco, Mexico.** Each landrace has a different name, planting time, and preferred location.

The pure line selection process can be modified in a variety of ways. One is to transfer genes between existing pure lines through artificial cross-pollination in an effort to produce a new line with a new combination of characteristics. Sometimes this is accomplished by repeatedly backcrossing the progeny of an artificial cross with a parent having a specific desired characteristic.

### Production of Synthetic Varieties

In cross-pollinated plants, an analog to a self-pollinated pure line, called a *synthetic variety* or synthetic cultivar, can be created through a variety of techniques. The underlying principle is to limit the parental genotypes to a few that are known to have superior characteristics and to cross well. In alfalfa, for example, this can be done by planting seed from only a few specific sources (such as two or three clonal lines) in an isolated field and allowing natural crossing to occur. Seed produced from this field is then distributed as a synthetic variety. Synthetic varieties have greater genetic variability than self-pollinated pure line varieties, but far less variability than mass-selected, open-pollinated varieties.

### Hybridization

The primary method of directed selection today in many important crop plants — especially corn — is the production of hybrid varieties. A hybrid is a cross between two very different parents, each from a different pure-breeding line. The process of creating a hybrid variety involves two basic steps.

First, the two distinct pure-breeding lines are produced. (Pure-breeding means that the genomes are largely homozygous at most gene loci.) In cross-pollinated plants (and self-pollinated plants that cross-pollinate frequently), this step involves artificial inbreeding, which is accomplished in a variety of ways.

Second, the two pure-breeding lines are crossed to produce the hybrid seed that is planted by farmers for production of the crop. Neither self-pollination nor cross-pollination between plants of the same line can occur in this step, necessitating the use of certain techniques. One technique, used in corn, is to plant the pollen-donor parental line and the seed-producing parental line in alternating rows or strips and to detassel the seed-producing plants before the tassels produce pollen (the tassels contain only male flowers). An alternative technique, used extensively in self-pollinated plants such as sorghum, is to introduce genetically controlled male sterility, called *cytosterility*, into one of the inbred parental lines. This line is then used as the seed-producing parental line, because it can be pollinated only by pollen from the other, nonsterile parental line.

The hybrid offspring of two selectively inbred parents are usually quite different from either parent. They are often larger and produce larger seeds or fruits, or have some other desirable characteristic not possessed by either parent. This response, known as *hybrid vigor*, or *heterosis*, is one of the great advantages of a hybrid variety. Another desirable characteristic (from the standpoint of conventional agriculture) is genetic uniformity: all the hybrid seeds of a particular cross will have the same genotype.

Hybrid varieties, however, have an inherent disadvantage. Seeds produced by hybrid plants — through either self- or cross-pollination — are usually undesirable for planting since sexual recombination will produce a variety of new gene combinations, most of which will not exhibit the hybrid vigor of the parents. Therefore, farmers must purchase hybrid seed each year from seed producers.

In crop types with tubers or other means of asexual reproduction, such as potatoes and asparagus, once a hybrid is produced with a suite of desirable traits, it is then propagated asexually as a *clone*. With advances in techniques of tissue culture, this method of propagating hybrids without seed has been applied more widely. Small amounts of tissue from different parts of important hybrid cultivars can be used to rapidly reproduce clones under strictly controlled conditions.

### Induced Polyploidy

Many of today's important crop types, such as wheat, corn, coffee, and cotton, arose long ago as natural polyploids. Since polyploid plants are often more robust and have larger fruits or seeds than their normal diploid parents, people found them desirable when they occurred in early cropping systems, and they were selected, even though farmers were not aware of what made them different.

When it was discovered by modern cytologists that many favorable traits in crop plants were the result of polyploidy, methods were developed to artificially induce it. Through the use of colchicine or other chemical stimulators during the first steps of meiosis, artificial multiplication of the number of chromosomes has become possible. Induced polyploidy has produced some of the most useful lines of wheat, for example, such as the hexaploid *Triticum aestivum*. Once produced, polyploids themselves can then be used to perpetuate pure lines or develop new hybrids.

## Transgenic Modification

Plant breeding using the techniques described above is tedious, time-consuming, and dependent to some extent on luck. Genes occur in the company of many other thousands or millions of genes on chromosomes, and the plant breeder cannot determine how a few genes of interest are distributed and recombined in each generation. Moreover, these techniques are restricted to breeding parents that are closely related — usually within the same species.

No such limitations exist for genetic engineers. Using various techniques developed in recent decades, they can transfer single genes from one organism — a bacterium, for example — to another completely unrelated organism, such as a higher plant. *Genetic engineering* enables crop geneticists to introduce specific traits such as resistance to freezing or herbivory, into a crop species, and to create customized organisms, each with its own unique suite of traits.

As noted in Chapter 1, the end results of genetic engineering are called transgenic, genetically modified, or genetically engineered (GE) organisms. Transgenic crops being planted today on a commercial basis include strains of corn, soybeans, wheat, rice, cotton, canola seed, sugar beets, tomatoes, lettuce, peanuts, and potatoes. The area planted to these and other GE crops has increased regularly every year since the mid-1990s; in 2004, they covered an estimated 200 million acres (81 million ha). Throughout this period of rapid growth in the planting of biotech crops, one company — Monsanto Corporation — has maintained an 80 to 90% share of the market.

GE crops are created with a variety of goals in mind. Some are intended to be resistant to attack by a particular pest, some to create food with better nutritive value, some to resist the application of herbicides. Because of these and other characteristics, genetic engineering has been touted as the technological answer to many of the challenges faced by agriculture: producing more food, producing better food, reducing the need for pesticides and herbicides, and growing crops on marginal land.

Transgenic modification of crop organisms has been controversial ever since it began to be practiced on a commercial basis in the 1990s, and for good reason. Growing GE crops poses a variety of potential problems and serious risks (Table 14.1). As just one example, several researchers have recently cautioned against the use of transgenic rice on the grounds of its potential effects on food safety, cross pollination with "wild rice" varieties and other types of environmental impact on nontarget organisms (Bottrell, 1996; Lu and Snow, 2005; Saito and Miyata, 2005).

Some of the potential drawbacks of growing transgenic crops are listed below. They are not all hypothetical; most have already been documented to occur.

**TABLE 14.1**
**Traits of Transgenically Modified Crops**

| Desired Trait | Claimed Benefits | Examples |
|---|---|---|
| Disease resistance | Higher yield due to reduced crop loss | Inserting an artificial bacterial chromosome into strains of potato to confer resistance to late blight |
| Pest resistance | Lower pesticide use; higher yield due to reduced crop loss | Introducing toxin-producing genes from the bacterium *Bacillus thuringiensis* into cotton |
| Improved food quality | Less malnutrition in developing countries | Engineering the Vitamin-A production pathway in rice ("golden rice") |
| Tolerance for abiotic stresses (e.g., drought, salinity) | Higher food yield on marginal land; less irrigation | Introducing genes allowing biosynthesis of citric acid in sugar beets, to increase tolerance of aluminum and uptake of phosphorus in acidic soil |
| Herbicide resistance | Higher yield due to reduced weed competition when crop treated with herbicide | Roundup Ready® soybeans, engineered to resist the herbicide glyphosate (farmers buy the GE seed and the herbicide from Monsanto) |
| Production of a particular useful compound | Lower product cost | Plant-made pharmaceuticals, such as a drug for the treatment of cystic fibrosis produced by GE corn |

- *Unintended and hidden effects on the expression of the genome.* Although geneticists can insert specific genes and create GE organisms that express desired traits, they have little control over the unpredictable interaction of inserted genes with the organism's own genes. A GE organism could, for example, exhibit resistance to a particular fungal disease but have a hidden vulnerability to a bacterial disease.
- *Accelerated evolution of pesticide-resistant pests.* When a pest species is confronted with an environment consisting entirely of a crop producing a specific deterrence compound, natural selection will favor the evolution of resistance to that compound.
- *Creation of "super weeds."* It may be possible for pest-resistance, herbicide-resistance, or improved-vigor genes to move from a GE crop species to a closely related noncrop species or variety, creating weeds even more resistant to human attempts to control them, or capable of disrupting natural ecosystems.
- *Introduction of toxic agents and allergens into the food supplies.* The compounds produced by genes imported into GE organisms may harm human consumers, in addition to deterring pests. Even if the genes are incorporated into varieties only meant for animal feed, they may find their way into the human food supply.
- *Genetic pollution of the environment.* Genes from GE crops may jump to related native species, with unpredictable consequences for natural ecosystems.
- *Harm to wildlife and beneficial species.* Toxins produced by GE crops may kill beneficial insects, pollinators, birds, and other animals, in addition to the targeted pests. Herbicide-resistant GE crops, used in concert with herbicides, can also have a negative effect on beneficials and wildlife species by reducing the number of weeds on which the species depend for food and cover.
- Consolidation of agribusiness control of genetic resources. GE organisms are protected by patents and intellectual property laws. Their increased use reduces agrobiodiversity, makes farmers more dependent on off-farm inputs, and perpetuates the economic divide between developed and developing countries.

In addition to all these problems, a broader objection to transgenic engineering of crop plants (and livestock) is that it has all the pitfalls — potentially magnified — of other modern plant breeding techniques. These are discussed below.

## TRENDS IN THE USE OF GENETIC RESOURCES IN AGRICULTURE

Large-scale conventional agriculture, aided by advances in our knowledge of genetics, has marshaled the genetic resources of domesticated organisms to help create the dramatic yield increases of the twentieth century. But because the creation and deployment of new agricultural varieties and breeds has been directed primarily toward the goal of increasing the profits of agribusiness conglomerates, conventional agriculture has also threatened the foundation of the food system by tending to centralize the control over genetic resources, promote genetic uniformity, and narrow the diversity pool of our crop and livestock species. These trends undermine agriculture's

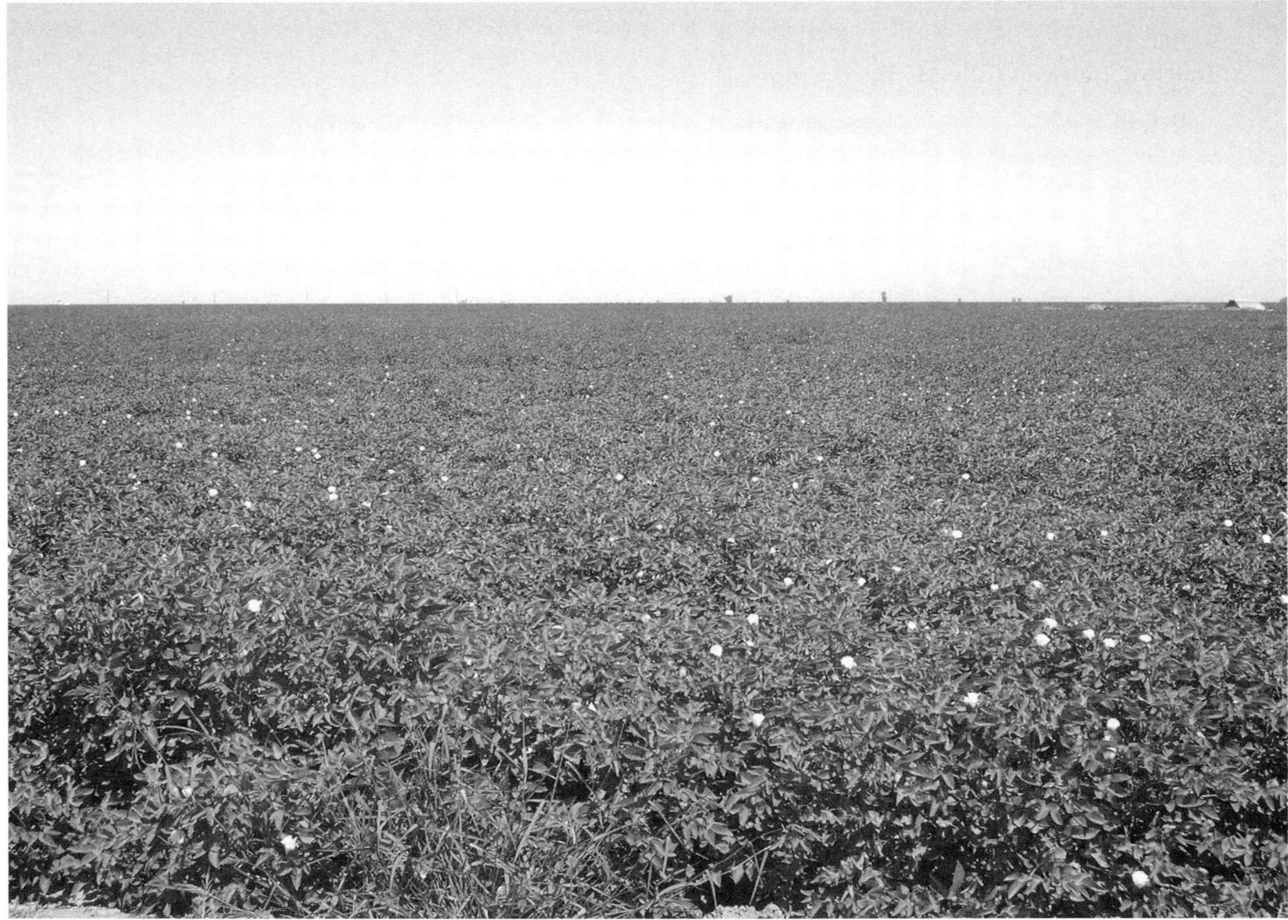

**FIGURE 14.6 Genetically engineered cotton in California's Central Valley.** A few transgenic varieties account for a large percentage of the state's crop.

long-term sustainability by reducing genetic diversity at many levels, making domesticated species more vulnerable to pests, diseases, and environmental changes, and increasing the dependence of cropping and livestock production systems on human intervention and external inputs (Figure 14.6).

### 14.3.2 Loss of Genetic Diversity

Genetic diversity in agriculture, or *agrobiodiversity*, matters in two ways: in the differences among organisms — what can be called diversity's *genetic* component — and in how these differences are arrayed spatially in actual on-the-ground use — what we can term the *geographic* component of diversity (Brookfield, 2001). And for each component, diversity matters at three distinct scales. Geographically, diversity is important at a worldwide scale, a regional or national scale, and a farm scale. Genetically, we can focus on the diversity of food types, the diversity within a species, or the diversity that exists within a particular breed or variety. Since these components of diversity are independent, they combine to create nine different facets of agrobiodiversity, from food diversity worldwide to the genetic diversity of a crop variety on a particular farm. These facets of agrobiodiversity, somewhat simplified, are shown in Table 14.2.

As a result of the ways that conventional agriculture has been exploiting the genetic resources at its disposal over the last century or so, agrobiodiversity is being lost in all nine ways. These trends of loss are also shown in Table 14.2.

There is no shortage of evidence that agrobiodiversity is declining at every geographic scale and every genetic level. This decline is seen in two interrelated ways: fewer and more uniform varieties and breeds are in widespread use, and more varieties and breeds are disappearing from use and being lost altogether. Here are a few telling facts:

- There are perhaps as many as 300,000 edible plant species on earth, but now more than 60% of the world's dietary energy comes from just four of these plant species — wheat, rice,

**TABLE 14.2**
**Facets of Agrobiodiversity, with Trends for Each**

| Genetic Component | Geographic Component: World Food System | Geographic Component: Region or Country | Geographic Component: Farm or Field |
|---|---|---|---|
| Food diversity: number of food types and species grown or raised for food | Fewer species are satisfying food needs globally. *Example*: about 60% of the world's dietary energy comes from four plant species — wheat, rice, corn, and potatoes | Regions and countries are increasingly likely to specialize in a few crops or livestock types | It is increasingly common for an individual farming operation to raise one type of livestock or one type of crop (monoculture) |
| Species diversity: number of breeds or varieties of each food species | Fewer varieties and breeds are being grown, and many of the others are going extinct. *Examples*: three varieties of oranges make up 90% of Florida's orange crop; four varieties of potatoes produce over 70% of the world crop | | It is increasingly common for an individual farming operation to grow or raise one genetic line |
| Variety or breed diversity: number of unique genomes in the plant variety, or degree of uniformity in the livestock breed | Pure line, synthetic, hybrid, and transgenic varieties — all highly uniform — make up an increasing percentage of crops grown worldwide | | It is increasingly common for an individual farming operation to plant a single genome. *Example:* a farm that grows a strain of hybrid corn |

corn, and potatoes (FAO, 1999; Nierenberg and Halweil, 2004).

- In 1993, 71% of the commercial corn crop in the U.S. came from six varieties, 65% of the rice from only four varieties, and 50% of the wheat from nine varieties (Raeburn, 1995).
- 70% of the U.S. dairy herd is Holstein, and almost all chicken eggs sold (more than 90%) are laid by one breed, the white leghorn (Halweil, 2004).
- Since 1900, more than 6000 known varieties of apples (86% of those ever recorded) have become extinct (Fowler and Mooney, 1990), as have half of the domesticated animal breeds in Europe, and about 1000 breeds of poultry and cattle worldwide (Hall and Ruane, 1993).
- In Iran, one of the cradles of agrobiodiversity, only about one quarter of the total number of varieties of wheat, rice, and sorghum account for 70 to 85% of the total area planted in these crops (Koocheki et al., 2006).
- Worldwide, the FAO estimates that 28 to 43% of animal breeds are in danger of extinction, and that somewhere in the world at least one breed of traditional livestock dies out each week (Thrupp, 2004; FAO, 1999).
- About four fifths of the maize varieties known in Mexico in 1930 have been lost.
- During the 20th century, about 75% of plant genetic diversity has been lost as farmers worldwide have abandoned their local varieties and landraces for genetically uniform, high-yielding varieties (HYVs) (FAO, 1999; Nierenberg and Halweil, 2004).

The loss of agrobiodiversity is a cause for concern because it represents the loss of potentially valuable information. If the accumulated genetic resources of thousands of years of plant and livestock breeding and domestication can be likened to a library full of books, old and new, on a vast array of subjects, then the impact of trends in conventional agriculture can be c ompared to replacing that library with one that loans only the current best-selling paperbacks.

The genetic information we are losing has a variety of proven and potential values:

- Genetic diversity in general is the raw material for plant and animal breeding. Loss of this diversity may restrict opportunities for future breeding efforts.
- Genetic diversity in a crop or livestock species, as manifested by the existence of many local landraces and breeds, allows the use of genetic lines that are well adapted to the particular conditions of specific localities. Locally adapted genetic lines, both in crops and in livestock, require fewer external inputs and are therefore a basis for sustainable systems.
- Genetic diversity in a crop variety or livestock breed is an important component of environmental resistance in the field. The broader the

genetic basis of a crop or breed, the greater the chance that some individuals will have innate resistance to disease, unusual variations in environmental conditions, or herbivore attack in the case of crops, preventing total crop loss or herd decimation if one of these events should occur.

- Genetic diversity is also a reservoir of potential environmental resistance. A few individuals in a genetically diverse crop variety or livestock breed may have genes or gene combinations that may confer resistance to future events or conditions, such as the spread of a new disease. These genes may be selected in a population to provide it with resistance.
- Genetic diversity in crops ensures a reservoir of traits with potential value for satisfying (in developed countries) the growing consumer demand for organically grown foods with higher nutritive value. Varieties with innate disease and pest resistance are much easier to grow in more-sustainable low-input organic systems.
- Genetic diversity gives a system overall long-term flexibility, the ability to adjust and adapt to changes in conditions from season to season and from decade to decade.

Some farmers, geneticists, plant breeders, and others saw, several decades ago, the dangers of losing genetic diversity in our food crops. One response was the establishment of "gene banks," where the seeds of varieties and cultivars not in general use would be stored for possible later use. These gene banks serve an important purpose (Qualset and Shands, 2005), but are limited in what they can do to stem genetic erosion. First, the vast majority of current gene banks only maintain stocks of crops that have national and international research programs supporting them, and even then, only a fraction of the genetic diversity of protected crops has been collected. Second, management and evaluation of genetic resources within gene banks are often lacking, so that deterioration of material occurs. Third, germplasm collections are really static, with no incorporation of the processes that maintain and create genetic diversity in the first place, including both environmental and cultural selection pressures. Unfortunately, we may never know how many varieties have already been lost, especially for the large number of minor crops that meet local needs around the world, but are not part of current germplasm preservation efforts.

## Greater Genetic Uniformity in Crop Varieties

The erosion of diversity at the level of the variety or breed deserves a closer look. Therefore, we examine this subject in greater detail as it applies to crop plants.

**FIGURE 14.7 An endangered variety of corn from rural Mexico.** Many pressures have pushed farmers away from using their local varieties, and many of those that are left are being contaminated by genetic material from genetically engineered varieties.

All higher organisms have very complex genetic structures. A great many genes — a single plant can contain upward of 10 million — all work together in complex ways to control the way the organism functions and interacts with its environment. Some genes act alone, but most appear to act in complex combinations with others. In nature, each species' genetic totality, or genome, is the product of a very long evolutionary process, as described above. The genome as a whole is typically very diverse because it is made up of many individual genotypes, many, or all of them unique.

Traditional methods of mass selection, though changing the content of a species' genome, tend to preserve much of its genetically rich structure (Figure 14.7). Modern crop breeding, in contrast, tends to both alter and narrow a crop variety's genome by focusing on the optimization of one or a few genotypes of the variety. Although this process creates plants that perform exceedingly well in specific, highly altered modern agricultural environments, it also greatly restricts a variety's genetic basis. At the most uniform end of the scale, the genetic diversity of a crop variety is restricted to a single genome — that of the hybrid seed of that variety. At the most diverse end of the scale, the genetic diversity of a mass-selected, open-pollinated variety is the product of countless unique individual genomes. Figure 14.8 illustrates this contrast in the structure of genetic diversity.

Commercially produced, hybrid, high-yielding varieties (HYVs), have captured the seed market and are now planted over large areas in genetically uniform fields. Their dominance is challenged only by equally uniform

Mass-selected, open-pollinated variety

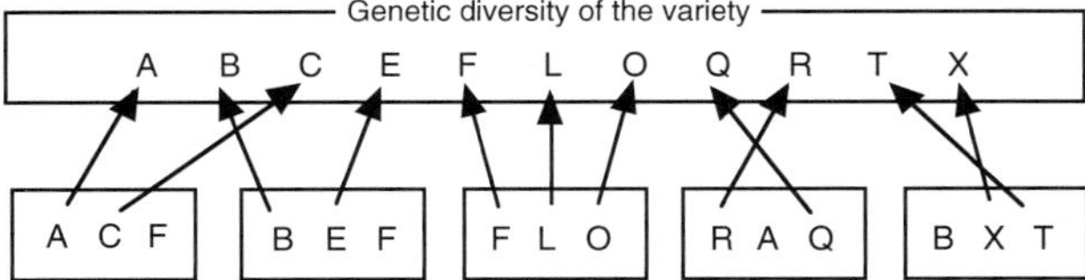


Genes and gene combinations in individuals' genotypes

Hybrid variety

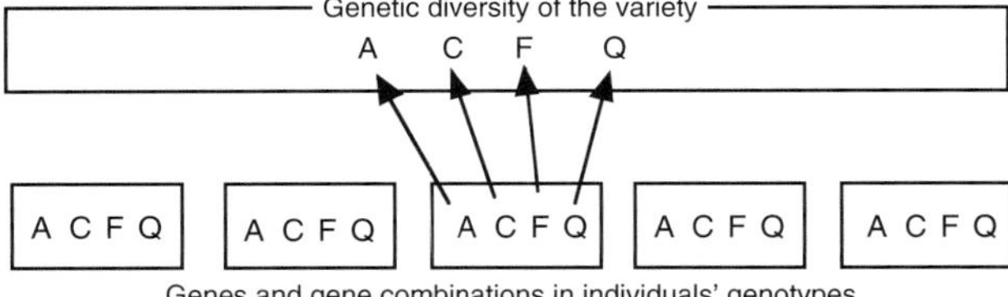


Genes and gene combinations in individuals' genotypes

**FIGURE 14.8 Genetic diversity in a mass-selected crop variety and a hybrid crop variety.** In a mass-selected variety, overall genetic diversity is much greater than that of any individual; in a hybrid variety, any individual contains all the genetic diversity of the variety.

GE crops. This situation, along with the other types of loss of diversity, makes our crops increasingly vulnerable to the age-old enemies of agriculture — pests, diseases, and unusual weather.

## Genetic Vulnerability

This consequence of the loss of genetic diversity in crop plants and livestock deserves further discussion. *Genetic vulnerability* is the susceptibility of the narrowed genetic stock of plants and animals to attack by pests and diseases, or to losses caused by extremes in the weather. The basic problem is that when a crop variety or livestock breed is genetically uniform over a large area, the ideal conditions for the rapid outbreak of a pest or disease population are in place.

Pest and disease populations evolve at a relatively rapid rate, in part, because of their short generation time. With this capacity for rapid genetic change, they can adapt quickly to changes in their hosts' defenses — or to factors (such as pesticides) introduced into the environment by humans. For this reason, pests and diseases in agriculture have been able to (and might always be able to) overcome just about everything agricultural science has thrown at them, from pesticides and antibiotics to resistant varieties to new practices.

In traditional agroecosystems, where crop plants are subjected to both natural and human-imposed selection pressures and the system retains many of the characteristics of a natural ecosystem, crop plants have a fighting chance to stay one step ahead of pathogens and herbivores. But with modern plant breeding, large-scale monocultures, and uniformity of farming practices, we have given pests and diseases the advantage. We strive to change both the genetic and environmental mechanisms of resistance by breeding crops of specific traits rather than general fitness, and by planting crops in large single-species populations at the same time in the same place. This creates an environment that is more uniform and predictable than it might otherwise be, setting the stage for outbreaks to occur.

Moreover, changes independent of agriculture are increasing the threat of serious outbreaks of disease and pests. The interconnections of global commerce give pathogens even more vectors for expansion into new areas, and climate change threatens to both allow pathogens to move into areas where the climate formerly excluded them, and to put stresses on crop plants that make them more vulnerable to pest and pathogen attack.

The increasing uniformity of the genetic base in livestock production also makes poultry, cattle, swine, goats, and sheep more vulnerable to the spread of disease. In more traditional pastoral systems, livestock breeds vary regionally and are well adapted to local conditions, providing good resistance to disease. In contrast, modern confined animal feeding operations (CAFOs) pack large numbers of genetically similar animals together in an artificial context that is the perfect setting for the rapid spread of a pathogen.

One of the most well known examples of the dangers of genetic uniformity is the Irish Potato Blight. In 1846 the late-blight fungus (*Phytophthora infestans*) destroyed half of Ireland's potato crop, causing widespread famine and forcing a quarter of the population to emigrate. The blight occurred because Irish potato farmers had developed a dependence on only two potato genotypes that had been brought to the country over 300 years before and then vegetatively propagated; the blight had such a profound impact because the country had become overly dependent on the carbohydrate-rich potato as a food source. The fungus was well adapted to the cool, moist conditions of the region, and once the disease arrived and got established, there was no stopping it. Interestingly, the same fungus is also found in the place of origin of the potato, the Andes of South America, but the great genetic diversity of potatoes there, combined with ongoing natural selection, ensures that a large proportion of the crop will be resistant.

Another well-known example is the 1970 to 1971 outbreak of southern corn leaf blight (*Helminthosporium maydis*), which destroyed almost the entire corn crop in areas of Illinois and Indiana and resulted in the loss of more than 15% of the corn crop in the U.S. as a whole (Ullstrup, 1972). This outbreak was linked to the genetic factors for cytosterility bred into the lines of corn used to produce hybrid seed. These factors produced male sterility and eliminated the need for expensive hand detasseling, but they also increased the hybrid's susceptibility to southern corn leaf blight. When a new strain of the blight appeared, therefore, it spread quickly. Seed producers and crop breeders were able to respond quickly and altered the combination of susceptible factors by the 1972 season.

Similar problems have been encountered with rice in Southeast Asia. In the 1970s, the International Rice Research Institute began releasing rice varieties with resistance to specific pests, promoted them for planting over broad geographic ranges. A short time after each variety was adopted and planted widely, newly evolved biotypes of the pests, such as the brown leafhopper, overcame the resistance and decimated the crops. Each new variety of rice lasted only 2 or 3 years before its level of resistance was overcome by the rapid evolution by the pest (Chang, 1984). The lesson is clear: as long as only a few varieties dominate, pests will be able to take advantage of the low genetic diversity of the crop and overcome its resistance. When failure occurs, farmers are totally dependent on the infrastructure that produces new resistant varieties (or provides chemical pesticides) since they no longer have access to the genetic variability that used to be present in their own fields (Altieri and Merrick, 1987).

In many ways, the overall success of agriculture in developed countries over the past 3 decades has masked the problem of genetic vulnerability. Surplus yields in some regions can compensate for failures elsewhere. But the regional failures are still happening, and the potential exists for failures on a larger scale (Qualset and Shands, 2005).

### Increased Dependence on Human Intervention

There is an important link between conventional agriculture's control of genetic resources and its dependence on external inputs, mechanization, and off-the-farm technological expertise. The dramatic reduction of the genetic diversity of our crops and livestock breeds has been closely paralleled by the dramatic increases in pesticide and fertilizer production, irrigation and water use, mechanization, and agricultural use of fossil fuels.

This link is very clear in the widespread use of hybrid crop seeds. A modern hybrid crop variety is virtually helpless outside the confines of the farm — it usually cannot even reproduce itself from its own seed. At the greatest extreme, the crop cannot succeed in a farming system without very specific kinds of intensive, technology-based human modification and control of the farm environment.

When a farmer abandons local crop varieties for hybrids, it is more than the hybrid seed that has to be purchased. Every hybrid has a "package" of inputs and practices that go along with the seed: soil cultivation equipment, irrigation systems, soil amendments and fertilizers, pest control materials, and other on-farm inputs. The package also includes changes in many other aspects of the farm organization and management as well. In order to recover the investment necessary to pay for these new inputs and equipment, farmers often must intensify the production of more profitable crops. This usually requires a concentration of production in fewer and fewer crops, a dependence on centralized market structures, a differently skilled labor force, and further intensification of inputs to reduce risk and the chance of crop failure. Technical advice is relied upon (and usually paid for) from sources outside the farm environment. The entire farm is forced to change.

These changes too often result in farmers losing the important local, traditional knowledge they have about crops, the farm, and the farming process and relying on genetic information that was developed under highly uniform, highly modified conditions. Cumulatively, the end result is the loss of the local genetic diversity and cultural experience that characterized farms before modernization.

The link between the erosion of genetic diversity in livestock breeds and the input-intensiveness of conventional livestock production, especially in CAFOs, has already been touched upon in Chapter 1. The whole rationale of large-scale livestock production — to produce large quantities of animal-derived food products at the lowest cost — depends both on the management efficiency gained through genetic uniformity and the tightly controlled, input-dependent environment in which the animals are raised. This topic will be discussed in greater depth in Chapter 19.

### Loss of Other Genetic Resources

Agriculture depends on more than just the genetic diversity of crop plants and domesticated animals. Also important is the genetic diversity of an array of other organisms: (1) organisms in the natural ecosystems surrounding agroecosystems, especially the wild relatives of crop plants; (2) crops and animal breeds of minor economic importance; and (3) beneficial noncrop organisms such as parasitoids, allelopathic weeds, trees, and soil organisms.

Wild relatives of crops are an important source of new or novel variation in the directed selection process. They have been important sources of new or stronger genetic material, especially in the event of epidemics of the type mentioned above. However, wild relatives, such as the wild cotton in Figure 14.9, are disappearing rapidly in many parts of the world because of deforestation and other forms of habitat modification.

A similar kind of organism with potential value is the natural cross between an escaped agricultural variety and its wild relative. Such crosses are endangered as well because the habitats where crops and wild relatives can exchange genetic material are becoming rarer, mainly due to the spread of hybrid seed into even the most remote agricultural parts of the world, the simplification of the

**FIGURE 14.9 Wild perennial cotton (Gossypium sp.), Tabasco, Mexico.** Wild relatives of crops can still be found *in situ* in traditional farming systems.

farming environment that accompanies the use of improved varieties, and the increasing separation between agricultural and natural ecosystems.

Diverse agricultural habitats also contain many minor crop species that are of considerable importance for the entire system. Besides providing an array of harvestable useful products, these crops contribute to the ecological diversity of the system. They are part of the whole-system energy flow and nutrient cycling process. Minor crops of little or no current commercial value are preserved in many traditional cropping systems, especially in the developing world. They could have promising value for future use, but they too are disappearing as traditional systems give way to modernization.

Apart from crops and crop relatives, agroecosystems are also made up of a diversity of noncrop plants and animals, including predators and parasites of pests, allelopathic weeds, and beneficial soil organisms. Many of these can play very important roles in maintaining overall system diversity and stability (Chapter 16). Since their presence and genetic diversity depends to a great extent on the overall diversity of the system, they are threatened by the tendency toward agroecosystem uniformity.

More generally, attention needs to be paid to the overall genetic diversity of agroecosystems. A fully functioning crop and farm system preserves all the genetic, ecological, and cultural processes that produce diversity in the first place. Biological control information, plant defenses, symbionts, and competitors are all actively interacting and preserving genetically based information that is of great agroecological value. And since only a fraction of all this information is in the germplasm of the key crop, loss of farming habitats can be even more devastating than narrowing of the crop gene pool itself.

## PRESERVING AGROBIODIVERSITY

Sustainability requires a fundamental shift in how we manage and manipulate the genetic resources in agroecosystems. The key theme in this shift is genetic diversity. Sustainable agroecosystems are genetically diverse at every level, from the genome of the individual organisms to the system as a whole. And this diversity should be a product of coevolution — genetic changes should have occurred in an environment of interaction among the various populations. In this way, all of the component organisms — crop plants, animals, noncrop associates, beneficial organisms, and so on — are adapted to local conditions and the local variability of the environment, in addition to possessing traits that make them specifically useful to humans.

Traditional, indigenous, and local agroecosystems contain many of the genetic elements of sustainability, and we can learn from their example. In particular, they have higher genetic diversity within populations as well as in the cropping community as a whole. Intercropping is much more common, noncrop species and wild relatives occur within and around cropping fields, and opportunities for genetic diversification are abundant at the field level. In such systems, resistance to environmental stress and biotic pressures has a much broader genetic base, genetic vulnerability is lower, and while pests and diseases occur, catastrophic outbreak is rare. In essence, genetic change in such systems takes place much like it does in natural ecosystems.

## Breeding for Durable Resistance in Crop Plants

Agricultural plant breeding has focused mainly on creating resistance to limiting factors of the environment, be they physical factors such as drought, poor soils, and temperature extremes, or biological factors such as herbivory, disease, and competition from weeds. Remarkable gains in yield have been achieved as a result of these breeding programs, but as we have already noted, another result is increased vulnerability to crop failure and increasing reliance on nonrenewable inputs.

As each problem presents itself, crop breeders screen the genetic variability of a crop until they find a resistant genotype. This resistance is often provided by a single gene. The gene-transfer and backcrossing techniques described above are employed to incorporate the gene into a specific crop pedigree. The result is sometimes called vertical resistance. It has two weaknesses. First, the resistance will continue to function only as long as the limiting factor does not change. Unfortunately, in the case of pests, diseases, and weeds, the limiting factor is never static for very long because of continual natural selection. So, the problem organism eventually develops "resistance to the resistance," and an outbreak or epidemic occurs. This dynamic is the basis of the well-known crop breeders' treadmill. Second, in the process of breeding for vertical resistance, genes providing partial resistance to the wider spectrum of pathogens are lost.

A more durable type of resistance is needed that does not break down easily in the face of new strains of pests, diseases, or weeds. Rather than directing breeding programs towards the development of specific resistances, the idea is to manage the whole crop system. Selection for durable resistance requires the accumulation of many resistance characters using population-level breeding methods, and relies on an understanding of the simultaneous nature of the interaction between a crop, pests, the environment, and the human managers. Selection takes place at all levels at the same time, rather than for single specific characters. The more durable type of resistance that results is termed *horizontal resistance* (Robinson, 1996). An example of this type of breeding program is described in the case study *Breeding for Horizontal Resistance in Beans*.

Breeding methods that provide the most durable resistance rely on the use of open-pollinated, locally adapted landraces. Open-pollinated crops are generally lower yielding when compared to hybrid varieties, but they are very responsive to local selection pressures because of their genetic diversity. They also have the best average performance in the face of the combination of all the local environmental factors, including pests, diseases, and weeds.

The importance of system-level resistance is accepted more easily by ecologists than by agricultural scientists. The study of selection in natural ecosystems has repeatedly demonstrated the ways a wild ecotype responds to either positive or negative selection pressures when it is introduced into an ecosystem different from the one in which it evolved. Selection operates simultaneously at the level of all of the factors, biotic and abiotic, that the organism encounters. Seen in this light, the problems associated with genetic uniformity in crops become more apparent. resistant v

## On-Site Selection and Conservation of Plant Genetic Resources

The concern for the erosion and loss of genetic resources led to the establishment, in 1974, of the International Board of Plant Genetic Resources (IBPGR). An international network of *ex situ* (off-site) crop germplasm repositories was established and genetic material from the major crop gene centers was collected in order to establish the IBPGR system of genebanks. Plant breeders have since relied heavily on these genetic resources for the conventional development of higher-yielding and resistant varieties, and the number of genebanks of all types has increased to an FAO-estimated 1460 worldwide, which together hold more than 5.4 million samples. In 2004, the FAO and the 15 Future Harvest Centers of the Consultative Group on International Agricultural Research partnered to create the Crop Diversity Trust, an independent international organization charged with assuring the long-term security of our most important collections of crop diversity. In particular, the Trust seeks to salvage collections that are at risk and to assist developing countries with managing their collections.

Although *ex situ* conservation is important, it cannot by itself stem the erosion of agricultural genetic diversity. Limited funding for genebanks has restricted the range of crops and regions from which material is collected, leaving much of the world's crop genetic diversity out of these reservoirs. Corn, wheat, beans, rice, and potatoes have received the most attention, excluding a very large number of much of the world's food crops. An added problem is that these *ex situ* genetic conservation efforts remove crops from their original cultural-ecological context, severing the adaptive tie between genome and environment (Hamilton, 1994; Nevo, 1998).

To achieve sustainability, conservation of genetic resources must also occur *in situ* or in the setting of the crop community (Brush, 2004). *In situ* conservation involves ongoing selection and genetic change, rather than static preservation. It allows genetic screening to occur, maintaining and strengthening local landraces. It attempts to mimic all the conditions — location, timing, and cultivation techniques — under which future cultivation of the crop will occur. As a result, cultivars remain well adapted to (1) the conditions of the local environment, (2) the cultural conditions of the local environment (such as irrigation, cultivation, and fertilization), and (3) all the locally important biotic crop problems (such as pests, diseases, and weeds).

*In situ* conservation requires that farms be the repositories of genetic information and farmers the repositories of the cultural knowledge of how crops are cared for and managed. At one extreme, therefore, the principle of *in situ* conservation argues for each farm having its own breeding and preservation program. Indeed, farmers ought to be able to select and preserve their own locally adapted landraces, where feasible. But a more practical approach focuses at the regional level (Figure 14.1). Because regional characteristics of a farming region establish important selection criteria, screening programs can be centralized to a certain extent for a particular geographically- and ecologically-defined region, as long as constant exchange of crop genetic material takes place among farmers of that region (Brush, 1995; Cunningham, 2001).

Ultimately, *in situ* and *ex situ* genetic resource conservation efforts must be integrated. Already, partnerships between nonprofit groups and farmers show that the two kinds of programs can complement each other and

## BREEDING FOR HORIZONTAL RESISTANCE IN BEANS

In the Mixteca bean-growing region of Mexico, farmers confront a variety of potent disease organisms. Bean common mosaic virus (BCMV), the common blight bacterium (*Xanthomonas* sp.), and the fungus *Macrophomina* sp., a soil-borne root pathogen, are three of the worst among a large cast of viruses, bacteria, and fungi. Using traditional breeding techniques, farmers have stayed a half-step ahead of these diseases — generally avoiding disastrous epidemics but always losing some potential production because of disease-weakened plants.

More than 10 years ago, Dr. Roberto García-Espinosa at the Colegio de Postgraduados (College of Postgraduates), in Montecillos, Mexico, began a bean-breeding program designed to help Mixtecan farmers increase bean yields without using fungicides. His approach was to develop bean varieties with horizontal resistance — the ability to resist the pathogen community as a whole, or what he calls the "pathosystem."

García-Espinosa's trials show that he has succeeded remarkably well. More than 5 years of data have consistently demonstrated that his varieties are more disease resistant than their conventional counterparts, and produce yields up to four times larger without the use of fungicides (García-Espinosa et al., 2003; Lotter, 2004).

García-Espinosa's breeding process is a radical departure from that of mainstream plant breeders, who focus on developing complete resistance to specific races of pathogens, or vertical resistance. His first step is to eliminate vertical resistance from the parental lines, because plants with vertical resistance tend to lose their horizontal resistance. He collects as wide a spectrum of germplasm as possible, plants these different genetic lines, and inoculates the plants with pathogens at a level intended to be sublethal. Then among the surviving plants, he collects seeds from those that were *most susceptible* to the diseases. If he were a mainstream breeder, he would be doing the exact opposite.

After this susceptibility screening, García-Espinosa has a handful of genetic lines with no vertical resistance, but as-yet-unexpressed horizontal resistance. He hand-cross-pollinates these lines, plants out the progeny, and subjects them to the same diseases. This time, however, he selects seeds based on resistance to the diseases. This process, called recurrent mass selection, is repeated at least three times, with only 1 to 10% of plants selected at each round. This allows the selected cultivars to accumulate polygenes for broad resistance. Then, the selected cultivars are planted out in the bean-growing region, where dozens of diseases and pest insects exist. In these additional on-site breeding cycles, resistance to all the locally active pests and diseases is looked out for.

This breeding process produces varieties that yield 2000 to 2400 kg/ha, compared to 1500 kg/ha for commercial and green revolution varieties. Despite this success, García-Espinosa has trouble attracting funding for his work, both in breeding the resistant varieties and in disseminating them to farmers (Figure 14.10). García-Espinosa attributes this to several factors. A big problem is that his breeding process does not fit into conventional notions of plant breeding. But perhaps more importantly, his horizontally-resistant cultivars cannot help agribusiness corporations turn a bigger profit, because they are difficult to patent and they don't require agrochemical inputs (Lotter, 2004).

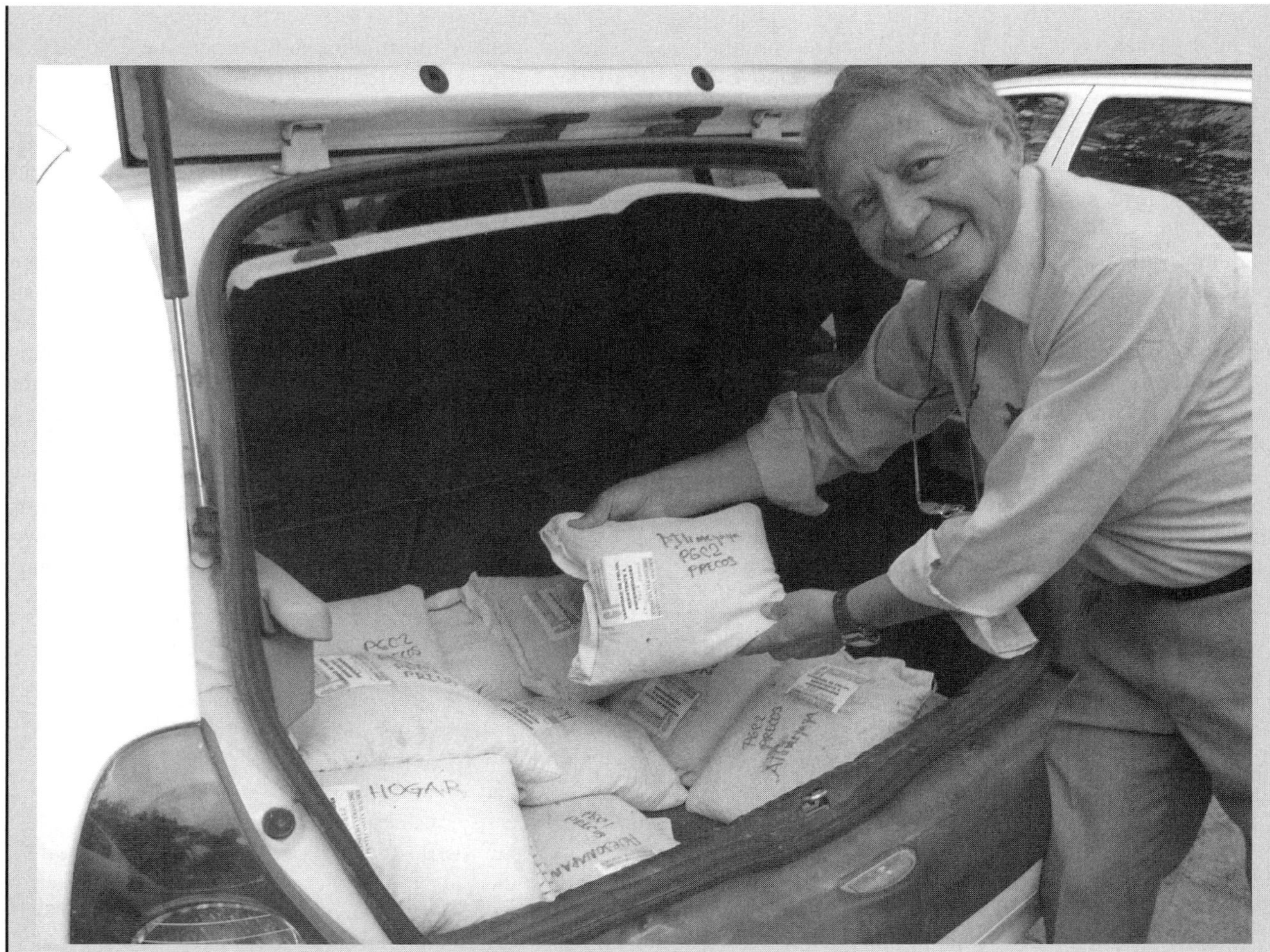


**FIGURE 14.10 Dr. Roberto García-Espinosa, with seeds of some of the horizontally resistant beans he has bred.** Lacking funds for distributing the seeds to farmers, he does the job himself from the back of his car. (Photo courtesy of Dr. Don Lotter.)

promote more effective and equitable conservation. The organization Native Seeds/SEARCH in Tucson, Arizona, for example, complements its *ex situ* seed collection and storage activities by encouraging farmers to grow local and traditional varieties of crops. The organization provides seeds for farmers who have lost varieties, and then purchases the farmers' excess production. The farmers' own fields, then, become the sites for both the retention of traditional genetic resources as well as the screening grounds for the varieties of the future. When these fields also use local knowledge, local resources, and limited industrial inputs, breeding for sustainability can take place (Tuxill and Nabhan, 2001).

## Preserving Minor Crops and Noncrop Resources

Genetic resources in agroecosystems extend beyond the relatively few crop species that today provide the bulk of the food consumed by much of the human population. Locally important, minor, or underutilized crops, as well as a range of noncrop species with potential as new crops, all form part of the genetic resources available for breeding programs for sustainable agriculture. They also form part of the whole-system, horizontal resistance process that is essential for maintaining a genetic basis for sustainable agricultural systems. It is important, therefore, to extend genetic conservation efforts to include all these other types of crop, noncrop, and wild-relative species. This goal is best achieved by preserving the traditional agroecosystems in which these species occur (Altieri and Nicholls, 2004a).

## Valuing Genetic Diversity in Livestock

Genetic diversity in livestock is valuable for the same reasons as genetic diversity in crops: a diversity of breeds (and diversity within a breed) gives farmers the raw material for selecting stocks or developing new breeds in response to environmental changes, new disease threats,

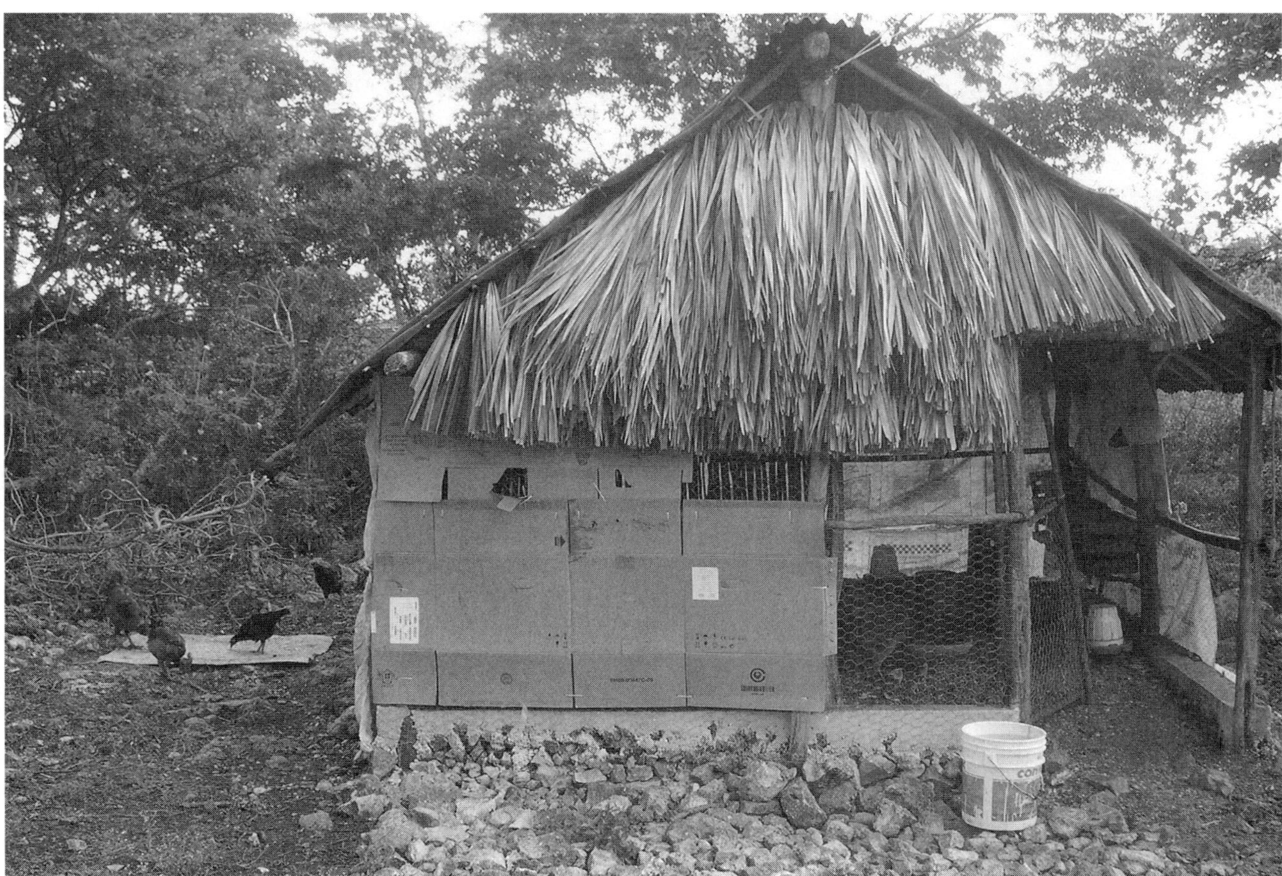

**FIGURE 14.11 Chickens of an endangered locally adapted breed in Mexico's Yucatan Peninsula and their rustic chicken house.** School children rear and promote the chickens as part of a school Forest Garden project in the town of Cepeda. Unlike high-yielding modern varieties, these chickens are free ranging, resistant to the hot Mayan lowlands, and are able to subsist with minimal external inputs while providing an excellent protein complement to local diets.

and changing market conditions or consumer preferences. Livestock breeds indigenous to particular locales often have valuable traits such as high fertility, good maternal instincts, disease resistance, ability to thrive on poor quality feed, adaptation to harsh conditions, longevity, and unique product characteristics. All of these traits are desirable — and necessary — for low-input livestock production, and are often missing in the widespread commercial breeds. Thus, indigenous breeds of poultry, swine, cattle, and other livestock types are crucial for sustainable livestock production.

Despite the value of genetic diversity in livestock, however, it is at greater risk than the genetic diversity of our crop plants, in part, because an animal's genome cannot easily be stored in a genebank. The FAO has estimated that as many as 43% of the livestock breeds in the world are threatened with extinction.

Recognizing the multiple roles that animals can play in sustainability offers a chance to reverse this trend. As we will see in Chapter 19, local and traditional breeds can play a critical role in the process of reintegrating livestock animals into crop production systems — a necessary step in the creation of sustainable food systems.

## CONCLUSIONS

Growing concern about the negative impacts of external human inputs on agroecosystem sustainability, coupled with regulations limiting the types of inputs farmers can use, is bringing renewed interest in breeding defenses and resistance back into crop organisms. This may result in changes in the genetic basis of adaptations, but we must also change the environmental background. If we continue to plant large monocultures, and focus only on resistance to particular stresses or problems, without determining why those problems occur in the agroecosystem in the first place, we will continue to select the very problems we are trying to avoid (Table 14.3).

The goal of agroecology is to apply ecological knowledge to the design and management of sustainable agroecosystems. If we are to pursue this goal, we need to broaden the context of our plant and livestock breeding

**TABLE 14.3**
**Genetic Resources and Processes of Importance in Sustainable Agriculture**

| Resource or Process | Advantage for Sustainability |
|---|---|
| Broad genetic base in the form of many landraces and developed varieties | Reduces genetic vulnerability; allows continued production of genetic variation |
| Variable gene frequency within and among landraces | Reduces genetic vulnerability |
| Gene flow within and between landraces, occasionally from wild relatives | Maintains variability, diversity, and environmental resistance |
| Selection for diversity of local adaptations | Maintains local flexibility in environmental resistance |
| Relatively small populations | Promotes diversity due to genetic drift |
| Open pollination breeding systems | Promotes outcrossing; maintains variability |
| Longer life cycles | Promotes outcrossing |
| Regional, patchy distribution | Promotes diversity |
| Presence of wild relatives | Can lead to spontaneous hybrids and variation |
| Local breeding | Promotes diversity and adaptability; maintains environmental resistance |
| Flexible and diverse environmental conditions on the farm (e.g., intercropping) | Provides microsites for retention of variable genetic lines |
| High overall diversity in the agroecosystem | Allows for interaction and development of more complex interdependencies and coevolution |

*Source*: Adapted from Salick J, and Merrick LC. In C. R. Carroll, J. H. Vandermeer, and P. M. Rosset (eds.), *Agroecology*. pp. 517–548. McGraw-Hill: New York. 1990.

efforts so that they work with all the multiple levels of the farm system. We need to reduce vulnerability and dependence on human interference through a strategy of diversifying the agricultural landscape, the crop species in agroecosystems, the varietal composition within species, and the resistance mechanisms within varieties. Otherwise, we end up stuck on the crop breeder's treadmill, trying to stay a short step ahead of the problems created by the very systems we have designed.

## FOOD FOR THOUGHT

1. What are the similarities and differences between an obligate mutualism in a natural ecosystem and the relationship between humans and their domesticated organisms?
2. What can we learn from traditional farming systems in developing countries about applying directed selection in a way that promotes sustainability?
3. What are the weaknesses of a germplasm preservation program that focuses only on the key crops and the storing of genetic material in large environmentally controlled germplasm banks isolated from the field situation?
4. How do your own personal choices at the market exert pressure on the selection of the genetic material used by farmers?
5. What is meant by "agroecosystem selection" in the directed selection process?

## INTERNET RESOURCES

Biological Diversity in Food and Agriculture
http://www.fao.org/biodiversity
The agrobiodiversity section of the United Nations' Food and Agriculture Organization site. A portal to a great deal of information and data about agrobiodiversity.

Center for Food Safety
www.centerforfoodsafety.org
This organization's site has a great deal of good information related to the hazards of genetically engineered crops.

International Plant Genetics Resources Institute
www.ipgri.cgiar.org
The world's largest nonprofit agricultural research and training organization devoted to promoting agrobiodiversity.

Native Seeds/SEARCH
www.nativeseeds.org
This organization works to preserve the many locally adapted plant varieties used by indigenous groups in the Americas.

People and Plants International
www.peopleandplants.org
An organization devoted to sustainable resource management, and focused on the preservation of plant biodiversity worldwide.

Union of Concerned Scientists, genetic engineering section
www.ucusa.org/food_and_environment/genetic_engineering
A wealth of information and research on the risks of genetic engineering.

## RECOMMENDED READING


Bains, W. 1998. *Biotechnology from A to Z.* 2nd ed. Oxford University Press: New York. An introduction to the field of biotechnology in all of its forms and approaches.

Bellwood, P. 2004. *First Farmers: the Origins of Agricultural Societies.* Blackwell Science: London. An archaeological perspective on the origins and history of agriculture and crop domestication.

Brookfield, H. 2001. *Exploring Agrodiversity.* Columbia University Press: New York. An integrated overview of the concept of diversity in agriculture, focusing as much on the choice of crops as on the management of land, water, and biota as a whole.

Brush, 2004. *Farmer's Bounty: Locating Diversity in the Contemporary World.* Yale University Press: New Haven, CT. A thorough assessment of the present state of crop diversity worldwide, written from the standpoint of an anthropologist but with a wide scope that includes the work of ecologists, geneticists, and ethnobotanists.

Chrispeels, M.J. and D.E. Sadava (eds.), 2003. *Plants, Genes and Crop Biotechnology.* Jones and Bartlett: Sudbury, MA. An interdisciplinary overview of the pressing issues surrounding agricultural biotechnology and farm management.

Doyle, J. 1985. *Altered Harvest: Agriculture, Genetics, and the Fate of the World's Food Supply.* Viking Penguin: New York. A review of the social and economic factors that have switched the emphasis from genetic diversity to chemical inputs for increased production, while limiting access to seed for farmers of the world.

Fox, C.W., D.A. Roff, and D.J. Fairbairn. (eds.), 2001. *Evolutionary Ecology: Concepts and Case Studies.* Oxford University Press: New York. A synthetic view of the field of evolutionary ecology, viewed as an integration of ecology and evolutionary biology.

Gaston, K.J. and J.I. Spicer. 2004. *Biodiversity: an Introduction.* 2nd ed. Blackwell Science: Malden, MA. An overview of what biodiversity is, its relevance to humanity and issues related to its conservation.

Gliessman, S.R. 1993. "Managing diversity in traditional agroecosystems of tropical Mexico." In: Potter, C.S., J.I. Cohen, and D. Janczewski (eds.) *Perspectives on Biodiversity: Cases Studies of Genetic Resource Conservation and Development.* American Association for the Advancement of Science Press: Washington, D.C. pp. 65–74. An example of how diversity is managed in agricultural settings that range from a specific farm field to the natural–agricultural landscape.

Gussow, J.D. 1991. *Chicken Little, Tomato Sauce and Agriculture: Who Will Produce Tomorrow's Food?* The Bootstrap Press: New York. A critical view of many of the problems and dangers inherent in modern breeding and biotechnology programs in today's agriculture.

National Academy of Sciences. 1972. *Genetic Vulnerability of Major Crops.* National Academy Press: Washington, D.C. An early call for concern about the potential risks of narrowing the gene pool for our major crop varieties.

National Academy of Sciences. 1975. Underexploited Tropical Plants with Promising Economic Value. National Academy of Sciences: Washington, D.C. An examination of the potential value and roles of minor crop species, especially those used by small or resource-limited farmers.

Plucknett, D.L., N. Smith, J. Williams, and N. Anisletty. 1987. *Gene Banks and the World's Food.* Princeton University Press: Princeton, New Jersey. A very thorough look at the strengths and weaknesses of gene banks as means for preserving agricultural genetic resources.

Qualset, C. and H. Shands. 2005. Safeguarding the Future of U.S. Agriculture: The Need to Conserve Threatened Collections of Crop Diversity Worldwide. University of California, Division of Agriculture and Natural Resources, Genetic Resources Conservation Program. Davis, CA, USA. Makes a good case for putting more resources into genebanks and emphasizes the value of crop plant diversity for fighting emerging diseases.

Reiss, M.J. and R. Straughan. 2000. *Improving Nature? The Science and Ethics of Genetic Engineering.* Cambridge University Press: Cambridge, UK. An evaluation of the scientific and ethical concerns surrounding genetic engineering, from the integrated visions of a biologist and a philosopher.

Ridley, M. 2004. *Evolution.* 3rd ed. Blackwell Science: Malden, MA. A text that covers the history of evolutionary theory from its origins until present times.

Rissler, J. and M. Mellon. 1996. *Perils Amidst the Promise: Ecological Risks of Transgenic Crops in a Global Market.* MIT Press: Cambridge, Massachusetts. A strong ecological critique of the risks involved in the use of genetically engineered crops for agriculture.

Schurman, R.A., and D.D.T. Kelso (eds.), 2003. *Engineering Trouble: Biotechnology and Its Discontents.* University of California Press: Berkeley. With examples from agriculture, food, forestry, and pharmaceuticals, this book critically examines some of the most contested issues of genetically engineered organisms, including its social and political consequences.

Silvertown, J. and D. Charlesworth. 2001. *Introduction to Plant Population Ecology.* 4th ed. Blackwell Science: Oxford, London, Vermont. The fundamentals of plant genetics presented from a population ecology perspective.

Simmonds, N.W. and J. Smartt. 1999. *Principles of Crop Improvement.* 2nd ed. Blackwell Science: London. A complete review of genetics and crop improvement.

Simpson, B.B. and M.C. Ogorzaly. 2001. *Economic Botany: Plants in Our World.* 3rd ed. McGraw-Hill, Inc.: New York. A very complete and well-illustrated coverage of the useful plants of the world, including aspects of history, morphology, taxonomy, chemistry, and modern use.

# 15 Species Interactions in Crop Communities

In ecological terms, a cropping system is a *community* formed by a complex of interacting populations of crops, weeds, microorganisms, insects, and, sometimes, other animals. The interactions among the populations of the crop community, which arise from the different kinds of interference, give the community characteristics, called emergent qualities, which exist only at the community level. These emergent qualities cannot be fully explained in terms of the properties of populations or individuals. In both natural ecosystems and agroecosystems, community-level phenomena are of critical importance in a system's stability, productivity, and dynamic functioning.

Agricultural researchers, however, normally focus their attention on the crop population of central importance in the farming system, rather than on the community of which it is a part. Because of this reductionist approach, they fail to understand cropping systems as communities, and thereby lose the ability to take advantage of community-level emergent qualities or to manipulate community interactions to the benefit of the cropping system.

To be sure, conventional agriculture has been greatly concerned with species interactions — in the sense that it has focused on the detrimental effects on crop yields, arising from the impacts of noncrop organisms such as weeds, pest herbivores, and diseases on the crop environment. Research for many years has been directed toward eliminating these detrimental effects. Noncrop organisms are said to "compete" with the crop or have a yield-reducing effect; they must therefore be eliminated from the cropping system. At the same time, considerable research has been done to determine the optimum densities for each crop (usually planted as a monoculture) in order to minimize intraspecific competition for resources and thereby obtain maximum yields.

By striving to eliminate and minimize interactions, the conventional approach tends to simplify the crop community. In a sense, the ultimate goal is to reduce it to a single-crop population growing in an otherwise sterile abiotic environment.

In contrast, the agroecological approach to cropping system management is to understand species interactions in the context of the larger community. The agroecologist recognizes the existence of beneficial species interactions, understands how they arise from the impacts of interference, and knows that a certain level of complexity is desirable. By paying attention to the ecology of the crop community, it is possible to create beneficial interactions and emergent qualities that not only reduce the need for external inputs, but also increase overall yields.

## INTERFERENCE AT THE COMMUNITY LEVEL

The basis for understanding species interactions in the context of community structure and function was developed in Chapter 11. There, we discussed how organism–organism interactions can be conceptualized as *interferences*, in which an organism has some kind of impact on its environment, and through this impact, another organism is affected. We identified two types of interferences: removal interferences, in which the environmental impact consists of the removal of some resource by one or both of the interacting organisms, and addition interferences, in which one or both organisms adds some substance or structure to the environment. Either kind of interference can have beneficial, detrimental, or neutral effects on neighboring organisms. As was discussed in Chapter 11, the advantage of the interference approach is that it allows a more complete understanding of the *mechanisms* of interaction.

At the level of the community, the existence of many populations means that many kinds of interferences may be going on at the same time. These many interferences may interact with and modify each other, creating complex relationships among the members of the community. Despite this complexity, we can understand both the individual types of interference that exist between populations and the overall effect of the complex of interferences on the community as a whole because the interference concept allows analysis of the mechanisms of interaction.

Some of the ways in which interferences may combine to affect the crop community are described in Figure 15.1. Direct removal of something from the environment leads to interactions such as competition or herbivory, whereas additions can lead to allelopathy or the production of food for beneficial organisms in the crop community. Both removal and addition interferences may go on simultaneously, leading to different types of interactions. Many mutualisms, for example, arise from combined addition/removal interferences. Examples are pollination (removal of nectar and addition of pollen) and biological nitrogen fixation (addition of fixed nitrogen by the bacteria and removal of nitrogen by the legume). Additionally,

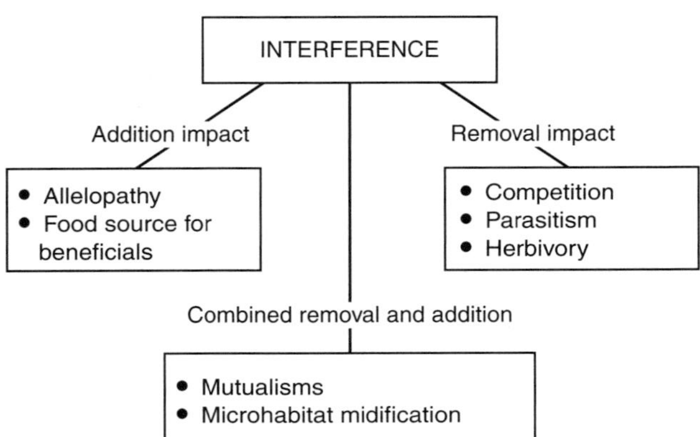


**FIGURE 15.1 Modes of interference underlying species interactions in communities.**

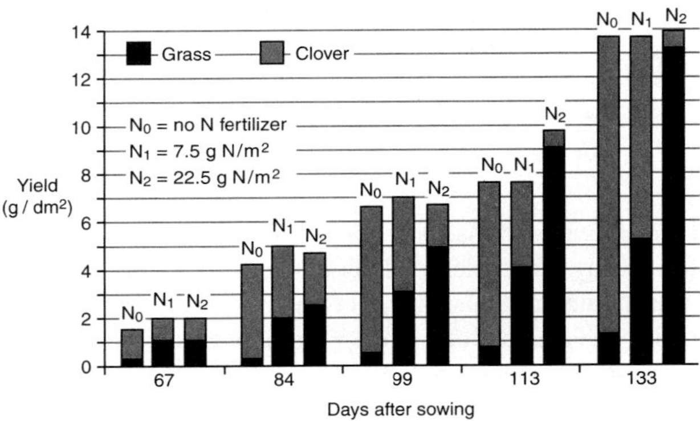


**FIGURE 15.2 Relative dominance of grass (Lolium rigidum) and clover (Trifolium subterraneum) in relation to levels of nitrogen fertilizer.** (Stern, W. R. and C. M. Donald. 1961. *Australian Journal of Agricultural Research* 13: 599–614. With permission.)

combined addition/removal interference between populations may modify the microclimate of a cropping system in ways that affect populations of other species. Shading, soil insulation, temperature and wind modification, and altered moisture relations can combine to create a microclimate within the cropping system that is conducive to the presence of organisms that are beneficial for the entire crop community.

## COMPLEXITY OF INTERACTIONS

The ways in which the various populations of a crop community influence the community as a whole through their interferences may be complex and difficult to discern. An example will help illustrate this point.

Canopy development over time was studied in a grass and clover mixture. The data from this study are shown in Figure 15.2. When the interaction between grass and clover is considered without any nitrogen being added, it appears that competition for limited light under the canopy of the crop mixture takes place. Shading by the clover appears to inhibit the grass. We could conclude from these data that due to its mutualism with nitrogen-fixing bacteria, the clover is able to avoid nitrogen competition and establish dominance. But the data obtained from adding different amounts of nitrogen fertilizer to the mixture alter the picture of community dynamics. The effect of adding nitrogen is to shift the balance of species dominance; by the last sample date the mixture at low nitrogen levels is dominated by clover, but the mixture at high nitrogen levels is dominated by grass. The advantage of one crop over the other is altered by the availability of nitrogen, with grass becoming more dominant as nitrogen supply is increased. These data lead to somewhat different conclusions; perhaps competition for light is the key factor, or perhaps some complex interaction of light, nitrogen availability, and some other factor (e.g., allelopathic chemicals added to the soil by the grass) is at work in the crop mixture.

These data raise other questions. For example, what would happen in a crop mixture where the two species involved had very similar nitrogen needs and procurement abilities? Under conditions of limited nitrogen supply, competition would probably result, and both species might suffer, but eventually one would begin to dominate the other. But another outcome is possible. The two different species could have complementary ways of using nitrogen when it is in limited supply; their timing of growth might be different, or their root systems might occupy different regions in the soil. They could thus avoid competition and coexist in the same system.

### Coexistence

In complex natural communities, populations of ecologically similar organisms often share the same habitat without significant apparent competitive interference, even though their niches overlap to a considerable degree. Similarly, it is often the case in natural communities that more than one species shares the role of dominant species. It would appear, then, that the principle of competitive exclusion, which implies that two species with similar needs cannot occupy the same niche or place in the environment, does not fully apply in many communities.

The ability to "avoid" competition and instead coexist in mixed communities leads to advantages for all involved members of the community. Therefore, this ability may well provide significant selective advantage in an evolutionary sense. Although selection for competitive ability has undoubtedly been very important in evolution, ecologists who study evolutionary biology now more widely accept the idea that selection for coexistence may be more the rule than the exception, especially in more mature communities (Pianka, 2000).

It is possible, too, that many domesticated species have undergone directed selection for coexistence by being grown most commonly in polycultures for many thousands of years. In this context, the plants would have coevolved, each developing adaptations for coexistence. (The traditional corn–bean–squash polyculture discussed later in this chapter is a possible example.)

Mixed populations are able to coexist due to many different mechanisms, such as resource partitioning, niche diversification, or specific physiological, behavioral, or genetic changes that reduce direct competition and allow for its avoidance. Understanding the mechanisms of interference that make coexistence possible could form an important foundation for the design of multiple crop communities.

In agroecosystems, combining species with slightly different physiological characteristics or resource needs is an important way of allowing for the coexistence of species in a multiple cropping community. Such an approach to designing the cropping community has much greater potential than trying to maintain single-species dominance in a monocultural field where considerable human intervention is needed to keep out potentially competing noncrop weeds or herbivorous pest insects. Successful mixed crop communities around the world offer fruitful ground for research on how avoidance of competition, or coexistence, plays an important ecological role in cropping systems.

## MUTUALISMS

Species with a mutualistic relationship are not only able to coexist, but are dependent on each other for optimal development. Mutualisms are likely the result of coexisting species continuing in the same evolutionary direction, coevolving adaptations for achieving mutual benefit through some kind of close association. Ecologists now know that mutualistic relationships among organisms of different species are relatively common in complex natural communities, creating intricate interdependencies among community members. Their prevalence is another factor explaining the observed complexity and diversity of many communities and their food webs. The same coevolutionary process has undoubtedly also occurred during domestication in agriculture, either by deliberate human selection or coincidentally in the context of multiple cropping systems.

The types of mutualisms that are most commonly recognized include the following:

- *Inhabitational mutualisms*: One mutualist lives wholly or partly inside the other. A classic example is the relationship between Rhizobium bacteria and leguminous plants. The nitrogen-fixing bacteria in this relationship cannot function outside of the nodules formed on the plant roots. This mutualism is the cornerstone of many of the most sustainable farming systems around the world.
- *Exhabitational mutualisms:* The organisms involved are relatively independent physically, but interact directly. An example is the relationship between a flowering plant and its pollinating insect. Many crop plants are unable to produce fertile seed without pollination from bees, and the bees depend on the crop plants for their main source of food in the form of nectar or pollen.
- *Indirect mutualisms:* The interactions among a set of species modify the environment in which they all live to the benefit of the mixture. An example is a polyculture agroecosystem. A tall crop species can modify conditions of the microclimate to the benefit of associated crop species, and the presence of several crops attracts a range of beneficial arthropods that facilitate the biological management of potential pests. Unlike the first two types of mutualisms, indirect mutualisms involve more than two species. Indirect mutualisms can also include both inhabitational and exhabitational mutualisms.

Some mutualisms are obligate for all involved members, while in others, only one of the members may require the relationship. In other cases, called facultative mutualisms, all members of the mutualism may be able to survive quite well without the interaction, but definitely do better when in relationship. Often, the mutualism functions not so much because of some stimulus or direct benefit to the organisms involved, but because it helps the species avoid some negative impact or impacts.

The expansion of the theory of mutualism in ecology has begun to find ready application in the development of more diverse cropping communities in which mutualistic relationships can occur. Making these relationships an integral part of crop communities is key to establishing sustainable systems that require fewer external inputs and less human intervention.

By contributing beneficial interactions, mutualisms in agroecosystems increase the resistance of the entire system to the negative impacts of pests, diseases, and weeds. At the same time, the efficiency of energy capture, nutrient uptake, and recycling in the system may be improved. Whenever mutualistic relationships can be incorporated into the organization of the cropping community, sustainability is much easier to achieve and maintain.

## HISTORY OF THE STUDY OF MUTUALISM

The idea that organisms may relate to each other in mutually beneficial ways has a very long history (Boucher, 1985). The ancient Greeks and Romans recognized that nature was full of examples of plants and animals helping each other. In his *History*, for example, the historian Herodotus describes such a relationship between a plover and a crocodile. The bird helps the crocodile by picking and eating leeches from the crocodile's mouth, and the crocodile never harms the bird even though a simple snap of its jaws would provide it with lunch.

In the 1600s, the theory of natural theology promoted the view that plants and animals sometimes selflessly aided each other in concert with the natural order of things. Divine Providence, it was believed, gave each organism a specific role to play in the larger "society" of the natural world and that some organisms had the role of guardian or helper.

As the industrial revolution progressed during the 18th and 19th centuries, the idea that competition among organisms was the driving force in nature gained prominence in science. The publication of Charles Darwin's *Origin of Species* was pivotal in bringing emphasis to competition, because it posited that the "struggle for existence" was the primary selective pressure in the evolutionary process. Interpretations and popularizations of Darwin's work went even further in casting nature as "red in tooth and claw."

Soon after the publication of the *Origin of Species*, however, interest in the concept of mutualism was revived. The term itself was officially introduced in 1873 by Pierre Van Beneden in a lecture to the Royal Academy of Belgium, and in 1877, Alfred Espinas' doctoral thesis documented multiple examples of mutualisms. Then, in an important 1893 paper, Roscoe Pound finally challenged the romantic notion of mutualism as help given freely and selflessly between organisms, explaining that each organism in a mutualism is simply acting in its own self-interest. The plover, for example, is obtaining food, and the crocodile is being relieved of parasites. The fact that such an interaction is mutually beneficial makes it a mutualism; the individual organism's intent is irrelevant.

As ecology developed into a science in the 20th century, interest in mutualisms remained at the fringes of the discipline, with most research on community-level interaction focusing on competition. Mutualism did not emerge as an important area of study until the 1970s.

Mutualisms have historically been important to agriculture, which in itself can be viewed as an obligate mutualism between humans and the crop plants and livestock we have domesticated. Traditional agroecosystems developed around facilitating mutualisms such as the Rhizobium–legume relationship (described in the next chapter), and coordinating the influences of beneficial insects and noncrop species. Conventional agriculture tends to eliminate these beneficial interactions and replace them with human-derived inputs.

## Mutually Beneficial Interferences at Work in Agroecosystems

Many sustainable traditional agroecosystems, upon analysis, reveal species interactions and modes of interference that benefit the community as a whole. Similar agroecosystems have been developed out of agroecological research and practical experimentation by farmers. These systems are based on the purposeful combining of various crop and noncrop species — including cover crops with crops, weeds with crops, and crops with other crops — in order to allow coexistence and take advantage of mutualistic relationships.

### Beneficial Interferences of Cover Crops

In a crop community, cover crops are plant species (usually grasses or legumes) grown in pure or mixed stands to cover the soil of the crop community for part or all of the year. They are often planted after the harvest of the primary crop to cover the soil during the fallow season, but they can also be planted in alternating years with the primary crop or grown in association with the primary crop. The cover crop plants may be incorporated into the soil by tillage in seasonal cover crop systems, or retained as live or dead plants on the soil surface for several seasons. When cover crops are tilled into the soil, the organic matter added to the soil is called *green manure*. When the cover crops are grown directly in association with other crops, they are called living mulch (Figure 15.3).

No matter how they are incorporated into the crop community, cover crops have important impacts on the environment, many of which can be highly beneficial. These impacts arise from the ability of cover crops to modify the soil–atmosphere interface, to offer physical protection of the soil from sunlight, wind, and rain, and to engage in a variety of addition and removal interferences. The benefits that accrue to the crop community — known to agriculture for a long time — include reduced soil erosion, improved soil structure, enhanced soil fertility, and suppression of weeds, insects, and pathogens. Some cover crops can even be used for animal feed or grazing, with the animals adding manure that is reincorporated back into

**FIGURE 15.3 Cover crop of bell bean (Vicia faba) and barley (Hordeum vulgare), Watsonville, CA.** This mixed cover crop inhibits weed growth, and when its biomasss is returned to the soil, it adds organic matter and fixed nitrogen.

the soil along with any residual plant matter. When cover crops can fulfill these roles in the crop community, there is less need for human interference and external inputs (Figure 15.3). Table 15.1 lists many of the benefits of cover crops along with the interferences (environmental impacts) that make them possible.

Despite the proven benefit of cover crops in general, their use must be tailored to the individual agroecosystem. The farmer needs to know how a cover crop species will interact with other organisms in the crop system, as well as how it will impact the conditions of the environment in which they all live. In addition, it must be remembered that forms of interference between members of the crop community, which may be of benefit at one time may be a liability at another. If resources in the crop system are limiting, the cover crop can create competitive interference. If allowed to become too dense, some cover crop species may be allelopathic to the crop. Residues or breakdown products of incorporated cover crops may produce growth-suppressing substances. Damaging herbivores or disease organisms may find the cover crop species to be an ideal alternate host, later moving onto the crop. Cover crop residue may also interfere with cultivation, weeding, harvesting, or other farming activities.

The case study *Cover cropping with Rye and Bell-beans* describes a study that demonstrates the ability of cover crops, especially those that are made up of mixed species, to control weeds and increase the yield of the subsequent main crop.

### Beneficial Interferences of Weeds

Weeds in cropping systems are most often considered to be detrimental, competing with the crop species and thereby reducing yields. Although weeds do often have negative effects on crops, it has been clearly shown that in many circumstances, weeds and other noncrop plants may benefit the crop community through their reactions on the environment (Radosevich et al., 1997; Chacón and Gliessman, 1982). Weeds exert their beneficial influences in much the same way as cover crops and often fill the same ecological roles; with proper management based on an understanding of the mechanisms of weeds' interactions, farmers can take advantage of their positive effects.

**TABLE 15.1**
**Potential Benefits of Cover Crops**

| | Interferences | Benefits to Crop Community |
|---|---|---|
| Impacts on soil structure | Enhanced root penetration in upper soil layers; shielding of soil surface from sunlight, wind, and the physical impact of raindrops; addition of organic matter to soil; enhanced biological activity in root zone | Improved water infiltration<br>Reduced soil crust formation<br>Decreased runoff<br>Less soil erosion<br>More stable soil aggregates<br>Increased percentage of macropores<br>Decreased soil compaction<br>Decreased bulk density |
| Impacts on soil fertility | Creation of cooler, moister surface and subsurface habitat; fixation of nitrogen by Rhizobium bacteria; carbon fixation (greater biomass); capture of nutrients by roots | Increased organic matter content<br>Retention of nutrients in system<br>Prevention of leaching loss<br>Increased nitrogen content<br>Greater diversity of beneficial biota in soil |
| Impacts on pest organisms | Addition of allelopathic compounds; removal of resources (light and nutrients) needed by weeds; creation of habitat for beneficial predators, parasites, and parasitoids; modification of microclimate | Inhibition of weeds by allelopathy<br>Competitive suppression of weeds<br>Control of soil pathogens by allelochemicals<br>Increased presence of beneficial organisms<br>Suppression of pest organisms |

*Source*: Lal, R., E. Regnier, D. J. Exkert, W. M. Edwards, and R. Hammond. 1991. In W. L. Hargrove (ed.), *Cover Crops for Clean Water*. pp. 1–14. Soil and Water Conservation Society: Iowa.; Altieri, M. A. 1995a. In M. A. Altieri, *Agroecology: The Science of Sustainable Agriculture*. 2nd ed., pp. 219–232. Westview Press: Boulder, CO.; Magdoff, F. and R. R. Weil. 2004. *Soil Organic Matter in Sustainable Agriculture*. Advances in Agroecology Series. CRC Press: Boca Raton, Florida.

## COVER CROPPING WITH RYE AND BELLBEANS

Multispecies cover crop systems often confer greater benefits to the agroecosystem than a cover crop of just one of the component species. These benefits arise from interactions between the species in the mixture.

One such system has been studied at the Center for Agroecology's farm facility at the University of California, Santa Cruz (UCSC). A legume (bellbean) is mixed with a grass (cereal rye) as a winter cover crop for vegetable fields. This multispecies cover crop has been used by local farmers since the turn of the century. Farmers plant the grass–legume mixture following the harvest of the summer crop, before winter rains begin. It is allowed to grow through the cool, wet months of winter and is disked into the soil in March or early April. The summer vegetable crop is then planted toward the end of May. The UCSC study used cabbage as the vegetable crop.

Rye produces significant amounts of biomass and limits weed growth in the plots, possibly through the release of allelopathic chemicals (Putnam and DeFrank, 1983). Bellbeans bring nitrogen into the system through their symbiotic relationship with nitrogen-fixing bacteria, but they produce limited biomass and have only a minor effect on weed growth. When bellbeans and rye are planted together, the advantages of both are combined: the mixture suppresses weed growth, is highly productive, and adds nitrogen to the system. But that is not all. The mixed cover crop does a better job of increasing nitrogen levels in the soil than does a legume-only crop, even when the legume-only crop has a higher legume biomass. It is possible that the increased organic matter being disked in with the bellbeans slows decomposition, retaining more nitrogen in the soil.

The mixed cover crop also proves to be of benefit to the vegetable crop that follows it. Although cabbage yield was highest in the bellbean-only treatment, it was not statistically different from the high cabbage yield of the rye–bellbean treatment, and both yields were significantly higher than those for rye alone and the control. Because of the greater bulk of organic matter it adds to the soil, the mixed cover crop would probably show the greatest benefits over a period of many years.

**TABLE 15.2**
**Impact of Bellbeans (*Vicia faba*) and Cereal Rye (*Secale cereale*) on Various Factors of the Crop Environment**

| | Total Biomass, g/m² | | | Weed Biomass, g/m² | | Cabbage Yield, kg/100m² |
|---|---|---|---|---|---|---|
| **Cover crop** | **1985** | **1986** | **1987** | **1986** | **1987** | **1987** |
| Bellbeans | 138 | 325 | 403 | 17.4 | 80.7 | 849.0 |
| Rye | 502 | 696 | 671 | 0.7 | 9.7 | 327.8 |
| Rye/bellbeans | 464 | 692 | 448 | 0.3 | 3.9 | 718.0 |
| None (Control) | n.d . | 130 | 305 | 112.3 | 305.1 | 611.0 |

Data from Gliessman 1989; n.d. = data not determined.
*Source:* Gliessman, S. R. 1989. In C. H. Chou and G. R. Waller (eds.), *Phytochemical Ecology: Allelochemicals, Mycotoxins and Insect Pheromones and Allomones*. pp. 69–80. Institute of Botany: Taipei, Taiwan.

## Modification of the Cropping System Environment

Weeds can protect the soil surface from erosion through root and foliar cover, take up nutrients that might otherwise be leached from the system, add organic matter to the soil, and selectively inhibit the development of more noxious species through allelopathy. Most of these benefits of weeds stem from the fact that ecologically weeds are pioneer species, invading open or disturbed habitats, and, through their reactions on the environment, initiate the process of succession toward more complex communities. Most crop communities, especially those composed of predominantly annual species, are simplified, disturbed habitats. Weeds are especially well adapted to such conditions. When we gain an understanding of the ecological basis of the reactions of weeds on the crop environment, we can utilize their interference in ways that reduce the need for inputs from outside the crop community.

## Control of Insect Pests by Promotion of Beneficial Insects

Agriculture is usually concerned with keeping both weeds and insects out of the production system. This takes large amounts of external inputs to accomplish and does not always provide the hoped-for results. When interactions between weeds and insects are examined from an ecological point of view, however, the possibility of retaining weeds in the system in order to control the unwanted insects emerges as an option. A body of literature is accumulating, which supports the hypothesis that certain weeds should be regarded as important components of the crop community because of the positive effects they can have on populations of beneficial insects (Altieri and Nicholls, 2004b). Depending on the type of beneficial insect, weeds can modify the microenvironment in ways that provide habitat for the insect, and they can provide alternative food sources such as pollen, nectar, foliage, or prey (Marshall et al., 2003).

In a study where weed species were planted as narrow border strips (0.25 m wide) around 5 × 5 m plots of cauliflower, it was found that certain pest insects were reduced as a result of the increase in predatory or parasitic beneficials (Ruiz-Rosado, 1984). For example, with the weeds *Spergula arvensis* (corn spurry) and *Chenopodium album* (lamb's quarters) planted in pure borders around the crop, larvae and eggs of the common imported cabbage worm (*Pieris rapae*) and the cabbage looper (*Trichoplusia ni*) were much more heavily parasitized by beneficials such as the tachinid fly *Madremyia saundersii*. The adult tachinids are attracted to the food sources provided by the weeds, then search out prey on which to lay their eggs in the crop nearby (Figure 15.5).

In another study, with the weed *Spergula arvensis* planted in 1.0-m strips around a Brussels sprouts field, the numbers of aphid-controlling beneficial insects collected among the corn spurry rose considerably when the spurry flowered (Linn, 1984). Presumably, the flowers provided a nectar and pollen source for the beneficials. In addition, predatory and parasitic wasps and Syrphid flies were commonly found feeding on the corn spurry flowers. Larger numbers of beneficials were also found in sweep samples on the Brussels sprouts, but only for a distance up to 5.0 m into the crop field. Other studies have shown similar effects. Significant reductions in aphid populations extended throughout the field when Spergula was a more evenly distributed member of the weed/insect/crop complex (Theunissen and van Duden, 1980). Leaving weedy borders with grasses and legumes on the margins of corn and soybean fields in Michigan was shown to greatly increase the presence of predatory ground-dwelling Carabid beetles in the crops (Landis et al., 2005).

## MUSTARD COVER CROP FOR FUJI APPLES

Using cover crops to suppress the growth of invasive weeds can help reduce the need for herbicides in an agroecosystem. To be useful, however, a cover crop must exclude other weeds without inhibiting the growth of the crop plant. Wild mustard (*Brassica kaber*) appears to be a cover crop that meets these requirements well, when planted in fruit orchards (Figure 15.4).

In a study of the conversion from conventional to organic management of young Fuji semidwarf apple trees, James Paulus, a graduate student researcher at the University of California, Santa Cruz, demonstrated the potential use of mustard as a cover crop species (Paulus, 1994). He grew several different types of cover crops between the trees in different plots and examined their effectiveness at weed control. The cover crop treatments were compared to conventional management with herbicides and to an organic conversion plot using plastic tarp for weed control.

Mustard was the only cover crop tested that controlled weeds as effectively as conventional herbicides or plastic tarp. Forty-five days after mustard emerged, it had displaced nearly all of the other weed plants in the plot and accounted for 99% of the total weed biomass present. Other covercrops only achieved partial dominance, accounting for no more than 42% of the total weed biomass in their respective plots.

It appears that mustard achieves this level of dominance through allelopathic inhibition of other weeds. Many members of the genus Brassica, including mustard, have been observed to inhibit weed growth in the field, and research has shown they contain potentially allelopathic chemicals called glucosinolates, which inhibit seed germination in the laboratory (Gliessman, 1987). Seeds of monocot grasses — often a problem as weeds — are the most strongly inhibited.

Paulus found that the mustard not only inhibited weeds effectively, but actually helped increase apple production. Trees in the plots with a mustard cover crop produced more than three times as many apples per tree than trees in the conventional plots. Moreover, the trees grown with mustard increased in girth more rapidly, showing diameters as much as 50% larger than trees in the conventional plots after 2 years.

At least part of the yield advantage in the mustard–cover cropped plots was due to improved nutrient cycling. Analysis showed that the weed cover took up significant amounts of nitrogen during the winter, lowering its concentrations in the soil. When the winter rains came, nitrogen in the bare soil treatments was leached out and lost from the system, whereas the nitrogen in the covercrop treatments was immobilized in the weed biomass. When the cover crop was cut down in the spring, the nitrogen was made available to the trees to use for spring and summer growth.

**FIGURE 15.4 Wild mustard cover crop in an apple orchard.** Wild mustard (*Brassica kaber*) adds an array of species interactions to an apple agroecosystem by attracting beneficial insects to its flowers and allelopathically inhibiting other weedy plants.

**FIGURE 15.5 A border of corn spurry (Spergula arvensis) around a cauliflower crop.** The weed's flowers attract beneficial insects.

## INTERCROPPING

Whenever two or more crops are planted together in the same cropping system, the resulting interactions can have mutually beneficial effects and effectively reduce the need for external inputs. The body of information documenting these interactions has grown considerably in recent years (Francis, 1986; van Noordwijk et al., 2004), and several authors have discussed how an ecological approach to multiple cropping research can provide an understanding of how the benefits of intercropping come about (Hart, 1984, 1986; Vandermeer, 1989; Ong et al., 2004).

The most successful intercropping systems are known from the tropics, where a high percentage of agricultural production still is grown in mixtures. Because smaller-scale farmers in the tropics have limited access to purchased inputs, they have developed intercropping combinations that are adapted to low external-input management (Gliessman et al., 1981; Innis, 1997; Joshi et al., 2004).

The traditional corn, bean, and squash polyculture cropping system of Central America and Mexico, with roots in the pre-Hispanic period, has been studied in some detail. Both removal and addition interferences occur in this system, leading to habitat modifications and mutualistic relationships of benefit to all three crops (Figure 15.6).

In a series of studies of the corn–bean–squash polyculture, done in Tabasco, Mexico, it was shown that corn yields could be stimulated as much as 50% beyond monoculture yields when planted with beans and squash using the techniques of local farmers and planting on land that had only been managed using local traditional practices (Amador and Gliessman, 1990). There was significant yield reduction for the two associated crop species, but the total yields for the three crops together were higher than what would have been obtained in an equivalent area planted to monocultures of the three crops. As shown in Table 15.3, this comparison is made using the concept of *land equivalent ratio*, explained in greater detail in Chapter 16. A land equivalent ratio greater than 1 indicates that an intercropping system is *overyielding* in relation to monocultures of its component crops.

Additional research has identified some of the ecological mechanisms of these yield increases:

- In a polyculture with corn, beans nodulate more and are potentially more active in biological fixation of nitrogen (Boucher and Espinosa, 1982; Santalla et al., 2001).
- Fixed nitrogen is made directly available to the corn through mycorrhizal fungi connections between root systems (Bethlenfalvay, et al., 1991; Hauggaard-Nielsen and Jensen, 2005).
- Net gains of nitrogen in the soil have been observed when the crops are associated, despite its removal, with the harvest (Gliessman, 1982; Maingi et al., 2001).
- The squash helps control weeds; the thick, broad, horizontal leaves block sunlight, preventing weed germination and growth, while leachates in rains washing the leaves contain allelopathic compounds that can inhibit weeds (Gliessman, 1983; Fujiyoshi et al., 2002).

**FIGURE 15.6 The traditional corn–bean–squash intercrop system from Mesoamerica.** Complex species interactions are key to the success of this cropping system.

- Herbivorous insects are at a disadvantage in theintercrop system because food sources are less concentrated and more difficult to find in themixture (Risch, 1980; Verkerk et al., 1998).
- The presence of beneficial insects is promoted due to such factors as the availability of more attractive microclimatic conditions and the presence of more diverse pollen and nectar sources (Letourneau, 1986; Verkerk et al., 1998).

Interestingly, when the same varieties of corn, bean, and squash were simultaneously planted in the same way in a nearby soil that had at least 10 years of management history involving mechanical cultivation, synthetic chemical fertilizers, and modern pesticides, the yield advantages disappeared. Apparently, the positive interactions that occurred in the traditional farm field were inhibited by the alteration of the soil ecosystem, which occurred with conventional inputs and practices. This result points to an important link between cultural practices and ecological conditions.

The corn–bean–squash intercrop is only one of many crop combinations that either exist or could be developed. Our knowledge of the ecological mechanisms of interference that function in this crop community provides a tantalizing indication of what we can look for in mixtures anywhere farming occurs.

An enormous number of polycultures exist, reflecting the wide variety of crops and management practices that farmers around the world use to meet their requirements for food, fiber, feed, fuel, forage, cash, and other needs. Intercrop communities can include mixtures of annuals, annuals with perennials, or perennials with perennials. Legumes can be grown with an array of cereals, and vegetable crops may be grown in between rows of fruit trees. The patterns of planting such mixtures can range from alternating rows of two crops to complex assortments of annual herbs, shrubs, and trees, as found in home garden agroecosystems (Chapter 17). Planting and harvesting in polycultures can be distributed in both time and space to provide advantage to the farmer throughout the year. The integration of animals helps form even more fully integrated mixed-crop communities (Chapter 19). Understanding the ecological foundation of the interactions that take place in these crop communities is the key to returning polyculture to prominence in agricultural practice.

**TABLE 15.3**
**Yield of a Corn–Bean–Squash Polyculture Compared to Yields of the Same Crops Grown as Monocultures in Tabasco, Mexico**

| | Low-Density Monoculture[a] | High-Density Monoculture[a] | Polyculture |
|---|---|---|---|
| Corn density (plants/ha) | 40,000 | 66,000 | 50,000 |
| Corn yield (kg/ha)[b] | 1150 | 1230 | 1720 |
| Bean density (plants/ha) | 64,000 | 100,000 | 40,000 |
| Bean yield (kg/ha)[b] | 740 | 610 | 110 |
| Squash density (plants/ha) | 1875 | 7500 | 3330 |
| Squash yield (kg/ha)[b] | 250 | 430 | 80 |
| Land equivalent ratio (LER) | | | 1.97[c]<br>1.77[d] |

[a] The monoculture densities were designed to represent levels just above and below the normal monoculture planting densities.
[b] Yields for corn and beans expressed as dried grain, squash as fresh fruits.
[c] Compared to low-density monoculture.
[d] Compared to high-density monoculture.

*Source*: Amador, M. F. 1980. *Comportamiento de tres especies (Maiz, Frijol, Calabaza) en policultivos en la Chontalpa, Tabasco, Mexico.* Tesis Profesional, CSAT, Cardenas, Tabasco, Mexico.

## USING SPECIES INTERACTIONS FOR SUSTAINABILITY

In natural ecosystems, organisms occur in communities of mixed species assemblages. Our ability to understand the complexity of interactions going on in such mixtures has benefited greatly from a growing body of ecological knowledge focused at each of the four levels of organization in ecosystems. The community ecology level discussed in this chapter is based on an understanding of the individual organism level and the population level. At the community level of organization, unique qualities begin to emerge as a result of multispecies interactions. These qualities have importance at the ecosystem level, as we will see in following chapters.

The challenge for agroecologists, then, is to put this ecological understanding into the context of sustainability. It is important that we combine the agronomists' extensive knowledge of the ecology and management of single species populations of crops with the ecologists' extensive knowledge of species interactions and community processes. It is time to redirect a large portion of the resources that have generated all of the knowledge about single-species cropping systems toward the integration of both ecological and agronomic knowledge, and to do so with the broader goal of developing the ability to manage the entire community of interacting organisms, both crop and noncrop, and understand how each species contributes to the sustainability of the whole system. This is an extremely complex process, requiring a systems-level approach and the interaction of many disciplines, but the end result will be a better understanding of how effective change in agriculture can come about.

## FOOD FOR THOUGHT

1. What are some of the primary impediments to convincing conventional farmers of the potential advantages of managing complex, multispecies cropping systems?
2. Give an example of a complex cropping community where competition and mutualisms may play different but equally important roles in the success of the entire crop system.
3. Describe an example of how coexistence and mutualisms in a crop community can be essential to the success of a biological control mechanism for a particular crop pest.
4. A noncrop organism can have either positive or negative impacts on the rest of the crop community of which it is a member. Explain how this is possible.
5. Describe a complex cropping community of crop and noncrop populations that allows for a significant reduction in the use of nonrenewable synthetic agricultural chemicals. Be sure to explain the contribution made by each member of the crop community.
6. What are several "emergent qualities" of a crop community that are not evident at the population or single individual level in an agroecosystem?

## INTERNET RESOURCES

Agroecology in Action
www.agroeco.org
The website of Professor Miguel Altieri, at University of California, Berkeley, with extensive material on agroecological pest and habitat management.

Community Ecology Group, NERC Center for Population Biology (Great Britain)
www.cpb.bio.ic.ac.uk/communityecology/communityecology.html

Contains descriptions of current projects and references.

Department of Community Ecology, Center for Environmental Research (Germany)
www.ufz.de/index.php?en = 798
Focuses on the analysis and assessment of natural and anthropogenic structural changes in biological communities, and thus on the development of a scientific basis for understanding and managing biodiversity.

Natural Systems Agriculture Group, University of Manitoba
www.umanitoba.ca/outreach/naturalagriculture/articles/intercrop.html
This research group makes extensive use of intercropping in its approach to sustainable agriculture.

International Allelopathy Society
www-ias.uca.es
Information on researchers, events, and publications in allelopathic research.

## RECOMMENDED READING

Beets, W.C. 1990. *Raising and Sustaining Productivity of Small-Holder Farming Systems in the Tropics*. AgBe Publishing: Alkmaar, Holland. A very thorough and practical review of sustainable agriculture; most appropriate for much of the tropics, with good sections on multiple cropping systems.

Eilittä, M., J. Mureithi, and R. Derpsch (eds.), 2004. *Green Manure Cover Crop Systems of Smallholder Farmers: Experiences From Tropical and Subtropical Regions*. Springer: New York. A volume providing 12 in-depth case studies of smallholder intercropping strategies, analyzed from an interdisciplinary perspective.

Francis, C.A. (ed.). 1986. *Multiple Cropping Systems*. New York: Macmillan. A very thorough treatment of the agronomy and ecology of the great diversity of multiple cropping systems from around the world.

Francis, C.A. 1990. *Sustainable Agriculture in Temperate Zones*. Wiley and Sons: New York. An overview of sustainable agriculture for the developed world, with good examples of how multiple cropping may play a role.

Hajek, A.E. 2004. *Natural Enemies: an Introduction to Biological Control*. Cambridge University Press: Cambridge UK. An in-depth review of biological control of arthropods, vertebrates, weeds, and plant pathogens through use of natural enemies.

Huffaker, C.B. and P.S. Messenger. 1976. *Theory and Practice of Biological Control*. Academic Press: New York. The classic reference on biological control, with emphasis on the management of crop communities as a foundation.

Innis, D.Q. 1997. Intercropping and the Scientific Basis of Traditional Agriculture. Intermediate Technology Development Group: London, UK. Compares the practice and science of intercropping in traditional agricultural systems of several developing countries.

Koul, O., G.S. Dhaliwal, and G.W. Cuperus. 2004. *Integrated Pest Management: Potential, Constraints and Challenges*. CABI Publishing: Cambridge, MA. This volume covers key issues surrounding IPM, with an emphasis on insects. Topics include the pesticide paradox in IPM, risk-benefit analysis, IPM and sustainable development, and consumer response to IPM.

Liebman, M., C.L. Mohler, and C.P. Staver. 2001. *Ecological Management of Agricultural Weeds*. Cambridge University Press: Cambridge, UK. A very complete review of the principles and applications of ecological weed management in a range of temperate and tropical farming systems with several chapters that emphasize community ecology aspects.

Morin, P.J. 1999. *Community Ecology*. Cambridge University Press: Cambridge UK. An introduction to community ecology, with examples of interactions between plants and animals in aquatic and terrestrial habitats.

Narwal, S.S., R.E. Hoagland, R.H. Dilday, and M.R. Roger. 2000. *Allelopathy in Ecological Agriculture and Forestry*. Springer: New York. A review of current allelopathy research, with case studies from several countries.

Rice, E.L. 1995. *Biological Control of Weeds and Plant Diseases: Advances in Applied Allelopathy*. Norman, OK: University of Oklahoma Press. An excellent review of allelopathy as a means of managing weed and disease populations in crop or forest communities.

van Noordwijk, M., G. Cadish, and C.K. Ong. (eds.), 2004. *Below-ground Interactions in Tropical Agroecosystems: Concepts and Models with Multiple Plant Components*. CABI Publishing: Cambridge, MA. A synthesis of plant–soil–plant interactions in agroforestry and intercropping, with a focus on agroecological processes in multiple cropping systems.

Wardle, D.A. 2002. *Communities and Ecosystems: Linking the Aboveground and Belowground Components*. Princeton University Press: Princeton, NJ. A thorough review of the field of community ecology, focusing on the influence of interactions between above- and belowground components on ecosystem structure and function.

# 16 Agroecosystem Diversity and Stability

Both agroecosystems and natural ecosystems are made up of organisms and the nonliving physical environment in which the organisms live. The three preceding chapters have been concerned primarily with the organismal, or biotic, components of these systems, at the level of populations and communities. In this chapter, we begin to add the abiotic components of ecosystems to the picture, thereby reaching the ecosystem level of study. At this level, we look at systems as wholes, gaining a more complete picture of their structure and functioning.

The complexity that characterizes whole systems is the basis for ecological interactions that are a crucial foundation for sustainable agroecosystem design. These interactions are largely a function of the *diversity* of a system.

Diversity is at once a product, a measure, and a foundation of a system's complexity — and therefore, of its ability to support sustainable functioning. From one perspective, ecosystem diversity comes about as a result of the ways that the different living and nonliving components of the system are organized and interact. From another perspective, diversity — as manifested by the complex of biogeochemical cycles and the variety of living organisms — is what makes the organization and interactions of the system possible.

In this chapter, we first explore what it means to manage agroecosystems as whole systems, taking advantage of their emergent qualities. We then examine biodiversity in natural ecosystems, the value of diversity in an agroecosystem setting, how diversity is evaluated, and the possible role of island biogeography theory in managing diversity. Finally, we explore the connections between ecological diversity, stability, and sustainability in terms of developing a framework for agroecosystem design and management.

## WHOLE-SYSTEM APPROACHES AND OPPORTUNITIES

In Chapter 15, we saw how the interactions among the populations of a crop community lead to emergent qualities that exist only at the community level. At the ecosystem level, another set of emergent qualities exists that makes the agroecosystem much greater than the sum of its parts (or the farm much greater that the sum of the crop plants in its fields). Management that works at this level can take advantage of a huge array of beneficial interactions and processes.

### Managing the Whole System

Agroecology emphasizes the need to study both the parts and the whole. Although the concept of the whole being greater than the sum of its parts is widely recognized, it has been ignored for a long time by modern agronomy and technology, which emphasize the detailed study of the individual crop plant or animal as a way of dealing with the complex issues of farm production and viability. We have learned a great deal from specialization and a narrow focus on the yield of the crop components of farming systems, but an understanding of the entire farm (and the whole food system) must also be developed to fully understand agricultural sustainability and implement sustainable management practices.

When agroecosystem management considers the opportunities presented by the emergent qualities of whole systems, the paradigm of *controlling* conditions and populations is replaced by the paradigm of *managing* them. Under the management paradigm, we are always striving to consider the effects on the whole system of any action or practice, and we deliberately design practices that build on whole-system functioning and emergent qualities.

Under the conventional approach, the attempt to rigidly control and homogenize all the conditions separately too often results in the elimination of beneficial relationships and interferences, leaving only negative interferences and interactions. Conventional management practices work primarily at the individual or population level of the system, rather than at the community and ecosystem levels, where more complex interactions can take place.

The problems inherent in the population level, control-oriented conventional approach are readily seen in the way it has been applied to pest, weed, and pathogen control during the past several decades. Based on the principle that the only good bug or weed is a dead one, an incredible array of technologies have been developed to remove or eliminate each target pest from the cropping system. These technologies have simplified agroecosystems in various ways — for example, by eliminating the predators of the target pests. In simplified agroecosystems, however, pest invasions become more common and pernicious, and the

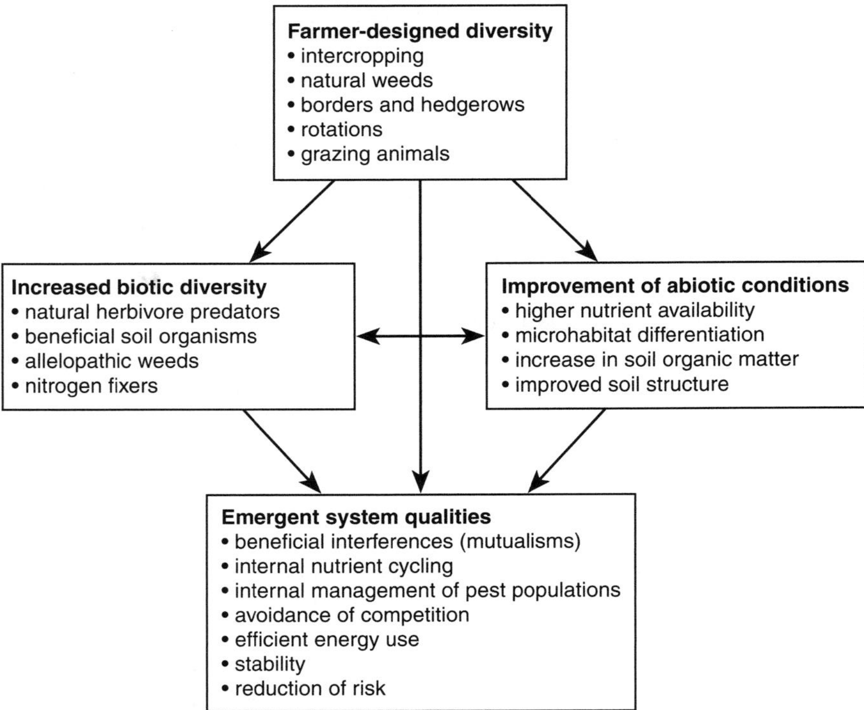


**FIGURE 16.1** **System dynamics in diverse agroecosystems.**

use of external inputs must increase to deal with the resulting problems.

### Building on Diversity

The central priority in whole-system management is creating a more complex, diverse agroecosystem, because only with high diversity is there a potential for beneficial interactions. The farmer begins by increasing the number of plant species in the system, through a variety of planting practices discussed in more detail below. Then livestock may be integrated with the crops, as discussed in Chapter 19. This diversification leads to positive changes in the abiotic conditions and attracts populations of beneficial arthropods and other animals. Emergent qualities develop that allow the system — with appropriate management of its specific components — to function in ways that maintain fertility and productivity and regulate pest populations. This very general conceptualization of the dynamics of managing a diverse agroecosystem is sketched out in Figure 16.1.

In a diverse and complex system, all the challenges facing farmers can be met with appropriate management of system components and interactions, making the addition of external inputs largely unnecessary. In the area of pest management, for example, pest populations can be controlled by system interactions intentionally set up by the agroecosystem manager. In the area of nutrient cycling, as another example, animals can convert plant matter, which humans cannot consume, into manure for use on the farm.

The many methods of "alternative" pest management developed by organic farmers and agroecologists are a good example of diversity-based whole-systems management. These methods rely on increasing agroecosystem diversity and complexity as a foundation for establishing beneficial interactions that keep pest populations in check. Descriptions of several of these methods, as applied in specific agroecosystems, are presented in Table 16.1.

## ECOLOGICAL DIVERSITY

In ecology, the concept of diversity tends to be applied mainly at the community level; diversity is understood as the number of different species making up a community in a particular location. Ecosystems, however, have other kinds of variety and heterogeneity beyond that encompassed by the number of species. They have diversity in the spatial arrangement of their components, for example, as shown by the different canopy levels in a forest. They have diversity in their functional processes and diversity in the genomes of their biota. And since they change in various ways over time, both cyclically and directionally, they have what could be called temporal diversity.

Diversity, therefore, has a variety of different *dimensions*. When these dimensions are recognized and defined, the concept of diversity itself is broadened and

**TABLE 16.1**
**Representative Examples of Alternative Pest Management Based on System Interactions**

| Pest Problem | Alternative Management Practice | Mechanism(s) of Action |
|---|---|---|
| Flea beetle (*Phyllotreta cruciferae*) damage on broccoli | Intercropping weedy mustard (*Brassica* spp.) | Trap crop attracts the pest away from the crop |
| Grape leafhopper (*Erythroneura elegantula*) damage on grape vines | Border plantings of weedy blackberries (*Rubus* spp.) | Increases abundance of alternate hosts for parasitic wasp *Anagrus epos* |
| Aphid (*Rhopalosiphum maidis*) damage on sugar cane | Border plantings of aggressive grassy weeds | Grassy weeds displace other plants that harbor the aphid |
| Corn earworm (*Heliothis zea*) damage | Allowing development of a natural weed complex in the corn | Enhances presence and effectiveness of predators of pest eggs and larvae |
| Fall armyworm (*Spodoptera frugiperda*) damage in corn | Intercropping with beans | Increases beneficial insect abundance and activity |
| Whitefly (*Aleurotrachelus socialis*) damage on cassava | Intercropping with cowpeas | Increases plant vigor and abundance of natural whitefly enemies |
| Webworm (*Antigostra* sp.) damage on sesame | Intercropping with corn or sorghum | Shading by the taller companion crops repels the pest |
| Diamondback moth (*Plutella xylostella*) damage on cabbage | Intercropping with tomato | Repels moth chemically, or masks presence of cabbage |
| Codling moth (*Cydia pomonella*) damage in apple orchards | Cover cropping with specific plant species | Provides additional food and habitat for natural enemies of codling moths |
| Pacific mite (*Eotetranychus willamette*) damage in vineyards | Cover cropping with grass | Promotes presence of predatory mites by providing winter habitat for alternative prey |
| Sugarbeet cyst nematode (*Heterodera schachtii*) damage on sugarbeet roots | Rotations with alfalfa | Provide "biological break" when no host plant is present |
| Western flower thrip (*Frankliniella occidentalis*) damage in flowering grapes | Flowering corridors | Provide a biological highway for predators to disperse into the center of the vineyard |

*Source*: Adapted from Altieri, M. A. and C.I. Nicholls. 2004a. An agroecological basis for designing diversified cropping systems in the tropics. In D. Clements and A. Shrestha (eds.) *New Dimensions in Agroecology*. Food Products Press/The Haworth Press: New York.; Altieri, M.A. and C.I. Nicholls. 2004b. *Biodiversity and Pest Management in Agroecosystems*. 2nd ed. Howarth Press: Binghamton, NY.; Andow, D. A. 1991. *Annual Review of Entomology* 36: 561–586.

complexified — it becomes what we will call *ecological diversity*.

Some of the possible dimensions of ecological diversity are listed in Table 16.2. Other dimensions may be recognized and defined, but these seven are the dimensions that will be used in this text. (The term "biodiversity" is commonly used to refer to a combination of species diversity and genetic diversity.) These different dimensions of ecological diversity are useful tools for fully understanding diversity in both natural ecosystems and agroecosystems.

## Diversity in Natural Ecosystems

Diversity seems to be an inherent characteristic of most natural ecosystems. Although the degree of diversity among different ecosystems varies greatly, ecosystems in general tend to express as great a diversity as possible given the constraints of their abiotic environments.

Diversity is in part a function of evolutionary dynamics. As discussed in Chapter 14, mutation, genetic recombination, and natural selection combine to produce variability, innovation, and differentiation among earth's biota. Once diversity is generated, it tends to be self-reinforcing. Greater species diversity leads to greater differentiation of habitats and greater productivity, which in turn allow even greater species diversity.

Diversity has an important role in maintaining ecosystem structure and function. Ever since Tansley (1935) coined the term "ecosystem" to refer to the combination of plant and animal communities and their physical environment, ecologists have attempted to demonstrate the relationship between diversity and system stability. Natural ecosystems generally conform to the principle that greater diversity allows greater resistance to perturbation and disturbance. Ecosystems with high diversity tend to be able to recover from disturbance and restore balance in their processes of material cycling and energy flow; in ecosystems with low diversity, disturbance can more easily cause permanent shifts in functioning, resulting in the loss of resources from the ecosystem and changes in its species makeup.

### Scale of Diversity

The size of the area being considered has an impact on how diversity (species diversity in particular) is measured. The species diversity of a single location in a river valley

## RHIZOBIUM BACTERIA, LEGUMES, AND THE NITROGEN CYCLE

One important way of taking advantage of ecological diversity is to introduce nitrogen-fixing legumes into the agroecosystem. As a result of the mutualistic relationship between the leguminous plants and bacteria of the genus Rhizobium, nitrogen derived from the atmosphere is made available to all the biotic members of the system. The ability of a system to supply its needs for nitrogen in this way is an emergent quality made possible by biotic diversity.

Rhizobium bacteria possess the ability to capture atmospheric nitrogen from the air in the soil and convert it to a form that is usable by the bacteria and also by plants. These bacteria can live freely in the soil; however, when legume plants are present, the bacteria infect the plants' root structure. A bacterium moves into an internal root cell, causing it to differentiate and form a nodule in which the bacterium can reproduce. The bacteria in a root nodule begin to receive all the sugars they need from the host plant, giving up their ability to live independently; they reciprocate by making the nitrogen they fix available to the host. The interaction provides an advantage to both organisms: the plant is able to obtain nitrogen that would otherwise not be available to it, and the bacteria are able to maintain a much higher population level than they can in the soil. A great deal more nitrogen fixation occurs with nodulated legumes, therefore, than with free-living Rhizobium alone. When the host plant dies, the bacteria can revert to an autotrophic lifestyle and reenter the soil community.

Because nitrogen is often a limiting nutrient, a legume's relationship with Rhizobium allows it to survive in soil that may contain too little nitrogen to support other plants. And if the legume is returned to the soil after it dies, the bacterially fixed nitrogen it incorporated into its biomass during its life becomes part of the soil, available for other plants to use.

This mutualism has been historically important in agriculture. The legume–Rhizobium symbiosis is the primary source of nitrogen addition in many traditional agroecosystems, and was one of the only methods used to incorporate environmental nitrogen into many crop systems before the development of nitrogen fertilizer. Legume crops have been intercropped with nonlegumes, as in the corn–bean-squash polyculture common in Latin America, and legumes are used as cover crops and green manure crops in the U.S. and other regions to improve soil quality and nitrogen content. Legumes have also been an important part of managed fallow systems. All of these systems take advantage of the legume–Rhizobium symbiosis, using biological nitrogen fixation to make usable nitrogen available to the entire plant community, and ultimately to humans.

**TABLE 16.2**
**Dimensions of Ecological Diversity in An Ecosystem**

| Dimension | Description |
|---|---|
| Species | Number of different species in the system |
| Genetic | Degree of variability of genetic information in the system (within each species and among different species) |
| Vertical | Number of distinct horizontal layers or levels in the system |
| Horizontal | Pattern of spatial distribution of organisms in the system |
| Structural | Number of locations (niches, trophic roles) in the system organization |
| Functional | Complexity of interaction, energy flow, and material cycling among system components |
| Temporal | Degree of heterogeneity of cyclical changes (daily, seasonal, etc.) in the system |

forest is different from the species diversity across the river valley's different communities.

Species diversity in a single location is often called *alpha diversity.* This is simply the variety of species in a relatively small area of one community. Species diversity across communities or habitats — the variety of species from one location to another — is called *beta diversity.* On a still larger scale is *gamma diversity,* which is a measurement of the species diversity of a region such as a mountain range or river valley.

The difference between the three types of diversity can be illustrated with a hypothetical 5-km transect. It is possible to measure alpha diversity at any location along the transect by counting the number of species within, say, 10 m of a specified point. A measure of beta diversity, in contrast, includes at least two points along the transect in different but adjacent habitats. If the species makeup of these two locations is very different, beta diversity is high; if the species makeup changes little as one moves between the two habitats, beta diversity is low. A measure of gamma diversity is made along the entire length of the transect, taking into account both the total number of species and the variation in their distribution. In principle, the distinction between alpha, beta, and gamma diversity can be extended to other dimensions of ecological diversity, such as structural and functional diversity.

Alpha, beta, and gamma diversity are helpful conceptual distinctions because they allow us to describe how different ecosystems and landscapes vary in the structure of their diversity. For example, a highly diverse natural grassland that stretches for hundreds of kilometers in every direction is likely to have high alpha

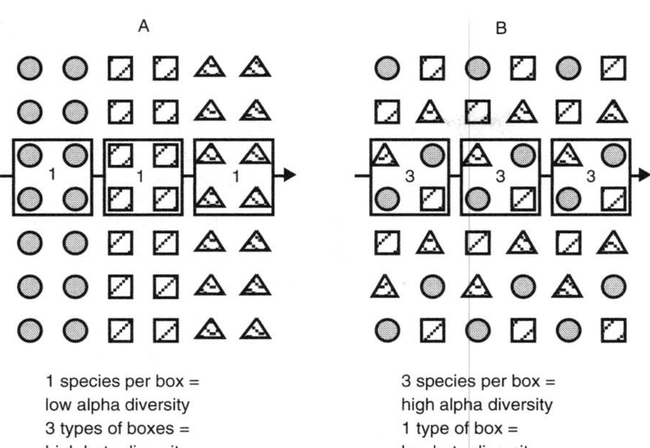


**FIGURE 16.2 Alpha diversity vs. beta diversity in an agroecosystem context.** For the sake of simplicity, each shape represents a crop plant and each box a locality. This scale is somewhat arbitrary in that a locality could comprise many more crop plants; the point of the diagram is to show the contrast between the two arrangements, which might represent (A) three crops planted in strips, and (B) an intercrop of the three crops.

diversity, but since the same species in the same relative proportions are found at all locations over a wide area, the grassland's beta and gamma diversity are relatively low. As a contrasting example, consider a landscape made up of a complex mosaic of simple communities, such as nonnative grassland, a forest community dominated by a single species, and a scrub community growing on steep slopes. Alpha diversity is relatively low in each of the communities, but any transect across the area crosses a variety of species groupings, making beta and gamma diversity relatively high.

The alpha and beta scales of diversity in particular have useful application in agroecosystems. A cropping system with high beta diversity, for example, can often provide the same advantages as one with high alpha diversity while offering greater ease of management (Figure 16.2).

## Successional Processes and Changes in Diversity

Studies of natural ecosystems in early stages of development or following disturbance have shown that all the dimensions of diversity tend to increase over time. This process takes place through niche diversification, habitat modification, competitive displacement, resource partitioning, and the development of coexistence, mutualisms, and other forms of interference. Variability and fluctuation in ecosystem processes are damped by this diversification, giving the system the appearance of greater stability as diversity increases.

When an ecosystem is disturbed, each of the dimensions of its ecological diversity is simplified, or set back to an earlier stage of development. The number of species is reduced, vertical stratification decreases, fewer interactions occur. Following the disturbance, the ecosystem begins the recovery process that is called secondary succession (Chapter 17). During this process, the system begins to restore the diversity of species, interactions, and processes that existed before disturbance.

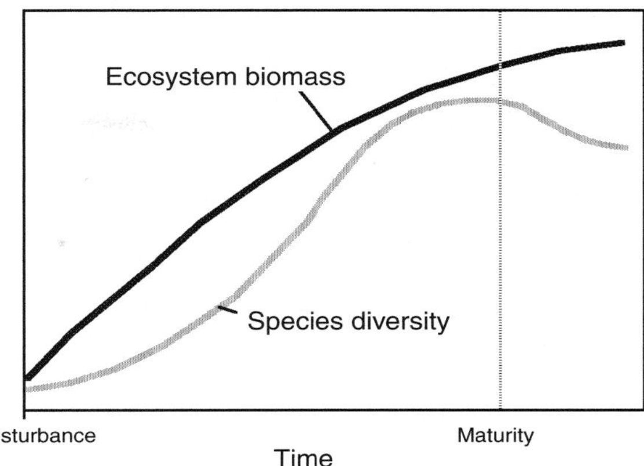


**FIGURE 16.3 Changes in species diversity and biomass during secondary succession.**

Eventually the system reaches something called maturity, which might be defined as the successional condition in which the full potential for energy flow, nutrient cycling, and population dynamics in that physical environment can be realized. The structural and functional diversity of the ecosystem at maturity provides resistance to change in the face of further minor disturbance.

Even though diversity tends to increase through the stages of succession, recent research in ecology indicates that maturity may not represent the stage with the greatest diversity, at least in terms of species. Rather, the greatest diversity is achieved as a system approaches maturity, with diversity declining slightly thereafter as full maturity is attained. Biomass continues to increase at maturity, though at a slower rate (Figure 16.3).

## Diversity and Stability

There has been considerable discussion in ecology about the relationship between diversity and stability. There appears to be some correlation between the two — that is, the greater the diversity of an ecosystem, the more resistant it is to change, and the better able it is to recover from disturbance — but there is disagreement over the degree and strength of the correlation.

Much of the problem arises from the restricted nature of the accepted definition of stability. "Stability" usually refers to the relative absence of fluctuations in the populations of organisms in the system, implying a steady-state condition, or a lack of change. This notion of stability is inadequate, especially in relation to describing the ecological results of diversity. What we need is an expanded definition of stability (or a new term) based on system characteristics, a definition that focuses on the *robustness* of an ecosystem, its ability to sustain complex levels of interaction and self-regulating processes of energy flow and material cycling. Such an expanded notion of stability is required, in particular, for understanding the value and use of diversity in agroecosystems, which are anything but "stable" in the conventional sense of the term.

To gain a better sense of what "stability" really is, we need more research into possible causal relationships among the different forms of ecological diversity and specific ecosystem processes and characteristics. Some important work in this area has already been done. It has been found, for example, that higher bird species diversity is correlated with more complex community structure, because it supports a greater variety of feeding and nesting behaviors. Similarly, predator–prey diversity and a more complex food web are correlated both with actual species numbers as well as habitat diversity.

We must continue to be cautious of falling into a trap of circular reasoning, where we begin to believe that diversity always leads to stability, and once we have more stability, that this will lead to more diversity. For the concepts of diversity and stability to be of application in agriculture, we need studies that correlate the different types of diversity with the process of productivity, and from there, to sustainability.

## Ecological Diversity in Agroecosystems

In most agroecosystems, disturbance occurs much more frequently, regularly, and with greater intensity than it does in natural ecosystems. Rarely can agroecosystems proceed very far in their successional development. As a result, diversity in an agroecosystem is difficult to maintain.

The loss of diversity greatly weakens the tight functional links between species that characterize natural ecosystems. Nutrient cycling rates and efficiency change, energy flow is altered, and dependence on human interference and inputs increases. For these reasons, an agroecosystem is considered ecologically unstable.

Nevertheless, agroecosystems need not be as simplified and diversity-poor as conventional agroecosystems typically are. Within the constraints imposed by the need for harvesting biomass, agroecosystems can approach the level of diversity exhibited by natural ecosystems, and enjoy the benefits of the increased stability allowed by greater diversity. Managing the complexity of interactions that are possible when more of the elements of diversity are present in the farm system is a key part of reducing the need for external inputs and moving toward sustainability.

### The Value of Agroecosystem Diversity

A key strategy in sustainable agriculture is to reincorporate diversity into the agricultural landscape and manage it more effectively. Increasing diversity is contrary to the focus of much of present-day conventional agriculture, which reaches its extreme form in large-scale monocultures. It would appear that diversity is seen more as a liability in such systems, especially when we consider all of the inputs and practices that have been developed to limit diversity and maintain uniformity.

Recent research on multiple cropping systems underscores the great importance of diversity in an agricultural setting (Francis, 1986; Vandermeer, 1989; Altieri, 1995b; Innis, 1997; Ong et al. 2004). Diversity is of value in agroecosystems for a variety of reasons:

- With higher diversity, there is greater microhabitat differentiation, allowing the component species of the system to become "habitat specialists." Each crop can be grown in an environment ideally suited to its unique requirements.
- As diversity increases, so do opportunities for coexistence and beneficial interference between species that can enhance agroecosystem sustainability. The relationships between nitrogen-fixing legumes and associated crop plants discussed earlier are a prime example.
- In a diverse agroecosystem, the disturbed environments associated with agricultural situations can be better taken advantage of. Open habitats can be colonized by useful species that already occur in the system, rather than by weedy, noxious pioneer invaders from outside.
- High diversity makes possible various kinds of beneficial population dynamics among herbivores and their predators. For example, a diverse system may encourage the presence of several populations of herbivores, only some of which are pests, as well as the presence of a predator species that preys on all the herbivores. The predator enhances diversity among the herbivore species by keeping in check the populations of individual herbivore species. With greater herbivore diversity, the pest herbivore cannot become dominant and threaten any crop.
- Greater diversity often allows better resource-use efficiency in an agroecosystem. There is better system-level adaptation to habitat

heterogeneity, leading to complementarity in crop species needs, diversification of the niche, overlap of species niches, and partitioning of resources. The traditional corn–bean–squash intercrop, for example, brings together three different but complementary crop types. When all three are planted in a heterogeneous field, soil conditions at any one site are likely to adequately meet the needs of at least one of the three crops. When planted in a uniform field, each crop will occupy a slightly different niche and make different demands on the soil's nutrients.

- Diversity reduces risk for a farmer, especially in areas with more unpredictable environmental conditions. If one crop does not do well, income from others can compensate.
- When livestock animals are integrated into an agroecosystem, many opportunities arise for beneficial interactions. Grazing, for example, can allow better nutrient cycling, increase the numbers of the beneficial arthropods that occupy the microsites provided by perennial pasture plants, and shift the dominance of noncrop species. These and other interactions are discussed in more detail in Chapter 19.
- A diverse crop assemblage can create a diversity of microclimates within the cropping system that can be occupied by a range of noncrop organisms — including beneficial predators, parasites, and antagonists — that are of importance for the entire system, and who would not be attracted to a very uniform and simplified system.
- Diversity in the agricultural landscape can contribute to the conservation of biodiversity in surrounding natural ecosystems, an issue that will be discussed in Chapter 22.

Diversity — especially that of the belowground part of the system — performs a varie-ty of ecological services that have impacts both on and off the farm, such as nutrient recycling, regulation of local hydrological processes, and detoxification of noxious chemicals.

When our understanding of diversity extends beyond the crop species to include noncrop plants (commonly called weeds, but of potential ecological or human value), animals (especially beneficial enemies of pests or animals useful to humans), and microorganisms (belowground diversity of bacteria and fungi are essential for maintaining many agroecosystem processes), we then begin to see the range of ecological processes that are promoted by greater diversity.

## Methods of Increasing Diversity in Agricultural Systems

A range of options and alternatives are available for adding the benefits of diversity discussed above to the agricultural landscape. These alternatives can involve (1) adding new species to existing cropping systems, (2) reorganizing or restructuring the species already present, (3) adding diversity-enhancing practices or inputs, and (4) eliminating diversity-reducing or diversity-restricting inputs or practices.

### *Intercropping*

A primary and direct way of increasing the alpha diversity of an agroecosystem is to grow two or more crops together in mixtures that allow interaction between the individuals of the different crops. Intercropping is a common form of multiple cropping, which is defined as "the intensification and diversification of cropping in time and space dimensions" (Francis, 1986). Intercropping can add temporal diversity through the sequential planting of different crops during the same season, and the presence of more than one crop adds horizontal, vertical, structural, and functional diversity. Best developed in traditional farming systems in rural or developing areas, especially in the tropics, intercropping or polyculture systems vary from relatively simple mixtures of two or three crop plants to the very complex mixtures of crops found in agroforestry or home garden agroecosystems (Figure 16.4) (discussed in more detail in Chapter 17).

### *Strip Cropping*

Another form of multiple cropping is to plant different crops in adjacent strips, creating what may be called a polyculture of monocultures. This practice, which increases beta diversity instead of alpha diversity, can provide many of the diversity benefits of multiple cropping. For some crops and crop mixtures, it is a more practical method of increasing diversity because it presents fewer management and harvest challenges than multiple cropping.

### *Hedgerows and Buffer Vegetation*

Trees or shrubs planted around the perimeter of fields, along pathways of a farm, or to mark boundaries, can have many useful functions. In practical terms, they can provide protection from wind, exclude (or enclose) animals, and produce an array of tree products (firewood, construction materials, fruit, etc.). Ecologically, hedgerows and buffer strips increase the beta diversity of the farm, and can serve to attract and provide habitat for beneficial organisms. When planted as wider strips, especially between farmlands and adjacent natural ecosystems, they form buffer zones that can mitigate a range of potential impacts of one system on the other, as well as increase the overall biodiversity of the region (Figure 16.5).

**FIGURE 16.4 Two examples of multiple cropping.** Carrots, beets, and onions are grown together in Witzenhausen, Germany (top); annual and perennial crops are combined to form a diverse home garden in Riva de Garda, Italy (bottom).

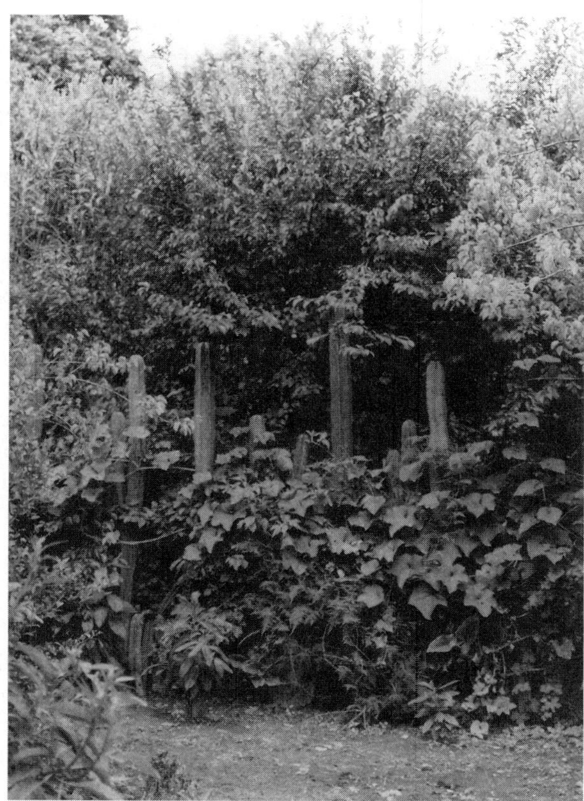

**FIGURE 16.5 A multiple-use hedgerow around a home garden in Tepeyanco, Tlaxcala, Mexico.** Cactus form a barrier to animals, and chayote squash and apricot trees provide food.

### Cover Cropping

A cover crop is a noncrop species planted in a field to provide soil cover, usually in-between cropping cycles. Cover crops range from annuals to perennials, and include many different taxonomic groups, although grasses and legumes are used predominantly. Increasing the diversity of a system by planting one or more cover crop species has a variety of important benefits. Cover cropping enhances soil organic matter, stimulates soil biological activity and diversity of the soil biota, traps nutrients in the soil left over from previous crops, reduces soil erosion, contributes biologically fixed nitrogen (if one of the cover crop species is a legume), and provides alternate hosts for beneficial enemies of crop pests. In some systems, such as orchards, cover crops may serve the additional purpose of inhibiting weed development (Sullivan, 2003).

### Rotations

Growing crops in rotation is an important method of increasing the diversity of a system over the dimension of time. Rotations usually involve planting different crops in succession or in a recurring sequence. The greater the differences between the rotated crops in their ecological impact on the soil, the greater the benefits of the method. Alternating crops can create what is known as a rotational effect, where a crop grown after another does better than when grown in continuous monoculture. By adding residues of different species of plants to the soil, rotations help maintain the biological diversity of soil microorganisms. Each residue type varies chemically and biologically, stimulating and/or inhibiting different soil organisms. In some cases, the residue from one crop is able to promote the activity of organisms that are antagonistic to pests or diseases for a subsequent crop. Rotations also tend to improve soil fertility and soil physical properties, reduce soil erosion, and add more organic matter. The well-known advantages of soybean/corn/legume–hay rotations in the Midwestern U.S. are based in part on the way that greater temporal diversity aids nutrient and disease management. Research on the impact of rotations on the dimensions of diversity can improve the effectiveness of this important practice (Altieri, 1999).

### Fallows

A variation of the rotation practice is to allow a period in the cropping sequence where the land is simply left uncultivated, or fallow. The introduction of a fallow period allows the soil to "rest," a process that involves secondary succession and the recovery of diversity in many parts of the system, especially the soil. Shifting cultivation, discussed in Chapter 10, is probably the most well-known fallow system; the long rest period allows the reintroduction of native plant and animal diversity and the recovery of soil fertility. In some systems, the fallow principle is used to create a mosaic of plots in different stages of succession, from farmed fields to second growth native vegetation. In dry-farmed regions, fallow may occur in alternate years to allow rainfall to recharge soil moisture reserves, while at the same time promoting the recovery of diversity in the soil ecosystem during the uncultivated cycle. Another variation on the use of the fallow is to make it productive in addition to being protective; in swidden-fallow agroforestry, specific crop plants are introduced just before the fallow begins, or intentionally allowed to reestablish, so that harvestable products can be obtained during the fallow period (Denevan and Padoch, 1987). Wherever a fallow period is incorporated into the cropping cycle, it is the lack of human-induced disturbance, not just the absence of a crop, which allows the diversity recovery process.

### Reduced or Minimum Tillage

Since disturbance in an agroecosystem has a major role in limiting successional development, diversity, and stability, a practice that minimizes disturbance may help enhance diversity. Reducing the intensity of soil cultivation and leaving residues on the surface of the soil is a primary method of effecting reduced system disturbance. The many advantages to be gained from reducing both the frequency and intensity of tillage were discussed in

Chapter 8. Comparisons of conventional tillage and no-till practices show increased earthworm abundance and activity, diversification in soil-organic-matter–consuming and –decomposing organisms, and an accompanying improvement in soil structure, nutrient holding capacity, internal nutrient cycling, and organic matter content (House and Stinner, 1983; Stinner et al., 1984; Hendrix et al., 1986). Even when the aboveground diversity of the cropping system remains low, the species diversity of the decomposer subsystem of the soil increases with reduced soil disturbance. Increasing plant diversity aboveground as well can only enhance this subsystem.

### *High Organic Matter Inputs*

High levels of organic matter are crucial for stimulating species diversification of the belowground subsystem, involving the same type of stimulation of structural and functional diversity noted above for reduced tillage systems. Long seen as a key component of organic agriculture, high organic matter inputs have an array of benefits that were reviewed in Chapter 8. The organic-matter content of the soil can be increased by applying composts, incorporating crop residues into the soil, cover cropping, diversifying crops, and using other diversity-enhancing cropping practices.

### *Reduction in Use of Chemical Inputs*

It has long been known that many agricultural pesticides either harm or kill many nontarget organisms in crop systems, or leave residues that can limit the abundance and diversity of many other organisms. Thus, eliminating or reducing the use of pesticides removes a major impediment to the rediversification of the agroecosystem. The recolonization process involved in this rediversification is discussed later in this chapter. It must be acknowledged, however, that removing pesticides from a system that has become dependent on them is a challenging task. The first response may be a dramatic increase in the pest population; only with time and the reestablishment of diversity can internal mechanisms develop for keeping the pest in check. An experimental framework for examining the changes in diversity that occurs when a "pesticide stress-free background" is established in a transitional agroecosystem is presented in Chapter 20.

### *Integration of Livestock*

Integrating animals back into the agricultural landscape increases the overall biodiversity of the agroecosystem. In addition, animal activity, such as grazing, crop residue consumption, and manure deposition can alter aspects of structural diversity, species dominance, and system function. Additional benefits accrue in the diversification of the farm enterprise itself. Livestock integration is discussed in more detail in Chapter 19.

## Managing Diversification

Moving from a uniform, monoculture agroecosystem to a more diverse system supporting beneficial processes and interactions is a multistep process. Initially, all of the above ways of introducing diversity into the agricultural landscape help mitigate the negative impacts of agricultural activities. Then the introduction of more species, either as a direct or indirect effect, broadens the opportunities for integrated agroecosystem structure and function, allowing built-in buffers and system dynamics to dampen variability of system response. Finally, the kinds and forms of interference in the diversifying landscape make possible more types of interactions, ranging from competitive exclusion to symbiotic mutualisms (summarized in Table 16.3).

Managing diversity at the farm level is a big challenge. Compared to conventional management, it can involve more work, more risk, and more uncertainty. It also

**TABLE 16.3**
**Methods of Increasing Ecological Diversity in An Agroecosystem**

| | Dimensions of Ecological Diversity Affected | | | | | | |
|---|---|---|---|---|---|---|---|
| **Method** | **Species** | **Genetic** | **Vertical** | **Horizontal** | **Structural** | **Functional** | **Temporal** |
| Intercropping | • | o | • | • | • | • | o |
| Strip cropping | • | o | o | • | o | o | o |
| Hedgerows and buffers | • | o | • | • | o | o | • |
| Cover cropping | • | o | • | • | • | • | o |
| Rotations | o | o | — | — | o | o | • |
| Fallows | o | o | — | — | o | o | • |
| Minimum tillage | • | o | — | — | o | • | o |
| High inputs of organic matter | • | o | — | — | o | • | — |
| Reduction of chemical use | o | o | — | — | o | • | — |
| Integration of livestock | • | o | — | o | o | • | o |

• denotes direct or primary effect; o denotes Indirect, secondary, or potential effect; — denotes little or no effect requires more knowledge. Ultimately, however, understanding the ecological basis for how diversity operates in agroecosystems, and taking advantage of complexity rather than striving to eliminate it, is the only strategy leading to sustainability.

## EVALUATING AGROECOSYSTEM DIVERSITY AND ITS BENEFITS

To manage diversity most effectively, we need means of measuring diversity and evaluating how increases in diversity actually impact the performance and functioning of an agroecosystem. We need to be able to recognize the presence of diversity and the patterns of its distribution on the landscape, and we need to know if, and to what extent, the presence of that diversity is of benefit to the performance of the agroecosystem, especially from the farmer's point of view. Several approaches can be taken to analyze and research the presence and impacts of diversity.

### Indices of Species Diversity

It is obvious that any kind of intercrop is more diverse than a monoculture. Comparing the diversity of two different intercropping systems, however — varying in both species numbers and planting ratios — requires that we measure the diversity of each. To do so, we can borrow tools and concepts developed by ecologists for natural ecosystems.

Ecologists recognize that the diversity of an ecosystem or community is determined by more than just the number of species. A community made up of 50 redwood trees, 50 tanbark oaks, and 50 Douglas firs is more diverse than one made up of 130 redwood trees, 10 tanbark oaks, and 10 Douglas firs. Both have the same number of species and total individuals, but the individuals in the first community are distributed more evenly among the species than those in the second community, where redwood trees dominate.

This example demonstrates that there are two components of species diversity: the number of species, called *species richness*, and the evenness of the distribution of the individuals in the system among the different species, called *species evenness*. Both components must be considered in any comprehensive measurement of diversity, in both natural ecosystems and agroecosystems.

How these concepts can be applied in analyzing the diversity of agroecosystems is demonstrated in Table 16.4, where four different hypothetical systems, each with the same number of individual crop plants, are compared. Among these systems, the even polyculture of three crops is the most diverse, since it is the only one in which both species richness and species evenness are high in relation to the other systems.

**TABLE 16.4**
**Diversity Measures of Four Hypothetical Agroecosystems**

| | Monoculture | Even Polyculture of Two Crops | Even Polyculture of Three Crops | Uneven Polyculture of Three Crops |
|---|---|---|---|---|
| Corn plants | 300 | 150 | 100 | 250 |
| Squash plants | 0 | 150 | 100 | 25 |
| Bean plants | 0 | 0 | 100 | 25 |
| Number of species ($s$) | 1 | 2 | 3 | 3 |
| Number of individuals ($N$) | 300 | 300 | 300 | 300 |
| Relative species richness | Low | Medium | High | High |
| Relative species evenness | High | High | High | Low |

Instead of using the number of individuals of each species as a basis for measuring a system's species diversity, it is possible to use some other species characteristic, such as biomass or productivity. This may be more appropriate, for example, when the biomass of a typical individual of one species is very different from the biomasses of the individuals of the other species. Number of individuals, biomass, and productivity are all examples of *importance values* for a particular species.

Ecology offers various ways of quantifying the species diversity of a system. The simplest method is to ignore species evenness, and to measure the number of species in terms of the number of individuals. Such a measure is provided by Margalef's index of diversity:

$$\text{diversity} = (s - 1)/\log N$$

where $s$ is the number species and $N$ is the number of individuals. The usefulness of Margalef's index is limited because it cannot distinguish the varying diversity of systems with the same $s$ and $N$, such as the even and uneven three-crop polycultures in Table 16.4.

There are two other diversity indices that do take species evenness into account, and are therefore more useful. The *Shannon index* is an application of information theory, based on the idea that greater diversity corresponds to greater uncertainty in picking at random an individual of a particular species. It is given by the following formula:

$$H = -\sum_{i=1}^{S}\left(\frac{n_i}{N}\right)\left(\log_e \frac{n_i}{N}\right)$$

where $n_i$ is the number of individuals in the system (or sample) belonging to the $i$th species.

The *Simpson index* of diversity is the inverse of an index of community dominance with the same name. It is based on the principle that a system is most diverse when none of its component species can be considered any more dominant than any of the others. It is given by the following formula:

$$\text{diversity} = \frac{N(N-1)}{\sum n_i(n_i-1)}$$

For the Simpson index, the minimum value is 1; for the Shannon index it is 0. Both minimums indicate the absence of diversity, the condition that exists in a monoculture. In theory, the maximum value for each index is limited only by the number of species and how evenly distributed they are in the ecosystem. Relatively diverse natural ecosystems have Simpson indices of 5 or greater, and Shannon indices of 3 to 4.

Calculations of Margalef, Simpson, and Shannon index values for the hypothetical systems in Table 16.4 are given in Table 16.5. The Shannon and Simpson values both show that the even polyculture of two crops is more diverse than the uneven polyculture of three crops, underscoring the importance of species evenness in agroecosystem diversity.

More detailed descriptions of the Shannon and Simpson indices, including the theory on which they are based and the ways they can be applied, can be found in the ecology texts cited in the recommended readings at the end of the chapter.

## Assessing the Benefits of Intercrop Diversity

On a farm, a way of measuring the value gained from greater diversity in the cropping system will be very useful in helping the farmer evaluate the advantages or disadvantages of different cropping arrangements. The diversity indices described above can quantify diversity, but they do not tell us how that diversity translates into performance, or what the ecological basis of any improved performance is. In cropping systems where two or more crop species are in close enough proximity to each other, various kinds of between-species interference are possible (as described in Chapter 11 and Chapter 13) that can provide clear benefits in improved yield, nutrient cycling, and so on.

Despite the fact that researchers have accumulated a great deal of evidence that intercropping can provide substantial yield advantages over monocropping, it is important to remember that there can also be disadvantages to intercropping. There may be practical difficulties in the management of the intercrop, and yield decreases may occur because of the effects of adverse interference. Such cases should not be used as arguments against intercropping, but rather as a means of determining where research needs to be focused to avoid such problems.

### The Land Equivalent Ratio

An important tool for the study and evaluation of intercropping systems is the land equivalent ratio (LER). LER provides an all-other-things-being-equal measure of the yield advantage obtained by growing two or more crops as an intercrop compared to growing the same crops as a collection of separate monocultures. LER thus allows us to go beyond a description of the pattern of diversity into an analysis of the advantages of intercropping.

The LER is calculated using the formula

$$\text{LER} = \sum \frac{\text{Yp}_i}{\text{Ym}_i}$$

where Yp is the yield of each crop in the intercrop or polyculture, and Ym is the yield of each crop in the sole crop or monoculture. For each crop (i) a ratio is calculated to determine the partial LER for that crop, then the partial LERs are summed to give the total LER for the intercrop. An example of how the LER is calculated is given in Table 16.6.

An LER value of 1.0 is the break-even point, indicating no difference in yield between the intercrop and the

**TABLE 16.5**
**Diversity Index Values for the Four Hypothetical Agroecosystems in Table 16.4**

| | Monoculture | Even Polyculture of Two Crops | Even Polyculture of Three Crops | Uneven Polyculture of Three Crops |
|---|---|---|---|---|
| Margalef diversity | 0 | 0.4 | 0.81 | 0.81 |
| Shannon diversity | 0 | 0.69 | 1.10 | 0.57 |
| Simpson diversity | 1.0 | 2.01 | 3.02 | 1.41 |

**TABLE 16.6**
**Representative Data for Calculation of LER**

| | Yield In Polyculture (Yp), kg/ha | Yield in Monoculture (Ym), kg/ha | Partial LER $\left(\frac{\text{Yp}_i}{\text{Ym}_i}\right)$ |
|---|---|---|---|
| Crop A | 1000 | 1200 | 0.83 |
| Crop B | 800 | 1000 | 0.80 |
| | | | $\Sigma \frac{\text{Yp}_i}{\text{Ym}_i} = 1.63$ |

collection of monocultures. Any value greater than 1 indicates a yield advantage for the intercrop, a result called *overyielding*. The extent of overyielding is given directly by the LER value: an LER of 1.2, for example, indicates that the area planted to monocultures would need to be 20% greater than the area planted to the intercrop for the two to produce the same combined yields. An LER of 2.0 means that twice as much land would be required for the monocultures.

### Application and Interpretation of the LER

Since the partial and total LER values are ratios, and not actual crop yields, they are useful for comparing diverse crop mixtures. In a sense, the LER measures the level of intercrop interference going on in the cropping system.

Theoretically, if the agroecological characteristics of each crop in a mixture are exactly the same, planting them together should lead to the same total yield as planting them apart, with each crop member contributing an equal proportion to that total yield. For example, if two similar crops are planted together, the total LER should be 1.0 and the partial LERs should be 0.5 for each. In many mixtures, however, we obtain a total LER greater than 1.0, and partial LERs proportionately greater than what would theoretically be obtained if each crop were agroecologically the same as the others. A total LER higher that 1.0 indicates the presence of positive interferences among the crop components of the mixture, and may also mean that any negative interspecific interference that exists in the mixture is not as intensive as the intraspecific interference that exists in the monocultures. Avoidance of competition or partitioning of resources is probably occurring in the mixture.

When the total LER is greater than 1.5, or when the partial LER of at least one member of the mixture is greater than 1.0, there is strong evidence that negative interference is minimal in the intercrop interactions and that positive interferences allow at least one of the members of the crop mixture to do better in the intercrop than it does when planted in monoculture.

The traditional corn–bean–squash intercrop discussed in chapter 15 — with a total LER of 1.97 — provides a good example of this situation (Table 15.3). The corn component of the system expressed a partial LER of 1.50, meaning that it actually produced better in the mixture than when planted alone. The positive interference responsible for this result may be mutualistic mycorrhizal connections between the corn and beans, or a habitat modification by the squash that enhanced the presence of a beneficial insect and reduction of a pest. Although partial LERs for beans and squash were very low (0.15 and 0.32, respectively), their presence obviously was important for the yield enhancement of the corn.

When the total LER is less than 1.0, negative interference has probably occurred, especially if the LERs of the component parts of the mixture are all lowered in a similar fashion. In this case, the intercrop provides a yield disadvantage compared to monocropping.

When analyzing LERs and partial LERs, confusion can often arise about what constitutes an advantage and what the magnitude of the advantage is. Avoiding confusion requires the recognition that different circumstances call for different criteria for evaluating an intercrop's advantage. There are at least three basic situations (Willey, 1981):

1. *When combined intercrop yield must exceed the yield of the higher-yielding sole crops.* This situation may exist when assessing mixtures of very similar crops, such as pasture forage mixes, or mixtures of genotypes within a crop, such as a multiline wheat crop. In such cases, partial LERs are not important in determining advantage as long as total LER is greater than 1.0, because the farmer's requirement is mostly for maximum yield, regardless of which part of the crop system it comes from. The quantitative advantage is the extent to which the combined intercrop yield is increased and total LER exceeds 1.0, as compared to the yield of the highest yielding sole-crop.
2. *When intercropping must give full yield of a "main" crop plus some additional yield of a second crop.* This situation occurs when the primary requirement is for some essential food crop or some particularly valuable cash crop. For there to be an advantage to the intercrop, total LER must exceed 1.0 *and* the partial LER of the primary crop should be close to 1.0 or even higher. With the emphasis on a key crop, the associated plants must provide some positive intercrop interference. The corn–bean–squash intercrop mentioned above is a good example of this situation because the farmer is mainly interested in the corn yield. If some additional yield is obtained from the beans and squash, even if their partial LERs are very low, it is seen as an additional bonus beyond the yield advantage gained by corn. The quantitative advantage is the extent to which the main crop is stimulated beyond its performance in monoculture.
3. *When the combined intercrop yield must exceed a combined sole-crop yield.* This situation occurs when a farmer needs to grow both (or all) the component crops, especially when there is limited land for planting. For the intercrop to be advantageous, total LER must be greater than 1.0, but no member of the mixture can suffer a great reduction in its partial LER in the

process. Negative interference definitely cannot be functioning for such a mixture to be beneficial. This situation can present problems in the use of the LER value since it is not always readily apparent what proportions of sole crops the total LER value should be based on. Comparison cannot be made only on sown proportions because interference in the intercrop situation can often produce yield values that are very different from the monocrop's proportions, leading to skewed partial LERs.

Recognizing these different situations is important for two reasons. First, it helps to ensure that research on a given combination is likely to be grounded in farming practice. Second, it should ensure that yield advantages are assessed in valid, quantitative terms that are appropriate to the situation being considered. Ultimately, the intercropping pattern that functions best is the one that meets the criteria of both the farmer and the researcher.

To put certain different crops on a more comparable basis, figures other than harvest yields can be used to calculate an LER (Andersen et al., 2004). These measurements include protein content, total biomass, energy content, digestible nutrient content, or monetary value. Such calculations allow the use of a similar indicator to evaluate different contributions the crop may make to the agroecosystem.

## COLONIZATION AND DIVERSITY

Up to this point we have explored how the farmer can directly increase diversity by planting more species, and how he or she can create the conditions that allow "natural" diversification to occur in an agroecosystem. We have ignored the question of how organisms not actually planted by the farmer enter the system and establish themselves there. This question concerns both the desirable organisms whose presence is encouraged — such as predators and parasites of herbivores, beneficial soil organisms, and helpful allelopathic weeds — and the undesirable ones, such as herbivores, that the farmer would like to exclude from the system.

To address this question of how an agroecosystem is colonized by organisms, it is helpful to think of a crop field as an "island" surrounded by an "ocean" that organisms have to cross in order to become part of the species diversity of the agroecosystem. In an ecological sense, any isolated ecosystem surrounded by distinctly different ecosystems is an island because the surrounding ecosystems set limits on the ability of organisms to reach and colonize the island. Building on our study of the dispersal and establishment process in Chapter 13, we will here explore how the study of the colonization of actual islands by organisms can be applied to understanding the colonization of agroecosystems and how this process is related to agroecosystem diversity.

### ISLAND BIOGEOGRAPHY THEORY

The body of ecological theory concerning islands is known as island biogeography (MacArthur and Wilson, 1967). It begins with the idea that island ecosystems are usually very isolated from other similar ecosystems. The sequence of events that allows an organism to reach an island sets in motion a set of responses that guide the development of the island ecosystem. A key characteristic of an island is that many of the interactions that eventually determine the actual niche of an organism after it reaches the island are very different from the conditions of the niche the organism left behind. This situation gives the organism an opportunity to occupy more of its potential niche, or even evolve characteristics that could allow it to expand into a new niche. This is especially true in the case of a newly formed island in the ocean — an environment very similar to that of a recently disturbed (e.g., plowed) farm field. The first pest to arrive in an "uncolonized" field has the opportunity to very rapidly fill its potential niche, especially if it is a specialist pest adapted to the conditions of the crop in that field.

Island biogeography theory offers methods of predicting the outcome of the species diversification process on an island. These methods take into account the size of the island, the effectiveness of the barriers limiting dispersal to the island, the variability of the habitats on the island, the distance of the island from sources of emigration, and the length of time the island has been isolated.

Experimental manipulation of island systems (Simberloff and Wilson, 1969) and studies of island diversity have provided the basis for the following principles:

- The smaller the island, the longer it takes for organisms to find it.
- The further an island is from the source of colonists, the longer it takes for the colonists to find it.
- Smaller and more distant islands have smaller and more depauperate flora and fauna.
- Many niches on islands can be unoccupied.
- Many of the organisms that reach islands occupy a much broader niche than the same or similar organisms on the mainland.
- Early colonizers often arrive ahead of limiting predators and parasites, and can experience very rapid population growth at first.

- As colonization proceeds, changes occur in the niche structure of the island, and extinction of earlier colonists can take place.
- The earliest arrivals are mostly *r*-selected.

Ultimately, the theory should be able to predict the colonization and extinction rates that are possible for a particular island. Such a prediction should then make it possible to understand the relationship between ecological conditions and potential species diversity, and what factors control the establishment of an equilibrium between extinction and further colonization.

### Agricultural Applications

The parallels between islands and crop fields allow researchers to apply island biogeography theory to agriculture. Experiments can be designed where either one crop field is completely surrounded by a different crop, or small plots are marked out in a larger field of the same crop. An early example was a study by Price (1976) of the rates that pests and natural enemies colonize soybean fields. The study was carried out using small plots in a field of soybeans as the experimental islands; the plots were surrounded by an "ocean" of soybeans, with natural forest abutting one side, and more soybean fields on the other sides. Small plots in the soybean field located at different distances from the various sources of colonization were monitored for the full crop season, allowing the measurement of the arrival rates, abundance, and diversity of both pests and their beneficial control agents. The more easily dispersed pests were the first ones to reach the interior plots of the field, and were followed later by some of their predators and parasites. The equilibrium between species and individuals of both pests and natural enemies that was predicted by island biography theory was not reached, probably due to the short life cycle of a soybean field. This study has encouraged other studies of a similar nature (Altieri and Nicholls, 2004b).

More recent research suggests that beneficial arthropods move more readily into a crop field from their refuges surrounding the field when they are provided with habitat highways — vegetated corridors providing food and refuge — that penetrate into the crop field (e.g., Nicholls et al., 2000). A study examining how this principle operates in a vineyard is described in *Using Flowering Plant Corridors to Increase Beneficial Insect Diversity in a Vineyard.*

Ma and others (2002) adapted the island biogeography approach to analyze floral richness in agricultural buffer zones — delimiting strips of vegetation used to prevent pesticide drift, nutrient leaching, and soil erosion — in the human-dominated landscape of Finland. They found that the width of the buffer zone was the factor that most affected and correlated with floral species richness at the landscape level. Denys and Tscharntke (2002) conducted a similar study focusing on insect species richness in different types of margin vegetation strips surrounding crop fields. Higher ratios of predatory to herbivorous insects were observed in larger strips, thus supporting the trophic-level hypothesis of island biogeography, which states that the role of predators and parasitoids tends to increase with area (Figure 16.6).

## USING FLOWERING PLANT CORRIDORS TO INCREASE BENEFICIAL INSECT DIVERSITY IN A VINEYARD

In many of the grape-growing regions of California, large-scale monoculture vineyards dominate the landscape. The numbers of natural insect predators and parasitoids that might otherwise exist in these landscapes are greatly reduced because of the relative lack of important food resources and overwintering sites offered by natural and noncrop vegetation.

In contrast, where viticulturalists have retained or created a more diverse landscape by keeping vineyards smaller and maintaining natural vegetation patches and riparian corridors at vineyard perimeters, they have encouraged the presence of natural predators and parasitoids. The positive effect of landscape diversification practices in increasing the diversity of beneficial insects has been demonstrated in a variety of agroecosystems (Altieri, 1994a; Coombes and Sotherton, 1986; Corbett and Plant, 1993; Thomas, Wratter, and Sotherton, 1991).

In these more diverse viticultural areas, where strips and patches of natural and other noncrop vegetation are interspersed among monoculture vineyards, analysis of the dynamics of insect predator and herbivore populations is a good application of island biogeography theory. The grape monocultures in these landscapes are "islands" in the sense that beneficial insects do not live in them year-round but instead disperse into them from the adjacent noncrop vegetation when their prey and hosts are present.

A study by Clara Nicholls, Michael Parrella, and Miguel A. Altieri (2000) has shown that where noncrop vegetation already exists adjacent to a vineyard, its positive effect on beneficial insect biodiversity can be greatly enhanced by a relatively simple practice: penetrate the vineyard with corridors of flowering plants contiguous with the adjacent natural vegetation. The corridors serve beneficials both as a habitat and a "biological highway," allowing them to move from their refugia in nonagricultural areas deep into the vineyard.

The researchers compared two adjacent vineyard blocks that differed in only one respect: block A was bisected by a 600-meter-long corridor of noncrop vegetation contiguous with a bordering riparian forest; block B had the bordering forest but no analogous corridor. The corridor in block A supported 65 species of locally

**FIGURE 16.6 Wild mustard (Brassica campestris) forming a barrier around "islands" of cauliflower.** The mustard can attract beneficial insects and retard the movement of herbivorous insect pests to the crop.

adapted flowering plants, including fennel (*Foeniculum vulgare*), yarrow (*Achillea millefolium*), daisy fleabane (*Erigeron annuus*), and butterfly bush (*Buddleia* spp.). Most of these plants were nonnative but not particularly weedy (an exception is fennel; care should be taken in using it for such corridors).

Various sampling methods allowed the researchers to observe the following patterns:

- The corridor supported a healthy diversity of arthropod predators including green lacewings, minute pirate bugs, big-eyed bugs, damsel bugs, and several species of hoverflies, ladybugs, tumbling flower beetles, and spiders.
- Diversity of generalist predators overall was higher in the vineyard block with the plant corridor.
- In the vineyard block with the corridor, the numbers of the two major grape herbivores present (western grape leafhoppers and western flower thrips) were lowest near the plant corridor and highest in the central areas. In the other vineyard block, these herbivores were distributed evenly throughout the block.
- Most generalist predators showed a density gradient in the block with the corridor, reaching their greatest numbers near the plant corridor. In the other block, these generalist predators were more evenly distributed.
- The rate of parasitization of leafhopper eggs by *Anagrus epos* wasps was roughly the same throughout both vineyard blocks.

These results showed that the positive effect of the adjacent riparian forest on the biodiversity of beneficials was — with the exception of *A. epos* — amplified by the flowering plant corridor in block A. For ladybugs and lacewings, the corridor provided food in the form of aphids and other homoptera; for hoverflies it supplied nectar and pollen; for predatory insects such as minute pirate bugs it offered neutral insect prey. By providing these food resources, the corridor allowed beneficials to move more deeply into the vineyard. In island biogeography terms, the corridor effectively reduced the size of the monoculture "islands," facilitating their "colonization" by beneficials.

In addition to demonstrating the applications of island biogeography theory and the value of diversity, this study highlights the importance of looking at diversity and ecological processes at the scale of the landscape. Agricultural practices that allow, create, or retain a more diverse agricultural landscape that includes remnants of natural vegetation and noncrop areas are to be encouraged for a variety of reasons, a concept we will explore in more detail in Chapter 22 (Table 16.7).

## USING FLOWERING PLANT CORRIDORS TO INCREASE BENEFICIAL INSECT DIVERSITY IN A VINEYARD

In many of the grape-growing regions of California, large-scale monoculture vineyards dominate the landscape. The numbers of natural insect predators and parasitoids that might otherwise exist in these landscapes are greatly reduced because of the relative lack of important food resources and overwintering sites offered by natural and noncrop vegetation.

In contrast, where viticulturalists have retained or created a more diverse landscape by keeping vineyards smaller and maintaining natural vegetation patches and riparian corridors at vineyard perimeters, they have encouraged the presence of natural predators and parasitoids. The positive effect of landscape diversification practices in increasing the diversity of beneficial insects has been demonstrated in a variety of agroecosystems (Altieri 1994a; Coombes and Sotherton 1986; Corbett and Plant 1993; Thomas, Wratter, and Sotherton 1991).

In these more diverse viticultural areas, where strips and patches of natural and other noncrop vegetation are interspersed among monoculture vineyards, analysis of the dynamics of insect predator and herbivore populations is a good application of island biogeography theory. The grape monocultures in these landscapes are "islands" in the sense that beneficial insects don't live in them year-round but instead disperse into them from the adjacent noncrop vegetation when their prey and hosts are present.

A study by Clara Nicholls, Michael Parrella, and Miguel A. Altieri (2000) has shown that where noncrop vegetation already exists adjacent to a vineyard, its positive effect on beneficial insect biodiversity can be greatly enhanced by a relatively simple practice: penetrate the vineyard with corridors of flowering plants contiguous with the adjacent natural vegetation. The corridors serve beneficials both as a habitat and a "biological highway," allowing them to move from their refugia in nonagricultural areas deep into the vineyard (Figure 16.7).

The researchers compared two adjacent vineyard blocks that differed in only one respect: block A was bisected by a 600-meter-long corridor of noncrop vegetation contiguous with a bordering riparian forest; block B had the bordering forest but no analogous corridor. The corridor in block A supported 65 species of locally adapted flowering plants, including fennel (*Foeniculum vulgare*), yarrow (*Achillea millefolium*), daisy fleabane (*Erigeron annuus*), and butterfly bush (*Buddleia* spp.). Most of these plants were nonnative but not particularly weedy (an exception is fennel; care should be taken in using it for such corridors).

**FIGURE 16.7 Corridor of flowering plants penetrating the interior of a vineyard in California.** The corridor facilitates the movement of beneficial insects into the vineyard from their refugia in the riparian forest (in the background).

Various sampling methods allowed the researchers to observe the following patterns:

- The corridor supported a healthy diversity of arthropod predators including green lacewings, minute pirate bugs, big-eyed bugs, damsel bugs, and several species of hoverflies, ladybugs, tumbling flower beetles, and spiders.
- Diversity of generalist predators overall was higher in the vineyard block with the plant corridor.
- In the vineyard block with the corridor, the numbers of the two major grape herbivores present (western grape leafhoppers and western flower thrips) were lowest near the plant corridor and highest in the central areas. In the other vineyard block, these herbivores were distributed evenly throughout the block.
- Most generalist predators showed a density gradient in the block with the corridor, reaching their greatest numbers near the plant corridor. In the other block, these generalist predators were more evenly distributed.
- The rate of parasitization of leafhopper eggs by *Anagrus epos* wasps was roughly the same throughout both vineyard blocks.

These results showed that the positive effect of the adjacent riparian forest on the biodiversity of beneficials was — with the exception of *A. epos* — amplified by the flowering plant corridor in block A. For ladybugs and lacewings, the corridor provided food in the form of aphids and other homoptera; for hoverflies it supplied nectar and pollen; for predatory insects such as minute pirate bugs it offered neutral insect prey. By providing these food resources, the corridor allowed beneficials to move more deeply into the vineyard. In island biogeography terms, the corridor effectively reduced the size of the monoculture "islands," facilitating their "colonization" by beneficials.

In addition to demonstrating the applications of island biogeography theory and the value of diversity, this study highlights the importance of looking at diversity and ecological processes at the scale of the landscape. Agricultural practices that allow, create, or retain a more diverse agricultural landscape that includes remnants of natural vegetation and noncrop areas are to be encouraged for a variety of reasons, a concept we will explore in more detail in Chapter 22.

**TABLE 16.7**
**Research Questions Related to the Colonization and Island Biogeography Theory**

| Type of Organism | Source | Barrier Variables | Island Variables | Research Question |
|---|---|---|---|---|
| Herbivore pest | Surrounding crop fields | Type of barrier | | What are effective barriers against the dispersal of the pest into the crop field? |
| Herbivore pest | Surrounding crop fields | Size of barrier | | What distance between fields of similar crops can best control the spread of the pest from one field to another? |
| Undesirable weed | Surrounding crop fields | Type, size, and nature of barrier (e.g., windbreak) | | What are effective barriers against dispersal of the weed into the crop field? |
| Predator on herbivores | Anywhere outside the system | | Habitat for alternate host | How can colonization by the predator be encouraged? |
| Disease organism | Surrounding crop fields | | Size of island | Is a small crop island more difficult for a disease organism to find or reach? |
| Undesirable weed | Surrounding crop fields | | Occupation of niches | Can an occupied niche resist the invasion of new colonizers? |
| Beneficial insects | Anywhere outside the system | Strip crops around the crop field | Corridors within the crop field | Can the area between crops be diversified in ways that attract and retain beneficials? |

## DIVERSITY, STABILITY, AND SUSTAINABILITY

Diversity in agroecosystems can take many forms, including the specific arrangement of crops in a field, the way that different fields are arranged, and the ways that different fields form part of the entire agricultural landscape of a farming region. With increased diversity, we can take advantage of the positive forms of interference that lead to interactions between the component parts of the agroecosystem, including both crop and noncrop elements. The challenge for the agroecologist is to demonstrate the advantages that can be gained from introducing diversity into farming systems, incorporating many of the components of ecosystem function that are important in nature, and managing such diversity for the long term.

In part, meeting this challenge means determining the relationships between the different kinds of diversity presented in this chapter and the stability of the agroecosystem over time. This stability should be understood as both the resistance of the system to change and the resilience of the system in response to change. Since each species in the agroecosystem brings something different to the processes that maintain both types of stability, an important part of agroecological research is focused on understanding the contribution each species makes and using this knowledge to integrate each species into the system in the optimal time and place. As this integration takes place, the emergent qualities of system stability appear, allowing the ultimate emergent quality — sustainability — to develop.

The most sustainable agroecosystems might be those that have some kind of mosaic pattern of structure and development, in which the system is a patchwork of levels of diversity, mixing annuals, perennials, shrubs, trees, and animals. Or the most sustainable systems might be those with several stages of development occurring at the same time as a result of the type of management applied. Such systems might incorporate minimum tillage to allow a more mature soil subsystem to develop, even with a simpler aboveground plant system, or use strip cropping or hedge rows to create a mosaic of levels of development and diversity across the farm landscape. Once the parameters of diversity are established, the issue becomes one of frequency and intensity of disturbance — which we will explore in the next chapter.

## FOOD FOR THOUGHT

1. Describe a pest management strategy that builds on the theory of island biogeography.
2. Explain a situation where lack of diversity in one component of an agroecosystem can be compensated for by greater diversity in some other component.
3. What is the connection between diversity and the avoidance of risk in agroecosystems? Give examples to support your viewpoint.
4. What are some possible mechanisms allowing a crop to produce a higher yield in an intercrop than when planted by itself in monoculture?
5. What are the main disincentives for farmers to shift into more diverse farming systems? What kinds of changes need to occur in order to provide the necessary incentives?
6. What are some of the forms of agroecosystem diversification that will best promote the successful use of Integrated Pest Management (IPM)?
7. Why are intercropping and agroforestry agroecosystems more common in the tropics than in the temperate parts of the world?

## INTERNET RESOURCES

Agroecology in Action
www.agroeco.org
The website of Professor Miguel Altieri, at the University of California, Berkeley, with extensive material on agroecological pest and habitat management.

Cedar Creek Natural History Area and LTER
www.lter.umn.edu
An important Long Term Ecological Research site (STER), directed by recognized diversity ecologist David Tilman, and with information on long-term vegetation diversity experiments and publications.

Farming Systems Research Unit, Center for Environmental Farming Systems
www.cefs.ncsu.edu/frsu.htm
An excellent site located at North Carolina State University, with projects and publications focusing on management of agroecosystem diversity.

The Land Institute
www.landinstitute.org
A well-known research and training center in Salina, Kansas, which has focused on agroecosystem diversity management through its natural systems agriculture approach.

## RECOMMENDED READING

Altieri, M.A., and C. Nicholls. 2004. *Biodiversity and Pest Management in Agroecosystems.* 2nd ed. Howarth Press: Binghamton, NY. A review of the role of vegetational diversity in insect pest management, combining an analysis of ecological mechanisms and design principles for sustainable agriculture.

Carlquist, S. 1974. *Island Biology.* Columbia University Press: New York. An excellent overview of the biological and evolutionary processes characteristic of island ecosystems.

Carson, R. 1962. *Silent Spring*. Houghton Mifflin: Boston. The classic alarm call about the negative impact of pesticides on biodiversity.

Gaston, K.J. and J.I. Spicer. 2004. *Biodiversity: An Introduction*. 2nd ed. Blackwell Science: Malden, MA. An overview of what biodiversity is, its relevance to humanity and issues related to its conservation.

Golley, F.B. 1994. *A History of the Ecosystem Concept in Ecology.* Yale University Press: New Haven, Connecticut. A full review of the development and importance of the ecosystem concept.

Loreau, M., S. Naeem, and P. Inchausti. 2002. *Biodiversity and Ecosystem Functioning: Synthesis and Perspectives.* Oxford University Press: New York. A comprehensive and critical overview of recent empirical and theoretical research on the relationship between biodiversity and ecosystem function.

Ricklefs, R.E. 2001. *The Economy of Nature.* W.H. Freeman and Company: New York. 5th ed. A very balanced review of the field of ecology that links basic principles with an understanding of environmental problems.

Smith, R.L. and T.M. Smith. 2001. *Ecology and Field Biology.* 6th ed. Prentice-Hall: New York. A text of general ecology that provides an overview of the discipline with an excellent focus on applications in the field.

# 17 Disturbance, Succession, and Agroecosystem Management

The ecological concepts of disturbance and recovery through succession have important application in agroecology. Agroecosystems are constantly undergoing disturbance in the form of cultivation, soil preparation, sowing, planting, irrigation, fertilizer application, pest management, pruning, harvesting, and burning. When disturbance is frequent, widespread, and intense — as it is in conventional agriculture — agroecosystems are limited to the earliest stages of succession. This condition enables high productivity but requires large inputs of fertilizer and pesticides, and tends to degrade the soil resource over time.

More sustainable food production can be achieved by moving away from dependency on continual and excessive disturbance and allowing successional processes to generate greater agroecosystem stability. Based on our understanding of disturbance and succession in natural ecosystems, we can enhance the ability of agroecosystems to maintain both fertility and productivity through appropriate management of disturbance and recovery.

## DISTURBANCE AND RECOVERY IN NATURAL ECOSYSTEMS

A long-standing tenet of ecology is that following a disturbance, an ecosystem immediately begins a process of recovery from that disturbance. Recovery takes place through the relatively orderly process of succession, which was introduced in Chapter 2. In the broadest sense, ecological succession is the process of ecosystem development, whereby distinct changes in community structure and function occur over time.

Ecologists distinguish two basic types of succession. **Primary succession** is ecosystem development on sites (such as bare rock, glaciated surfaces, or recently formed volcanic islands) that were not previously occupied by living organisms or subject to the changes that the biotic components can bring to bear on the abiotic components. **Secondary succession** is ecosystem development on sites that were previously occupied by living organisms, but had some or all of those organisms removed by fire, flooding, severe wind, intense grazing, or some other event. Depending on the intensity, frequency, and duration of the disturbance, the impact on the structure and function of the ecosystem will vary, as will the time required for recovery from the disturbance. Since the disturbance and recovery process that occurs in agriculture usually takes place in sites that formerly had other biotic components, we will focus our attention here on the secondary succession process.

## THE NATURE OF DISTURBANCE

Although natural ecosystems give the impression of being stable and unchanging, they are constantly being altered on some scale by events such as fire, wind storms, floods, extremes of temperature, epidemic outbreaks, falling trees, mudslides, and erosion. These events disturb ecosystems by killing organisms, destroying and modifying habitats, and changing abiotic conditions. Any of these impacts can change the structure of a natural ecosystem and cause changes in the population levels of the organisms present and the biomass they store.

Disturbance can vary in three dimensions:

- *Intensity of disturbance* can be measured by the amount of biomass removed or the number of individuals killed. The three types of fire described in Chapter 10 provide good examples of variation in disturbance intensity: surface fires usually create low-intensity disturbance, whereas crown fires cause high-intensity disturbance.
- *Frequency of disturbance* is the average amount of time between each disturbance event. The longer the time span between disturbances, the greater the ability of the ecosystem to fully recover after each disturbance.
- *Scale of disturbance* is the spatial scope of the disturbance, which can vary from a small, localized patch to the entire landscape. The small gap in the forest canopy created by an individual tree falling is a small-scale disturbance, whereas the massive destruction of a powerful hurricane is very large-scale.

All three characteristics of disturbance are often intertwined in complex ways. Fire, for example, may occur with varying frequency; it may be distributed over the landscape in a patchy manner; and where it does occur it may burn some areas very intensely and others hardly at all.

## The Recovery Process

Any change or alteration of the ecosystem by a disturbance is followed by a recovery process. Recovery occurs through the combined action of several ecosystem dynamics: (1) the biotic community as a whole modifies the physical environment through the many forms of interference described in previous chapters; (2) competition and coexistence between individual organisms and populations cause changes in the diversity and abundance of species; and (3) energy flow shifts from production to respiration as more and more energy in the system is needed to support the growing amount of standing biomass. The interaction of these processes directs a recovering ecosystem through a number of stages of development (originally called seral stages) that eventually lead to a structure and level of ecosystem complexity similar to what existed before the disturbance occurred.

During the recovery process, many important changes in ecosystem structure and function occur. These are most distinct following a relatively severe and extensive disturbance. A summary of some of the more important characteristics of the successional process that follows a major disturbance is presented in Table 17.1. The early or pioneer stages of succession are dominated by *r*-selected, easily dispersed weedy species, but as these early invaders either alter the conditions of the environment or are displaced by interference from later arrivals, *K*-selected species begin to dominate. The replacement of earlier species of plants and animals by others over time has been commonly observed during the recovery process (e.g., Keever, 1950; Bazzaz, 1996; Finegan, 1996).

Most of the components of ecological diversity (described in the previous chapter) increase during succession, especially in the early stages, often reaching their highest levels prior to full recovery. Of particular agroecological importance is the fact that gross photosynthesis during the early stages of succession normally greatly exceeds total respiration, resulting in high net primary productivity and high harvest potential. As the standing crop increases with successional development, however, a greater proportion of productivity is used for maintenance, creating the impression of greater stability.

**TABLE 17.1**
**Changes that Occur in Ecosystem Structure and Function During the Course of Secondary Succession Following a Major Disturbance**

| | Changes During Successional Process[a] | | |
|---|---|---|---|
| **Ecosystem Characteristic** | **Early Stages** | **Middle Stages** | **Maturity** |
| Species composition | Rapid replacement of species | Slower replacement of species | Little change |
| Species diversity | Low, with rapid increase | Medium, with rapid increase | High, with possible slight decline |
| Total biomass | Low, with rapid increase | Medium, with moderate increase | High, with slow rate of increase |
| Mass of non-living organic matter | Low, with rapid increase | Medium, with moderate increase | High, with slow rate of increase |
| Gross primary productivity | Increases rapidly | Declines slightly | |
| Net primary productivity | Increases rapidly | Declines slightly | |
| System respiration | Increases | Increases slowly | |
| Food chains/webs | Become increasingly complex | Remain complex | |
| Species interactions | Become increasingly complex | Remain complex | |
| Efficiency of overall nutrient and energy use | Increases | Remains efficient | |
| Cycling of nutrients | Flow-through; open cycles | -----------------> | Internal cycling; closed cycles |
| Retention of nutrients | Low retention, short turnover time | -----------------> | High retention, long turnover time |
| Growth form | *r*-selected, rapidly growing species | -----------------> | Long-lived *K*-selected species |
| Niche breadth | Generalists | -----------------> | Specialists |
| Life cycles | Annuals | -----------------> | Perennials |
| Interference | Mostly competitive | -----------------> | More mutualistic |

[a] Although some changes are presented in stepwise form, all occur as gradual transitions.

*Source:* Adapted from Odum, E. P. 1993. *Ecology and Our Endangered Life-Support Systems*. Sinauer Associates Incorporated: Sunderland, Massachusetts.

Another aspect of successional development that has important agroecological implications is the increase in biomass and the standing crop of organic matter with time, especially in the early stages of succession. Since biomass is eventually converted to detritus and humus as it passes through the decomposers, this increase in biomass results indirectly in an increase in soil organic matter.

During the early stages of recovery, nutrient availability is usually high and nutrient conservation relatively inefficient. Fast-growing, ruderal plant species quickly become dominant, and population interaction is limited to the few species present. As succession progresses, nutrient retention improves, colonizing species begin to occupy a greater diversity of niches in the system, population interaction intensifies (especially interactions that involve resource partitioning and mutualistic interference), and the structure of the ecosystem becomes more complex and interconnected.

If enough time is allowed to pass after a disturbance, an ecosystem eventually reaches a point (formerly referred to as the climax stage) at which most of the characteristics presented in Table 17.1 cease to change significantly in rate or character. In terms of species diversity, for example, new colonizing species equal the number of emigrating species or those going extinct. Nutrient losses from the system are balanced by inputs from outside. The population levels of species fluctuate seasonally, but do so around a fairly constant mean number. At this stage, the system is once again in a tenuous equilibrium with the regional climate and local conditions of soil, topography, and moisture availability. Change still occurs, but it is no longer directional, developmental change, but change oriented around an equilibrium point. In chapter 2, we described such a condition as one of *dynamic equilibrium*, a concept that takes into account the fact that all environments are constantly changing and evolving, with new disturbances occurring frequently on at least a small scale.

In the typical mature ecosystem, then, localized sites may be undergoing disturbance on a regular basis, but the characteristics listed in Table 17.1 are developed sufficiently enough for energy and nutrient utilization to be highly efficient, food webs complex, and mutualistic relationships prevalent. The system is relatively stable, in the double sense of being able to resist change and to be resilient when disturbance occurs. Thus, the disturbance events that do occur do not result in drastic change, but neither do they allow a steady-state condition.

### Intermediate Disturbance

In some ecosystems, the frequency, intensity, and scale of disturbance is such that the system never reaches full maturity, but is nevertheless able to maintain the species diversity, stability, and energy-use efficiency characteristic of a mature ecosystem. Where hurricanes occur, for example, these high-intensity disturbance events — as long as they are low in frequency — tend to generate forest systems with both high species diversity and high biomass (Vandermeer et al., 2000; Mascaro et al., 2005). Ecologists studying these systems have posited the **intermediate disturbance hypothesis**, which states that in natural ecosystems where environmental disturbances are neither too frequent nor too seldom (at some intermediate frequency) both diversity and productivity can be high (Connell, 1978; Connell and Slayter, 1977). The disturbance in these systems retains the early-successional characteristic of high productivity, while the system's overall stability allows the high species diversity more characteristic of mature ecosystems.

Some natural ecosystems for which the intermediate disturbance hypothesis may apply are presented in Table 17.2. An examination of these systems reveals that intermediate disturbance can come about through a great variety of different combinations of disturbance frequency, disturbance intensity, and disturbance scale. At an ecosystem level, relatively intense and frequent disturbance on a small scale, for example, can have an effect similar to that of low-intensity, low-frequency disturbance on a larger scale.

In many intermediate-disturbance situations, disturbance distributed irregularly over the landscape in time and space creates what is known as a **patchy landscape**, in which numerous stages of succession occur in a relatively small area. The variation in developmental stage from patch to patch contributes to the maintenance of considerable diversity at the ecosystem level. Successional **patchiness** can therefore be seen as an important aspect of the ecological dynamics of ecosystems. Patch size, variation in patch development, and the nature of the interfaces between patches all become important variables, and ecologists have invested considerable study attempting to understand their role in natural ecosystems (Pickett and White 1985; Hubbell et al., 1999). The inherent patchiness of many agricultural landscapes points out the potential application of intermediate disturbance and patchiness to agroecosystem management (Bruun, 2000). As we will see in more detail in Chapter 22, the concept of patchiness has become especially important in approaches that seek to conserve biodiversity and ecosystem services in agricultural landscapes (McIntyre and Hobbs, 1999; Swift et al., 2004).

## APPLICATIONS TO AGROECOSYSTEM MANAGEMENT

Modern agriculture has developed practices, technologies, and inputs that allow farmers to ignore most successional processes. In place of natural recovery, farmers use inputs and materials that replace what is removed at harvest or altered with cultivation. Constant disturbance keeps the

**TABLE 17.2**
**Some Examples of Intermediate Disturbance in Natural Ecosystems**

| Frequency | Scale | Intensity | Nature of disturbance |
|---|---|---|---|
| High | Small | low | Natural wind-felling of trees in forests |
| Low | Large | high | Hurricane damage to coral reef or coastal tropical forest |
| High | Medium | low | Removal of above-ground biomass by grazing herbivores in grasslands |
| Medium | Medium | medium | Ice and sleet damage to trees in temperate forests |
| Medium | Medium | low | Surface fires in dry-summer tropical forests |

**TABLE 17.3**
**Desirable Ecological Characteristics of Agroecosystems in Relation to Successional Development**

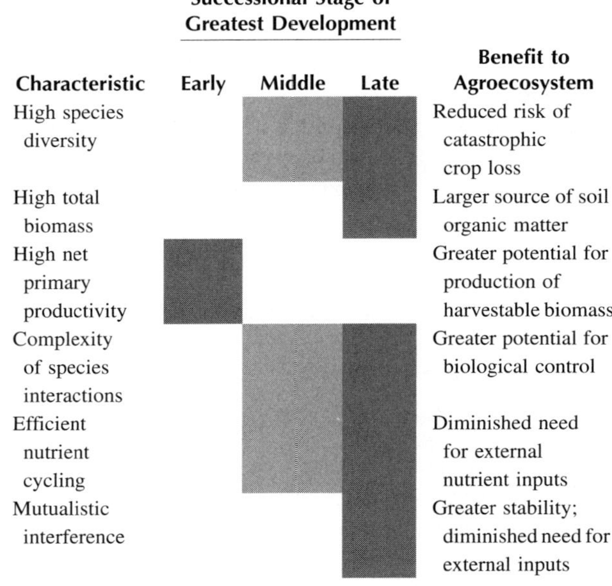

| Characteristic | Successional Stage of Greatest Development: Early | Middle | Late | Benefit to Agroecosystem |
|---|---|---|---|---|
| High species diversity | | | | Reduced risk of catastrophic crop loss |
| High total biomass | | | | Larger source of soil organic matter |
| High net primary productivity | | | | Greater potential for production of harvestable biomass |
| Complexity of species interactions | | | | Greater potential for biological control |
| Efficient nutrient cycling | | | | Diminished need for external nutrient inputs |
| Mutualistic interference | | | | Greater stability; diminished need for external inputs |

agroecosystem at the early stages of succession, where a greater proportion of gross productivity is available as net productivity or harvestable biomass. But in order to develop more stable systems that are much less dependent on human interventions and polluting, non-renewable inputs, we must do much more to take advantage of natural ecosystem recovery processes. Our knowledge of the successional process in natural ecosystems can be used both to aid agroecosystems in their recovery from the impacts of human-induced disturbance and to introduce disturbances in a planned manner.

Simply stated, the task is to design agroecosystems that on the one hand take advantage of some of the beneficial attributes of the early stages of succession, yet on the other hand incorporate some of the advantages gained by allowing the system to reach the later stages of succession. As shown in Table 17.3, only one desirable ecological characteristic of agroecosystems — high net primary productivity — occurs in the early stages of successional development; all the others do not become manifest until the later stages of development.

The challenge for research, then, is to develop ways of integrating disturbance and development so as to take best advantage of both extremes. This involves learning how to use successional processes for installing and developing an agroecosystem, as well as for re-introducing disturbance and recovery at appropriate times in the life of the system.

### Allowing Successional Development

Agriculture has long taken advantage of disturbance to keep farming systems in the earlier stages of succession. This is especially true for annual cropping systems, where no part of the ecosystem is allowed to progress beyond the early pioneer stage of development. In this stage, the system can produce large amounts of harvestable material, but keeping an agroecosystem at this high output level takes its toll on other developmental processes and makes stability impossible.

Another approach to agroecosystem management is to "mimic nature" by installing a farming system that uses as a model the successional processes that go on naturally in that location (Soule and Piper, 1992; Ewel, 1999; Jackson, 2002). Through such an approach — sometimes called the "analog model" or "natural systems agriculture" — we can establish agroecosystems that are both stable and productive.

Under a scheme of managed succession, natural successional stages are mimicked by intentionally introducing plants, animals, practices, and inputs that promote the development of interactions and connections between component parts of the agroecosystem. Plant species (both crop and non-crop) are planted that capture and retain nutrients in the system and promote good soil development. These plants include legumes, with their nitrogen-fixing bacteria, and plants with phosphorus-trapping mycorrhizae. As the system develops, increasing diversity, food web complexity, and level of mutualistic interactions all lead to more effective feedback mechanisms for pest and disease management. The emphasis during the development process is on building a complex and integrated agroecosystem.

Such a strategy may require more intensive human management, but because processes and interactions are internalized within the agroecosystem, it should lead to less dependence on human-derived inputs from outside of the system and greater stability.

There are many ways that a farmer, beginning with a recently cultivated field of bare soil, can allow successional development to proceed beyond the early stages. One general model, beginning with an annual monoculture and progressing to a perennial tree crop system, is illustrated in Figure 17.3 and described below.

1–2. The farmer begins by planting a single annual crop that grows rapidly, captures soil nutrients, gives an early yield, and acts as a pioneer species in the developmental process. The farmer could also choose to introduce other less aggressive annuals into the initial planting, mimicking the early successional process.

3. As a next step (or instead of the previous one), the farmer can plant a polyculture of annuals that represent different components of the pioneer stage. The species would differ in their nutrient needs, attract different insects, have different rooting depths, and return a different proportion of their biomass to the soil. One might be a nitrogen-fixing legume. Small livestock such as ducks or geese might be allowed to graze on weeds or feed on snails that might be common colonizers. All of these early species would contribute to the initiation of the recovery process, and they would modify the environment so that non-crop plants and animals — especially the macro- and microorganisms necessary for developing the soil ecosystem — can also begin to colonize.

4. Following the initial stage of development (towards the end of the first season or at the beginning of the second or third season), short-lived perennial crops might begin to be introduced. Taking advantage of the soil cover created by the pioneer crops, these species can diversify the agroecosystem in important ecological aspects. Deeper root systems, more organic matter stored in standing biomass, and greater habitat microclimate diversity all combine to advance the successional development of the agroecosystem (Figure 17.1).

**FIGURE 17.1 The short-lived perennial yuca (Manihot esculenta) growing in an annual corn crop, Turrialba, Costa Rica.** The yuca is introduced after the corn is established.

5. Once soil conditions improve sufficiently, the ground is prepared for planting longer-lived perennials, especially orchard or tree crops, with annual and short-lived perennial crops maintained in the areas between them. While the trees are in their early growth, they have limited impact on the environment around them. At the same time, they benefit from having annual crops around them, because in the early stages of growth they are often more susceptible to interference from the more aggressive weedy *r*-selected non-crop species that would otherwise occupy the area (Figure 17.2).
6. As the tree crops develop, the space in between them can continue to be managed with annuals and short-lived perennials, using the agroforesty approach described below. Larger livestock can be introduced at this point for vegetation management, enterprise diversification, and better nutrient cycling (Chapter 19).
7. Eventually, once the trees reach full development, the end point in the developmental process is achieved. This end point could be modeled after the structure of natural ecosystems of the region. Once it has been achieved, the farmer has the choice of maintaining it (possibly as an integrated system with livestock) or introducing controlled disturbance in ways that return the agroecosystem, or selected parts of it, to earlier stages of succession.

It is useful to examine how net primary productivity and standing biomass change over time when an agroecosystem is allowed to progress through the stages described in Figure 17.3. These changes will be similar to those that occur in a natural ecosystem as it undergoes succession after disturbance; a general model for these changes over time is presented in Figure 17.4. NPP increases rapidly during the earliest stages of agroecosystem development, with most of that increase being available as harvestable products. A time interval in the early stages of successional development (e.g., stages 2 and 3 in Figure 17.3) will show the most rapid increase in net primary productivity available during the

**FIGURE 17.2 Seedlings of the tree Gmelia arborea intercropped into a corn–squash planting in southern Campeche, Mexico.** The practice of initiating a tree crop system in an annual system is called *taungya*.

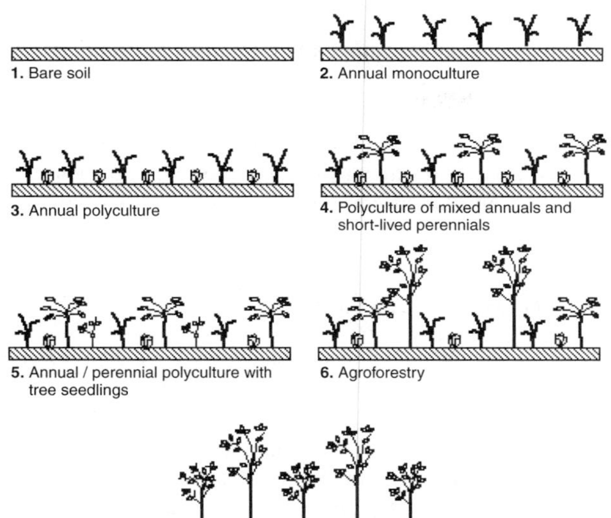


**FIGURE 17.3 Steps in the successional development of an agroecosystem.** At any stage in the process, disturbance can be introduced to bring all or part of the system back to an earlier stage of development.

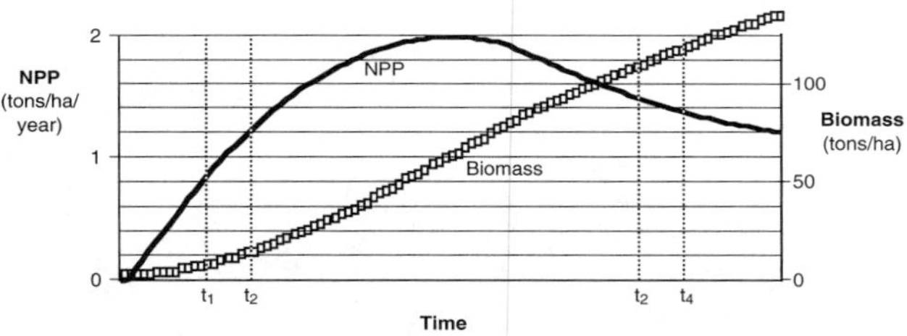


**FIGURE 17.4 Change over time in the relationship between annual net primary productivity (NPP) and accumulated living and dead biomass in a representative successionally developing ecosystem.** A time interval (e.g., one season) in the early stages of succession (such as $t_2$–$t_1$) will witness a rapid increase in NPP, whereas NPP will decline slightly during a time interval of similar length (such as $t_4$–$t_3$) during the latter stages of succession. (Modified from Whittaker, R. H. 1975. *Communities and Ecosystems*. 2nd ed. MacMillan: New York.; Odum, E. P. 1993. *Ecology and Our Endangered Life-Support Systems*. Sinauer Associates Incorporated: Sunderland, Massachusetts. With permission.)

developmental process, and provide the greatest amount of harvestable material in the shortest time. This could also be the point at which the most biomass is available for grazing animals. In the later stages of development (e.g., Stage 7 in Figure 17.3), when the rate of NPP begins to decrease, standing biomass (in the form of accumulated perennial biomass) is relatively high, but the actual amount of new harvestable material produced in each time interval begins to drop.

The changing relationship between NPP and biomass over time determines what management and production strategies can be used at each stage of agroecosystem development. The trade-offs and constraints change. In the early stages of development, for example, constant removal of NPP restricts the accumulation of biomass, whereas restricted harvest of NPP forces a farmer to wait several years for harvest. Grazing animals can help accelerate biomass turnover as long as their manures are kept in the system. At the intermediate stages of development, NPP is high enough for part of it to be harvested as fruit or nuts and part allowed to accumulate as biomass. By the later stages (e.g., stage 7 in Figure 17.3), NPP declines to a low enough level that a workable strategy is to allow all new NPP to accumulate as biomass, and to harvest the biomass selectively for fuel, timber, forage, paper pulp, or even food.

## Managing Successionally Developed Agroecosystems

Once a successionally developed agroecosystem has been created, the problem becomes one of how to manage it. The farmer has three basic options:

- Return the entire system to the initial stages of succession by introducing a major disturbance, such as clear-cutting the trees in the perennial system. Many of the ecological advantages that have been achieved will be lost and the process must begin anew.
- Maintain the system as a perennial or tree crop agroecosystem, with or without livestock.
- Reintroduce disturbance into the agroecosystem in a controlled and localized manner, taking advantage of the intermediate disturbance hypothesis and the dynamics that such patchiness introduces into an ecosystem. Small areas in the system can be cleared, returning those areas to earlier stages in succession, and allowing a return to the planting of annual or short-lived crops. If care is taken in the disturbance process, the below-ground ecosystem can be kept at a later stage of development, whereas the above-ground system can be made up of highly productive species that are available for harvest removal. Such a mixture of early and later stages of development leads to the formation of a **successional mosaic.** This mosaic can be adjusted and managed according to the ecological conditions of the area, as well as the needs of the farmer and changes in market conditions. It can also incorporate livestock.

The latter option provides the most advantages and offers the greatest flexibility to the farmer. Within the constraints imposed by the ecological limits of the cropping region, the final mixture of annual and perennial plants and grazing animals can be tailored to the needs of the farmer and farm community and adjusted to fit market demand, the distance to market, the ability to enter into the market, and the farmer's ability to purchase and transport inputs. The closer the farm is to inputs, labor, and markets, the heavier the emphasis can be on the annual component.

The biggest challenge in managing a successionally developed system is to learn how to introduce disturbance in ways that stimulate system productivity on the one hand, and provide resistance to change and variation within the ecosystem on the other. This can be done in many different ways depending on local environmental conditions, the structure of mature natural ecosystems normally present, and the feasibility of maintaining modifications of those conditions over the long term.

For example, in the prairie region of the United States, where a large percentage of the country's annual grain production currently takes place, the use of a successional model for designing a tree-less perennial grain system (discussed in Chapter 13) might be the focus. Another example applies to the rice-growing regions of the Yangtze River valley of China, where the long-term maintenance of paddy systems is based on knowledge of wetland ecosystems, periodic flooding, and human alteration of paddy soil. A successionally developed paddy rice agroecosystem could incorporate a perennial component by using trees that tolerate wet, flooded conditions, such as willows, bald cypress, and other riparian or wetland species, and by adding an animal component consisting of waterfowl and fish (Figure 17.5).

## AGROFORESTRY SYSTEMS

Although the perennial components of a successionally-developed agroecosystem do not have to be trees, systems with trees provide some of the best examples of how successional development can be managed. The term **agroforestry** has been given to practices that intentionally retain or plant trees on land used for crop production or grazing (Wiersum, 1981; Nair, 1983). Such systems combine elements of crop or animal agriculture with elements of forestry, either at the same time or in sequence, building on the unique productive and protective value of trees. There are many variations in practices that fall into the category

FIGURE 17.5(a) **Variations in the mixture of annuals and perennials in successionally developed agroecosystems.** Corn and beans grown for the local market are surrounded by persimmon trees in the urban fringe around Beijing, China (above). At a greater distance from any markets, a rural farm in southern Costa Rica (right) concentrates on perennial shrub and tree crops.

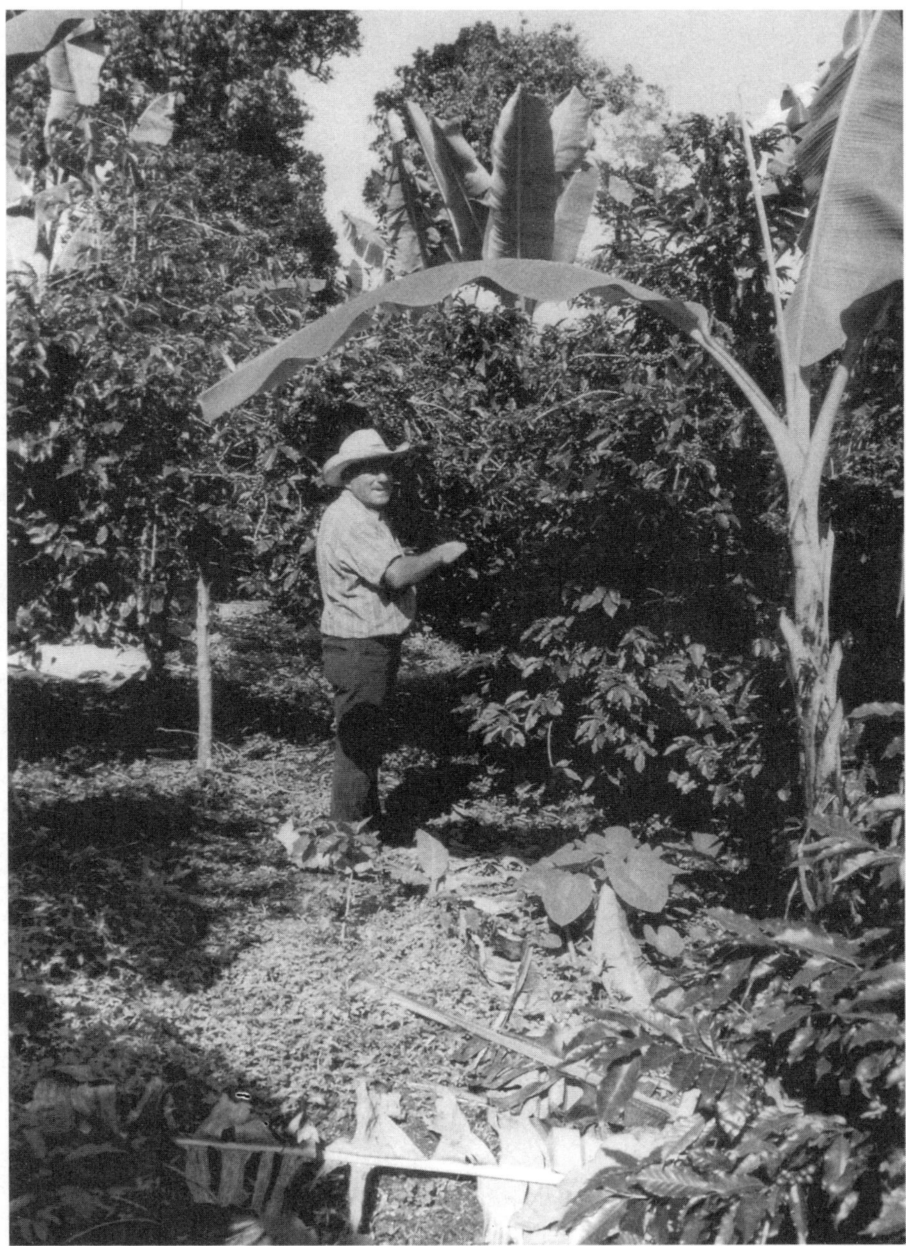

**FIGURE 17.5(b)** continued.

of agroforestry: in agrosilviculture, trees are combined with crops; in silvopastoral systems, trees are combined with animal production; and in agrosilvopastoral systems, the farmer manages a complex mixture of trees, crops, and animals. All agroforestry systems are good examples of taking advantage of diversity and successional development for production of food and other farm products.

Incorporating trees into agroecosystems is a practice with a long history. This is especially true in the tropical and subtropical regions of the world, where farmers have long planted trees along with other agricultural crops and animals to help provide for the basic needs of food, wood products, and fodder, and to help conserve and protect their often limited resources (Nair, 1983). Agroforestry systems in temperate regions of the world are also well known (Gordon and Newman, 1997) (Figure 17.6).

The objective of most agroforestry systems is to optimize the beneficial effects of the interactions that occur among the woody components and the crop or animal components in order to obtain more diversity of products, lessen the need for outside inputs, and lower the negative environmental impacts of farming practices. In many respects, agroforestry systems create the same ecological benefits as multiple cropping systems, and the research

**FIGURE 17.6 Cows crowding into the shade of a lone Ceiba pentandra left in a tropical lowland pasture in Tabasco, Mexico.** Trees can provide a number of benefits to pasture and grazing systems.

methods used to analyze multiple cropping systems apply equally well to agroforestry systems.

## THE ECOLOGICAL ROLE OF TREES IN AGROFORESTRY

Trees are capable of altering dramatically the conditions of the ecosystem of which they are part (Reifsnyder and Darnhofer, 1989; Farrell, 1990). The sustainable productivity of agroforestry systems is due in large part to this capability of trees.

Below ground, a tree's roots penetrate deeper than those of annual crops, affecting soil structure, nutrient cycling, and soil moisture relations. Above ground, a tree alters the light environment by shading, which in turn affects humidity and evapotranspiration. Its branches and leaves provide habitats for an array of animal life and modify the local effects of wind. Shed leaves provide soil cover and modify the soil environment; as they decay they become an important source of organic matter. These and other ecological effects of trees are summarized in Figure 17.7.

Because of these effects, trees in agroecosystems are a good foundation for developing the emergent qualities of more complex ecosystems. They allow more efficient capture of solar energy; enhance nutrient uptake, retention, and cycling; and maintain the system in dynamic equilibrium. By providing permanent microsites and resources, they make possible a more stable population of both pests and their predators. In an agroforestry system, all of these factor interactions can be managed to the benefit of the associated crop plant and animals, while at the same time lessening the dependence of the system on outside inputs.

## DESIGN AND MANAGEMENT OF AGROFORESTRY SYSTEMS

In an agroforestry system, farmers have the choice of how many trees to include, how frequently and in what patterns to remove them, and what kind of pattern of successional mosaic to maintain. These management decisions depend on the local environment and culture, as well as the nature and proximity of markets.

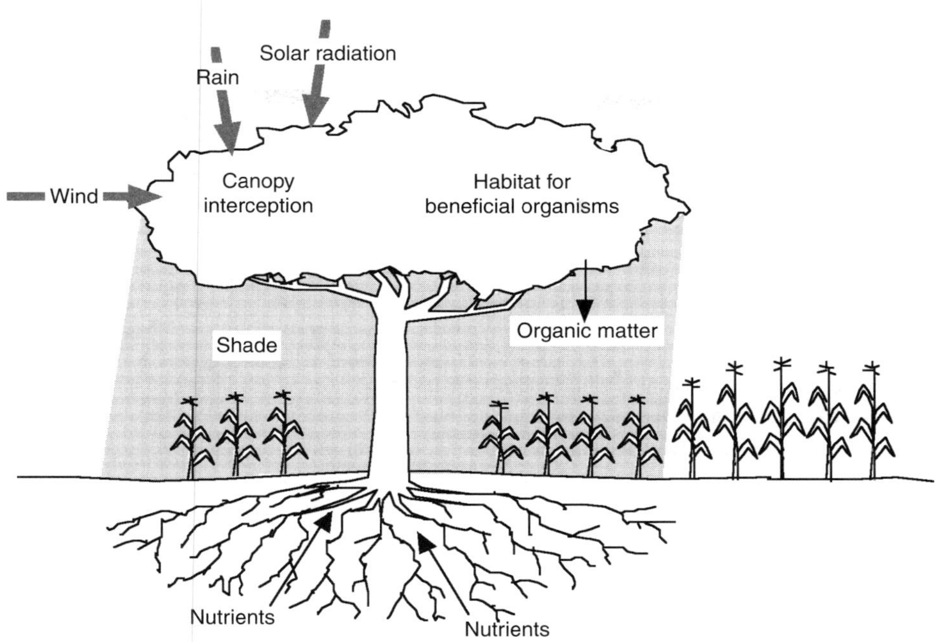


**FIGURE 17.7 Effects of a tree on the surrounding agroecosystem.** Because of its size, root depth, and perennial nature, a tree has significant effects on the abiotic conditions of an agroecosystem and takes part in many biotic interactions. In addition to the effects and interactions shown, a tree can limit wind and water erosion, provide shade and browse for animals, form mycorrhizal associations, moderate soil temperature, and reduce evapotranspiration. Leguminous trees can contribute nitrogen to the system through their association with nitrogen-fixing bacteria. (Adapted from Nair, P. K. R. 1984. *Soil Productivity Aspects of Agroforestry: Science and Practice in Agroforestry.* International Council for Research in Agroforestry (ICRAF): Nairobi, Kenya.; Farrell, J. 1990. In S. R. Gliessman (ed.), *Agroecology: Researching the Ecological Basis for Sustainable Agriculture*. pp. 169–183. Springer-Verlag: New York. With permission.)

## EFFECT OF TREES ON SOIL IN TLAXCALA, MEXICO

Trees affect the environment of an agroforestry system in a variety of ways. The specific effects vary from system to system, depending on factors such as altitude, annual rainfall, wind patterns, geography, soil type — and, of course, the species of the tree. To effectively use trees in an agroecosystem, it is important to consider all these factors, as well as the farmer's needs.

In the low-lying areas of Tlaxcala, Mexico, farmers typically maintain some combination of five different types of trees, either scattered in their fields or arranged as borders. Researcher John Farrell chose to study the two trees that are most commonly associated with agricultural fields in Tlaxcala, *Prunus capuli* and *Juniperus deppeana* (Farrell, 1990). For each species, he studied the conditions directly under the crown of the tree, in the shade zone of the tree, in the zone affected by the tree root system, and in the zone outside the direct influence of the tree.

Farrell found that soil conditions were consistently improved by the presence of the trees. Carbon, nitrogen, and phosphorus content of the soil were significantly higher in the zone of influence of the trees; other beneficial effects included a higher soil pH, increased moisture content and lower soil temperature. All these effects decreased with distance from the tree.

On the negative side, harvest yields were reduced directly under the canopy of the tree; corn planted in this area was shorter and produced approximately half as much grain as corn outside the zone. However, corn in the partially shaded areas within the zone of root influence produced just as well as corn grown outside the influence of the tree. Farrell concluded that the lower yield of the shaded corn was due solely to shading and not to competition for nutrients.

The shading of crop plants under a tree's canopy demonstrates that using trees in agroecosystems always involves trade-offs. However, with proper management, farmers can maximize the substantial benefits of trees while minimizing their negative impacts on harvest yield.

### Optimizing Positive Impacts of Trees

Knowledge of both the positive and negative impacts of trees on the rest of the agroecosystem is essential to fully and effectively integrating trees into the system. The positive impacts discussed above need to be balanced with the possible negative impacts of trees. These include competitive or allelopathic interference between trees and other crops, microclimate modification that creates conditions favoring disease or pest outbreak, and damage to crop quality caused by branches or fruits falling from mature trees. These negative effects of trees can usually be avoided or mitigated by appropriate spatial arrangement of the trees, choice of tree species, choice of annual species, timing of planting, and pruning. Integration of trees takes extensive knowledge of the full range of ecological interactions that can occur.

### Managing Interdependency

As our knowledge of the ecological processes taking place in complex agroforestry systems becomes more complete, we can begin to see how the different components of such systems become interdependent. An annual cropping component can become dependent on the trees for habitat modification, nutrient capture from deeper depths in the soil, and harboring of beneficial insects. The presence of the cropping component in the system can displace invasive non-crop plants that might interfere with the growth of the trees. Animals benefit from the high net primary productivity of the annual or short-lived crop or forage part of the system, and return nutrients to the soil in the form of urine and manures (for further discussion of the role of animals in agroforestry systems, Chapter 19). Management of agroforesty systems should focus on maximizing the benefits of these complex sets of ecological interdependencies.

We must also remember that ecological interdependencies are only part of the picture. Humans are dependent on trees in agroecosystems for such items as fire wood, construction material, browse for animals, fruits and nuts, spice, and medicinals. Agroforestry systems can be designed and managed with these needs in mind, so that the trees serve important roles both ecologically and economically. When this occurs, an interdependency can develop between the farming community and its farms.

### Spatial Arrangement of Trees

Trees can be arranged in an agroforestry system in a variety of different ways. The pattern used will depend on the needs of the farmer, the nature of the agroecosystem, and the local environmental and economic conditions. As an example, Figure 17.8 shows six different ways that the same percentage of ground in an agroecosystem can be covered by trees.

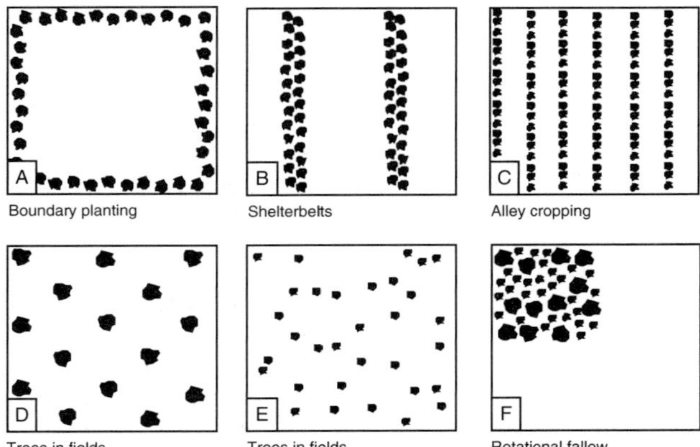


**FIGURE 17.8 Models for the arrangement of trees in agroforestry systems** (Adapted from Young, A. 1989. In W. S. Reifsnyder and T. O. Darnhofer (eds.), *Meteorology and Agroforestry*. pp. 29–48. International Council for Research in Agroforestry: Nairobi, Kenya. With permission.)

If the primary emphasis of the farmer is on silvopastoral activities, with trees intended to provide living fences, wind breaks, occasional forage from prunings, and harvestable products such as firewood or fruit, then a boundary planting of trees around areas of pasture (a) may be the best design. If, in another case, wind is a problem, but the focus is on crop production, a shelterbelt or windbreak system (b) may be best. When the tree component is intended to provide mulch from leaf fall or prunings to enhance crop production, shelterbelts can be narrow tree rows between alleys used for agriculture (c). When the trees also have agricultural value, they may be dispersed amongst the cropping system or pasture, either uniformly (d) or more randomly (e). Finally, if soil conditions are so poor that permanent cropping or grazing is not feasible, a rotational design (f) can be employed where the successional period during tree development is determined by a range of factors similar to those used to determine the length of fallow needed in shifting cultivation. A thorough understanding of the interaction, integration, and interdependency of all components of the system will ultimately help in determining trees' spatial arrangement and how it may change over time.

## TROPICAL HOME GARDENS

An agroforestry system with great complexity and diversity, as well as opportunities for maintaining a mosaic of stages of succession, is the tropical home garden system. It is probably one of the most complex and interesting types of agroecosystems, and one we have much to learn from regarding resource management for a sustainable agriculture (Méndez, 2000; Nair, 2001; Kumar and Nair, 2004) (Figure 17.9).

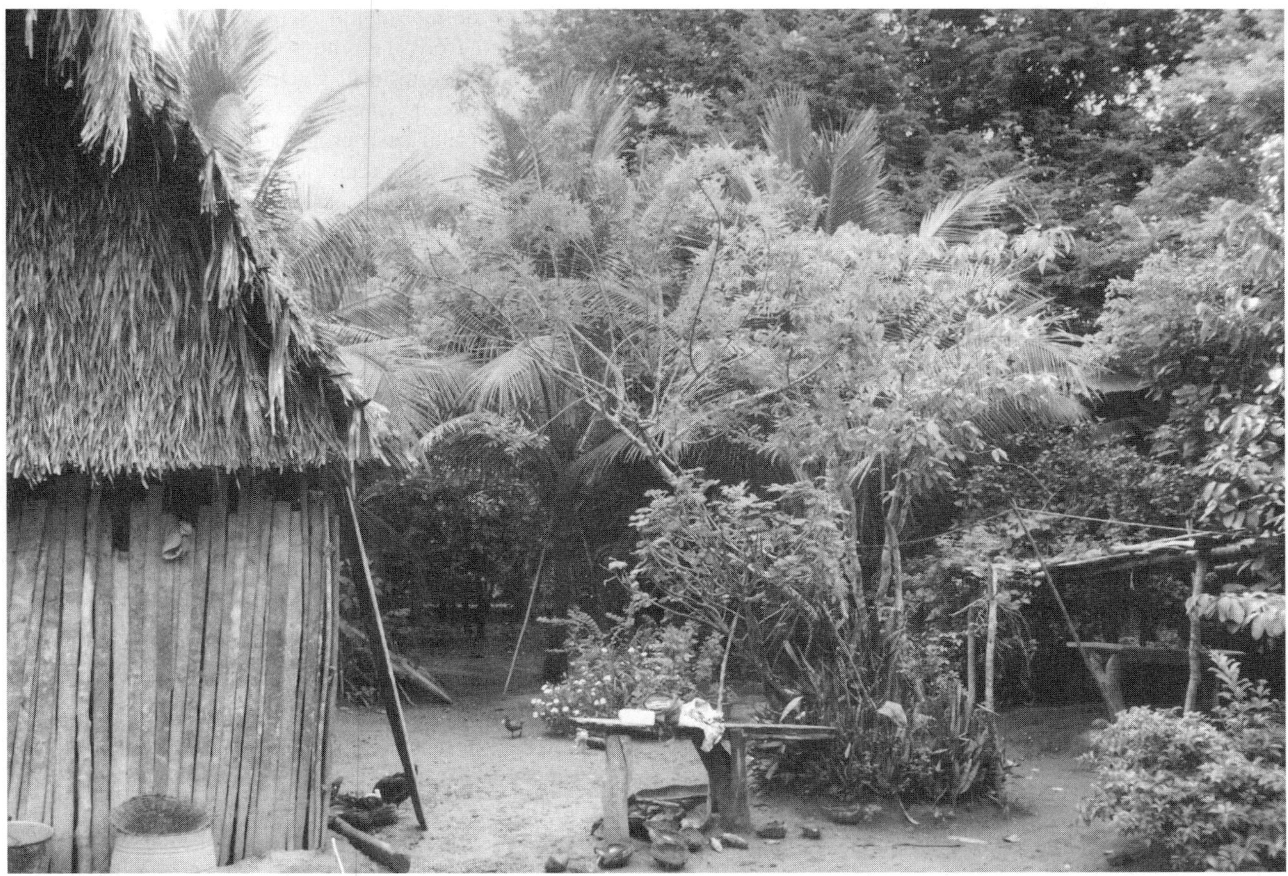

**FIGURE 17.9 A traditional tropical home garden in Cupilco, Tabasco, Mexico.** A diverse mixture of useful herbs, shrubs, and trees are associated with the area close to the dwelling.

The home garden is an integrated ecosystem of humans, plants, animals, soils, and water, with trees playing key ecological roles. It usually occupies a well-defined area, between 0.5 and 2.0 ha in size, in close proximity to a dwelling. Rich in plant species, home gardens are usually dominated by woody perennials; a mixture of annuals and perennials of different heights forms layers of vegetation resembling a natural forest structure. The high diversity of species permits year-round harvesting of food products and a wide range of other useful products, such as firewood, medicinal plants, spices, and ornamentals. Tropical home gardens also provide good opportunities for incorporating domestic animals such as chickens (Méndez et al., 2001; Del Angel-Perez and Mendoza, 2004; Kehlenbeck and Maass, 2005).

## High Diversity

The ecological diversity of home gardens — including diversity of species, structure, function, and vertical and horizontal arrangement — is remarkably high. Two examples serve as illustrations.

In a study of home gardens in both upland and lowland sites in Mexico, it was found that in quite small areas (between 0.3 and 0.7 ha) high diversity permitted the maintenance of gardens that in many aspects were similar to the local natural ecosystems (Allison, 1983). The gardens studied had relatively high indices of diversity for cropping systems (Table 17.4), and had leaf area indices and cover levels that approximated the much more complex natural ecosystems of the surrounding regions.

In another study (Ewel et al., 1982), in which nine different tropical ecosystems were analyzed for a series of ecosystem characteristics, a 40-yr-old home garden was found to have the most evenly distributed canopy — one that was fairly uniformly stratified from ground level to more than 14 m in height. Its leaf area index was 3.9 and its percent cover 100%, and the leaf biomass per square meter (307 g/m2) was the next-to-highest among all of the ecosystems examined. Total root biomass per square meter down to a depth of 25 cm was identical to leaf biomass. Perhaps most importantly, of the nine systems tested, the first 25 cm of the home garden's soil had the highest small-diameter (< 5 cm) root surface area per area of ground surface. These traits are indicative of an

**TABLE 17.4**
**Characteristics of Home Garden Systems at Two Sites in Mexico**

| Characteristics | Lowland Site (Cupilco, $n$ = 3) | Upland Site (Tepeyanco, $n$ = 4) |
|---|---|---|
| Garden size | 0.70 ha | 0.34 ha |
| Useful species per garden | 55 | 33 |
| Diversity (Shannon index) | 3.84 | 2.43 |
| Leaf area index | 4.5 | 3.2 |
| % Cover | 96.7 | 85.3 |
| % Light transmission | 21.5 | 30.5 |
| Perennial species (%) | 52.3 | 24.5 |
| Tree species (%) | 30.7 | 12.3 |
| Ornamental plants (%) | 7 | 9 |
| Medicinal plants (%) | 2 | 2.8 |

*Source:* Allison, J. 1983. M.A. Thesis, Univ. of Calif., Santa Cruz.

ecologically efficient system, especially in its ability to capture light, garner nutrients in the upper layers of the soil, store nutrients in the above-ground biomass, and reduce the impact of rain and sun on the soil.

The trees in a home garden — and the way in which they are managed by their human caretakers — make possible much of the garden's diversity, complexity, and efficient functioning. Carbon dioxide trapped between canopy layers might be able to stimulate photosynthetic activity, and the layers themselves may increase habitat diversity for birds and insects useful for maintaining biological control in the system. The trees' roots prevent nutrients from leaching out of the system, and the trees' leaf litter recycles nutrients back into the rest of the system.

## Multiple Uses and Functions

An important characteristic of home gardens is their multifaceted usefulness. The trees can produce food, such as coconuts, that can serve as either subsistence food or a cash crop. The woody parts of trees can be used for both firewood and construction material. The diversity of food types from both plants and animals provides a varied diet balanced in carbohydrates, protein, vitamins, and minerals (Dewey, 1979; Dharmasena and Wijeratne, 1996). Due to the mixture of species and their variability in flowering time and fruit maturity, there is always something ready to be harvested, ensuring sources of food or income throughout the entire year (Gliessman, 1990a).

The home garden can have such social or aesthetic functions as serving as an indication of the social status of the owner or beautifying or improving the environment directly associated with the house. At the same time, the gardens have an important economic function for rural families. In studies in Java it was found that between 20 and 30% of the annual income of many households was obtained from their home gardens (Hisyam et al., 1979). Production in local gardens fell considerably during the rice harvest when labor was concentrated on this essential food and cash crop, but during the rest of the year, activity in the gardens was quite high.

A case study in Nicaragua found that agroforestry homegardens were important to household livelihoods for both income generation and products for consumption (Méndez, 2000; Méndez et al., 2001). On average, households derived 34% of their income from sales of homegarden products, and in three cases out of the 20 studied, it was the only source of income. In addition, families reported obtaining at least 40 different types of plant products from their gardens, including firewood, fruit, timber, and medicinal plants. The authors found a relationship between the level of dependence on the homegardens for products and income and the number of plant species and management zones. Although dependence on homegardens varied, they represent a reliable and flexible resource that is held in high esteem by the families that maintain them.

## Dynamic Change

The few long-term studies of home gardens that have been carried out have shown that the gardens are dynamic and changing. In a study in Costa Rica, a home garden near Puerto Viejo was shown to be in the process of change due to a need for cash income, as well as the limited availability of both land and labor (Flietner, 1985). The tree stratum in approximately half of the 3264-m2 garden was in the process of being replaced with coconuts planted in evenly spaced rows, and the understory had been planted to pure stands of yuca (*Manihot esculenta*) and pineapple (*Ananas comosus*). With the construction of an all-weather road to the region, trucks had become much more available for hauling produce to distant urban markets, creating a demand for crops such as coconut and pineapple that a few years before did not exist. Farmers were adjusting their agroecosystems to meet this demand. Also, the farmer of the study garden had recently taken a job off the farm and was much less able to meet the management needs that a more diverse home garden would require.

As the coconuts mature and generate a much shadier environment on the ground below them, the farmer will have to decide what shifts will be necessary in the understory plants. He may shift to the malanga (*Colocasia esculenta*), common already in the shadier parts of the garden. He may also decide to clear out part of the tree crop in order to reintegrate more of the annual crops and short-lived perennials that were common earlier in the development of the system (Figure 17.10).

In a home garden system studied in Cañas, Guanacaste, Costa Rica, interesting shifts in diversity and organization

**FIGURE 17.10 The home garden near Puerto Viejo, Costa Rica, undergoing a transition to market crops.** A new road opened up market opportunities and prompted the changes in the species mix of the garden.

of the garden were observed to take place from one year to the next (Gliessman, 1990a), as can be seen from the data in Table 17.5. The total number of species in the garden increased by 12, but more impressive is the major increase in total number of individual plants. A large part of this increase came primarily from the greater predominance of ornamental species the second year, along with more medicinal and spice species. Some of the food species that had been very common the year before, such as squash, were not present in 1986 due to a drought that had eliminated seedlings planted earlier.

Some of the changes in the garden can be traced to changes in the household's economic situation. In 1986, the woman of the household had less time to care for the garden since she and her daughters had begun a small-scale baking business making bread for sale in the local community. If the baking business fails, food crops will probably once again receive greater emphasis.

Even though socio-economic factors account for part of the change in the garden, some of it occurs for ecological reasons. Change in home gardens is ongoing and sometimes quite rapid because of the shifting dynamics of the disturbance–recovery process.

## Links With the Social System

As indicated by the studies described above, social and economic factors can have significant impacts on home garden systems and the way they are managed. A long-term study of traditional agriculture in Tlaxcala, Mexico (González, 1985) found that changes took place in home garden diversity, structure, and management in response to industrialization and population increase. In general, farmers reduced the number of species in their home gardens, used more orderly and easily managed cropping patterns, and planted species that could more easily enter the cash economy. However, because Tlaxcala has gone through several "boom and bust" cycles over a longer time period, where off-farm employment has been alternately available and then limited, farmers have a certain mistrust of job security off the farm. As a result, relatively diverse agroecosystems have been maintained even in times of off-farm employment as insurance against the probable loss of the outside income.

Regional population growth has had a mixed impact on home garden structure. Since Tlaxcala is close to the large and expanding urban-industrial centers of Puebla and

**TABLE 17.5**
**Comparison of Plant Species in a 1240-square-meter Home Garden over 2 Years in Cañas, Guanacaste, Costa Rica**

| | 1985 | 1986 |
|---|---|---|
| Species | 71 | 83 |
| Individuals | 940 | 1870 |
| Tree species | 17 | 16 |
| Food species | 21 | 18 |
| Ornamental species | 23 | 31 |
| Medicinal species | 7 | 9 |
| Firewood species | 3 | 5 |
| Spice species | 0 | 4 |

*Source:* Gliessman, S. R. (ed.) 1990a. *Agroecology: Researching the Ecological Basis for Sustainable Agriculture.* Springer-Verlag: New York.

Mexico City, there is considerable demand and market for a large variety of agricultural products, from basic corn and beans to cut flowers. This demand is a stimulus to diversify the local cropping systems, but it also puts pressure on farmers to emphasize cash crops and abandon many subsistence species. Those families that see an advantage in combining both cash and subsistence crops maintain the most diverse home gardens, while others shift to mostly cash crops.

Although regional economic change has a clear impact on home gardens, the link between the two can go in the other direction as well. Where they exist, home gardens tend to stabilize the local economy and social structure by giving families a means of economic survival. They act as a bridge between the traditional local economy and the modern industrial economy, helping to buffer the forces that encourage migration to industrial centers and abandonment of traditional social ties. By offering the possibility of local autonomy, economic equity, and ecological sustainability, they provide important examples that can be adapted and applied around the world (Méndez et al., 2001; Major et al., 2005).

## DISTURBANCE, RECOVERY, AND SUSTAINABILITY

Agroforestry and home garden agroecosystems have been examined in this chapter because of their usefulness as models of sustainable agriculture. They incorporate a range of desirable characteristics applicable and adaptable to any agroecosystem. Manageable and productive, they have the ability to respond to different factors or conditions in the environment, to meet the needs of the inhabitants for a great diversity of products and materials, and to respond to external socio-economic demands. At the same time, they are not dependent on expensive imported agricultural inputs, and have very limited negative environmental impacts.

More information on existing types of successionally developed systems, especially those with perennial shrubs and trees, is desperately needed. Urbanization and the rapid move towards agroecosystem simplification and cash cropping is threatening the existence of these systems, especially in developing countries. We need to locate, describe, and monitor existing systems that incorporate traditional knowledge of the management of succession and disturbance with selected agroecologically-based improvements. Moreover, studies of such systems (e.g., Berkes et al., 2000; Altieri, 2002) require more institutional support.

Perhaps the greatest value of agroforestry systems is that they offer principles that can be applied to agroecosystems with few trees or none at all. By viewing all agroecosystems as successional systems in which we incorporate perennial species, appropriately introduce disturbance, and promote recovery from disturbance, we can make important steps toward sustainable food production. The limits are set only by the kind of mature ecosystems that would naturally occur in a region, and the human component in the design and management of sustainable alternatives that build upon such ecosystem models. Regardless of whether they are grain systems or home gardens, they must be dynamic, diverse, and flexible, incorporating the important ecosystem characteristics of resilience and resistance to disturbance, and the ability to constantly be renewed and regenerated by the recovery process of succession.

The more widespread implementation of practices based on disturbance and recovery will involve considerable research. But it can lead to the development of an agricultural landscape that is a mosaic of agroecosystems. The need for high harvestable yields could come from annual and short-lived perennial crops, grown in polycultures of several species that are ecologically complementary and interdependent. In such systems, animals could once again play important roles in nutrient cycling. Field structure and organization could change over time as succession leads to a gradual conversion to long-lived perennials. And incorporated into the disturbance cycle could be a patchwork of rotations in which areas are allowed to develop to maturity and their perennial or tree vegetation harvested or recycled to open up parts of the agroecosystem once again for annual cropping. In the end, a sustainable mosaic could be achieved.

## FOOD FOR THOUGHT

1. How similar or different are the ecological impacts of human-induced disturbances in

agroecosystems to those of disturbances in natural ecosystems?

2. Describe how the "analog model" for agroecosystem design and management might be applied in your own farming region. Be sure to clearly indicate the successional stages your system would need to go through and how they mirror what happens in the natural ecosystems that exist (or once existed) around your farm.
3. Give some examples of how agroforestry system design can be informed by knowledge about the ecological impact of trees on the environment, and how it can be shaped by the farmer's need for particular products.
4. How would you integrate both ecological balance and harvestability in the design of a home garden agroforestry system specifically suited to the location in which you live? Be sure to describe both the ecological and cultural background that affect your design determinations.
5. Why have trees disappeared from so many agricultural landscapes over the past several decades, especially in developed countries?
6. From an agroecological perspective, what are some of the most important relationships between diversity and disturbance in sustainable agriculture?
7. Describe how an agricultural landscape made up of a mosaic of successional patches might be described as a "polyculture of monocultures."

## INTERNET RESOURCES

World Agroforestry Centre
www.worldagroforestry.org
Considerable information about research and development partnerships in the area of agroforestry in the tropics, directed towards reducing poverty and environmental impacts.

Agroforestry Net
www.agroforestry.net
An organization based in Hawaii and focused on the Pacific Islands.

Association for Temperate Agroforestry
www.aftaweb.org
An excellent source of information on agroforestry systems suitable for more temperate regions of the world.

Edible Forest Gardens
www.edibleforestgardens.com
An organization dedicated to developing the vision, design, ecology, and stewardship of perennial polycultures of multipurpose plants in small-scale settings.

Holistic Management International
www.holisticmanagement.org
A natural-resource approach that uses ecological processes, including succession, for pasture and agroecosystem management.

The Overstory
www.agroforestry.net/overstory
An agroforestry "ejournal," focused on homegardens.

Tropical Homegardens
www.css.cornell.edu/ecf3/Web/new/AF/homeGardens.html
An interesting learning module on tropical homegardens by Professor Erick Fernandes at Cornell University.

## RECOMMENDED READING


Bazzaz, F.A. 1996. *Plants in Changing Environments: Linking Physiological, Population, and Community Ecology.* Cambridge University Press: Cambridge, UK. A global overview of the interactions between plants and different organisms as they participate in succession and ecosystem recovery.

Biotropica. 1980. Volume 12. Special Issue on Tropical Succession. A collection of research papers that cover an array of topics related to succession in tropical ecosystems and agroecosystems.

Buck, L., J.P. Lassoie, and E.C.M. Fernandes (eds.), 1999. *Agroforestry in Sustainable Agricultural Systems. Advances in Agroecology Series.* CRC Press: Boca Raton, FL. An excellent edited volume covering social, ecological, and production aspects of agroforestry research and practice in both tropical and temperate regions.

Huxley, P. 1999. *Tropical Agroforestry.* Blackwell Science: Oxford, UK. A comprehensive textbook covering general aspects of agroforestry, with a focus on the tropics.

Odum, E.P. 1969. The strategy of ecosystem development. Science 164:262–270. A key paper for understanding the relationship between succession and ecosystem development.

Ronnenberg, K.L., G.A. Bradshaw, and P.A. Marquet (eds.), 2003. *How Landscapes Change: Human Disturbance and Ecosystem Fragmentation in the Americas. Ecological Studies 162.* Springer: Berlin & New York. A multidisciplinary overview of the interactions between humans and ecosystem processes, and how they have affected the landscapes of North and South America.

Rundel, P.W., G. Montenegro, and F.M. Jaksic (eds.), 1999. *Landscape Disturbance and Biodiversity in Mediterranean-Type Ecosystems.* Ecological Studies 136. Springer: Berlin & New York. An analysis of the effects of disturbances and environmental degradation, at the landscape scale, on flora and fauna of the Mediterranean regions of the world.

Schelhas, J. and R. Greenberg. 1996. *Forest Patches in Tropical Landscapes.* Island Press: Washington, D.C. An assessment of the ecological and social value of tropical forest remnants, and the issues surrounding their management and conservation.

Schroth, G., G.A.B. da Foseca, C.A. Harvey, C. Gascon, H.L. Vasconcelos, and A.M.N. Izac. (eds.), 2004. *Agroforestry and Biodiversity Conservation in Tropical Landscapes.* Island Press: Washington, D.C. Rich in case studies, this edited volume critically analyzes the biodiversity conservation potential of tropical agroforestry systems at the landscape scale.

Soule, J. D. and J. K. Piper. 1992. *Farming in Nature's Image.* Island Press: Washington, DC. A review of what it means to use nature and our understanding of ecological processes as the model for designing and managing agroecosystems, with a focus on the midwestern U.S.

Walker, L.R. and R. del Moral. 2003. *Primary Succession and Ecosystem Rehabilitation.* Cambridge University Press: Cambridge, UK. A comprehensive discussion of how plant, animals, and microorganisms develop and interact following disturbances.

Watt, A. S. 1947. Pattern and process in the plant community. J. Ecol 35:1–22. A classic paper on the way succession works in plant communities.

# 18 The Energetics of Agroecosystems

Energy is the lifeblood of ecosystems and of the biosphere as a whole. At the most fundamental level, what ecosystems do is capture and transform energy.

Energy is constantly flowing through ecosystems in one direction. It enters as solar energy and is converted by photosynthesizing organisms (plants and algae) into potential energy, which is stored in the chemical bonds of organic molecules, or biomass. Whenever this potential energy is harvested by organisms to do work (e.g., grow, move, reproduce), much of it is transformed into heat energy that is no longer available for further work or transformation — it is lost from the ecosystem.

Agriculture, in essence, is the human manipulation of the capture and flow of energy in ecosystems. Humans use agroecosystems to convert solar energy into particular forms of biomass — forms that can be used as food, feed, fiber, and fuel.

All agroecosystems — from the simple, localized plantings and harvests of the earliest agriculture to the intensively altered agroecosystems of today — require an input of energy from their human stewards in addition to that provided by the sun. This input is necessary in part because of the heavy removal of energy from agroecosystems in the form of harvested material. But it is also necessary because an agroecosystem must to some extent deviate from, and be in opposition to, natural processes. Humans must intervene in a variety of ways — manage noncrop plants and herbivores, irrigate, cultivate soil, and so on — and doing so requires work.

The agricultural "modernization" of the last several decades has been largely a process of putting ever-greater amounts of energy into agriculture in order to increase yields. But most of this additional energy input comes directly or indirectly from nonrenewable fossil fuels. Moreover, the return on the energy investment in conventional agriculture is not very favorable: for many crops, we invest more energy than we get back as food. Our energy-intensive form of agriculture, therefore, cannot be sustained into the future without fundamental changes.

## ENERGY AND THE LAWS OF THERMODYNAMICS

An examination of the energy flows and inputs in agriculture requires a basic understanding of energy and the physical laws that govern it. First of all, what is energy? Energy is most commonly defined as the ability to do work. Work occurs when a force acts over some distance. When energy is actually doing work it is called kinetic energy. There is kinetic energy, for example, in a swinging hoe and a moving plow, and also in the light waves coming from the sun. Another form of energy is potential energy, which is energy at rest yet capable of doing work. When kinetic energy is doing work, some of it can be stored as potential energy. The energy in the chemical bonds of biomass is a form of potential energy.

In the physical world and in ecosystems, energy is constantly moving from one place to another and changing forms. Two laws of thermodynamics describe how this occurs. According to the first law of thermodynamics, energy is neither created nor destroyed regardless of what transfers or transformations occur. Energy changes from one form to another as it moves from one place to another or is used to do work, and all of it can be accounted for. For example, the heat energy and light energy created by the burning of wood (plus the potential energy of the remaining products) is equal to the potential energy of the unburned wood and oxygen.

The second law of thermodynamics states that when energy is transferred or transformed, part of the energy is converted to a form that cannot be passed on any further and is not available to do work. This degraded form of energy is heat, which is simply the disorganized movement of molecules. The second law of thermodynamics means that there is always a tendency toward greater disorder, or entropy. To counter entropy — to create order, in other words — energy must be expended.

The operation of the second law can be clearly seen in a natural ecosystem: as energy is transferred from one organism to another in the form of food, a large part of that energy is degraded to heat through metabolic activity, with a net increase in entropy. In another sense, biological systems don't appear to conform to the second law because they are able to create order out of disorder. They are only able to do this, however, because of the constant input of energy from outside the system in the form of solar energy.

Analysis of energy flows in any system requires measuring energy use. Many units are available for this purpose. In this chapter, we will use kilocalories (kcal) as the preferred unit because it is best oriented to linking human nutrition with energy inputs in food production. Other units and their equivalents are listed in Table 18.1.

**TABLE 18.1**
**Units of Energy Measure**

| Unit | Definition | Equivalents |
|---|---|---|
| Calorie (cal) | The amount of heat necessary to raise 1 g (1 ml) of water 1°C at 15°C | 0.001 kcal<br>4.187 J |
| Kilocalorie (kcal) | The amount of heat needed to raise 1 kg (l L) of water 1°C at 15°C | 1000 cal; 4187 J;<br>3.968 Btu |
| British thermal unit (Btu) | The amount of heat needed to raise 1 lb of water 1°F | 252 cal;<br>0.252 kcal |
| Joule (J) | The amount of work done in moving an object a distance of 1 m against a force of 1 N | 0.252 cal;<br>0.000252 kcal |

## CAPTURE OF SOLAR ENERGY

The starting point in the flow of energy through ecosystems and agroecosystems is the sun. The energy emitted by the sun is captured by plants and converted to stored chemical energy through the photosynthetic process discussed in Chapter 3 and Chapter 4. The energy accumulated by plants through photosynthesis is called *primary production* because it is the first and most basic form of energy storage in an ecosystem. Energy left after the respiration needed to maintain plants is net primary production (NPP) and remains as stored biomass. Through agriculture, we concentrate this stored energy in biomass that can be harvested and utilized, either by consuming it directly or by feeding it to animals that we can either consume or use to do work for us.

Plants vary in how efficiently they can capture solar energy and convert it to stored biomass. This variation is the result of differences in plant morphology (e.g., leaf area), photosynthetic efficiency, and physiology. It also depends on the conditions under which the plant is grown. Agricultural plants are some of the most efficient plants, but even in their case the efficiency of their conversion of sunlight to biomass rarely exceeds 1% (a 1% efficiency means that 1% of the solar energy reaching the plant is converted to biomass).

Corn, considered one of the most productive food and feed crops per unit of area of land, can produce as much as 15,000 kg/ha/season of dry biomass, divided fairly equally between grain and stover. This biomass represents about 0.5% of the solar energy reaching the cornfield during the year (or about 1% of the sunlight reaching the field during the growing season). A potato crop that yields 40,000 kg/ha of fresh tubers (the equivalent of 7000 kg/ha of dry matter) has a conversion efficiency of about 0.4%. Wheat, with a grain yield of 2700 kg/ha and a dry matter yield of 6750 kg/ha, has about 0.2% conversion efficiency. The conversion efficiency of sugarcane in tropical areas — about 4.0% — is one of the highest known.

Even though these efficiencies are relatively low, they are still several times greater than the average conversion efficiency of mature natural vegetation, which is estimated to be about 0.1% (Pimentel et al., 1978). We must also take into consideration the fact that little of the biomass in natural vegetation is available for human consumption, whereas a large portion of the stored energy in agricultural species is consumable (Figure 18.1).

Since much of the food consumed in developed countries is not plant biomass but animal biomass, we should also examine the efficiency of the conversion from plant-matter energy to meat, milk, and eggs. The production of animal biomass from plant biomass is inefficient because animals lose so much metabolic energy to maintenance and respiration.

Analysis of this conversion is normally done in terms of the energy content of the protein in the animal biomass, since meat, milk, and eggs are produced mainly for their protein. Feedlot or confined livestock need 20 to 120 units of plant food energy to produce each unit of protein energy, depending on the animal and the production system. This is equivalent to an efficiency of 0.8% at the low end and 5% at the high end. If these conversion efficiencies are combined with those for the production of the animals' feed, the inefficiency of animal production systems becomes evident. As an example, the plant products fed to feedlot cattle contain about 0.5% of the solar energy that reached the plants, and the protein in the consumed meat of the cattle contains 0.8% of the energy that was in

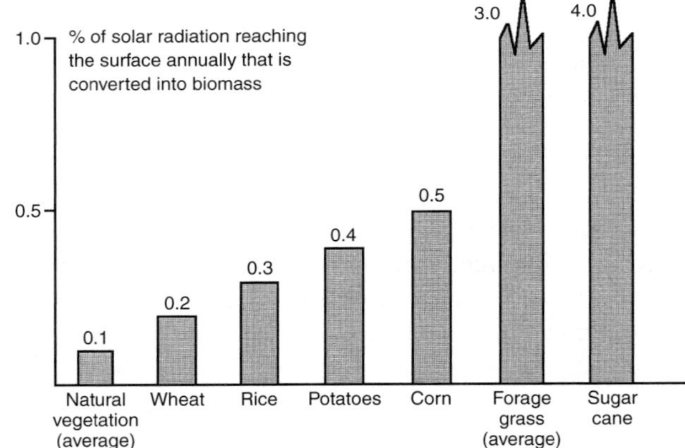


**FIGURE 18.1 Efficiency of solar energy-to-biomass conversion.** (Pimentel, D., D. Nafus, W. Vergara, D. Papaj, L. Jaconetta, M. Wulfe, L. Olsvig, K. French, M. Loye, and E. Mendoza. 1978. *Bioscience* 28: 376–382.; Pimentel, D., W. Dahzhong, and M. Giampietro. 1990. In S. R. Gliessman (ed.), *Agroecology: Researching the Ecological Basis for Sustainable Agriculture*. pp. 305–321. Springer-Verlag: New York.; Ludlow, M. M. 1985. *Australian Journal of Plant Physiology* 12: 557–572. With permission.)

**FIGURE 18.2 Dairy cows fed on concentrated diets to increase milk production.** Corn silage, pelletized alfalfa, and other supplements increase the energy cost of producing dairy products.

the feed, yielding an overall efficiency of only 0.004% (Figure 18.2).

Open-range livestock must be considered somewhat differently, since they can graze on land that might not be suitable for other forms of agriculture, and consume forage directly from a natural ecosystem or low energy-requiring pasture systems. They can transform the energy contained in biomass that humans cannot consume directly.

## ENERGY INPUTS IN FOOD PRODUCTION

Although all the energy in the food we consume comes originally from the sun, additional energy is needed to produce the food in the context of an agroecosystem. This additional energy comes in the form of human labor, animal labor, and the work done by machines. Energy is also required to produce the machines, tools, seed, and fertilizer, to provide irrigation, to process the food, and to transport it to market. We must examine all these energy inputs to understand the energy costs of agriculture and to develop a basis for more sustainable use of energy in agriculture.

It is helpful, first of all, to distinguish between the different types of energy inputs in agriculture. The primary distinction is between energy inputs from solar radiation, called *ecological energy inputs*, and those derived from human sources, called *cultural energy inputs*. Cultural energy inputs can be further divided into biological inputs and industrial inputs. Biological inputs come directly from organisms and include human labor, animal labor, and manure; industrial inputs of energy are derived from fossil fuels, radioactive fission, and geothermal and hydrological sources.

It is important to note that even though we are referring to all these sources of energy as "inputs," cultural energy of either form can be derived from sources within a particular agroecosystem, making it not an input at all in the sense that we have been using the term. Such "internal inputs" of energy include the labor of farm residents, the manure of on-farm animals, and energy from on-farm windmills or wind-driven turbines (Figure 18.3).

### Cultural Energy Inputs and Harvest Output

From the standpoint of sustainability, the key aspect of energy flow in agroecosystems is how cultural energy is used to direct the conversion of ecological energy to biomass. The greater the modification of natural processes that humans try to force on the environment in the production of food, the greater the amount of cultural energy required. Energy is needed to maintain a low-diversity system, to limit interference, and to modify the physical and chemical conditions of the system in order to maintain optimal growth and development of the crop organisms.

Larger inputs of cultural energy enable higher productivity. However, there is not a one-to-one relationship between the two. When the cultural energy input is very high, the "return" on the "investment" of cultural energy is often minimal. Since the output of an agroecosystem can be measured in terms of energy, we can evaluate the efficiency of energy use in the agroecosystem with a simple ratio: the amount of energy contained in the harvested biomass compared to the amount of cultural energy required to produce that biomass. Across all the world's agroecosystems, this ratio varies from one in which much more energy comes out than is put in to one in which the energy inputs are larger than the energy output.

Nonmechanized agroecosystems (e.g., pastoralism or shifting cultivation) that use only biological cultural energy in the form of human labor are able to realize returns that vary from 5 to nearly 40 cal of food energy for each calorie of cultural energy invested. Permanent farming systems using draft animals have a higher input of cultural energy, but because this greater energy investment

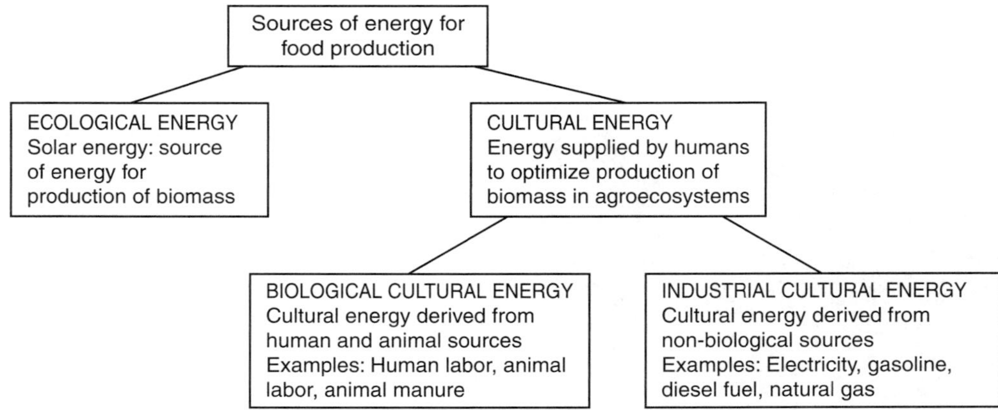


**FIGURE 18.3 Types of energy inputs in agriculture.** Biological cultural energy and industrial cultural energy can either come from outside a particular agroecosystem (in which case it is a form of external human input) or be derived from sources with the system.

enables higher yields, such systems still have favorable returns on their investment of cultural energy.

In mechanized agroecosystems, however, very large inputs of industrial cultural energy replace most of the biological cultural energy, enabling high levels of yield but greatly reducing energy-use efficiency. In the production of grains such as corn, wheat, and rice, these agroecosystems can yield 1 to 3 cal of food energy per calorie of cultural energy. In mechanized fruit and vegetable production, the energy return is at best slightly greater than the energy investment, and in most cases it is smaller (Pimentel and Pimentel, 1997). For production of animal food, the ratio is in most cases even less favorable. Beef production in the U.S., for example, requires about 5 cal of cultural energy for each calorie obtained, and pork requires as much as 10 cal (Pimentel and Pimentel, 1997).

Since animal foods are valued more for protein content than total energy content, we should also consider the energy efficiency of their production in terms of the energy in the protein of these foods compared to the energy in the feed consumed by the animals. In these terms, each calorie of protein in milk, pork, and feedlot beef requires between 30 and 80 cal of energy to produce. By comparison, a calorie of plant protein can be produced with as little as 3 cal of cultural energy (in the case of protein from grains). Even the production of concentrated plant protein (e.g., tofu from soybeans) takes no more than 20 cal of energy for each calorie of protein (Figure 18.4).

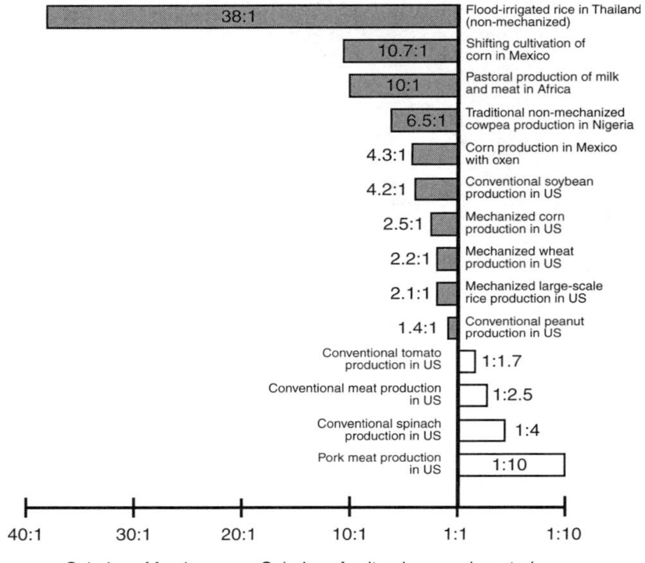


**FIGURE 18.4 Comparison of the returns on energy investment for various agroecosystems.** Bars extending to the left indicate systems in which the realized output is greater than the input; bars extending to the right indicate systems in which the energy input is greater than the energy value of the resulting food. (Cox, G. W. and M. D. Atkins. 1979. *Agricultural Ecology.* Freeman: San Francisco.; Pimentel, D. and M. Pimentel (eds.), 1997. *Food, Energy, and Society.* 2nd ed. University Press of Colorado: Niwot, Colorado.; Pimentel, D., M. Pimentel, and M. Karpenstein-Machan. 1998. International Commission of Agricultural Engineering Ejournal (cigr-ejournal.tamu.edu. With permission.)

The data presented in Figure 18.4 reinforce our claim that the cultural energy requirement in agriculture is closely related to the level of modification of natural ecosystem processes. The costs are small when humans leave the basic structure of the ecosystem intact. When certain minor modifications are made that increase the abundance of a specific crop species of interest, more cultural energy is required, but the return is still favorable. But when a complex natural ecosystem is replaced by a crop monoculture with a life form very different from that of the native species — as is the case with irrigated cotton in the former arid scrub lands of the western San Joaquin Valley of California — cultural energy costs rise steeply. When the

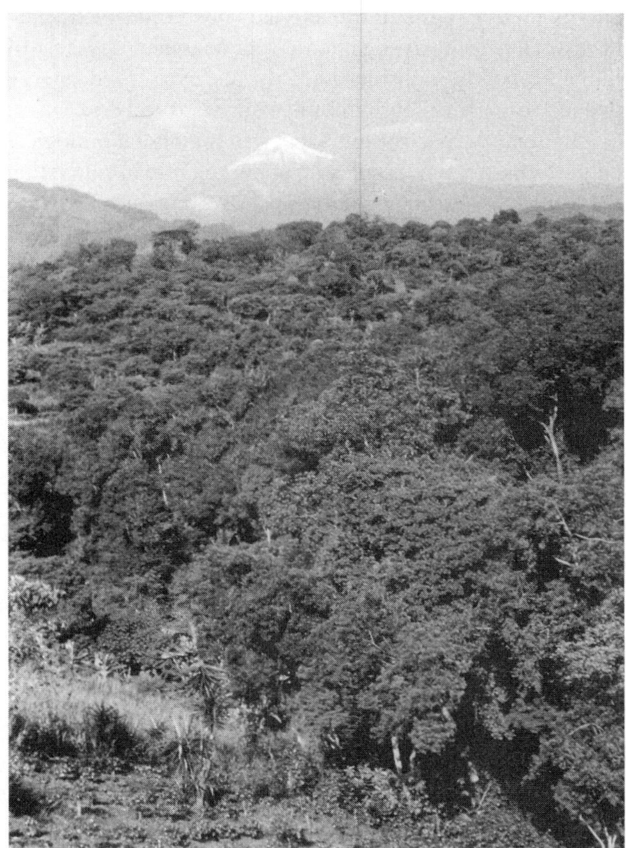

**FIGURE 18.5 Coffee grown under the shade of native trees in Veracruz, Mexico.** In this agroecosystem, coffee is substituted for understory species without major alteration of the upper canopy of native trees. Because the natural ecosystem is altered so little, only small inputs of cultural energy are required to maintain the productivity of the system.

goal is to also increase the level of solar energy capture (productivity) above that shown by the previous natural system, the levels of cultural energy required can be very high (Figure 18.5).

Figure 18.6 offers another perspective on the relative energy costs and energy benefits of different types of agroecosystems. Although using a large amount of cultural energy enables conventional agroecosystems to be more productive than others, such systems are not realizing a good return on their energy investment. Food production that is more energy efficient is possible if we decrease inputs of industrial cultural energy, increase the investment of biological cultural energy, and change how industrial cultural energy is used.

## Use of Biological Cultural Energy

Biological cultural energy is any energy input with a biological source under human control — this includes human labor, the labor of human-directed animals and

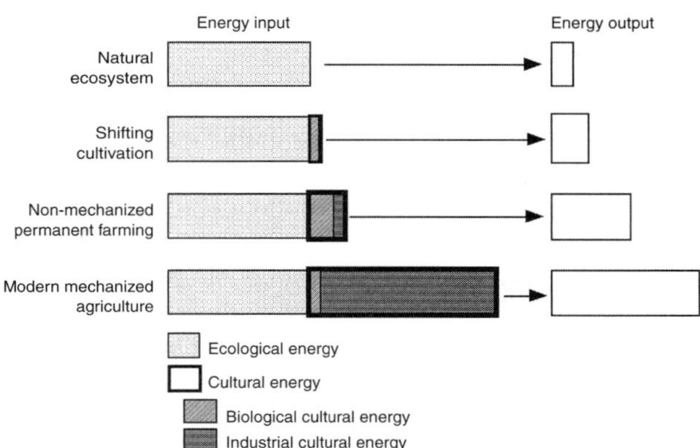


**FIGURE 18.6 Approximate relative size of energy inputs and outputs in four types of systems.** The actual size of the ecological energy input for each system is much larger than shown. Note that for modern mechanized agriculture, the total energy output is smaller than the input of cultural energy; this disparity is often more extreme than shown.

their by-products, and any human-directed biological activity or by-product. Some of the different forms of biological cultural energy, with their approximate energy values, are presented in Table 18.2.

Biological cultural energy is renewable in that it derives from food energy, the ultimate source of which is solar energy. Biological cultural energy is also efficient in facilitating the production of harvestable biomass. As we saw above, agroecosystems that rely mainly on biological cultural energy are able to obtain the most favorable ratios of energy output to input.

Human labor has been the key cultural energy input to agriculture ever since its beginning, and in many parts of the world today it continues to be the primary energy input, along with animal labor. In shifting cultivation systems, for example, human labor is practically the only form of energy added other than the energy captured through photosynthesis. These systems' high ratios of food energy produced to cultural energy invested, ranging from 10:1 to 40:1, is a reflection of how efficiently human labor can direct the conversion of solar energy into harvestable material (Rappaport 1971; Pimentel and Pimentel, 1997). As an example, the energy budget for a traditional shifting cultivation or swidden corn crop in Mexico is shown in Figure 18.7.

Many other types of traditional, nonmechanized food production systems, where biological cultural energy is the primary input, realize a very favorable return on their investment of cultural energy. In pastoral agroecosystems, in which herding and animal care are the main human activities, and animals gain their food energy from natural vegetation, the ratios of food energy produced to cultural energy invested range from 3:1 to 10:1. Even intensive, nonmechanized farming systems maintain a positive energy budget. Paddy rice production systems in parts of Southeast

**TABLE 18.2**
**Energy Content of Several Types of Biological Cultural Energy Inputs to Agriculture**

| Input Type | Energy Value |
|---|---|
| Human labor, heavy (clearing with a machete) | 400–500 kcal/hr |
| Human labor, light (driving a tractor) | 175–200 kcal/hr |
| Large draft animal labor | 2400 kcal/hr |
| Locally produced seed | 4000 kcal/kg |
| Cow manure | 1611 kcal/kg |
| Pig manure | 2403 kcal/kg |
| Commercial compost | 2000 kcal/kg |
| Biogas slurry | 1730 Kcal/kg |

*Source:* Cox, G.W. and M.D. Atkins. 1979. *Agricultural Ecology.* Freeman: San Francisco.; Pimentel, D. and M. Pimentel (eds.), 1997. *Food, Energy, and Society.* 2nd ed. University Press of Colorado: Niwot, Colorado.; Zhengfang, L. 1994. Energetic and ecological analysis of farming systems in Jiangsu Province, China. Presented at the 10th International Conference of the International Federation of Organic Agriculture Movements (IFOAM), Lincoln University, Lincoln, New Zealand, December 9–16.

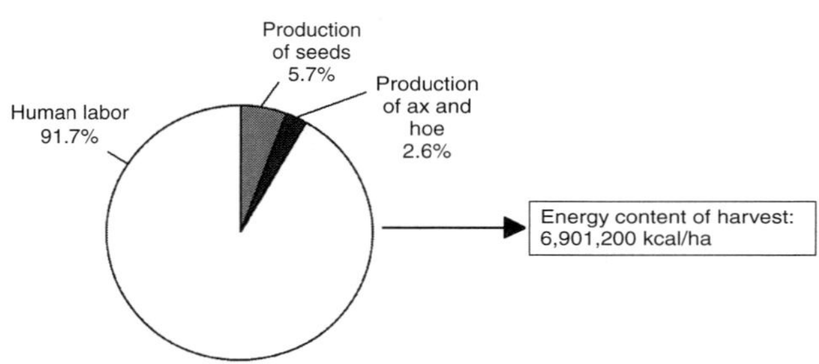


**FIGURE 18.7 Cultural energy inputs to a traditional shifting cultivation corn crop in Mexico.** The ratio of the food energy output to the cultural energy input for this system is 10.7:1. Only the axe and hoe (used for clearing and seed planting) required an input of industrial cultural energy. (Pimentel, D. and M. Pimentel (eds.), 1997. *Food, Energy, and Society.* 2nd ed. University Press of Colorado: Niwot, Colorado.)

Asia, for example, are able to gain up to 38 cal of food energy for every calorie of cultural energy invested.

The energy value of the human labor in these systems is calculated by determining how many food calories a person burns while working. Although this technique provides good baseline data, it does not take into account a variety of other factors. One could also consider the energy required to grow the food that is metabolized while working, and the energy needed to provide for all the other basic needs of the human workers when they aren't working. Such additions would increase the energy value of human labor. On the other hand, people's basic needs must be provided for whether or not their labor serves as an energy input in agriculture, and they need food even when at rest. On this basis, one could reduce the energy cost of human labor by considering only the extra food energy needed to perform agricultural work (Figure 18.8).

In many agroecosystems that rely mainly on biological cultural energy, animals play an important role in cultivating the soil, transporting materials, converting biomass into manure, and producing protein-rich foods such as milk and meat. Animal use increased considerably in agriculture when the transition from shifting cultivation to permanent agriculture began to occur.

Although the use of animal labor increases the total biological cultural energy input and lowers the ratio of harvested energy to invested energy to the neighborhood of 3:1, it allows for permanent instead of shifting agriculture, increases the area that can be planted, produces manures for enriching the soil, and allows for the harvest of meat, milk, and animal products. In addition, animals consume biomass that cannot be consumed directly by humans, which lowers their relative energy cost. An example of the energy efficiency of corn production using animal traction is presented in Figure 18.9.

Biological cultural energy is an important component of sustainable agriculture. Energy inputs from humans and their animals are generally renewable, providing energy that helps transform a greater proportion of solar energy into harvestable food energy. The use of human and animal labor takes advantage of the first law of thermodynamics by altering natural ecosystem processes in ways that concentrate energy in useful products, but still obeys the second law by always returning to ecological inputs of energy from the sun in order to maintain the agroecosystem over the long term. When doing an energetic analysis of biological cultural energy, it must be remembered that this form of energy is more than an economic cost for agriculture — it is an integral part of a sustainable production process.

## Use of Industrial Cultural Energy

Once agriculture began to mechanize, the use of energy from industrial cultural sources increased dramatically. Mechanization and industrial cultural energy greatly increased productivity, but they also changed the nature of agricultural production. Human and animal labor was displaced, and farming became tied to fossil fuel production and consumption.

Present-day conventional agroecosystems have come to rely heavily on industrial cultural energy inputs. Corn production in the U.S. is a good example of an agroecosystem where almost all of the energy inputs to the system come from industrial sources. Figure 18.10 shows the total energy inputs per hectare in corn production, and how this energy is distributed among the various input types. Biological cultural energy in the form of human labor is a minimal part of this system.

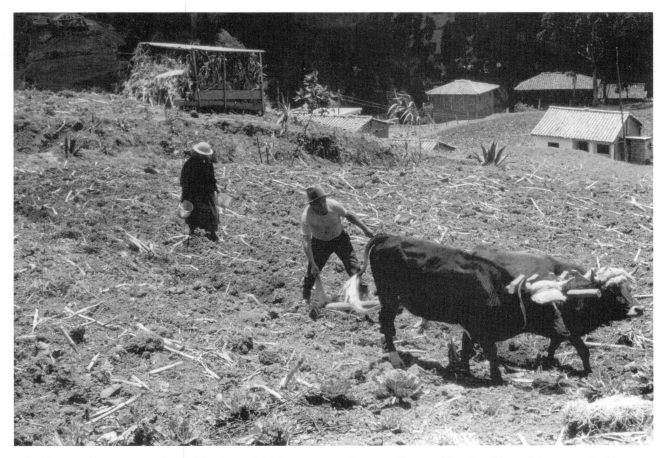

**FIGURE 18.8 Oxen-drawn plow cultivating a field for corn planting near Cuenca, Ecuador.** Most of the energy in this system is from renewable local sources.

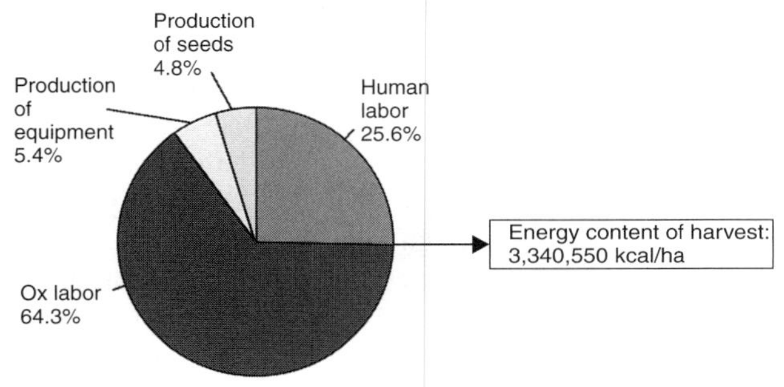


**FIGURE 18.9 Cultural energy inputs into a traditional corn production system using animal labor.** The ratio of the food energy output to the cultural energy input for this system is 4.34:1. The energy in the cover crop and fallow plants that were incorporated into the soil is not included in the calculations. Animal manures returned to the soil are included in the energy input from the oxen. (Data from Cox, G. W. and M. D. Atkins. 1979. *Agricultural Ecology.* Freeman: San Francisco.; Pimentel, D. and M. Pimentel (eds.), 1997. *Food, Energy, and Society.* 2nd ed. University Press of Colorado: Niwot, Colorado.)

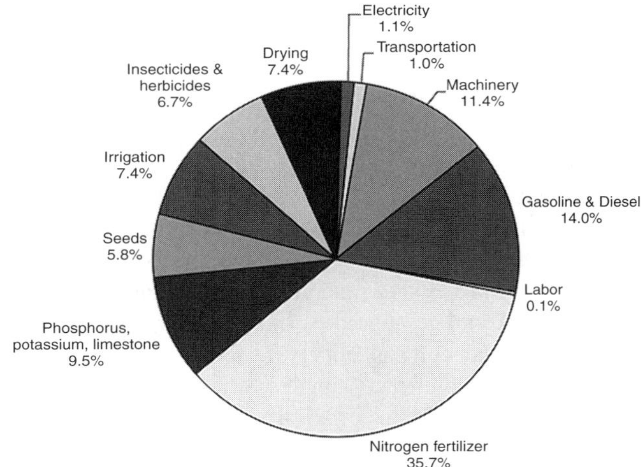


**FIGURE 18.10 Components of the 10,535,650 kcal/ha of cultural energy used for corn production in the U.S.** Total grain yield averages 7500 kg/ha and the kcal output-to-input ratio is 2.5:1 (Data from Pimentel, D. and W. Dazhong 1990. Technological changes in energy use in U.S. agricultural production. In Carroll, C.R., J.H. Vandermeer, and P.M. Rosset (eds.). *Agroecology* McGraw-Hill: New York.)

The changes that have occurred since World War II in the way cultural energy is used to produce corn is a good example of how energy use has changed in agriculture in general. Between 1945 and 1983, corn yields in the U.S. increased threefold, but energy inputs increased more than five-fold. In 1945, the estimated ratio of energy output to energy input in corn was between 3.5:1 and 5.5:1. By 1975, this ratio had declined to between 3.2:1 and 4.1:1, and in the early 1990s, it stood at 2.53:1 (Pimentel and Pimentel, 1997). During the last decade, this ratio of return has probably remained about the same, with the continued intensification of inputs to agriculture balanced by tailoring of inputs to measured crop needs ("precision agriculture").

Energetically speaking, industrial cultural energy is of a higher quality than both solar energy and biological cultural energy. It is more concentrated — calorie for calorie, it has a greater capacity for doing work than solar energy or biological cultural energy; 1 kcal of energy in the form of fossil fuel, for example, is able to do about 2000 times as much work as 1 kcal of solar radiation.

But even though industrial cultural energy is generally of very high quality in terms of the work it can do, each form of this energy varies in the amount of energy that was required to give it this higher quality state. A kilocalorie of electricity, for example, can do four times the work of a kilocalorie of petroleum fuel, but much more energy was expended to create the electricity. As the laws of thermodynamics tell us, humans must expend energy in order to concentrate energy, and no new energy can be created in the process. So we are as much concerned with the absolute amount of work that can be done by each kilocalorie of a certain form of energy as we are with the total amount of energy that is expended to transform it into that energy form. To compare industrial cultural energy inputs in these terms, we can calculate their energy costs. Table 18.3 presents a range of energy costs for some commonly used industrial energy inputs.

Industrial cultural energy is used either directly or indirectly in agriculture. Direct use occurs when industrial cultural energy is used to power tractors and transport vehicles, run processing machinery and irrigation pumps, and heat and cool greenhouses. Indirect energy use occurs when industrial cultural energy is used off the farm to produce the machinery, vehicles, chemical inputs, and other goods and services that are then employed in the farming operation. This energy is sometimes referred to as embodied energy, or emergy, in order to emphasize the energy costs that are often overlooked when we calculate the direct energy consumed in a farming system (Odum, 1996). In the typical conventional farming system, about one third of energy use is direct, and two thirds is indirect.

The production of fertilizers — especial nitrogen fertilizer — accounts for the great majority of indirect energy use in agriculture. Nearly one third of *all* the energy used in modern agriculture is consumed in the production of nitrogen fertilizer. This energy cost is high because nitrogen fertilizer is used so intensively and because a large amount of energy is required to produce it. In corn production, for example, about 152 kg/ha of nitrogen fertilizer is applied to the field, which represents 30% of the total energy input per hectare (Pimentel and Pimentel, 1997). This energy input could be reduced greatly by using manures, biological nitrogen fixation, and recycling.

**TABLE 18.3**
**Approximate Energy Costs of Commonly Used Industrial Cultural Inputs**

| | |
|---|---|
| Machinery (average for trucks and tractors) | 18,000 kcal/kg |
| Gasoline (including refining and shipping) | 16,500 kcal/l |
| Diesel (including refining and shipping) | 11,450 kcal/l |
| LP gas (including refining and shipping) | 7700 kcal/l |
| Electricity (including generation and transmission) | 3,100 kcal/kwh |
| Nitrogen (as ammonium nitrate) | 14,700 kcal/kg |
| Phosphorus (as triple superphosphate) | 3000 kcal/kg |
| Potassium (as potash) | 1860 kcal/kg |
| Lime (including mining and processing) | 295 kcal/kg |
| Insecticides (including manufacturing) | 85,680 kcal/kg |
| Herbicides (including manufacturing) | 111,070 kcal/kg |

*Source:* Fluck, R.C. (ed.) 1992. *Energy in Farm Production*. Vol. 6. Elsevier: Amsterdam.

Another 15% of indirect energy use occurs in the production of pesticides. When formulation, packaging, and transport to the farm are included, the energy cost is somewhat higher. Although newer pesticides are usually applied in smaller quantities than those common a few decades ago, they are typically higher in energy content.

Most of the industrial cultural energy inputs in agriculture, both direct and indirect, come from fossil fuels or are dependent on fossil fuels for their manufacture. Other sources of industrial energy play a very small role in agriculture overall, even though they may be significant on a local basis. An analysis of the energy budget for corn production in the Midwestern U.S. showed that more than 90% of the industrial energy inputs came from fossil fuels, and less than 1% of the total energy needed for production came from renewable biological cultural energy in the form of labor (Pimentel and W. Dazhong, 1990). When crop production depends so fully on fossil fuels, anything that affects the cost or availability of such energy can have dramatic impacts on agriculture.

Current trends indicate that fossil fuel use in agriculture will continue to increase to meet growing production needs (Pimentel and Pimentel, 1997), resulting in more rapid depletion of world petroleum reserves, and competition with other uses for fossil fuels.

## TOWARD SUSTAINABLE USE OF ENERGY IN AGROECOSYSTEMS

Examining conventional agriculture through the lens of energy reveals a critical source of unsustainability. Conventional agriculture is today using more energy to produce, process, transport, and market food than the food itself contains, and most of this invested energy comes from sources with a finite supply. We have come to depend on fossil fuels to produce our food, yet fossil fuels will not always be available in abundant supply. Moreover, dependence on fossil fuel use in agriculture is linked with virtually every other source of unsustainability in our food production systems.

### PROBLEMS WITH INTENSIVE FOSSIL FUEL USE

Growing levels of energy inputs to agriculture have played an important role in increasing yield levels in many of the world's agricultural ecosystems over the past several decades. However, as described above, most of these energy inputs come from industrial sources, and most are based on the use of fossil fuels. If the strategy for meeting the food demands of the growing population of the world continues to depend on these sources, several critical problems will begin to emerge. Some of these problems are ecological, but others are economic and social.

As has been noted throughout the chapters of this book, when ecological processes are ignored, environmental degradation begins to appear in the agroecosystem. The use of intensive cultural energy inputs is what has permitted us to ignore ecological processes. The application of inorganic fertilizers masks declines in soil fertility; pesticides contribute to and hide declines in agricultural biodiversity.

However, the consequences of ignoring ecological processes are now becoming more evident. At the farm level, a shift to heavy mechanization and high use of fossil fuel–derived chemical inputs have led to problems of organic matter loss, nutrient leaching, soil degradation, and increased soil erosion. Water supplies have become polluted, and excessive pumping of the ground water has led to exhaustion of aquifers and accompanying water shortages. Pests and diseases have developed resistance to inundative use of pesticides, and pesticides have contaminated both farm environments and natural ecosystems, causing health problems for farmers and farm workers and destroying populations of beneficial insects and microorganisms.

Off the farm, the wind and water erosion of soil associated with mechanized agriculture has had negative impacts on other systems, especially downstream. Recent work on gaseous emissions from nitrogenous fertilizers ($N_2O$ and NO) has shown that the addition of these materials to the atmosphere is beginning to impact the global nitrogen cycle, further damage the ozone layer, and exacerbate the global warming problem. The simplification of farming systems, which always accompanies high industrial energy inputs to agriculture, is causing greater loss of regional biodiversity.

From an economic and social perspective, the problems with excessive dependence on fossil fuel energy in agriculture go much beyond the issue of the efficiency of return on investment for the energy that is used. Dependence on fossil fuel use means greater vulnerability to changes in the price and supply of petroleum. As was seen in the oil crisis of 1973, and then periodically since then, petroleum prices can suddenly rise, increasing the costs of agricultural production. With fossil fuel consumption continuing to rise worldwide, the risks to fossil fuel-based agriculture become even greater. The problem will become even more critical as developing countries are forced to intensify their own agricultural output to meet the growing demand for food.

A final problem with fossil fuel-based agriculture is that it is linked to a certain kind of agricultural development: it enables large-scale, mechanized agriculture, which all over the world is displacing traditional agriculture and thus forcing migration to cities, disrupting cultural ties, and undermining self-reliance.

## FUTURE ENERGY DIRECTIONS

Clearly, sustainable food production depends to a large extent on more efficient use of energy, as well as less reliance on industrial cultural energy inputs and fossil fuels in particular. As suggested in this chapter, a key to more sustainable use of energy in agriculture lies in expanding the use of biological cultural energy. Biological inputs are not only renewable, they have the advantages of being locally available and locally controlled, environmentally benign, and able to contribute to the ecological soundness of agroecosystems. Also important is the conversion to alternative energy sources and appropriate technologies that lessen dependence on fossil fuel.

Many agroecosystems currently in use point the way toward the future. The typical organic farming system, in which animals and legumes replace some of the fossil fuel-derived inputs, consumes 28 to 32% less energy than an equivalent conventional system (Pimentel et al., 2005). A Danish study found that a grass–clover integrated organic dairy farm was able to reduce total energy use 37.5% over its conventional counterpart, and systems using legume rotations in organic cereals and row crops reduced total energy use by 81.5 and 75%, respectively, compared to conventional systems (Dalgaard et al., 2001).

Many of the ecologically based options and approaches presented throughout this book relate directly to improving energy efficiency. They suggest a number of strategies for fashioning food production systems that use energy in a more sustainable manner:

- Reduce the use of industrial cultural energy, especially nonrenewable or contaminating sources such as fossil fuels.
- Use minimum or reduced tillage systems that require less mechanized cultivation.
- Employ practices that reduce water use and water loss in order to reduce the amount of energy expended for irrigation.
- Use appropriate crop rotations and associations that stimulate recovery from the disturbance caused by each cropping cycle without the need for artificial inputs.
- Develop renewable, energy-efficient industrial cultural sources and uses of energy to replace fossil fuels and their uses.
- Develop on-farm sources of industrial cultural energy (e.g., photovoltaic electricity, wind energy, small-scale hydropower, biofuels) where possible.
- Use industrial cultural energy more efficiently by reducing waste and making more appropriate matches between the energy's quality and its use.
- Reduce the consumption of animal products overall, and for the animal products that are consumed, rely more on livestock that are range- or grass-fed or raised on agricultural plant biomass that would otherwise be waste.
- Reduce energy use in the agricultural sector by regionalizing production, and putting consumers and producers more directly in contact both seasonally and geographically.
- Increase the use of biological cultural energy.
- View human energy as an integral part of energy flow in agriculture rather than as an economic cost that must be reduced or eliminated.
- Return harvested nutrients to the farmland from which they came.
- Make more extensive use of manures and plant by-products to maintain soil fertility and quality.
- Design and implement integrated livestock-and-crop systems that harness the ability of livestock to supply work, recycle nutrients on the farm, and provide other ecosystem services.
- Increase the local and on-farm use of agricultural products in order to lessen the energy costs of long distance transport.
- Expand the use of biological control and integrated pest management.
- Encourage the presence of mycorrhizal relationships in the roots of crops in order to lessen the need for external inputs.
- Design agroecosystems in which biological and ecological relationships provide more of the nutrient and biomass inputs and population-regulating processes, and that, therefore, require lower levels of cultural energy inputs.
- Make greater use of nitrogen-fixing crops, green manures, and fallows.
- Make greater use of biological pest management through cover cropping, intercropping, encouragement of beneficials, well-designed livestock integration, etc.
- Introduce crops that are appropriate or adapted to the local environment rather than trying to alter the environment to meet the needs of the crop.
- Incorporate windbreaks, hedgerows, and non-crop areas into cropping systems for habitat and microclimate management.
- Design agroecosystems using local natural ecosystems as a model.
- Maximize the use of successional development in the cropping system (e.g., through agroforestry) in order to maintain better agroecosystem regeneration capacity.
- Diversify rather than simplify farming systems.
- Emphasize agroecosystem design and management approaches that store carbon in biomass or soil organic matter in order to make agriculture a net sink for carbon, and hence, a force for counteracting global warming and climate change.
- Develop energy-related indicators of sustainability that incorporate the parallel goals of efficiency, productivity, and renewability.

Too often we hear the argument that without the continued intensive use of fossil fuels, agriculture will not be able to meet the growing demand for food around the globe. Although this point of view highlights the main challenge we will face in the coming decades, it ignores both the seriousness of the problems caused by our present methods of food production and the very real and practical alternatives that exist and that can be developed if research is directed toward whole-system analysis of agroecosystems.

Nor can we rely on biofuel substitutes for fossil fuels. The current push to develop biofuels has considerable risk because biofuel production can divert biomass and food products away from direct human consumption and use in agriculture (Hunt et al., 2006). Moreover, biofuels don't always have a positive energy balance. For example, producing 1000 L of ethanol requires 8.3 million kcal of energy (much of it from fossil fuels) but that same 1000 L of ethanol has an energy value of only 5.0 million kcal (Pimentel et al., 1998). Although biofuels have their place in developing more sustainable agroecosystems, they are not the easy solution that some claim.

The rapid increase in energy use in agriculture during the 20th century radically changed the nature of farming. With an understanding of energy as an ecological factor in agriculture, and its use and flow as an emergent quality of the entire agroecosystem, better means of evaluating current practices can be developed, contributing at the same time to the development of practices and policies that establish a more sustainable basis for the world's food production systems in the 21st century. The longer it takes to develop alternative, ecologically sound energy use and conversion systems, the more vulnerable our current energy-dependent systems will become.

## THE SUNSHINE FARM PROJECT

Before the middle of the 1900s, many farms ran mostly on sunlight. They used crop rotations and farm-produced manure to maintain soil fertility, and work was done by draft horses and people fed by on-farm production. With these farms of 100 yr ago in mind, Marty Bender at the Land Institute set out in the early 1990s to create a modern farm that could provide its own fuel and fertility. The result was the Sunshine Farm, a 10-yr-long demonstration project consisting of 50 acres of conventional crops and 100 acres of prairie pasture grazed by cattle near Salina, Kansas.

As the farm took shape, it showed many similarities to farms of the early 1900s and before. Livestock and crops were integrated, draft horses performed work, a variety of crops were grown, and at any one time about 40% of the cropland was planted in legumes. Unlike a farmer in the 1920s, however, Bender had, at his disposal, some newer renewable energy technologies.

He had a 4.5-kW photovoltaic array installed to provide for all of the farm's electricity needs, which included running the workshop tools, charging the electric fencing, running the water pumps, heating the chick brooders, and providing electricity for the farmhouse. A pair of Percheron draft horses and a biodiesel tractor provided motive power for field operations. Bender planted about one quarter of the farm's cropland in soybeans and sunflowers to provide the raw material for the tractor's biodiesel fuel; however, since on-farm processing was not feasible, the oilseed was sold to a local cooperative, and an equivalent amount of biodiesel fuel purchased.

The livestock side of the farm's commercial enterprises consisted of a beef cattle operation, along with poultry raised to produce eggs and broilers. About three fourths of the feed for these animals (and the draft horses) was produced on the farm. On the crop side, wheat was grown for sale, and excess oilseed meal was also sold. The major components of the farm operation are listed in Table 18.4.

Energy accounting was a crucial facet of the Sunshine Farm project. Bender and colleagues carefully measured the weight of every farm input and output, using energy factors published in the academic literature to derive equivalent energy values. These data were painstakingly entered into a database, and used to generate energy budgets for the farm as a whole and for its constituent enterprises. These budgets included both direct and indirect energy costs.

The energy accounting showed that over the course of the demonstration, about 90% of the farm's energy needs — not counting the energy embodied in capital outlays and human labor — were supplied by on-farm inputs. The remaining 10% was the energy embodied in purchased seed and feed, and in the phosphorus and potassium removed in the marketed crops (Bender, 2002).

The Sunshine Farm project served many purposes. Primarily, it demonstrated that farming operations can come close to attaining energy self-sufficiency without sacrificing yields. It showed that many traditional farming practices — rotations, green manuring, livestock integration, crop diversity, use of draft animals — can be essential components of energy-efficient agroecosystems, and that modern alternative energy technologies can also play an important role. In addition, it showed that increasing the energy self-sufficiency of individual farms is not the only means of reducing agriculture's dependence on fossil fuels. Farms may also need to be integrated into a local renewable energy economy, as the Sunshine Farm did in growing oilseed but leaving biodiesel fuel production to a larger-scale cooperative, and in tying its photovoltaic array into the local power grid.

**TABLE 18.4**
**Components of the Sunshine Farm, with Their Energy Sources and Functions**

| Energy Source | Component | Function |
|---|---|---|
| Grain produced on the farm, plus some purchased feed | Draft horses | Field operations |
| Sunlight | 4.5-kW photovoltaic array | Electricity for workshop tools, water pumping, electric fencing, chick brooding |
| Purchased biodiesel from local cooperative, with raw-material contribution from the farm | Biodiesel tractor | Field operations |
| Grain produced on the farm, plus some purchased feed | Texas longhorn beef cattle | Marketing |
| Grain produced on the farm, plus some purchased feed | Poultry | Marketing (eggs and broilers) |
| Primary production, animal manure | Grain crops | Marketing (wheat) and animal feed (alfalfa, sorghum, oats) |
| Primary production, animal manure | Oilseed crops | Biodiesel production (pressed oil) and animal feed (leftover meal) |
| Primary production | Leguminous crops | Nitrogen fixation, forage, animal feed |

## FOOD FOR THOUGHT

1. How do biological cultural energy inputs and industrial cultural energy inputs differ with respect to ecological impacts?
2. What are some of the types of industrial cultural energy inputs to agriculture that can come from renewable sources?
3. How can we use renewable energy sources to replace nonrenewable sources, yet still meet the increasing demand for food?
4. What roles can animals play in improving the efficiency and effectiveness of energy concentration and transfer in agroecosystems?
5. What is your definition of sustainable energy use in agriculture?
6. How has the use of fossil fuels masked the environmental costs of conventional agriculture?
7. How has our "faith in technology" influenced the development of ecologically based, sustainable sources of energy for agriculture?
8. What are some of the limitations to "growing" energy crops on the farms where the energy will be used?

## INTERNET RESOURCES

Alternative Fuels Data Center
www.eere.energy.gov/afdc
A vast collection of information on alternative fuels and the vehicles that use them.

Energy Bulletin
www.energybulletin.net
An online news bulletin on energy issues, with a section dedicated to agriculture

National Sustainable Agriculture Information Service: Energy in Agriculture
www.attra.ncat.org/energy.html
This private, non-profit organization helps people by championing small-scale, local, and sustainable solutions to reduce poverty, promote healthy communities, and protect natural resources.

The Land Institute
www.landinstitute.org
A nonprofit research and education organization that promotes Natural Systems Agriculture, in which nature is the model for reconnecting people, land, and community.

Windustry: Wind Farmers Network
www.windustry.org
A nonprofit organization working to create an understanding of wind energy opportunities for rural economic benefit.

## RECOMMENDED READING

El Bassam, N. and P. Maegaard. 2004. *Integrated Renewable Energy for Rural Communities*. Elsevier: Amsterdam, The Netherlands. A comprehensive overview of how to meet the energy needs of rural and agricultural communities in a more sustainable way.

Fluck, R.C. (ed.), 1992. *Energy in Farm Production. Energy in World Agriculture* Volume 6. Elsevier: Amsterdam. A very comprehensive review of the basic principles of energy use in agriculture; includes data on energy use efficiency and potential alternative energy sources.

Odum, H.T. 1983. *Systems Ecology: An Introduction.* Wiley: New York. A key work on the systems view in ecology that analyzes how energy flows through natural ecosystems and examines how this knowledge can be linked to the sustainability of human-managed systems.

Outlaw, J.L., K.J. Collins, and J.A. Duffield. 2005. *Agriculture as a Producer and Consumer of Energy.* CABI Publishing. An examination of agriculture's role as a producer and consumer of energy, including recent research on issues related to efficiency, alternative fuels, and environmental impact.

Pimentel, D. and M. Pimentel (eds.) 1997. *Food, Energy, and Society.* 2nd ed. University Press of Colorado: Niwot, Colorado. A review of the problems inherent in an agriculture that is dependent on nonrenewable sources of energy and the complex issues involved in developing alternatives.

van Ierland, E.C. and A.O. Lansink (eds.), 2002. *Economics of Sustainable Energy in Agriculture*. Springer: Berlin and New York. A collection of case studies on energy efficiency improvement and the use of biomass for more sustainable agricultural systems.

# 19 Animals in Agroecosystems

Livestock animals figure prominently among the many reasons given in Chapter 1 for the unsustainability of conventional agriculture. Confined animal feeding operations pollute the air and water, turning manure into a problem instead of a resource; the meat industry stands as a prime example of economic concentration, vertical integration, and enemy of family farming; production of soybeans and corn for animal feed takes up too high a percentage of the world's arable land; concentrated production of meat and animal products for human consumption is energetically inefficient and ecologically harmful; factory farming of meat tends to undermine the economic base of rural farmers in developing countries who rely on small-scale livestock production; trend of diets toward more meat consumption accentuate income disparities between rich and poor; and diseases of livestock such as mad cow disease and avian flu threaten the human population. Combined with the risks to human health presented by antibiotic- and hormone-laden meat and diets too high in animal fat, these problems put livestock animals in a bad light, making them a target for criticism among many critics of conventional agriculture, advocates for sustainability, and consumer activists, as well as vegetarians, animal rights activists, and the like.

**FIGURE 19.1 An integrated farming system with organic walnuts and chickens near Tres Pinos, CA.** The mobile chicken houses are relocated daily so chickens can feed, help in weed management, and add manure to the soil. The walnut trees provide shade in the hot summer. The chickens are marketed directly to consumers, who come to the ranch to pick up their freshly slaughtered orders. Walnuts are harvested in the fall. A covercrop is grown during the winter.

Certainly some of the criticism is well deserved. But the problems lie not so much with the animals themselves or their use as food as they do with the ways the animals are incorporated into today's agroecosystems and food systems. Animals can play many beneficial roles in agroecosystems, and therefore make strong contributions to sustainability. Indeed, as we will see in this chapter, the inclusion of animals in an agroecosystem can often make the difference in realizing ecological sustainability and economic viability.

Relatively recently in agricultural history — around the turn of the century in the U.S. — farms included both livestock animals and crops as a matter of course. To use the central concept in this chapter, crops and livestock were integrated. The separation between crops and livestock that has occurred since then represents a literal dis-integration of agriculture. This disintegration not only threatens the ecological foundations of our food system, but it has also fundamentally altered the terms of the millennia-long mutualistic relationship we have developed with our domesticated animals.

Sustainability today depends in part on reintegrating animals and crops. It demands not the rejection of animal protein in our food system, but a more sensible and integrated approach to raising livestock for food that uses agroecological concepts and principles to adapt the best aspects of preindustrial agriculture into the postindustrial age. In this chapter we will explore the ways in which this reintegration can take place. The focus is not on how to do animal husbandry sustainably, but rather on the synergisms that derive from mixing crops and animals and their role in moving us toward sustainability (Figure 19.1).

## THE ROLE OF ANIMALS IN ECOSYSTEMS

Animals — defined broadly as heterotrophs — are essential components of all ecosystems on earth. They consume autotrophs (plants), transforming their biomass into animal biomass, which is eventually cycled back to autotrophs in the form of nutrient-rich waste and once-living organic matter. Since agroecosystems are modified natural ecosystems, managed for the purpose of harvesting biomass, they too require animals. Of course, as the ultimate consumers of the biomass harvested from agroecosystems, humans fill the animal role in all agroecosystems. But there are many reasons why we shouldn't be the only species in that role. As natural ecosystems demonstrate, there is plenty of room in an agroecosystem for a variety of animal species.

As we explore the ways in which reintegrating nonhuman animals into crop-based agroecosystems can help us achieve more sustainable food production systems, we need to begin, as always, with natural ecosystems. They show us how animals can enhance ecological integrity and stability, rather than disrupting or degrading it.

The role that animals play in the structure and function of natural ecosystems was discussed in some detail in Chapter 2. Here, we review and expand on some of the concepts presented there, with a view toward applying that knowledge to the design and management of agroecosystems that incorporate livestock animals.

## Shaping Vegetation

As heterotrophic organisms in the trophic structure of an ecosystem, herbivorous animals consume the biomass produced by plants, and this feeding behavior impacts the abundance, distribution, and diversity of the plant species, which in a terrestrial ecosystem are collectively termed the vegetation. Herbivorous animals are therefore key components — and determinants — of the structural makeup of most terrestrial ecosystems.

Because of the tight association between herbivores and the plants they consume, many ecosystems show a strong correlation between herbivore diversity and floral diversity. For example, multispecies grazing by a diverse ungulate fauna in the Serengeti of East Africa is intimately connected to a striking richness in both predatory animals and plants (McNaughton, 1985, 1990; Murray and Illius, 1996). Ten or more species of grazing ungulate may be found in close proximity, with several species occurring together in mixed groups. These herbivores eat different plant species, different parts of plants (leaves, stems, flowers), and plants in different life stages (green or dry); in addition, because of migration cycles, they put their herbivorous pressure on plants at different times of the year. These variations in the dietary specialization of each herbivore species coevolved with the vegetation, the components of which followed diverse strategies for coping with herbivory and for using it to minimize interspecific competition with other plants. The resulting structural diversity of the system allowed niche overlap, coexistence, and mutualisms to evolve as well and contribute to further diversity (Figure 19.2).

**FIGURE 19.2 Bison grazing at Konza Prairie Biological Station, near Manhattan, Kansas.** Bison have been a key element in shaping the prairie ecosystems of much of the Midwestern U.S. Photo courtesy of Catherine Burns.

The perennial grass prairie ecosystem of the North American Great Plains — in its aboriginal form — was another example of herbivore diversity coupled with floral diversity. Bison, elk, antelope, and other grazers selectively consumed different plant species, different plant parts, at different times in the season, coevolving with a prairie vegetation comprised of shortgrass, midgrass, and tallgrass species. The prairie ecosystem also demonstrated the direct influence of herbivores on ecosystem structure. The proportion of shortgrass, midgrass, and tallgrass species was determined primarily by grazing behavior and fire, with shifts in one direction or the other due to abiotic factors such as soil type and rainfall (Briske, 1996). Following the severe reduction in wild herds of the prairie herbivores, species composition of the native plants changed as well.

Recent restoration programs of native prairie ecosystems face the challenge of how to restore this native grazing pressure, or face the alternative of having to simulate the natural grazing with fire, mowing, or the use of domestic animals. At the Tallgrass Prairie Reserve in Oklahoma, The Nature Conservancy is using herds of 2000-plus bison and a "patch-burn" management tool to restore the prairie ecosystem and promote its original native plant and animal diversity. In another case, the World Wildlife Fund has a goal of restoring 17 to 27 million acres of the Northern Great Plains ecoregion so that it can support two bison herds of at least 10,000 animals each, as well as much of the accompanying plant and animal biodiversity characteristic of these systems.

### Enabling Energy Flow

When herbivores eat plants, and are in turn eaten by carnivorous animals, energy is flowing from one trophic level to another. You will recall that the energy flow between trophic levels is inefficient. A relatively small percentage of the solar energy fixed by photosynthesis and stored in plant biomass is preserved when that biomass is converted into animal biomass at the next trophic level. The vast majority of the energy (up to 90%) moving from one trophic level to another is given off as metabolic heat by the animals or deposited as manures back into the soil (Odum and Barrett, 2005). The energy contained in the manures of animals is not lost, however, because it is an essential driver of soil organism activity.

The loss of energy at each jump in trophic level means that the biomass at each higher level must be progressively smaller — thus the shape of the familiar "energy pyramid" in basic ecology texts. Since plants can occupy only the bottom level in the energy pyramid, the energy stored in animal biomass at any level is essential to the secondary consumers at each higher trophic level. Thus animal diversity in an ecosystem is a primary determinant of the number of trophic levels through which energy can be transferred — that is, the height of the energy pyramid and the diversity of the fauna generally. Returning to the example of the Serengeti: the diversity of herbivores is what makes possible the relatively high diversity of predators and other secondary consumers, including cheetah, lions, hyenas, aardwolves, leopards, wild dogs, jackals, eagles, vultures, crocodiles, and a variety of smaller carnivores and omnivores.

### Cycling Nutrients

In all natural ecosystems, herbivorous animals play an essential role in the dynamic process by which matter is cycled through the system. The emergent properties of efficiency, productivity, and stability are all related to this fundamental ecosystem process.

Ecosystems are dependent on animals, decomposers, and detritivores to release nutrients from their storage in plant material. Animals are therefore an important part of the nitrogen cycle, the carbon cycle, and the phosphorus cycle (all discussed in Chapter 2). Whether the nutrients are released back into geologic or atmospheric reservoirs, the initial step is consumption of plant tissue, followed by digestion, excretion, and decomposition (Figure 19.3).

### Influencing Community Dynamics

As discussed above, herbivory has a direct effect on the vegetation of an ecosystem. This was noted in a structural sense, but it can also be understood in a functional sense, as a factor affecting interspecific interactions and ecosystem complexity. Grazing or foraging by herbivores involves selective removal of certain species or plant parts, which affects how populations of each species in the community interact. Grazing pressure, for example, is often a key factor preventing a particular plant species from dominating an ecosystem through competitive exclusion and thereby reducing diversity and complexity. When grazing patterns change — due, for example, to removal of native grazers, changes in herbivore populations, or introduction of non-native herbivores — shifts in plant species dominance inevitably occur.

An example of the important role herbivores play in community dynamics is provided by ecosystems dominated by introduced species. In many parts of the world, invasive non-native plant species have established dominance in association with introduced nonnative grazing animals, causing changes in the native ecosystems that can persist even after the exotic herbivores are removed (Colvin and Gliessman, 2000). Conversely, invasion by non-native plants can become problematic because of the absence of herbivores able to control the aliens through consumption.

An awareness of how animals, especially larger herbivores, function as part of community dynamics and the other ecosystem processes discussed above can guide us as we consider how livestock may be integrated into crop production. As heterotrophic consumers of plants (and in some cases arthropod and molluscan pests), livestock

**FIGURE 19.3 Bison and bison manure.** The consumption of plant biomass by animals contributes to the recycling of nutrients in most natural ecosystems in the world. Photo courtesy of Catherine Burns.

animals can play a role in managing species interactions (Chapter 15), increasing agroecosystem diversity and stability (Chapter 16), taking advantage of successional processes (Chapter 17), and maximizing the efficiency of energy capture and use (Chapter 18).

## COEVOLUTION OF LIVESTOCK ANIMALS AND AGRICULTURE

In the earliest human cultures, people made a living off the land as hunters and gatherers, exploiting both the animals and the plants available in the ecosystems around them. Therefore, it is not surprising that as some human societies developed economies that could support larger populations and ensure more reliable food supplies, they domesticated animals at about the same time as they domesticated plants.

Domestication was a coevolutionary process in two senses. First, as discussed in Chapter 14, domesticated species changed in concert with human cultures, each becoming dependent on the other. Second, the domestication of plants — that is, the development of agriculture — proceeded parallel with, and was often directly connected with, the domestication of animals and the development of grazing and pasture systems.

The vastly different environments and ecosystems around the world offered very different opportunities for, and placed different constraints upon, the development of more-settled socioeconomic modes based on domestication of wild species. Some environments were too cold and arid to support any kind of agriculture, but offered native ungulates suitable for domestication. Other environments were conducive to both agriculture and raising of livestock. Of the prehistoric human cultures inclined to develop toward agricultural societies, some created animal-based systems and others, crop-based systems. In some cases the two were directly coupled.

However, while there are many examples of systems relying almost exclusively on domesticated animals, there are very few crop-based systems that lack domesticated animals entirely. In this sense, animals are truly a hallmark of agriculture.

## HOLISTIC RESOURCE MANAGEMENT

In rangelands around the world, range managers and their herds of livestock are a relatively new presence. Not very long ago, most of these lands were natural grassland, savannah, or shrubland grazed by herds of wild herbivores preyed upon by large carnivores. In too many cases, conversion of these ecosystems to human-managed rangeland has resulted in ecological degradation of various kinds — extensive soil erosion, loss of wetland habitat, replacement of native plant species with exotics, simplification of community structure, and outright desertification.

Many ranchers, institutions, and researchers have been grappling with the problems of range management. One of the more promising and agroecologically oriented approaches is that of Holistic Resource Management (HRM), developed over the last several decades by Allan Savory and colleagues associated with the Savory Center as a framework for creating sustainable grazing systems.

A foundation of HRM is the idea that we must mirror how animals work in natural grazing systems. When treated as an integral part of the rangeland ecosystem, livestock animals can actually improve the functioning of the system, increasing its productivity and stability.

In the typical rangeland environment, where precipitation and humidity are erratic, periodic disturbance is needed to maintain soil cover, promote floral diversity, and promote fresh growth — plants neither grazed nor trampled are suppressed or die under old leaves and stems that take many years to decay. In the past, this disturbance was provided by large herds of grazing animals — bison, elk, deer, zebra, wildebeest, buffalo, kangaroos, and so on — that constantly moved over the landscape, bunched tightly together to ward off predators that hunted in packs. The key to sustainable grazing, therefore, is to use livestock animals to simulate the herbivory and trampling of native grazers.

For the range manager or rancher, this amounts to "getting animals to the right place at the right time and for the right reasons." At all times, the herd is affecting the soil, plants, and wildlife. If left in any one place too long, or if returned to a place too soon, the animals will overgraze plants and pulverize soils. Therefore, livestock movement is a key aspect of HRM, just as in natural grazing systems.

Under HRM, the land can produce the maximum amount of high quality forage in the growing season, on an increasing or sustained basis, insuring that in the nongrowing months there is adequate forage and cover for livestock and wildlife. In essence, HRM uses the herd to benefit the whole system — which includes the rangeland ecosystem, its wildlife, its riparian corridors, and the economic enterprise of the rancher.

**FIGURE 19.4 Cows improving woodland understory in southern Spain.** Animals are moved through the system at key times to manage herbaceous cover, promote tree development, and produce animal products.

Wherever animals were domesticated, they became an integral part of human societies, receiving both care and respect. In this way the mutualistic sense of "coevolution" was carried through. The raising of livestock is often called *animal husbandry*, and in the older meaning of the term *husbandry*, the concept of caretaking is clear. Husbandry is defined in Webster's 1913 dictionary as "care of domestic affairs; economy; domestic management; thrift." Thus "animal husbandry" links the stewardship of domestic animals with the welfare of humans and their households.

## Grazing and Pasture Systems

First, we will examine the development of systems based on the domestication of grazing herbivores. This strand of agricultural evolution resulted in animal-only systems that survive to this day, but it also played a direct role in the evolution of integrated systems employing both plants and animals.

As explained in Chapter 14, humans transitioned from observant hunter-gatherers, to careful managers of wild populations, to caretakers of livestock domesticates. During this process, animals became dependent on humans for protection, feed, and reproduction, and humans came to depend on animals for a range of services and products.

Depending on local environmental constraints and the availability of native mammals and birds suitable for domestication, various types of food production systems incorporating animals were developed by human societies. These systems evolved over time in different ways, but overall it is possible to describe a general process of coevolution in which the animals became more thoroughly domesticated, humans intensified their management practices, and the plant species eaten by the animals developed more desirable characteristics in response to management.

The earliest form of animal husbandry was pastoral nomadism, in which humans accompanied animals as they made their way across the landscape in search of feed and water (Koocheki and Gliessman, 2004). Pastoral nomadism still exists in some very arid and mountainous lands, where it is doubtful that human communities would be able to survive without their herds of domestic animals. In regions where crop agriculture would be extremely difficult or even impossible, at least without considerable technological intervention, these animals are able to forage for scarce resources and turn vegetation that humans cannot consume directly into harvestable animal products. As the caretakers of these systems, humans must respect the limits of the carrying capacity of the landscape for grazing, understand the seasonal and regional variations in resource availability, and develop social structures around the needs of their animals. There are examples of well-managed present-day nomadic systems that date back to the early times of animal domestication, with some of the most notable in the arid regions of the Middle East (Figure 19.5).

**FIGURE 19.5 Sheep being herded by pastoralists in a nomadic system in the Negev Desert of Israel.** Rainfall in this region rarely exceeds 30 mm per year, too little for crop agriculture.

In parts of the world with more rainfall and more access to water resources, pastoral nomadism evolved into a type of managed grazing. People established permanent settlements, and animals were taken out for periods of time to forage on well-defined grazing areas. Good husbandry evolved into not just caretaking the animals, but also maintaining the health of the range lands. As discussed in Chapter 10, fire was most likely one of the earliest tools used for pasture and range improvement in these systems. They proved to be sustainable when the human managers, using natural ecosystems as the benchmark, developed and maintained a thorough knowledge of vegetative structure, species composition, forage quality, and other indicators of healthy range or forage lands. Managed grazing systems exist today in most parts of the world, in arid to humid rainfall regimes, from warm to cold climates, and across most soil types and conditions, in ecosystems that include natural grasslands, shrublands with forage and grasses, savannahs or open woodlands with trees interspersed in grassland, or forests with understory vegetation appropriate for animal consumption (Hodgson and Illius, 1996).

Ultimately, the coevolution of the human–livestock relationship reached another stage, in which humans planted and managed pasture species for improved feed quality and quantity. The transition from managing natural grazing ecosystems to the direct sowing of edible forage and pasture species probably occurred hand-in-hand with the domestication of livestock that were capable of pulling cultivation implements and producing manures that could be applied as soil improving amendments. Obviously a parallel coevolution was taking place as well on the plant side of the equation, as grain size, forage quality, and growth vigor all increased. Grasses, grains, and legumes all became part of the pasture mix, each providing complementary nutrition for livestock, and balanced nutrient inputs to the soil. In places where there was an extended time of the year when the planted pasture would not grow, such as during a rainless summer or cold winter, systems developed whereby the pasture biomass was harvested, dried, and stored for feed to be used during the time of scarcity, at which time the animals were often kept in confinement.

Such pasture systems are still very common today all over the world. Many agricultural universities and colleges have entire departments and programs devoted to the study of pasture design, management, and improvement, especially where animal production systems are most prevalent.

### Mixed Crop–Livestock Systems

While the coevolution of animals and forage plants was taking place, humans in some parts of the world were also developing crops for their own consumption. Animals were nearly always part of this crop development, since they provided the cultivation and transport power, as well as manures for fertilization of crops, and played a part in the diversification of farm landscapes that must have come about as humans balanced the needs of themselves, their animals, and the environment upon which both depended.

All early agricultural societies employed domesticated animals to some extent. Ancient cultures in the Indus Valley domesticated the chicken. The cultures of Southeast Asia raised fowl and water buffalo. In Mesopotamia, cattle, sheep, and goats were important. Even in the New World, where domesticable wild species were less abundant, domesticated animals such as the turkey and hairless dog played important roles in agricultural societies. The Anasazi of the American Southwest, for example, grew corn, beans, and squash, but raised domesticated turkeys for feathers, emergency food, and fertilizer (Figure 19.6).

The degree of integration of plants and animals varied in early agricultural systems. In some societies, crop production systems developed alongside livestock pasture systems; in other cases, food derived from animal domesticates supplemented a crop-based system. Either way, the pattern was set for integrated crop–livestock systems to develop along with the major centers of human civilization.

These integrated systems involved a diverse mixture of different activities, managed together as a working whole to take advantage of the ecological complementarity of each component or enterprise. In many temperate parts of the world with adequate rainfall, including the Middle East, Europe, northern Africa, and southern Asia, integrated systems reached a level of considerable complexity. In early modern Europe, for example, a typical integrated farm had pasture for harvestable feed or forage (annual and perennial), crops (annuals and perennials), animal grazing areas (with some possible improvement in plant species used as forage), corrals, forest or woodlot, often some sort of wetland, stream or well, and places for human habitation and activity, along with rotations and fallows involving multiple combinations of each component.

**FIGURE 19.6 Domesticated sheep in ancient Mesopotamia.** Domesticated animals, such as the sheep depicted on this fragment of a stone carving in The Louvre, were important in the early form of agriculture practiced in the "fertile crescent."

This style of integrated farm — and the associated cultural values of animal husbandry — was imported to the U.S., where it became the model system. Until the beginning of the 20th Century, most farms in the U.S. showed this integration of multiple enterprises. Integrated farms still exist today, but they are greatly outnumbered by specialized and industrial-scale operations that completely separate livestock and crop production.

The disintegration of livestock and crops came about with the widespread introduction of specialized machinery, fertilizers, and pesticides following World War II, but specialization in U.S. agriculture began many decades before that (Gregson, 1996). In order to respond to uniform market signals and distant markets, farmers began to rely on production inputs that had the effect of standardizing both growing conditions and response to management and climate. Diversity seemed to be less necessary, and farms began to simplify. Ready access to effective and cheap chemical fertilizers encouraged the perception that farmers no longer had to depend on biological nitrogen fixation and nutrient recycling through livestock to maintain soil health. Government support programs and academic research institutions further promoted the value of specialization, and by the end of the 1980s, the separation of livestock and crops was fairly complete (Gregson, 1996).

Yet, as we saw in Chapter 1, the problems associated with the disassociation of livestock and crops — and the concomitant growth of large-scale confinement systems for livestock and monoculture for crops — have come back to haunt us (Nierenberg, 2005). Now that sustainability is a primary goal of agriculture, and the costs of the inputs that promoted specialization in the first place are rising faster than the value of the crops they produce, the idea of integrating crops and livestock has gained prominence once again.

## ANIMALS IN INTEGRATED FARMING SYSTEMS

An "integrated farm" is one in which livestock are incorporated into farm operations "specifically to capture positive synergies among enterprises — to perform tasks and supply services to other enterprises — not just as a marketable commodity" (Clark, 2004). In this definition, "enterprise" refers to any focus or purpose of the farm system, from saleable products to weed management to soil health.

The positive synergies that arise from integrating animals into agroecosystems come about in large part because of the ecological complementarity of livestock animals and crop and forage plants. Plants feed animals, and animal excrement provides, in concentrated form, the nutrients that plants require. Thus, an integrated system — as opposed to one that is merely diversified — harnesses this complementarity to move energy and nutrients between the crop component and the animal component. When animals are integrated into agroecosystems in this way, more of the ecosystem processes operating in natural systems can be incorporated into the functioning of the agroecosystem, increasing its stability and sustainability.

### EXAMPLES OF INTEGRATED SYSTEMS

The basic concept of integrating the raising of animals and the growing of crops in the same agroecosystem finds a variety of expressions around the world. The livestock component can include cattle, sheep, goats, pigs, rabbits, horses, oxen, yaks, water buffalo, poultry, waterfowl, fish, shellfish, bees, silkworms, or a variety of other species that can provide food, work, manure, ecosystem services, or some combination of these. The crop component can include grains, pulses, oil seeds, grazed forages, vegetables, potatoes, fruit or nut trees, fruit vines, and other food crops. Given these options, the possibilities for integration are nearly endless. Four of the most important types of systems are described below.

#### Crop Rotations with a Grazed Forage Phase

The most widespread and widely adaptable method of integrating livestock animals into cropping systems is the grazed forage rotation. The specifics of the practice vary greatly, but its essence is to rotate a field between crops grown for human consumption (often grains) and a forage crop grazed by livestock. A variation on this theme is to grow the forage crop without grazing and then harvest it as feed for confined livestock animals. The grazed forage rotation was once very common all over the world, with the type of livestock, forage species, crop species, and timing of the rotation all adapted to local conditions.

As has been discussed elsewhere in this text, crop rotations in general have many benefits for overall agroecosystem sustainability. Weed growth (Chapter 11), agroecosystem diversity (Chapter 16), and water availability (Chapter 9) are just a few factors that are positively affected. When an animal component is included, both the options for rotational sequences and the potential benefits of the rotation are increased.

#### Agropastoral Systems

In some mountainous areas of the world, particularly in Pakistan, India, China, Nepal, and Bhutan, the most common traditional agricultural system is agropastoral in nature. Crop production in warmer valleys is combined with the grazing of livestock animals on highland pastures during the summer. Usually the livestock provide the major source of income and food. Despite the spatial segregation in the summer, integration occurs in the use of animal manure for fertilizer and the growing of forage crops for winter animal feed.

#### Livestock in Agroforestry Systems

An agroforestry system, as discussed in Chapter 17, is a system that integrates trees with crops, animals, or both. When the focus is on the tree–animal combination, it is referred to as a *silvopastoral* system, and when all three (crops, animals, and trees) are integrated, the term *agrosilvopastoral* system is used.

The practice of silvopastoral agroforestry is best known in the tropics, where trees can mitigate the impacts of heavy rainfall, nutrient leaching, and intense solar gain. Some of the most common silvopastoral agroecosystems involve the use of trees as an overstory above natural or improved pasture (Buck et al., 1999). Typically, forest is cleared and specific trees are left to form the shade over the pasture, and often some additional tree species of ecological or economic value are planted as well, in patterns that ensure good tree development. The management of the animals in livestock agroforestry systems is the key, because it must meet the needs of the trees — and the crops, if they are present — and the animals at the same time.

Both trees (Chapter 17) and livestock animals can have many positive impacts on an agroecosystem; so when the two are combined many components of sustainability can be brought together. The journal *Agroforestry Systems*

**FIGURE 19.7 A simple silvopastoral type of agroforestry system in Puerto Viejo, Costa Rica.** The leguminous tree *Glyricidia sepium*, used as a living fence, is pruned three to four times per year, and the cattle eat the protein-rich prunings as a complement to their diet of pasture grasses.

## SPAIN'S *DEHESA* SYSTEM

In mountainous regions of southern Europe, especially in the region of Andalucia, Spain, there exists a traditional agrosilvopastoral system that integrates livestock, crops, native herbaceous vegetation, and oak forest. Known in Spain as *Dehesa*, this integrated system shows the level of complexity and stability that can be achieved by combining careful management of domestic grazing animals and limited crop agriculture in the context of the natural landscape (Sevilla Guzman, 1999).

The term *"Dehesa"* was originally used to refer to parcels of land that were located at the margin of a community's common grazing areas, meant to be used by specific community members for the pasturing and resting of the animals used for meeting the farm labor needs of the community. Today, the term describes the management system practiced on the forested lands surrounding communities, which were *Dehesa* in the older sense. These areas are vegetated by an open oak forest of several *Quercus* species, with an herbaceous understory that germinates with the first rains of the Mediterranean fall, grows through the winter into spring, and is dry during the rainless summer.

The basis of the *Dehesa* system is rotational and mixed grazing by the traditional race of Iberian pig (Figure 19.8), sheep, and occasional horses in the open oak forest. In addition, the people of the community gather firewood and some cork from the forest, as well as a vast number of native plant species for use as food, medicine, and spices. In open areas with better soils they grow small plots of forage grasses or legumes. More recently, cattle have been added to the mix of grazers (Figure 19.4), and native wildlife species are hunted for sport.

The key aspect of the *Dehesa* is the maintenance of the oak forest ecosystem. This is only possible with the rational and careful integration of the animal component. Sheep, cattle, and horses graze the natural herbaceous cover during the winter and spring, after which time they are either sold, moved to lowland areas with better pasture, or kept in limited numbers on stored forage. The Iberian pig brood-stock and the current year's offspring are kept in large pens under oaks where they are able to move freely, and are fed grain from the small production areas. In the late summer when the acorns begin to fall from the trees, the young pigs are released from the pens and allowed to range freely and feed on the acorns. In a period of less than three months, the pigs gain more than 50% of their

their weight on the acorn diet, producing flesh used for the unique and highly sought-after ham they are famous for (*jamón Ibérico*). Cattle prefer the green herbaceous growth of the winter months, and the sheep are able to do well on the green biomass of winter as well as dry plant material that persists into the summer.

The *Dehesa* represents an example of the sustainable use of resources in a marginal environment. It is sustainable only because its management is based on optimizing the natural productivity of the landscape through careful management of livestock, not on maximizing yields. Its base is diversification, complementarity of plants and animals, extensive rather than intensive use of the fragile natural resources of the oak forest, local animal breeds, and management knowledge built up over centuries.

Competition from factory-farmed pork, the desire for higher returns per unit area of land, and the movement of rural people to the cities, however, all put pressure on this remarkable system. An agroecological understanding of its value is needed to preserve it into the future.

**FIGURE 19.8 Iberian pigs.** This locally adapted breed is the most important animal component of the *Dehesa* system in Spain.

contains many examples of research on animal-oriented agroforestry systems that demonstrate many characteristics of ecological sustainability and provide many economic and social benefits as well.

## Aquaculture and Crop Production

A variety of systems in use around the world incorporate the raising of either waterfowl or aquatic species such as fish or shellfish with crop production. As one might expect, these systems are most common in areas with abundant moisture, where wetlands would predominate in the absence of agriculture.

Integrated rice and duck farming is practiced in parts of Japan and China (Furuno, 2001). Weeds and insects are consumed by the ducks, manure from the ducks is returned to fertilize the rice, and humans harvest both ducks and rice at the end of the crop cycle. When the aquatic fern *Azolla,* with its nitrogen-fixing algal mutualist, is added to the system, fertility is maintained and yields are improved. Similar results are achieved with the integrated fish and rice systems of southern China (Guo and Bradshaw, 1993). By allowing fish to occupy the irrigation channels and flooded rice paddies during the cropping season, nutrients are captured that might otherwise be lost from the system, especially in systems where the fish are algae feeders and the algae thrive on nutrients in the water. Even when part of the rice paddy is removed from rice production in order to dig ponds that allow for year-round presence of the fish, the ecological and economic benefits more than compensate.

Fish can be such an agroecologically and economically beneficial part of an integrated cropping system that some systems have developed that combine fish, crops, and livestock. In parts of Asia, for example, systems exist that integrate fish, silkworms, mulberries, pigs, sugarcane, vegetables, and grass in intensively managed wetlands. In some localities, farmers are adding an aquaculture component to an already-integrated crop–livestock system. Livestock manure is used to stimulate the growth of algae or plankton, which are consumed by the fish. Waste from the fish can then serve as a nutrient source for the crops, and the fish themselves are a marketable product.

Integrated crop–aquaculture systems offer a clear contrast to intensive, industrial-scale, single-species aquaculture systems. In many respects, these systems are not much different from the livestock confinement systems used for cattle, pigs, and poultry. Feed — different from what the animals consume in the wild — is often grown a large distance from the place of animal production, antibiotics and growth stimulants are often employed, and waste food and excrement contaminate the water.

## Beneficial Roles of Animals on an Integrated Farm

As they pursue their ecological role, mimicking the herbivores in natural systems, livestock animals transform the energy and matter contained in plant biomass into three agroecologically useful streams, as shown in Figure 19.9. The first stream, animal biomass, has value as food, fiber, fertilizer, and raw material. The second stream is the biological cultural energy represented by the ability of livestock to do work. Draft animals, which once performed all the work on farms that humans did not do, are the obvious workers among the many types of livestock, but sheep, goats, chickens, ducks, and other animals can also perform valuable "work" in the form of vegetation and weed management and pest control. The third stream, manure, is rich in plant nutrients and provides soil microorganisms with a key source of energy for their roles in the system.

Both the *products* of animal herbivory — animal biomass, work, and manure — and the *process* of herbivory itself combine to provide the farmer with an array of potential on-farm benefits. These benefits are discussed separately below, in the context of actual farm practices. The benefits are interrelated and overlapping, but by pulling them apart conceptually it is easier to see how various forms of integration can be combined to further the overall goal of establishing mutualistic synergies that improve agroecosystem structure and function and lessen the dependence on purchased external inputs. Table 19.1 summarizes many of the benefits of integration by comparing the conditions of an integrated system to those of a comparable nonintegrated crop-based system.

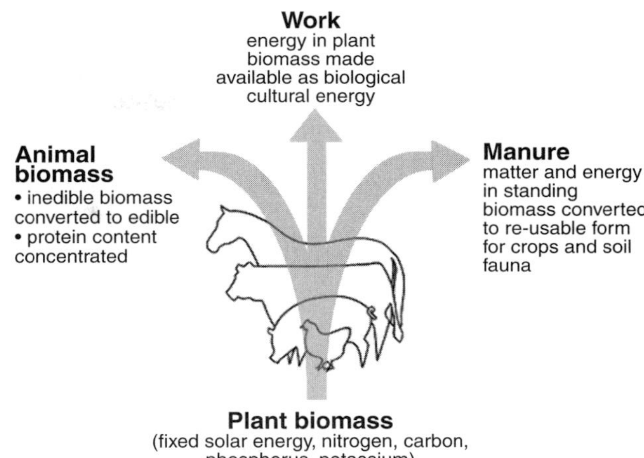


**FIGURE 19.9 The transformer role of livestock.** Animals transform plant biomass into useful forms of energy and matter.

### Producing Protein-rich Food and Other Products

Animal biomass is as important as food, both for subsistence and for market. Whether in the form of milk, meat, or eggs, it contains a much higher proportion of protein than plant biomass. Moreover, most livestock animals are able to obtain nutrition from types of plant biomass that humans can't eat — crop waste, food waste, plant tissues containing mostly cellulose — and convert it into various forms of animal biomass that humans can eat.

Animal biomass has many other economically valuable uses, of course. Sheep produce wool and waterfowl, feathers; and at the end of their lives, animals yield bones and other byproducts that can be used for a variety of purposes.

### Putting Crop Residue and By-products to Use

Since animals are able to consume much of the biomass left over after harvest, as well as many of the byproducts from agricultural processing; using such biomass as animal feed is an important way to produce harvestable animal products and convert a potential waste into recyclable nutrients at the same time. Maize, millet, wheat, oat, and

**TABLE 19.1**
**Benefits of Integrating Livestock into a Crop Production Agroecosystem**

| Aspect of Integrated System | Ecological Effects | Agricultural Benefits |
|---|---|---|
| Including grazed forages in crop rotation, especially perennial species | Higher diversity<br>Greater soil microbial activity<br>Maintenance of soil coverage<br>Less frequent disturbance; pioneer (weedy) species not encouraged<br>Provision of habitat for natural pest control agents | Soil organic matter allowed to increase<br>Soil not exposed to erosion<br>Reduction in weed pressure<br>Improved performance of subsequent crops<br>Elimination or reduction of need for biocides<br>Better retention of soil nutrients |
| Feeding marketed livestock on feedstuffs produced in the same system | Cyclical (vs. linear) nutrient flows<br>Less nutrient export | Reduction in dependence on purchased inputs<br>Enterprise diversification<br>Opportunities for productive use of crop wastes |
| Using livestock manure as a nutrient source instead of inorganic fertilizer | Improvement of soil structure<br>Higher soil biodiversity and microbial activity<br>Better nitrogen cycling | Reduction in dependence on purchased inputs<br>Improvement of soil organic matter content<br>Higher crop yields |
| Using livestock for vegetation and weed management | Mimics natural-system role of herbivores | Elimination or reduction of need for biocides<br>Reduction in manual human labor |

barley straw serve as supplemental feed for a range of animals in many agroecosystems around the world, with the greatest energy efficiency achieved when the site of straw production is as close as possible to the consuming animals. Allowing the animals to directly graze the straw remaining in the field is probably the most efficient method, although straw can be cut and hauled to a storage area in order to feed the animals when they are confined.

In California, where intensive vegetable production is so common, many crops produce residues that are used to supplement animal grazing or forage, such as culled Brussels sprouts, waste tomatoes, and carrot pulp after juice extraction. Pigs are excellent transformers of food and crop waste in many rural small-farm systems in the developing world. Since the animals in this case are not actually in the fields where the vegetable crops are raised, the manure the animals produce must be returned to the crop fields.

## Returning Nutrients to the Soil in Manure and Compost

Plants contain nutrients that they have taken up from the soil, and through their consumption in plant biomass, digestion, and deposition as manure, the nutrients are cycled back to the soil. Depending on the farming system, the manure can be collected, composted, and applied at any location on the farm where it is needed most. Returning manure to the soil is an important way to put both nutrients and organic matter back into the soil ecosystem (Chapter 8), as well as to reduce the need to import these materials from outside the farm operation.

The ecological and economic efficiencies of trying to import to cropping systems the massive amounts of manure and urine produced in large-scale livestock confinement systems have already been discussed in Chapter 1. Integrated livestock–cropping systems — in which forage is grown on the farm, fields are rotated between grazed forage and crops, and crop residues are incorporated into animal feed — can greatly increase the efficiency of manure and compost management. A study carried out in Denmark (Dalgaard et al., 2002) showed that a farm converted to mixed dairy and pig production, using an array of grain crops for harvest and grass and legume forage species for animal feed, could obtain total self sufficiency in animal fodder while reducing nitrogen contamination of local ground water systems to very low levels as compared to conventional systems and organic livestock operations more dependent on imported feed.

## Improving Soil Health

The key component of a healthy soil is soil organic matter. Many factors, organisms, and interactions drive organic matter quantity and quality, with soil health manifested in tilth, structure, water holding capacity, and resistance to both compaction and erosion. Long-term cultivation generally leads to the breakdown of soil organic matter, with accompanying degradation of the indicators of soil health. However, bringing livestock into the cropping system in the form of a rotated grazed forage not only reduces the need for cultivation, but it also adds nutrient- and energy-rich organic matter. Soil microbial activity increases, soil structure improves, and nutrient retention and availability favor better crop development. In some regions, especially the Midwestern areas of the U.S. and Canada, the inclusion of a perennial forage in cropping system rotations, with its accompanying respite from the negative impacts of tillage and restoration of soil health parameters, can easily justify the reduction in emphasis on cash grain crops (Clark 2004).

### Providing Work

Fueled by the matter and energy in the plant biomass they consume, animals are able to provide work in the form of cultivation and transport. This was discussed above, but it is important to note that the work performed by animals is a form of biological cultural energy. As such, it helps the farmer achieve more favorable ratios of energy input to energy output and reduce purchased energy inputs (Chapter 18). The use of draft animals may seem anachronistic in today's world, but the rising cost of fossil-fuel-derived energy makes it an increasingly attractive option. A different kind of "work" is provided by honeybees — when kept on the farm to produce honey as a marketable product, they also pollinate crops or fruit trees.

### Managing Vegetation and Controlling Weeds

Weed management appears to be a particularly important reason why farmers include forages in their crop rotations. In a farmer survey conducted in Canada, it was found that more than 80% of the 235 farmers contacted, reported reduction in weed pressure following forages (Entz et al., 1995). Many observed good control of several of the most problematic weeds such as wild oat (*Avena fatua* L.), Canada thistle (*Cirsium arvensis* L.), wild mustard (*Sinapis arvensis* L.), and green foxtail (*Setaria viridis* (L.) Beauv.).

Grazing animals can be used in other ways for landscape and vegetation management. Goats are used for poison oak (*Toxicodendron diversilobum*) control in many places of coastal California, or for weed suppression in crop systems at the end of harvest. Sheep are known to have been used for weed control in crops like corn before the advent of modern herbicides in parts of the corn belt of the U.S., and chickens are renowned for their ability to cultivate the soil, manage pests, and control weeds in home garden and small-scale cropping systems. Managed appropriately, cattle, sheep, and goats can be used to graze out undesirable species during reforestation, on young Christmas tree farms, and on rangelands. Obviously, the preference that most grazing livestock have for herbaceous rather than woody vegetation would be a key factor in the preferential removal of herb pressure in plantations of young trees or in regenerating forests following disturbance. The use of grazing livestock for vegetation management has strong resemblance to the impacts of natural grazing in places such as the Midwestern Prairies of North America and the Serengeti Plain mentioned earlier in this chapter (Figure 19.10).

**FIGURE 19.10 Ducks being used to graze out weeds and scavenge dropped seed in a rice paddy near Nanjing, China.** Waste matter is being converted into a resource in the form of animal products and manure.

### Increasing Subsequent Crop Yields in Rotations

One of the many benefits of including a grazed forage in crop rotations is that higher yields may be obtained from the crops planted after the forage. This effect is due to the other positive impacts of the grazed forage period: less soil disturbance, increased soil organic matter, and weed control. In the Canadian farmer survey noted above, over two-thirds of the surveyed farmers reported higher yields in the grain crops that were planted following the forage rotation (Entz et al., 1995).

### Providing Ecosystem Services

From an ecosystem perspective, animals on the agricultural landscape can provide many services beyond food production. Many of the benefits listed above contribute to various larger scale ecological processes:

- *Carbon sequestration.* Livestock animals are a part of putting cover back on the land as trees in silvopastoral systems or using perennial forage crops, two of the few known ways to produce a net increase in soil organic carbon, potentially making a contribution to reducing levels of carbon dioxide in the atmosphere.
- *Erosion control.* As integral parts of a grazed forage rotation, animals help improve the quality of vegetative cover, a crucial tool in soil erosion control.
- Maintenance of watershed health. The same factors that help in erosion control also promote the watershed processes of infiltration, percolation, and water retention discussed in Chapter 6.
- *Biodiversity protection.* The integration of animals back into the agricultural landscape — especially small livestock and locally adapted species and races — promotes the conservation

of agrobiodiversity. In addition, to the extent that animals provide other environmental services and lessen the negative off-farm impacts of agriculture, they enhance and protect biodiversity and the entire landscape.

## Social and Economic Benefits of Integration

Up to this point, we have focused mostly on the ecological benefits of mixed livestock–crop systems. As we have seen in many other cases, practices that have ecological benefits often have economic and social benefits as well, and livestock–crop integration is a good case in point. Many economic and social benefits, of course, are implicit in the points made above: increasing crop yields, improving soil health, and reducing costly purchased inputs all have direct positive impacts on the farmer's bottom line. But two socioeconomic benefits of integration deserve discussion on their own.

### Diversifying Enterprises and Reducing Economic Vulnerability

One of the original reasons why animals and crops were raised together on farms was that this mixing allowed a greater diversity of food types and agricultural products to be produced. This diversification had a simple economic logic: it increased economic security by spreading the risk of failure among more enterprises. While this was based on a self-sufficiency situation long gone in most parts of the world today, the same logic still applies in the context of producing food as a commodity. Raising animals in addition to crops provides the farmer with additional marketable products, whether they be eggs, milk, wool, honey, silk, lambs, or beef cattle. Depending on local market conditions, these animal-based enterprises can provide a valuable income stream and protect against crop failures and market fluctuations (Schierea et al., 2002).

Further, the various enterprises that may be based on an integrated farm are often ideal for marketing on a local or regional basis. By selling products at local stores, restaurants, and farmers markets, and through food cooperatives and community-supported agriculture organizations, the money that would otherwise go to distributors, wholesalers, transporters, and brokers goes to the farmer instead. We will discuss the importance of such localized food networks in more detail in Chapter 23.

### Alleviating Poverty in Developing Countries

Mixed crop–livestock systems are ideally suited to helping alleviate poverty in developing countries, the underlying cause of high infant mortality, chronic hunger, food insecurity, resource degradation, and the high societal costs that result from these problems. Integrated crop–livestock systems can be operated profitably on a small scale, comprise a multiplicity of income-producing activities, and require few off-farm inputs and relatively small capital investment; for these reasons they are effective and realistic ways of creating greater economic security for many people in developing countries (LEISA, 2005).

The animal portion of the system itself represents much of its economic value to poor farmers. A livestock animal is a living bank. It acts as a storehouse of capital, an investment in future productivity, and an insurance against crop production risks. Where diets tend to be protein- and calorie-deficient, livestock animals supply vital protein-rich food. In addition, since women often play an important role in animal husbandry activities, agricultural systems that incorporate livestock animals can promote gender equality, both socially and economically (Figure 19.11).

In many developing countries, mixed crop–livestock systems are already primary economic activities in rural areas, but these systems operate inefficiently, without taking advantage of all the potential synergies, and as a result, they don't realize their full potential for economic return. This is the case in the Indian state of Chhattisgarh, for example, where a study determined that several different alternative modes of structuring the typical small-holder agroecosystem could result in significant improvement of the socioeconomic status of the tribal farmers (Ramrao et al. 2005). In every case, the alternative system involved diversification and increase of the animal component and tighter integration of all system components.

Table 19.2 compares mixed crop–livestock systems, grazing systems, and industrial confinement systems in terms of three important characteristics. It demonstrates the high correlation between crop-livestock integration, ecological qualities, and potential for achieving social and economic equity.

**FIGURE 19.11 An agroforestry system in Tonga integrating cattle and coconut palms.** The palms provide coconut fruit, copra, and construction materials, and the cattle provide meat and milk. Systems such as this, combining agriculture, animal husbandry, and forestry, are especially appropriate for small holders in developing countries. Photo by Molly Wilson.

**TABLE 19.2**
**A Comparison of Three Livestock Production Systems in Ecological and Social terms**

| | Degree of plant–animal Integration | Need for External Inputs | Potential for Social Equity |
|---|---|---|---|
| Mixed crop–livestock systems | High | Low; system partially closed | High potential for poverty alleviation, reduction of risk, and gender equality |
| Extensive grazing systems | High | Low; system partially closed | Variable, depending on scale |
| Industrial confinement systems | Low | High; system open | Low |

## LIVESTOCK AND FOOD SYSTEM SUSTAINABILITY

Integrating livestock and crop production carries with it a wide variety of benefits to farms, farmers, developing countries, the agricultural landscape, and the environment in general, as the preceding section clearly demonstrates. It offers means of reversing many of the trends currently undermining the ecological and social-system foundations of agriculture. Increasing crop–livestock integration, therefore, is a key element of moving toward greater sustainability of the global food system. But in doing so, we face a huge challenge, because the momentum of change is in the other direction.

To better understand this challenge, we need to look at livestock production generally, in the context of the whole food system. This gets us into issues that go beyond farm-level integration of livestock and crops. It is not the intention of this chapter to delve into these issues in any detail, but noting them briefly here will prepare the ground for their further discussion in Section IV.

### Elements of a More Sustainable Animal Protein Economy

A thorough examination of the role of animals in the global food system would find serious problems at every level, from production to consumption. Livestock are raised in energetically wasteful and ecologically damaging ways; the products derived from them reach consumers only after traveling long distances; and at the consumer end, high demand for these products drives the whole system. To move toward a more sustainable animal protein economy, change must occur at the three levels of production, distribution, and consumption.

#### Production

Throughout, this chapter has supported the idea that producing meat, milk, and eggs in integrated farming systems is more sustainable than producing the same products in specialized, single-purpose, industrial systems. But it is unrealistic to expect all the animal protein in human diets worldwide to come from such systems. Therefore, it is important to mention two universally applicable principles for the more-sustainable production of animal protein:

- Energetically speaking, animals vary in the efficiency with which they convert plant food into animal protein. Producing chicken flesh, chicken eggs, and bovine milk are three of the most efficient ways to convert plant biomass into animal protein. Sustainability, therefore, depends on shifting the focus of animal production toward chicken flesh, eggs, and milk and away from the flesh of cattle.
- There is a significant difference between feeding beef and dairy cattle processed grain and feeding them plant biomass that humans can't eat. The former is an extremely inefficient use of arable land and requires a much larger fossil-fuel subsidy. In a sustainable food system, therefore, all ruminants used for meat or milk production would be range-fed. Similarly, all hogs would be fed mostly on food and crop wastes.

#### Consumption

The growth of confined animal feeding operations (CAFOs), and the disintegration of livestock and crop production generally, is driven largely by a rapidly increasing demand for meat and other animal products worldwide. People in developed countries are already eating more meat than they have ever eaten before, and every year, they eat even more. At the same time, people in developing countries are trying to match the prodigious meat consumption of their developed-country counterparts.

While it is beyond the scope of this chapter to address this issue, we must be aware that the trend toward more meat-intensive diets may be the single most serious barrier to creating a more sustainable food system. There is no question, therefore, that per capita meat consumption must be reduced. At the same time, the meat that is eaten must be produced in a way that minimizes its negative impacts. This means eating less beef and pork and relying more on poultry, milk, and eggs. It means preferring meat from animals fed in integrated and pasture systems over meat from animals raised in CAFOs (Figure 19.12).

**FIGURE 19.12 Locally produced eggs being sold by the grower at a farmer's market in Santa Cruz, California.** Eggs are among the most energy-efficient of all animal products, so the consumer buying them at a farmers' market is supporting a more sustainable food system in a variety of ways.

### Distribution

In the present food system, food commodities are typically transported long distances before they are finally consumed, using large amounts of fossil fuels and ensuring that most of each consumer dollar goes to processors, distributors, brokers, wholesalers and other "middlemen" instead of to the farmer. For alternatives to this distribution system — more localized food networks — to become stronger and more prevalent, there must be tighter geographic and economic connections between the producers of animal products and the consumers of those products.

Production of livestock on integrated farms is well suited to this transformation. Such production is necessarily smaller scale than production in CAFOs. High-volume, centralized CAFO production goes hand-in-hand with a high-volume, centralized processing and distribution system designed to distribute eggs and meat and dairy products to a national and even global market. Correspondingly, the low-volume, geographically dispersed production from individual integrated farms fits best with a more local processing and distribution system.

### Challenging Specialization in Agriculture

Not so long ago, the concept of specialization was unknown to farmers. Integration and enterprise diversification were the underlying principles of farm operation. As we lost this approach to food production, our communities lost their local food distribution systems and most of their family farms, and consumers lost the organic connection with both the people who produced their food and the animals from which it much of it came. Today, with livestock animals sequestered into CAFOs, fed with grains produced half a continent away, and their carcasses, eggs, and milk transported hundreds and thousands of miles to market, its not surprising that the typical consumer gives no thought to what it took to get the steak to his table.

As we have seen elsewhere in this book, specialization in agriculture is ill designed to meet the multiple needs of society for abundant, healthy food, produced in ecologically sound ways that provide sustainable livelihoods. Reintegration of livestock and crops helps reverse the trend toward specialization and economic concentration in agriculture, pointing the way toward more local food distribution networks, viability for smaller-scale, family-run farms, and more self-contained, closed-loop agroecosystems that don't rely so strongly on purchased inputs.

Reintegrating livestock and crop production really strikes at the heart of what's not sustainable in conventional agriculture. For this reason, supporting the integration of livestock and crops — in the marketplace, at research institutions, in the public policy arena — can go a long way toward making change happen. Such advocacy underlines the need for integration while increasing awareness of the huge social and environmental costs of specialization and concentration.

## FOOD FOR THOUGHT

1. What changes would consumers need to make in their diets in order to promote the reintegration of animals into farming systems?
2. Can vegetarianism and integrated livestock–crop production systems be combined?
3. What are some of the primary indicators of sustainability most appropriate for the analysis of integrated farming systems?
4. How can we reconcile production needs with the ethical treatment of animals?

## INTERNET RESOURCES

Alan Savory Center for Holistic Management
www.holisticmanagement.org

The Savory Center is an international not-for-profit organization established in 1984 to coordinate the development of holistic management, of which holistic resource management is a part.

American Forage and Grassland Council
www.afgc.org.

American Grassfed Association
www.americangrassfed.org

Center for Integrated Agricultural Systems, University of Wisconsin–Madison
www.cias.wisc.edu
Dedicated to the study of the relationships between farming practices, farm profitability, the environment, and rural vitality.

Eat Wild
www.eatwild.com
Information on pasture-based farming; lists farmers and ranchers who raise livestock on pasture and sell directly to consumers.

Heifer International
www.heifer.org
A nonprofit organization that helps communities in rural areas around the world integrate appropriate livestock technology, self-reliance, and sustainable development.

Livestock, Environment and Development (LEAD)
www.virtualcentre.org/en/frame.htm
Targets the protection and enhancement of natural resources that are affected by livestock production and processing; also focuses on poverty reduction and public health enhancement through appropriate forms of livestock development.

National Sustainable Agriculture Information Service, livestock section
www.attra.org/livestock.html
A project of the National Center for Appropriate Technology (NCAT).

Proceedings of the International Symposium on Silvopastoral Systems and Second Congress on Agroforestry and Livestock Production in Latin America
www.fao.org/WAIR OCS/LEAD/X6109E/x6109e00.htm
Research focusing on the theme of silvopastoral systems for restoration of degraded tropical pasture ecosystems.

The Nature Conservancy: Tallgrass Prairie Preserve
nature.org/wherewework/northameri a/states/oklahoma/preserves/tallgrass.html
Information on tallgrass prairie and bison conservation in the Great Plains of the Midwest.

World Wildlife Fund wild places: Northern Great Plains
www.worldwildlife.org/wildplaces/ngp/
Describes the conservation importance of the Northern Great Plains, and the WWF's conservation projects in this area. Current projects include the American Prairie Restoration project, which aims to restore the prairie so it can support large herds of bison and other native wildlife.

## RECOMMENDED READING

Cheeke, Peter R. 1999. *Contemporary Issues in Animal Agriculture* 2nd ed. Danville, IL: Interstate Publishers, Inc. Covers a wide range of ethical, environmental, and health issues related to animal agriculture.

de Haan, C., H. Steinfeld, and H. Blackburn. 1997. *Livestock and the Environment: Finding a Balance*. FAO/World Bank/USAID: Rome. An international assessment of the negative impacts of livestock production systems on the environment and an exploration on how to overcome them.

Furuno, T. 2001. *The Power of Duck*. Tagari Publications: Tasmania, Australia. A very readable and applicable example of how one animal in particular has played a critical role in sustainable agroecosystem design and management.

Hodgson, J. and A.W. Illius. 1996. *The Ecology and Management of Grazing Systems*. CAB International: Wallingford, UK. A unique look at our understanding and management of land resources used by grazing animals, combining perspectives from ecology, plant science, and animal science.

Little, D.C. and P. Edwards. 2003. *Integrated Livestock–Fish Farming Systems*. FAO: Rome. Describes the dynamic set of practices that constitutes integrated fish–livestock farming systems in Asia.

Nierenberg, D. 2005. Happier Meals: rethinking the global meat industry. WorldWatch Paper 171. WorldWatch Institute: Washington, D.C. A concise examination of the damaging effects of industrial animal agriculture, or factory farming, as well as proposals for alternative ways that farmers, processors, and consumers can help ensure that meat is made better for people, the environment, and the animals themselves.

Oltjen, J.W. and J.L. Beckett. 1996. The role of ruminant livestock in sustainable agricultural systems. *Journal of Animal Science* 74(6): 1406. Full text available online at jas.fass.org. Presents the argument that ruminants serve a valuable role in sustainable agricultural systems, and that they should be integrated in sustainable farming systems.

Schlosser, E. 2002. *Fast Food Nation*. Harper Collins Publishers: New York. An eye-opening exposé of the downside of the fast food industry and its dependence on the factory farm animal production model.

# Section IV

## The Transition to Sustainability

The appearance of the so-called Brundtland Report (WCED, 1987) in the late 1980s marked the emergence of sustainability as an issue of central concern in agriculture, rural development, natural resource use, and indeed every human endeavor. Since that time, a growing community of researchers and practitioners has made significant progress in developing useful systems for implementing and measuring sustainability, particularly in agriculture.

Although it has effectively lead the effort, the scientific community must still develop a much better understanding of what sustainability actually entails, so that the agenda for change is clear and actionable. A new field known as "sustainability science" has emerged that may help us meet this challenge (Kates et al., 2001; Turner et al., 2003).

From the perspective of sustainability science, food systems are so complex that many fields of inquiry must come together in understanding how to push their interdependent components toward more sustainable results. Understanding this complexity — and using it as the basis for change — is the goal of the sequence of chapters that make up Section IV.

Chapter 20 begins at a practical level, examining the issues surrounding the conversion to more-sustainable practices, and Chapter 21 follows with a focus on the indicators of sustainability. Chapter 22 describes how creating sustainable agroecosystems can aid in conserving biodiversity and ecosystem services, and vice-versa. Chapter 23 looks at food consumption and the necessity of reconstituting the economy and culture of food on a more local basis. With these building blocks in place, Chapter 24 broadens the agenda of sustainability to include the whole food system, which means integrating human society with knowledge of ecological sustainability. This brings us back to the understanding that agroecosystems are, after all, a product of the co-evolution between cultures and their environments. In this balance between social needs and ecological health lies true sustainability.

***An agricultural landscape in the mountains north of Quito, Ecuador.*** *This landscape shows many of the components of sustainability, including crop rotations, soil management techniques, diversity in and around fields, and equitable distribution of land, water, and local resources.*

# 20 Converting to Ecologically Based Management

Farmers have a reputation for being innovators and experimenters, willingly adopting new practices when they perceive that some benefit will be gained. Over the past 40 to 50 years, innovation in agriculture has been driven mainly by an emphasis on high yields and farm profit, not only resulting in remarkable returns but also an array of negative environmental and social side effects. Despite the continuation of this strong economic pressure on agriculture, however, many conventional farmers are choosing to make the transition to practices that are more environmentally sound and have the potential for contributing to long-term sustainability for agriculture. Others are starting agricultural enterprises from scratch that incorporate a variety of ecologically informed approaches. Both types of efforts represent "conversion" in the broad sense.

The remarkable growth of organic, alternative, and ecological agriculture in developed countries during the past several decades indicates that a transformation in the way we grow food is already underway. Between 1996 and 2001, the number of acres of organic cropland in the U.S. more than doubled, reaching 1.3 million. During this time, consumer demand for organic products rose about 20% annually, and continues to increase (USDA, 2003). Clearly, a more-sustainable approach to growing food, one that challenges conventional agricultural wisdom, is gaining ground both culturally and economically.

The conversion to ecologically based management is grounded in the principles discussed in the preceding chapters. In this chapter, we discuss how those principles can come into play in the actual process of changing the way food is grown. Farmers engaged in the conversion process know, through intuition, experience, and knowledge, what is unsustainable and what is, at the very least, more sustainable. Nevertheless, there is a clear need to study the process in more detail. This chapter makes a contribution toward that goal by proposing a protocol for converting conventional systems into more-sustainable systems (Figure 20.1). Determining what constitutes sustainability itself is the topic of Chapter 21

## FACTORS PROMOTING CONVERSION

Agriculture is always evolving and adopting new practices. In the 20th century, agriculture responded to a complex of economic and technological pressures that led to the development of the highly specialized and purchased-input-dependent systems that dominate agriculture today. Yield increasing technologies, farm support programs, and research developments helped push agriculture toward fewer larger farms. But some years ago, many farmers began to transition into what today we call "alternative agriculture" (National Research Council, 1989). The adoption of alternative practices has since accelerated, with several factors encouraging farmers to question conventional practices and manage agroecosystems in more sustainable ways.

- The cost of energy has risen dramatically and continues to rise.
- Crops produced with conventional practices have low profit margins.
- New practices with demonstrated potential for success have been and are being developed.
- Environmental awareness among consumers, producers, and regulators is increasing.
- There are new and stronger markets for alternatively grown and processed farm products.

Despite the fact that farmers often suffer a reduction in both yield and profit in the first year or two of the transition period, most of those that persist eventually realize both economic and ecological benefits from having made the conversion. Part of the success of the transition is based on a farmer's ability to adjust the economics of the farm operation to the new relationships of farming with a different set of input and management costs, as well as adjusting to different market systems and prices.

## GUIDING PRINCIPLES FOR CONVERSION

The conversion process can be complex, requiring changes in field practices, day-to-day management of the farming operation, planning, marketing, and philosophy. The following principles can serve as general guidelines for navigating the overall transformation:

- Shift from throughflow nutrient management to recycling of nutrients, with increased dependence on natural processes such as biological nitrogen fixation and mycorrhizal relationships.
- Use renewable sources of energy instead of nonrenewable sources.

**FIGURE 20.1 The experimental farm at the Center for Agroecology and Sustainable Food Systems, University of California, Santa Cruz.** Innovative research on the design and management of sustainable agroecosystems is carried out at this unique facility.

- Eliminate the use of nonrenewable off-farm human inputs that have the potential to harm the environment or the health of farmers, farm workers, or consumers.
- When materials must be added to the system, use naturally occurring materials instead of synthetic, manufactured inputs.
- Manage pests, diseases, and weeds instead of "controlling" them.
- Reestablish the biological relationships that can occur naturally on the farm instead of reducing and simplifying them.
- Make more appropriate matches between cropping patterns and the productive potential and physical limitations of the farm landscape.
- Use a strategy of adapting the biological and genetic potential of agricultural plant and animal species to the ecological conditions of the farm rather than modifying the farm to meet the needs of the crops and animals.
- Value most highly the overall health of the agroecosystem rather than the outcome of a particular crop system or season.
- Emphasize conservation of soil, water, energy, and biological resources.
- Incorporate the idea of long-term sustainability into overall agroecosystem design and management.

The integration of these principles creates a synergism of interactions and relationships on the farm that eventually leads to the development of the properties of sustainable agroecosystems that will be discussed in more detail in Chapter 21. Emphasis on particular principles will vary, but all of them can contribute greatly to the conversion process. We should neither be satisfied with an approach to conversion that only replaces conventional inputs and practices with environmentally benign alternatives; nor should we be satisfied with an approach dictated solely by market demands, or one that does not take into account the economic and social health of agricultural communities. Conversion must be part of ensuring long-term food security for everyone in all parts of the world.

## LEVELS OF CONVERSION

For many farmers, rapid conversion to sustainable agroecosystem design and practice is neither possible nor practical. As a result, many conversion efforts proceed in slower steps toward the ultimate goal of sustainability, or are simply focused on developing food production systems that are somewhat more environmentally sound. From the observed range of conversion efforts, three distinct levels of conversion have been discerned (MacRae et al., 1990; Gliessman, 2004), and a fourth is proposed here. These levels help us describe the steps that farmers actually take in converting from conventional agroecosystems, and they can serve as a map outlining a stepwise, evolutionary conversion process. They are also helpful for categorizing agricultural research as it relates to conversion, and for considering what additional steps might be needed to ensure that the conversion process promotes sustainability in food systems beyond the farm.

### LEVEL 1

*Increase the efficiency of conventional practices in order to reduce the use and consumption of costly, scarce, or environmentally damaging inputs.* The goal of this approach is to use inputs more efficiently so that fewer inputs will be needed and the negative impacts of their use will be reduced as well. This approach has been the primary emphasis of much conventional agricultural research, through which numerous agricultural technologies and practices have been developed. Examples include optimal crop spacing and density, improved machinery, pest monitoring for improved pesticide application, improved timing of operations, and precision farming for optimal fertilizer and water placement. Although these kinds of efforts reduce the negative impacts of conventional agriculture, they do not help break its dependence on external human inputs.

### LEVEL 2

*Substitute conventional inputs and practices with alternative practices.* The goal at this level of conversion is to replace resource-intensive and environment-degrading products and practices with those that are more environmentally benign. Organic farming and biological agriculture research has emphasized such an approach (Figure 20.2). Examples of alternative practices include the use of nitrogen-fixing cover crops and rotations to replace synthetic nitrogen fertilizers, the use of biological control agents rather than pesticides, and the shift to reduced or minimal tillage. At this level, the basic agroecosystem structure is not greatly altered; hence many of the same problems that occur in conventional systems also occur in those with input substitution.

**FIGURE 20.2 An on-farm study of a level-2 conversion process with strawberries on the central coast of California.** Conventional and organic practices are simultaneously compared for at least 3 years.

### Level 3

*Redesign the agroecosystem so that it functions on the basis of a new set of ecological processes.* At this level, overall system design eliminates the root causes of many of the problems that still exist at Levels 1 and 2. Thus rather than finding sounder ways of solving problems, the problems are prevented from arising in the first place. Whole-system conversion studies allow for an understanding of yield-limiting factors in the context of agroecosystem structure and function. Problems are recognized, and thereby, prevented by internal site- and time-specific design and management approaches, instead of by the application of external inputs. An example is the diversification of farm structure and management through the use of rotations, multiple cropping, and agroforestry.

### Level 4

*Reestablish a more direct connection between those who grow the food and those who consume it.* Conversion occurs within a cultural and economic context, and that context must support conversion to more sustainable practices. At a local level, this means consumers value locally grown food and support, with their food dollars, the farmers who are striving to move through conversion levels 1, 2, and 3. This support turns into a kind of "food citizenship" (Chapter 23) and becomes a force for food system change. The more this transformation occurs in communities around the world, the closer we move toward building a new culture and economy of sustainability (Hill, 1998, 2004).

In terms of research, agronomists and other agricultural researchers have done a good job of working on the transition from Level 1 to Level 2, and research on the transition to Level 3 has been underway for some time. Work on the ethics and economics of food system sustainability, however, has only just begun (Freyfogle, 2001). Agroecology provides the basis for the type of research that is needed. And eventually it will help us find answers to larger, more abstract questions, such as what sustainability is and how we will know we have achieved it.

## CONVERSION OF A STRAWBERRY PRODUCTION SYSTEM

The central coast of California, with its Mediterranean climate, is an important strawberry growing region. On approximately 12,100 acres, Monterey and Santa Cruz counties together produced more than $500 million worth of strawberries in 2004, about half of the total California crop. Strawberry production here, as in many other locales, is highly dependent on expensive, energy-intensive, and environmentally harmful off-farm inputs.

For more than 20 years, the Agroecology Research Group at the University of California, Santa Cruz has been carrying out a multifaceted research project centered on studying the process of converting these conventional strawberry production systems into more-sustainable agroecosystems. This project provides evidence that even systems strongly invested in conventional practices can be changed; it also exemplifies the difficulties and barriers inherent in conversion. The year-by-year evolution of the strawberry conversion research project is outlined in Table 20.1.

**TABLE 20.1**
**Chronology of Strawberry Conversion Research Activities[a]**

| Date | Activity or Milestone | Conversion Level |
|---|---|---|
| 1986 | Contact with first farmer in transition | Level 1 to Level 2 |
| 1987–1990 | On-farm comparative conversion study | Level 2 |
| 1990 | First conversion publication, *Calif. Agriculture* 44:4–7. | Level 2 |
| 1990–1995 | Refinement of organic management | Level 2 |
| 1995–1999 | Rotations and crop diversification | Initial level 3 |
| 1996 | Second conversion publication, *Calif. Agriculture* 50:24–31 | Level 2 |
| 1997–1999 | Alternatives to MeBr research projects | Level 2 |
| 1998 | BASIS (Biological Agriculture Systems in Strawberries) work group established | Levels 2 & 3 |
| 1999 | Soil health/crop rotation study initiated | Levels 2 & 3 |
| 2000–2006 | Strawberry agroecosystem health study | Levels 2 & 3 |
| 2002–2003 | Pathogen study, funded by NASGA (N. American Strawberry Growers Assn.) | Levels 2 & 3 |
| 2001–2005 | Poster/oral presentations at Amer. Soc. Agronomy meetings | Level 3 |
| 2003–2006 | Alfalfa trap crop project | Level 3 |
| 2004 | Organic Strawberry Production short course | Levels 2 & 3 |
| 2004–2008 | USDA–Organic Research Initiative project: Integrated network for organic vegetable and strawberry production | Levels 2, 3, & 4 |
| 2005–2006 | Local organic strawberries in UC Santa Cruz dining halls | Level 4 |
| 2006 | California Strawberry Commission and NASGA fund organic rotation system research | Level 3 |

[a] Carried out by the Agroecology Research Group at the University of California, Santa Cruz, CA.

The present system of conventional strawberry production in California can be traced back to the early 1960s, when the soil fumigant methyl bromide (MeBr) was introduced. Until that time, growers treated strawberries as a perennial crop, with each field requiring rotation out of strawberries for several years. Use of methyl bromide allowed growers to manage strawberries as an annual crop, planted year after year on the same piece of land. In the system used since the 1960s, strawberry plants are removed each year following the end of the season in late summer or early fall, then the soil is cultivated and fumigated before being replanted with new plants for the next season. Intensive systems of drip irrigation, plastic mulch, and soil manipulation are required (Figure 20.3).

**FIGURE 20.3 Conventional strawberry field fumigated with methyl bromide near Watsonville, California.** Vaporized MeBr is held under the plastic for several days. Conversion to organic management involves replacing this very toxic and expensive chemical with a variety of alternative inputs and practices.

*Level 1 conversion.* The first efforts related to conversion, carried out before the involvement of the Agroecology Research Group, were focused as much on increasing yields and profitability as on changing the nature of the production system. Extensive research was carried out to discover more effective ways of controlling pests and diseases so that inputs could be reduced and their environmental impacts lessened. For example, different miticides for control of the common pest two-spotted spider mite (*Tetranychus urticae*) were tested with the goal of overcoming the problems of evolving mite resistance to the pesticides, negative impacts on non target organisms, pollution of ground water, persistent residues on harvested berries, and health impacts for farm workers (Sances, 1982).

*Level 2 conversion.* In the early 1980s, as interest in organic food became a potential market force in agriculture and issues of pesticide safety and environmental quality came to the fore, farmers began to respond. It was in this environment that researchers at UC Santa Cruz and a local farmer formed a partnership for conversion. In 1987, this partnership became a comparative strawberry conversion research project. For three years, strawberries were grown in plots using conventional inputs and management side by side with strawberries grown under organic management. In the organic plots, each conventional input or practice was substituted with an organic equivalent. For example, rather than control the two-spotted spider mite with a miticide, beneficial predator mites (*Phytoseiulis persimilis*) were released into the organic plots. Over the three-year conversion period, population levels of the two-spotted spider mites were monitored, releases of the predator carried out, and responses quantified. By the end of the third year of the study, ideal rates and release amounts for the predator — now the norm for the industry — had been worked out (Gliessman et al., 1996).

After the 3-years comparison study, researchers continued to observe changes and the farmer continued to make adjustments in his input use and practices. This was especially true in regard to soilborne diseases. After a few years of organic management, diseases such as *Verticillium dahliae*, a source of root rot, began to occur with greater frequency. The response was to intensify research on input substitution. Initial experiments with mustard biofumigation took place, adjustments in organic fertility management occurred, and mycorrhizal soil inoculants were tested. But the agroecosystem was still basically a monoculture of strawberries, and problems with disease increased.

***Level 3 conversion.*** It was at this point that a whole-system approach began to come into play. Based on the concept that ecosystem stability comes about through the dynamic interaction of all the component parts of the system, the researchers and farmer conceived of ways to design resistance to the problems created by the simplified monoculture. The farmer realized he needed to partially return to the traditional practice of crop rotations that had been used before the appearance of MeBr. The researchers used their knowledge of ecological interactions to redesign the strawberry agroecosystem so that diversity and complexity could help make the rotations more effective, and in some cases, shorter. Testing of these ideas is ongoing. For example, mustard covercrops were tested for their ability to allelopathically reduce weeds and diseases through the release of toxic natural compounds. Broccoli is being tested as a rotation crop since it is not a host for the *Verticillium* disease organism, and broccoli residues incorporated into the soil release biofumigants that reduce the presence of disease organisms (Muramoto et al., 2005).

Rather than rely on biopesticides, which still have to be purchased outside the system and released, the researchers and farmer have undertaken redesign approaches intended to incorporate natural control agents into the system, keeping them present and active on a continuous basis. For example, they tested the idea that refugia for the *P. persimilis* predator mite could be provided, either on remnant strawberry plants or trap-crop rows around the fields. Perhaps the most novel redesign idea is the introduction of rows of alfalfa into the strawberry fields as trap crops for the western tarnished plant bug (*Lygus hesperus*). The pest can cause serious deformation of the strawberry fruit, and because it is a generalist pest, it is very difficult to control through input substitution. By replacing every 25th row in a strawberry field with a row of alfalfa (approximately 3% of the field), and then concentrating control strategies on that row (vacuuming, biopesticide application), it was possible to reduce Lygus damage to acceptable levels (Swezey et al., 2004). The ability of these alfalfa rows to also function as reservoirs of beneficial insects for better natural pest control is now being tested as well (Figure 20.4).

**FIGURE 20.4 Alfalfa rows used as a trap crop for pests and refugia for beneficials in a strawberry agroecosystem.** Such field-scale diversification is an example of level 3 of conversion.

***Level 4 conversion.*** Consumers have become a very important force in the conversion of agroecosystems to more sustainable design and management. The fourth level of conversion made its debut when students at the UC Santa Cruz campus convinced the campus dining service managers to begin integrating local, organic, and fair-trade items — including organic strawberries — into the meal service. There are other indicators that a culture of sustainability is beginning to take shape. Consumers are increasing the demand for organic produce, allowing organic farming to become increasingly important. In the two central coast counties, where so many strawberries are grown, there were a total of 46,775 organic-certified acres in 2004, more than ten times the organic acreage recorded in 1997. The total farm gate revenue from organic farming in these counties was $152 million in 2004, representing a dramatic increase of more than 1400% from 1997 (CDFA, 2004). A parallel increase in organic strawberry production occurred over this same time period, as can be seen in Table 20.2

Despite these positive trends, several sustainability issues are connected with this dramatic growth in strawberry production. For example, soil erosion and nutrient leaching have been observed in organic strawberries planted over a large area. What might be called "level-4 thinking" should include consideration of such issues, as part of a concern for the health of the entire system. And this includes more complex social issues. As can be seen in Table 20.2, the number of organic strawberry producers has recently declined, even as the acreage planted has increased. In addition, since organic strawberries usually require more labor, issues of worker health, safety, and pay equity must be also considered. The research carried out by the Agroecology Research Group over the last couple of decades has laid the foundation for making strawberries a crop for which many of these level-4 issues can be addressed (Table 20.2).

**TABLE 20.2**
**Changes in Organic Strawberry Production in California, 1997 to 2004**

| Year | Number of Organic Producers | Gross Declared Value ($ in Millions) | Number of Organic Producers |
|---|---|---|---|
| 1997 | n/a | n/a | n/a |
| 1998 | 82 | 2.5 | 82 |
| 1999 | 99 | 8.7 | 99 |
| 2000 | 119 | 9.7 | 119 |
| 2001 | 113 | 9.3 | 113 |
| 2002 | 105 | 12.5 | 105 |
| 2003 | 99 | 24.6 | 99 |
| 2004 | n/a | 28.4 | n/a |

[a] Acreage may tend to be an over-estimate since it may also include fallow or unplanted land set aside for future plantings.

*Source:* California Department of Food and Agriculture. 2004. County organic crop value and acreage reports. www.cdfa.ca.gov/is/fveqc/organic.htm

## EVALUATING THE CONVERSION EFFORT

Initially, the conversion to ecologically based agroecosystem management results in an array of ecological changes in the system (Gliessman, 2004). As the use of synthetic agrochemicals is reduced or eliminated, and nutrients and biomass are recycled within the system, agroecosystem structure and function change as well. A range of processes and relationships are transformed, beginning with aspects of basic soil structure, organic matter content, and diversity and activity of soil biota. Eventually major changes also occur in the activity of and relationships among weed, insect, and disease populations, and in the balance between beneficial and pest organisms. Ultimately, nutrient dynamics and cycling, energy use efficiency, and overall system productivity are impacted. Measuring and monitoring these changes during the conversion period helps the farmer evaluate the success of the conversion process, and provides a framework for determining the requirements for sustainability. This kind of evaluation will help convince a larger segment of the agricultural community that conversion to more-sustainable practices is possible and economically feasible.

For a researcher, the study of the process of conversion begins with identifying a study site. This should be a functioning, on-farm, commercial crop production unit whose owner-operator wishes to convert to a recognized alternative type of management, such as certified organic agriculture, and wants to participate in the design and management of the farm system during the conversion process (Gliessman, 2002b, 2004). Such a "farmer-first" approach is considered essential in the search for viable farming practices that eventually have the best chance of being adopted by other farmers.

The amount of time needed to complete the conversion process depends greatly on the type of crop or crops being farmed, the local ecological conditions where the farm is located, and the prior history of management and input use. For short-term annual crops, the time frame might be as short as 3 years, and for perennial crops and animal systems, the time period is probably at least 5 years or longer.

Study of the conversion process involves several levels of data collection and analysis:

1. Examine the changes in ecological factors and processes over time through monitoring and sampling.
2. Observe how yields change with changing practices, inputs, designs, and management.
3. Understand the changes in energy use, labor, and profitability that accompany the above changes.

4. Based on accumulated observations, identify key indicators of sustainability and continue to monitor them well into the future.
5. Identify indicators that are "farmer-friendly" and can be adapted to on-farm, farmer-based monitoring programs, but that are linked to our understanding of ecological sustainability.

Each season, research results, site-specific ecological factors, farmer skill and knowledge, and new techniques and practices can all be examined to determine if any modifications in management practices need to be made to overcome any identified yield-limiting factors. Ecological components of the sustainability of the system become identifiable at this time, and eventually can be combined with an analysis of economic sustainability as well.

The ultimate success of the conversion process will depend on changes in the attitudes, values, choices, and ethics of everyone in the food system. As these changes become manifest, a new culture of sustainability will emerge, encouraging the research and innovation that will move us beyond the mere substitution of inputs and practices to the redesigning of agroecosystems.

## CONVERSION TO ORGANIC APPLE PRODUCTION

Although organically managed agroecosystems may not be completely sustainable, they emphasize more sustainable practices than do conventional systems. Farmers considering converting from conventional to organic production, however, are concerned with more than just the ecological merits of certified organic agriculture. They want to know about the economic consequences of conversion — if they can support their families on the profits from an organic farming operation.

In recognition of such practical concerns, researchers study the conversion process and compare the economic viability of conventional and organic management. In one such study, a team of researchers from the Center for Agroecology and Sustainable Food Systems (CASFS) at the University of California, Santa Cruz analyzed the transition from Level-1 conventional to Level-2 organic management of Granny Smith apples at a farm in Watsonville, California (Swezey et al., 1994). The team monitored the ecological parameters of the transition, including nutrient content of the plants, weed species and abundance, pest damage, and the life cycle of the codling moth, the apple's primary pest. This careful monitoring allowed the team to adjust their management strategies as needed. These strategies included applying organic soil amendments and disrupting the mating cycle of the codling moth through the use of pheromone dispensers that confuse the moths.

The team also tracked economic costs and income over the study period. The organic system used 10% more labor than the conventional system, due to practices such as hand thinning of the apple fruit while immature, and the cost of materials was 17% higher than in the conventional system. However, the organic system produced a higher yield in terms of apple quantity and total apple mass. Overall, the organic system also yielded a higher economic return, due both to the higher harvest yield and to the higher price obtained on the market for premium organic apples.

This study demonstrates the organic production of apples can be profitable, even though the transition from conventional to certified organic takes careful planning and can be labor intensive. Similar studies have refined Level 2 conversion methods, leading to the publication of the first Organic Apple Production Manual for California (Swezey et al., 2000). The only Level 3 components of conversion mentioned in the manual are the use of permanent between-row legume and grass cover crops. The long-term sustainability of organic apple agroecosystems still needs to be addressed (Figure 20.5).

**FIGURE 20.5 Fuji apples on semidwarf rootstock under conversion to organic management, Corralitos, California.**

## FOOD FOR THOUGHT

1. What are some of the forces that are undercutting the long-term ecological sustainability of many traditional farming systems, and how might these forces be counteracted?
2. If you were to take over managing a farm in your community that has a long history of conventional management, what are some of the changes you would make first in order to begin the process of moving the farm to sustainable management?
3. How much time do you think is necessary for converting a farm from nonsustainable to sustainable management? What variables might influence the length of the conversion period?
4. What are some of the incentives that might be provided for farmers who are considering converting their farms to ecologically based management?
5. From an ecological perspective, why is the substitution phase of conversion not enough?

## INTERNET RESOURCES

Alternative Farming Systems Information Center
www.nal.usda.gov/afsic
An excellent source of information on alternative farming systems and practices, especially designed for farmers.

National Sustainable Agriculture Information Service
www.attra.org
A rich source of information especially designed to help small-scale and rural farmers and farm communities.

Sustainable Agriculture Research and Education (SARE)
www.sare.org
A good place to find research results about the transition to sustainable agriculture.

## RECOMMENDED READING

Edwards, C.A., R. Lal., P. Madden, R.H. Miller, and G. House. 1990. *Sustainable Agricultural Systems.* Soil and Water Conservation Society: Ankeny, Iowa. A diverse review of research from around the world on agroecosystems in the context of sustainability.

Filson, G.C. ed. 2004. *Intensive Agriculture and Sustainability: A Farming Systems Analysis.* University of British Columbia Press: Vancouver, Canada. A farming systems analysis for the issues associated with sustainable agriculture, including interactions between social, economic, and ecological indicators of sustainability.

Francis, C.A., C. Butler-Flora, and L.D. King. (eds.). 1990. *Sustainable Agriculture in Temperate Zones.* Wiley & Sons: New York. An in-depth examination of approaches to sustainability in temperate agricultural systems.

Gliessman, S.R. (ed.). 1990. *Agroecology: Researching the Ecological Basis for Sustainable Agriculture.* Springer Verlag Series in Ecological Studies #78: New York. An edited volume on the research approaches in the field of agroecology and sustainability.

Mason, J. 2003. *Sustainable Agriculture.* 2nd ed. Landlinks Press: Collington, Australia. Addresses some of the critical issues facing sustainable agriculture today, from an Australian perspective.

National Research Council. 1989. *Alternative Agriculture.* National Academy Press: Washington, D.C. An excellent review of the roots of the alternative agriculture movement in the U.S., its motivations, and its future.

Röling, N.G. and M.A.E. Wagemakers. 2000. *Facilitating Sustainable Agriculture: Participatory Learning and Adaptive Management in Times of Environmental Uncertainty.* Cambridge University Press: Cambridge, UK. Analyzes the implications of adopting sustainable agricultural practices, both at the farm and the landscape scale, with a focus on social aspects.

Uphoff, N. (ed.) 2002. *Agroecological Innovations: Increasing Food Production With Participatory Development.* Earthscan: London. A presentation of 12 case studies that demonstrate agroecology's potential to produce food in a socially and environmentally viable way

**COLOR PLATE 1** Example of the traditional Mesoamerican intercrop of corn, beans, and squash growing in Tabasco, Mexico. This co-evolved crop combination demonstrates many of the principles of agroecology at the autecological level.

**COLOR PLATE 2** A willow (*Salix* sp.) windbreak used to protect an apple orchard from wind damage near Lincoln, New Zealand. The farmer adds additional biodiversity to the system by managing permanent sod between the trees.

**COLOR PLATE 3** Ripe coffee berries are harvested from an organic, shade-covered agroecosystem in Aguabuena, Costa Rica. Small farms rely primarily on family members to provide the extra labor needed at peak times.

**COLOR PLATE 4** Canals and tree-lined platforms characterize the *Chinampas* of Xochimilco, Mexico. Canal sediments are periodically dredged by hand and used to replenish soil fertility on the platforms, where a diversity of crops are grown.

**COLOR PLATE 5** A small farmer coffee agroecosystem designed and managed for sustainability in Aguabuena, Costa Rica. Shade trees, mulch, grass hedges, and water catchment channels are integrated to maintain a healthy soil and high productivity.

**COLOR PLATE 6** *Oufei* compost being made in waterlogged pits using nearby canal sediments and Chinese milk vetch (*Lupinus* sp.) in Jiangsu, China. This traditional nitrogen-rich fertilizer, in use for many centuries, is rarely used today. Photo courtesy of Erle Ellis.

**COLOR PLATE 7** A home garden in American Samoa based on root and fruit crops. Taro is the principal crop, but papaya, bananas, coconut, and other crops are also grown.

**COLOR PLATE 8** Strawberries produced as part of a diverse rotation with other annual crops in Davenport, California. The cropping system is integrated into a diverse natural landscape.

**COLOR PLATE 9** Intercropped grapes and olives are grown in the traditional head-pruned and dry-farmed manner near New Cuyama, California. The crops form an integrated part of the surrounding landscape.

**COLOR PLATE 10** A diverse village landscape in the Guangdong Province of China. Paddy rice is the primary component in the lowest areas, but divers gardens surround homes, and tea and forest products are produced on the slopes and hills. Photo courtesy of Erle Ellis.

**COLOR PLATE 11** Free-range grassfed dairy cattle on a farm in Denmark. An extensive pasture rotation system is used to provide the majority of feed for these animals on a diverse landscape. Photo courtesy of Tommy Delgaard.

**COLOR PLATE 12** An agricultural landscape in the Imburra region of Ecuador. The communal management of hedgerows planted along irrigation channels is integrated with individual management of cropping fields.

**COLOR PLATE 13** A diverse subsistence farm in Cartegena, Colombia. Annual and short-lived perennial crops provide a diversity of ecological and social benefits for resource-limited farmers. Photo courtesy of Rose Cohen.

**COLOR PLATE 14** Plots used for studying the conversion of conventionally grown strawberries to organic management. In this level-2 study, more-sustainable inputs and practices are substituted for their conventional equivalents, an important step towards the level-three redesigned system in Figure 8.

**COLOR PLATE 15** A diverse landscape in the northern part of Andalucia, Spain. Annual crops, almonds, olives, sheep, and cattle are combined with areas of Mediterranean forest, allowing the landscape to provide an array of environmental services.

**COLOR PLATE 16** A river canyon in Veracruz, Mexico, supports a diversity of managed and natural ecosystems. Coffee, sugarcane, mangoes, avocados, cattle pasture, and annual corn/bean crops are integrated with natural forest ecosystems in this complex landscape.

COLOR PLATE 1

COLOR PLATE 2

COLOR PLATE 3

COLOR PLATE 4

COLOR PLATE 5

COLOR PLATE 6

COLOR PLATE 7

COLOR PLATE 8

COLOR PLATE 9

COLOR PLATE 10

COLOR PLATE 11

COLOR PLATE 12

COLOR PLATE 13

COLOR PLATE 14

COLOR PLATE 15

COLOR PLATE 16

# 21 Indicators of Sustainability

What is a sustainable agroecosystem? We answered this question in the abstract in Chapter 1. A sustainable agroecosystem is one that maintains the resource base upon which it depends, relies on a minimum of artificial inputs from outside the farm system, manages pests and diseases through internal regulating mechanisms, and is able to recover from the disturbances caused by cultivation and harvest.

It is a different matter, however, to point to an actually existing agroecosystem and identify it as sustainable or not and determine why, or to specify exactly how to build a sustainable system in a particular bioregion and sociocultural context. Generating the knowledge and expertise for doing so is one of the main tasks facing the science of agroecology today, and is the subject to which this chapter is devoted.

Ultimately, sustainability is a test of time: an agroecosystem that has continued to be productive and support local livelihoods for a long period of time without degrading its resource base — either locally or elsewhere — can be said to be sustainable. But just what constitutes "a long period of time"? How is it determined if degradation of resources has occurred? What tells us that all the components of the system are healthy and viable? How well integrated are the social and ecological components of sustainability? And how can a sustainable system be designed when the proof of its sustainability remains always in the future?

Despite these challenges, we need to determine what sustainability entails. In short, the task is to identify parameters of sustainability — specific characteristics of agroecosystems that play key parts in agroecosystem function — and to determine at what level or condition, and for how long, these parameters must be maintained for sustainable function to occur. Through this process, we can identify what we will call indicators of sustainability — agroecosystem-specific conditions necessary for and indicative of sustainability. With such knowledge it will be possible to predict whether or not a particular agroecosystem can be sustained over the long term, and to design agroecosystems that have the best chance of proving to be sustainable. This knowledge will also help us work to change the external forces that have kept most agroecosystems from being sustainable in the first place.

## LEARNING FROM EXISTING SUSTAINABLE SYSTEMS

The process of identifying the elements of sustainability begins with two kinds of existing systems: natural ecosystems and traditional agroecosystems. Both have stood the test of time in terms of maintaining productivity over long periods, and each offers a different kind of knowledge foundation. Natural ecosystems provide important reference points, or benchmarks, for understanding the ecological basis of sustainability; traditional agroecosystems offer abundant examples of actually sustainable agricultural practices as well as insights into how social systems — cultural, political, and economic — fit into the sustainability equation. Based on the knowledge gained from these systems, agroecological research can devise principles, practices, and designs that can be applied in converting unsustainable conventional agroecosystems into sustainable ones.

### Natural Ecosystems as Reference Points

As discussed in Chapter 2, natural ecosystems and conventional agroecosystems are very different. Conventional agroecosystems are generally more productive but far less diverse than natural systems. And unlike natural systems, conventional agroecosystems are far from self-sustaining. Their productivity can be maintained only with large additional inputs of energy and materials from external, human-produced sources; otherwise they quickly degrade to a much less productive level. In every respect, these two types of systems are at opposite ends of a spectrum.

The key to sustainability is to find a compromise between the two — a system that models the structure and function of natural ecosystems yet yields a harvest for human use. Such a system is manipulated to a high degree by humans for human ends, and is therefore not *self*-sustaining, but relies on natural processes for maintenance of its productivity. Its resemblance to natural systems allows the system to sustain, over the long term, human appropriation of its biomass without large subsidies of industrial cultural energy and without detrimental effects on the surrounding environment.

**TABLE 21.1**
**Properties of Natural Ecosystems, Sustainable Agroecosystems, and Conventional Agroecosystems**

| | Natural Ecosystems | Sustainable Agroeco-systems[a] | Conventiona l Agroeco-systems[a] |
|---|---|---|---|
| Production (yield) | Low | Low/medium | High |
| Productivity (process) | Medium | Medium/high | Low/medium |
| Diversity | High | Medium | Low |
| Resilience | High | Medium | Low |
| Output stability | Medium | Low/medium | High |
| Flexibility | High | Medium | Low |
| Human displacement of ecological processes | Low | Medium | High |
| Reliance on external human inputs | Low | Medium | High |
| Autonomy | High | High | Low |
| Interdependence | High | High | Low |
| Sustainability | High | High | Low |

[a] Properties given for these systems are most applicable to the farm scale and for the short- to medium-term time frame.

*Source:* Modified from Altieri M.A. 1995b, 2nd ed. Westview Press: Boulder, CO; Gliessman S.R. 2001, *Advances in Agroecology.* CRC Press: Boca Raton, FL.; Odum and Barrett, 2004. Fundamentaly of Ecology. 5th edition, Thomson Brooks/Cole:Belmont, CA.

Table 21.1 compares these three types of systems in terms of several ecological criteria. As the terms in the table indicate, sustainable agroecosystems model the high diversity, resilience, and autonomy of natural ecosystems. Compared to conventional systems, they have somewhat lower and more variable yields, a reflection of the variation that occurs from year to year in nature. These lower yields, however, are usually more than offset by the advantage gained in reduced dependence on external inputs and an accompanying reduction in adverse environmental impacts.

From this comparison, we can derive a general principle: *the greater the structural and functional similarity of an agroecosystem to the natural ecosystems in its biogeographic region, the greater the likelihood that the agroecosystem will be sustainable.* If this principle holds true, then observable and measurable values for a range of natural ecosystem processes, structures, and rates can provide threshold values, or benchmarks, that describe or delineate the ecological potential for the design and management of agroecosystems in a particular area. It is the task of research to determine how close an agroecosystem needs to be to these benchmark values to be sustainable (Gliessman, 2001).

## Traditional Agroecosystems as Examples of Sustainable Function

Throughout much of the rural world today, traditional agricultural practices and knowledge continue to form the basis for much of the primary food production. What distinguishes traditional and indigenous production systems from conventional systems is that the former developed primarily in times or places where inputs other than human labor and local resources were not available, or where alternatives have been found that reduce, eliminate, or replace the energy- and technology-intensive human inputs common to much of present-day conventional agriculture. The knowledge embodied in traditional systems reflects experience gained from past generations, yet continues to develop in the present as the ecological and cultural environment of the people involved go through the continual process of adaptation and change (Wilken, 1988; González Jácome and Del Amo, 1999) (Figure 21.1).

Many traditional farming systems can allow for the satisfaction of local needs while also contributing to food demands on the regional or national level. Production takes place in ways that focus more on the long-term sustainability of the system, rather than solely on maximizing yield and profit. Traditional agroecosystems have been in use for a long time, and during that time have gone through many changes and adaptations. The fact that they still are in use is strong evidence for a social and ecological stability that modern, mechanized systems could well envy (Klee, 1980).

Studies of traditional agroecosystems can contribute greatly to the development of ecologically sound management practices. Indeed, our understanding of sustainability in ecological terms comes mainly from knowledge generated from such study (Altieri, 1999).

What are the characteristics of traditional agroecosystems that make them sustainable? Despite the diversity of these agroecosystems across the globe, we can begin to answer this question by examining what most traditional systems have in common. Traditional agroecosystems:

- do not depend on external purchased inputs
- make extensive use of locally available and renewable resources
- emphasize the recycling of nutrients
- have beneficial or minimal negative impacts on both the on- and off-farm environment
- are adapted to or tolerant of local conditions, rather than dependent on massive alteration or control of the environment
- are able to take advantage of the full range of microenvironmental variation within the cropping system, farm, and region

**FIGURE 21.1 An example of the highly productive traditional corn-based agroecosystem of upland central Mexico.** This system, often integrating trees and crops, has flourished for hundreds of years.

- maximize yield without sacrificing the long-term productive capacity of the entire system and the ability of humans to use its resources optimally
- maintain spatial and temporal diversity and continuity
- conserve biological and cultural diversity
- rely on local crop varieties and often incorporate wild plants and animals
- use production to meet local needs first
- are relatively independent of external economic factors
- are built on the knowledge and culture of local inhabitants.

Traditional practices cannot be transplanted directly into regions of the world where agriculture has already been "modernized," nor can conventional agriculture be converted to fit the traditional mold exactly. Nevertheless, traditional practices and agroecosystems hold important lessons for how modern sustainable agroecosystems should be designed. A sustainable system need not have all the characteristics outlined above, but it must be designed so that all the functions of these characteristics are retained.

If we are to use traditional agroecosystems as a model for designing modern sustainable systems, we must understand the traditional agroecosystems at all levels of their organization, from the individual crop plants or animals in the field to the food production region or beyond. The examples of traditional practices and methods presented throughout this book provide an important starting point for the process of understanding how ecological sustainability is achieved.

Traditional agroecosystems can also provide important lessons about the role that social systems play in sustainability. For an agroecosystem to be sustainable, the cultural and economic systems in which its human participants are embedded must support and encourage sustainable practices and not create pressures that undermine them. The importance of this connection is revealed when formerly sustainable traditional systems undergo changes that make them unsustainable or environmentally destructive. In every case, the underlying cause is some kind of social, cultural, or economic pressure. For example, it is a common occurrence for traditional farmers to shorten fallow periods or increase their herds of grazing animals in response to higher rents or other economic pressures and to have these changes cause soil erosion or reduction in soil fertility. We will devote more attention to the link between social systems and sustainability in Chapter 23 and Chapter 24.

It is essential that traditional agroecosystems be recognized as examples of sophisticated, applied ecological knowledge. Otherwise, the so-called modernization process in agriculture will continue to destroy the time-tested knowledge they embody — knowledge that should serve as a starting point for the conversion to the more sustainable agroecosystems of the future.

## DEFINING AND MEASURING AGRICULTURAL SUSTAINABILITY

If we are concerned about maintaining the productivity of our food production systems over the long term, we need to be able to distinguish between systems that remain temporarily productive because of their high levels of inputs or external subsidies, and those that can remain productive indefinitely. This involves being able to *predict* where a system is headed — how its productivity will change in the future. We can do this through analysis of agroecosystem processes and conditions in the present.

A central question involves how a system's ecological parameters are changing over time. Are the ecological foundations of system productivity being maintained or enhanced, or are they being degraded in some way? An agroecosystem that will someday become unproductive gives us numerous hints of its future condition. Despite continuing to give acceptable yields, its underlying foundation is being destroyed. Its topsoil may be gradually eroding year by year; salts may be accumulating; the diversity of its soil biota may be declining. Inputs of fertilizers and pesticides may mask these signs of degradation, but they are there nonetheless for the farmer or agroecological researcher to detect. In contrast, a sustainable agroecosystem will show no signs of underlying degradation. Its topsoil depth will hold steady or increase; the diversity of its soil biota will remain consistently high.

Equally important is the question of the maintenance of farmer, farm family, and farm community livelihoods. Are the elements of social health and welfare being maintained so that farm families are able to enjoy a dignified, healthy life with opportunities for education, personal growth, and food security? Even if economic returns hold steady in a region, individual farmers may have to leave farming, children may be taken out of school to work on the farm, or local opportunities for employment may be reduced. Reducing the number of crops to meet market requirements or hiring undocumented labor at lower salaries and benefits may mask these signs, and an integrated analysis in necessary to detect them. A sustainable agroecosystem will show health and happiness in all segments of the social fabric of the food system.

In practice, distinguishing between systems that are degrading their foundations and those that are not is not as straightforward as it may seem. A multitude of ecological and social parameters, all interacting, determine sustainability — considering each one independently or relying on only a few may prove misleading. Moreover, some parameters are more critical than others, and the gains in one area may compensate for losses in another. A challenge for agroecological research is to learn how the parameters interact and to determine their relative importance (Gliessman, 1990, 1995, 2001; Giampietro, 2004).

In addition, analysis of agroecosystem sustainability or unsustainability can be applied in a variety of ways. Researchers or farmers may want to do any of the following, alone or in combination:

- Provide evidence of unsustainability on an individual farm in order to motivate changes in the practices on that farm.
- Provide evidence of the unsustainability of conventional practices or systems more generally to argue for changes in agricultural policy or societal values regarding agriculture.
- Predict how long a system can remain productive.
- Prescribe specific ways of averting productive collapse short of complete redesign of the agroecosystem.
- Prescribe ways of converting to a sustainable path through complete agroecosystem redesign.
- Develop supportive and equitable social relationships throughout the system.
- Suggest ways of restoring or regenerating a degraded agroecosystem.

Although these applications of sustainability analysis overlap, each represents a different focus and requires a different kind of research approach.

### Assessment of Soil Health

In Chapter 8 and Chapter 9, we discussed the many ways that farmers can manage soil factors. Depending on a farmer's skill and experience, this management can lead to improvement, degradation, or maintenance of the soil conditions needed to maintain both production and the qualities that promote it.

The overall picture of the condition of the soil — the soil's fitness to support crop growth without degradation — is called *soil health*. This term is used interchangeably with soil quality. The methods that soil scientists have developed to determine soil quality are usually fairly

**TABLE 21.2**
**Indicators of Soil Health**

| Indicator | Best Time to Test | Healthy Condition |
|---|---|---|
| Earthworm presence | Spring or Fall, when soil is moist | >10 worms per cubic ft; many castings and holes in tilled clods |
| Color of organic matter | When soil is moist | Topsoil distinctly darker than subsoil |
| Presence of plant residues | Anytime | Residue apparent on most of soil surface |
| Condition of plant roots | Late spring or during rapid growth | Roots extensively branched, white, extended into subsoil |
| Degree of subsurface compaction | Before tillage or after harvest | A stiff wire goes in easily to 2x plow depth |
| Soil tilth or friability | When soil is moist | Soil crumbles easily, feels spongy when walked on |
| Signs of erosion | After heavy rainfall | No gullies or rills; runoff from fields is clear |
| Water holding capacity | After rainfall during growing season | Soil holds moisture well more than a week w/o signs of drought stress |
| Degree of water infiltration | After rainfall | No ponding or runoff; soil surface does not remain excessively wet |
| PH | At same time each year | Near neutral and appropriate for crop |
| Nutrient holding capacity | At same time each year | N, P, and K trending up, but not into "very high" zone |

*Source*: Adapted from Magdoff F. and H. van Es, 2000, *Building Soils for Better Crops*. Second Edition. Sustainable Agriculture Network: Washington DC.

technical, costly, and laboratory-based. They tell us a great deal about the potential of any particular soil for farming, or the impacts of various farming practices on the soil, but they are impractical for farmers to use regularly. Farmers prefer to describe soil health subjectively and qualitatively, using words related to how the soil looks, feels, and smells. In this way they are able to assess characteristics such as ease of cultivation, water holding capacity, organic matter content, and potential for weed growth. Soil scientists have been able to correlate these subjective determinations with their quantitative analysis of soil quality, and they have developed score cards for assessing soil health on this basis (Magdoff and van Es, 2000).

Table 21.2 offers a fairly comprehensive set of soil health indicators that can be tested easily on the farm. Most are qualitative; only the last two require any testing equipment beyond a stiff wire and a shovel.

## The Productivity Index

One important aspect of sustainability analysis is to use a more holistic basis for analyzing an agroecosystem's most basic process — the production of biomass. Conventional agriculture is concerned with this process in terms of yield. How the harvest output, or *production*, is created is not important as long as the production is as high as possible. For sustainable agroecosystems, however, measurement of production alone is not adequate because the goal is sustainable production. Attention must be paid to the processes that enable production. This means focusing on *productivity* — the set of processes and structures actively chosen and maintained by the farmer to produce the harvest.

From an ecological perspective, productivity is a process in ecosystems that involves the capture of light energy and its transformation into biomass. Ultimately, it is this biomass that supports the processes of sustainable production. In a sustainable agroecosystem, therefore, the goal is to optimize the process of productivity so as to ensure the highest yield possible without causing environmental degradation, rather than to strive for maximum yields at all costs. If the processes of productivity are ecologically sound, sustainable production will follow.

One way of quantifying productivity is to measure the amount of biomass invested in the harvested product in relation to the total amount of standing biomass present in the rest of the system. This is done through the use of the *productivity index*, represented by the following formula:

$$\text{Productivity index (PI)} = \frac{\text{Total biomass accumulated in the system}}{\text{Net primary productivity (NPP)}}$$

The productivity index provides a way of measuring the potential for an agroecosystem to sustainably produce a harvestable yield. It can be a valuable tool in both the design and the evaluation of sustainable agroecosystems. A PI value can be used as an indicator of sustainability if we assume that there is a positive correlation between the return of biomass to an agroecosystem and the system's ability to provide harvestable yield.

The value of the productivity index will vary between a low of 1 for the most extractive annual cropping system, to a high of about 50 in some natural ecosystems, especially ecosystems in the early stages of succession. The higher the PI of a system, the greater its ability to maintain

**FIGURE 21.2 The traditional Chinese home garden agroecosystem, with pond, paddy, and vegetable beds.** The continual return of all forms of organic matter to the agroecosystem maintains a high productivity index.

a certain harvest output. For an intensive annual cropping system, the threshold value for sustainability is 2. At this level, the amount of biomass returned to the system each season is equal to what is removed as yield, which is the same as saying that half of the biomass produced during the season is harvested, and half returned to the system.

NPP does not vary much between system types (it ranges from 0 to 30 t/ha/yr); what really varies from system to system is standing biomass (it ranges between 0 and 800 t/ha). When a larger portion of NPP is allowed to accumulate as biomass or standing crop, the PI increases and so does the ability to harvest biomass without compromising sustainable system functioning. One way of increasing the standing biomass of the system is to combine annuals and perennials in some alternating pattern in time and space (Figure 21.2).

To be able to apply the PI in the most useful manner, we must find answers to a number of questions: How can higher ratios be sustained over time? How is the ratio of the return of biomass to the amount of biomass harvested connected to the process of productivity? What is the relationship between standing crop or biomass in an agroecosystem, and the ability to remove biomass as harvest or yield?

## Ecological Conditions of Sustainable Function

The ecological framework that has been described in this book provides us with a set of ecological parameters that can be studied and monitored over time to assess movement toward or away from sustainability. These parameters include such things as species diversity, organic matter content of the soil, and topsoil depth. For each parameter, agroecological theory suggests a general type of condition or quality that is necessary for sustainable functioning of the system — high diversity, high organic matter content, and thick topsoil. The specific rates, levels, values, and statuses of these parameters that together indicate a condition of sustainability, however, will vary for each agroecosystem because of differences in farm type, resources used, local climate, and other site-specific variables. Each system, therefore, must be studied separately to generate sets of system-specific indicators of sustainability.

The parameters listed in Table 21.3 provide a framework for research focusing on what is required for sustainable function of an agroecosystem from an ecological perspective. Explanations of the role of each parameter in a sustainable system are not provided here. The reader is

**TABLE 21.3**
**Ecological Parameters Related to Agroecosystem Sustainability**

**Characteristics of the Soil Resource**

***Over the long term***

Soil depth, especially that of the topsoil and the organic horizon
Percent of organic matter content in the topsoil and its quality
Bulk density and other measures of compaction in the plow zone
Water infiltration and percolation rates
Salinity and mineral levels
Cation-exchange capacity and pH
Ratios of nutrient levels, particularly C:N

***Over the Short Term***

Annual erosion rates
Efficiency of nutrient uptake
Availability and sources of essential nutrients

***Hydrogeological Factors***

***On-Farm Water Use Efficiency***

Infiltration rates of irrigation water or precipitation
Soil moisture-holding capacity
Rates of erosional losses
Amount of waterlogging, especially in the root zone
Drainage effectiveness
Distribution of soil moisture in relation to plant needs

***Surface Water Flow***

Sedimentation of water courses and nearby wetlands
Agrochemical levels and transport
Surface erosion rates and gully formation
Effectiveness of conservation systems in reducing non–point-source pollution

***Ground Water Quality***

Water movement downward into the soil profile
Leaching of nutrients, especially nitrates
Leaching of pesticides and other contaminants

***Biotic Factors***

***In the Soil***

Total microbial biomass in the soil
Rates of biomass turnover
Diversity of soil microorganisms
Nutrient cycling rates in relation to microbial activity
Amounts of nutrients or biomass stored in different agroecosystem pools
Balance of beneficial to pathogenic microorganisms
Rhizosphere structure and function

***Above the Soil***

Diversity and abundance of pest populations
Degree of resistance to pesticides
Diversity and abundance of natural enemies and beneficials
Niche diversity and overlap
Durability of control strategies
Diversity and abundance of native plants and animals

**Ecosystem-Level Characteristics**

Annual production output
Components of the productivity process
Diversity: structural, functional, vertical, horizontal, temporal
Stability and resistance to change
Resilience and recovery from disturbance
Intensity and origins of external inputs
Sources of energy and efficiency of use
Nutrient cycling efficiency and rates
Population growth rates
Community complexity and interactions

referred to the chapter in which each factor is discussed for more detail on the importance of that factor and how it might be measured.

## Social Conditions of Sustainable Function

Agriculture has been overly focused on the narrow economic goals of raising yields and increasing the returns on investments. When we use the criteria of sustainability, it is clear that the quality of life for the people involved in agriculture must also be taken into account, observed, and monitored over time. Social health, like soil health, is a composite picture of many factors, or parameters. These parameters include physical health and emotional well-being for individuals, and equity, participation, social function, and democratic expression for the family and community.

For each parameter, we can integrate agroecological concepts and social theory grounded in rural sociology to arrive at a general condition that reflects social health. For individuals, we can measure such factors as educational attainment, incidence of drug and alcohol use, and overall physical health. For families and communities, we can assess characteristics such as changes in the number of farms in the area, average income per farm, number of farm-related businesses, and level of participation in farmer networks. As for ecological parameters, social parameters have specific rates, levels, values and relations that together indicate a condition of sustainability; however, due to the great differences in culture, history, relationships, and belief systems, these indicators are more subjective and location-specific. Since the evaluator of sustainability cannot put his or her values on the people or communities being evaluated, participatory approaches to measurement are important (Bacon, 2005).

Some important social and economic parameters related to agroecosystem and regional food system sustainability are listed in Table 21.4. This is not an exhaustive list. The social framework for sustainability is discussed in more detail in Chapter 23 and Chapter 24, and the reader is referred to the recommended

**TABLE 21.4**
**Socioeconomic Parameters Related to Agroecosystem Sustainability**

| **Ecological Economics (Farm Profitability)** |
|---|
| Per-unit production costs and returns |
| Rate of investment in tangible assets and conservation |
| Debt loads and interest rates |
| Variance of economic returns over time |
| Reliance on subsidized inputs or price supports |
| Relative net return to ecologically based practices and investments |
| Off-farm externalities and costs that result from farming practices |
| Income stability and farming practice diversity |
| **The Social and Cultural Environment** |
| Equitability of return to farmer, farm laborer, and consumer |
| Autonomy and dependence on external forces |
| Degree of self-sufficiency and use of local resources |
| Social justice, especially cross-cultural and intergenerational |
| Equitability of involvement in the production process |
| Reproducibility of the farming culture |
| Extent of age, race, and gender empowerment |
| Stability of social organization and activity of social networking |
| Degree of sharing of agrarian values |

readings at the end of both chapters for more depth and information.

## RESEARCH ON SUSTAINABILITY

Research on the sustainability of agroecosystems has grown considerably in the past decade (Masera and López-Ridaura, 2000; Gliessman, 2001; Turner et al., 2003; Zhen and Routray, 2003; Liebig et al., 2004). The principles on which sustainability can be built are well established (and have been discussed in detail in this text), but we are just beginning to generate the more detailed knowledge needed to apply these principles to the design of sustainable systems and the global conversion of agriculture to sustainability.

Much more research still needs to be done because the resources and efforts of agricultural research have long been concentrated on other concerns. Research has focused on maximizing production, studying the component parts of systems, evaluating results based primarily on short-term economic return, answering questions involving immediate production problems, and serving the immediate needs and demands of agriculture as an independent industry (Pretty, 2002). The result has been the development of a high-yielding, industrial agriculture that is experiencing great difficulty responding to concerns about environmental quality, resource conservation, food safety, the quality of rural life, and the sustainability of agriculture itself.

In recent years, however, the emphasis in agriculture has begun to shift from maximizing yields and profit over the short term to valuing the ability to sustain productivity over the long term. Reflecting this shift, the number of university programs, nonprofit organizations, and development projects with a sustainability focus have grown in the past decade. Yet agriculture as a whole is just beginning to respond.

### Using an Agroecological Framework

The emerging agroecological approach permits research to apply an integrated system-level framework concerned with management for the long term (Gliessman, 2001; Rickerl and Francis, 2004). Agroecological research studies the environmental background of the agroecosystem, as well as the complex of processes involved in the maintenance of long-term productivity. It establishes the ecological basis of sustainability in terms of resource use and conservation, including soil, water, genetic resources, and air quality. Then it examines the interactions between the many organisms of the agroecosystem, beginning with interactions at the individual species level and culminating at the ecosystem level as the dynamics of the entire system are revealed.

The ecological concepts and principles on which agroecology is based establish a holistic perspective for the design and management of sustainable agricultural systems. The application of ecological methods is essential for determining (1) if a particular agricultural practice, input, or management decision is sustainable, and (2) what the ecological basis is for the functioning of the chosen management strategy over the long term.

The holistic perspective of agroecology means that instead of focusing research on very limited problems or single variables in a production system, these problems or variables are studied as part of a larger unit. There is little doubt that certain problems require research specialization. But in agroecological studies any necessary narrow focus is placed in the context of the larger system. Impacts that are felt outside of the production unit as a result of a particular management strategy (e.g., a reduction in local biodiversity) can be part of agroecological analysis. This broadening of the research context extends to the social realm as well — the final step in agroecological research is to understand ecological sustainability in the context of social and economic systems.

### Quantification of Sustainability

For agroecological research to contribute to making agriculture more sustainable, it must establish a framework for measuring and quantifying sustainability (Liverman et al., 1988; Gliessman, 2001). We need to be able to assess a particular system to determine how far from sustainability it is, which of its aspects is least sustainable, exactly how its sustainability is being undermined, and how it can be changed to move it toward sustainable functioning. And once a system is designed with the intent of being sustainable, we need to be able to monitor it to determine if sustainable functioning has been achieved.

The methodological tools for accomplishing this task can be borrowed from the science of ecology. Ecology has a well-developed set of methodologies for the quantification of ecosystem characteristics such as nutrient cycling, energy flow, population dynamics, species interactions, and habitat modification. Using these tools, agroecosystem characteristics — and how they are impacted by humans — can be studied from a level as specific as that of an individual species to a level as broad as that of the global environment.

We can also borrow methodological tools from rural and environmental sociologists who have developed a set of methodologies for evaluation of societal characteristics such as access to economic resources, social networks, political or economic status, and empowerment. Using these tools, broader agroecosystem characteristics — and how the are affected by political and economic structures and relationships — can be studied from a level as specific as a household to a level as broad as that of global markets and free trade agreements.

One approach is to analyze specific agroecosystems to quantify at what level a particular ecological or social parameter or set of parameters must be at for sustainable function to occur. Many researchers have begun work in this area, and some of their results are presented in Table 21.5. Even though the results are given individually, it is important to remember that such results must be used and interpreted in the context of the whole system and the complex of interacting factors of which they are only a part.

Another kind of approach is to begin with the whole system. Some researchers, for example, have been working on developing methods for determining the probability of an agroecosystem being sustainable over the long term (Fearnside, 1986; Hansen and Jones, 1996). Using a systems framework for measuring the carrying capacity of a particular landscape, they apply a methodology for integrating the rates of change of a range of parameters of sustainability and determine how quickly change is taking place toward or away from a specific goal. Such an analysis is limited by the difficulty of choosing which parameters to integrate into the model, but has the potential for becoming a tool allowing us to predict if a system will be able to continue indefinitely or not.

**TABLE 21.5**
**Selected Quantifiable Parameters and Their Approximate Minimum Values for Sustainable Function of Specific Agroecosystems**

| Parameter | Minimum Level for Sustainability | Agroecosystem | Ref. |
|---|---|---|---|
| Soil organic matter content | 2.9% | Strawberries in California | (Gliessman, et al., 1996) |
| Spores of the disease Verticillium wilt | Less than 1 spore per 100 grams of soil | Strawberries and vegetables in California | (Koike and Sabbarao, 2000) |
| Input harvest loss ratio for each macronutrient | Net positive balance over time | Mixed arable crops in Costa Rica | (Jansen et al., 1995) |
| Biocide Use Index[a] | Maintain at a level less than 15 | Mixed arable crops in Costa Rica | (Jansen et al., 1995) |
| Ecosystem biophysical capital[b] | GPP - NPP < 1 | Variable | (Giampietro, 2004) |
| Plant species diversity | Shannon index > 5.0 | Perennial pasture | (Risser, 1995) |
| Ratio of renewable energy input to total energy input | Should approach 1 | Mixed crops, forage, and animals in Central Italy | (Tellarini and Caporali, 2000) |
| Ratio of net energy output to total external input[c] | Maintain as far above 1 as possible | Mixed crops, forage, and animals in Central Italy | (Tellarini and Caporali, 2000) |
| Female participation in farm activities | Full acknowledgement of roles and activities | Small-scale traditional farms in NW Ethiopia | (Tsegaya, 1997) |
| Ratio of cost of all local inputs to cost of total inputs[d] | As close to 1 as possible | Mixed field crops in Bangladesh | (Rasul and Thapa, 2003) |

[a] Index based on several factors, including use rates, toxicity, and area sprayed; values above 50 are considered indicative of excessive biocide use.

[b] Defined as the capture of adequate solar energy to sustain cycles of matter in an ecosystem.

[c] An indicator of productivity.

[d] An indicator of input self-sufficiency.

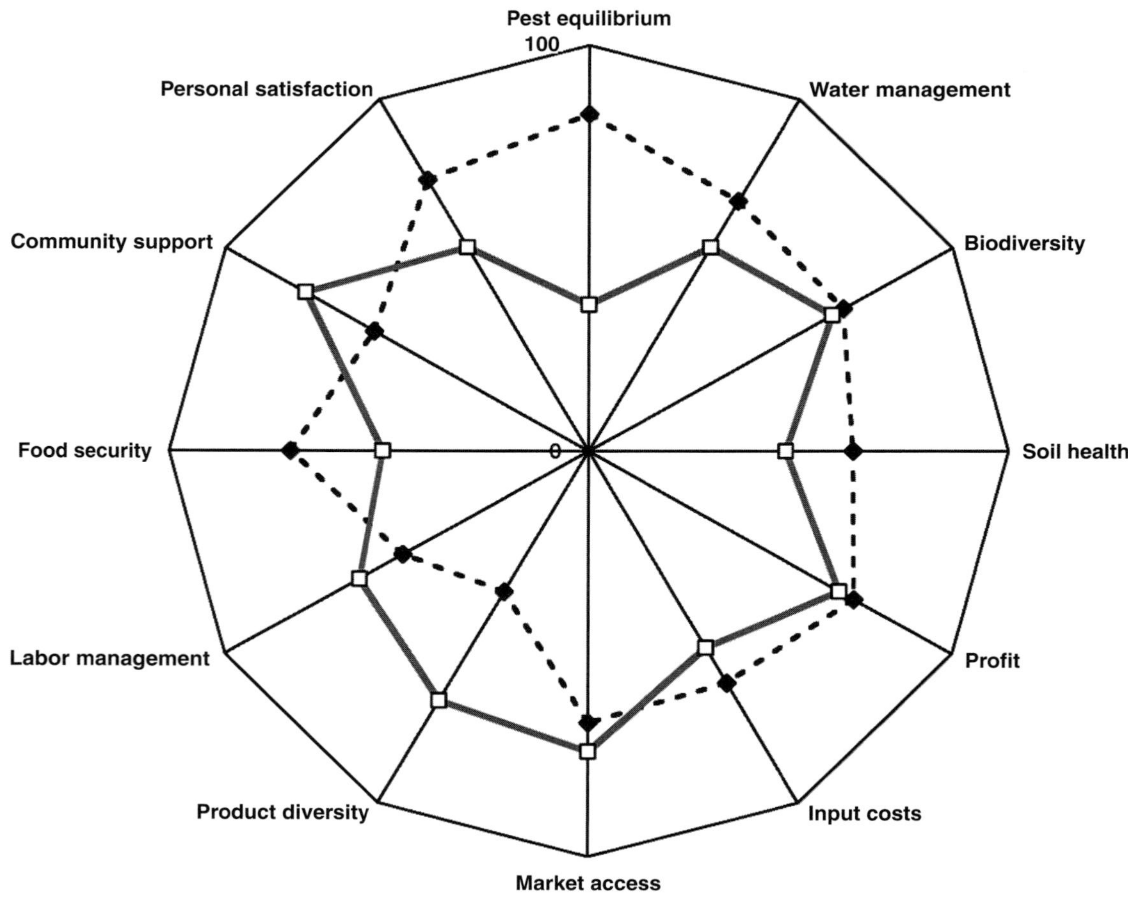


**FIGURE 21.3 An amoeba-type diagram comparing the sustainability of two organic farms in Santa Cruz County, California.** A combination of ecological, economic, and social indicators were used in the analysis.

Comparative analysis of multiple farms or farming systems is yet another means of assessing sustainability. Comparing a broad range of ecological and economic factors derived from the simultaneous study of contrasting farming systems over several years, especially when conventional and alternative practices are involved, will show factor differentiation through time (e.g., Gliessman et al., 1995). Correlating factor levels with crop performance can give indications of sustainability.

Survey instruments such as interviews and questionnaires can also be applied to multiple farms and farmers, with a set of parameters of sustainability being used to gain a bigger picture of the relationship between farm performance and farmer practice. For example, Pretty et al. (2000) carried out a survey of over 208 agroecologically based projects and initiatives in 52 developing countries involving almost 9 million farmers on 28.9 million ha of land in Africa, Asia, and Latin America. By using the promoted agroecologically based practices, yields were increased by 48 to 93% per hectare. The surveys were able to correlate these yield increases over several years with one of four mechanisms: (1) intensification of a single component of a farm system; (2) addition of a new productive element to a farm system; (3) better use of water and land to increase cropping intensity; or (4) introduction of new agroecological elements into farm systems and new locally appropriate crops varieties and animal breeds. The surveys allowed the farmers to tell their own stories of their experiments in sustainability, and give us a set of field-tested indicators.

Another valuable foundation for doing sustainability analysis of farming systems takes advantage of multiple, diverse, and often dispersed data sets that are connected to agriculture in a given region. For example, the New Mainstream Project of the Roots of Change in California (described in Chapter 24) has assembled an immense data set for current California food systems. Information that ranges from water use, number of farms, farmgate production values, number of farmers' markets, and pesticide use has been gathered from multiple agencies and organizations connected to agriculture and food issues. Data sets for some factors come from many years of data gathering and have considerable quantitative validity. They can be used to project forward into the future the kinds of changes that might be needed to offset negative trends, or to promote more sustainable activities, practices, or policies. These data sets, however, are limited in two ways: they have focused on the current food system, not the one we want to move toward, and they are not very well integrated and able to give a full view of how the component parts of a sustainable food system might be assembled.

An approach that can begin to overcome this problem and better integrate the separate parameters of sustainability is the system Marco para la Evaluación de Sistemas de Manejo de Recursos Naturales (MESMIS) for evaluating natural resource management systems, developed by Masera and others in Mexico (Masera and Lopéz-Ridaura, 2000). Indicators are chosen, ideal values for each are determined, and two or more systems are analyzed to determine how close, in percentage terms, each aspect of the system comes to the ideal value set for its indicator. The result is an "amoeba" diagram, or radar graph, like the one shown in Figure 21.3. The assumption is that the greater the percentage of the optimal area covered by an amoeba the higher the level of sustainability of the agroecosystem it represents. Areas of relative strength and weakness can thereby be compared. This system can be used to show how close each indicator is to a theoretically ideal value, offering a measure of progress toward sustainable function. Both qualitative and quantitative measures can be used. In addition, when applied in a

## SUSTAINABILITY IN A CHINESE VILLAGE AGROECOSYSTEM

Although it is often easy to identify processes that are degrading a system, it is much more difficult to determine what processes are necessary for sustainable productivity. Because the term "sustainable" describes a managed system that will maintain productivity over an indefinite period of time, it is difficult to find indicators of sustainability that can be measured over the short term.

One way to look for these indicators is to study systems that have a track record, systems that have sustained constant production of food for human consumption over a long period of time without degrading their ecological foundations. Many types of traditional agriculture around the world meet this requirement, but their relevance for the study of ecological sustainability is limited because their yields are much lower than those of modern systems. Not so for village agroecosystems in the Tai Lake region of China, located in the Yangtze River Delta. Sustained high yields under intensive human management have been documented in this area for more than nine centuries. The suitability of these systems for study of sustainability attracted researcher Erle Ellis in the 1990s.

Since traditional management practices have now in part been supplanted by modern practices, Ellis looked at the history of the region's agriculture, examining a multitude of factors, including landscape features, climate, soils, and human management practices. In order to elucidate the ecological mechanisms underlying the sustainability of the area's agriculture, Ellis studied the cycling of nutrients at the level of an entire village. This scale of study allowed him to compensate for the variability that exists between the practices of individual farmers and the variability of the landscape, and thereby draw more accurate conclusions. It also enabled him to discern overall processes that might be invisible at the field level.

With evidence suggesting that nitrogen was the limiting factor in traditional Chinese agroecosystems, Ellis made the cycling and management of this nutrient the focus of his research. He identified the specific practices and natural processes in the system that historically maintained adequate levels of soil nitrogen in the absence of inputs of inorganic fertilizer.

Ellis identified several aspects of traditional management practices that he believes were essential in maintaining nitrogen fertility (Ellis and Wang, 1997). One of the most important of these was the use of natural inputs, such as sediments from local waterways. Biological nitrogen fixation also played a significant role. A third factor, perhaps the most important, was the thorough recycling of organic matter. Nearly all organic wastes — including human excrement — were recycled in the village system, either by being returned to the fields directly or composted and then returned. Another important contributor to sustainability was the integration of animals to create a cyclical nutrient flow: farmers raised pigs specifically for their manure, and a portion of the animals' diet was food and agricultural waste.

Although these practices continue, they have been largely replaced by the application of inorganic fertilizers. This change, initiated in the 1960s, has made nitrogen a problematic source of pollution instead of a limiting nutrient. Although the use of inorganic inputs has boosted productivity even higher, feeding an ever-growing population, this change in management makes the continued sustainability of the region's agricultural systems an open question.

participatory action research setting, the farm community is a partner in selecting the galaxy of indicators that are of greatest concern or most interest for the comparison. The results are highly variable in terms of which factors are chosen to make the multiple axes, but this method enables a simple, yet comprehensive comparison of the systems being evaluated.

The most advanced, and probably the most complex analysis of indicators of agroecosystem sustainability is the one developed by Giampietro (2004). He employs all of the methodologies described throughout this chapter, and then some, to create what he calls Multi-Scale Integrated Analysis. This methodology applies complex system theory, integrates diverse components that cut across the ecological and social realms, and takes into account change through time over different scales. The results are layered with multiple scales of uncertainty, change, location, and cultural preference. Overall, this methodology calls attention to the need to move beyond the reductionist tendency of looking at single factors affecting sustainability.

### Moving to a Larger Context

An agroecological approach is more than just ecology applied to agriculture. It needs to take on a cultural perspective as it expands to include humans and their impacts on agricultural environments. Agricultural systems develop as a result of the coevolution that occurs between culture and environment, and a truly sustainable agriculture values the human as well as the ecological components, and the interdependence that can develop between the two.

One of the weaknesses of conventional agricultural research is the way in which the narrowness of its focus on production problems has ignored the social and economic impacts of agricultural modernization. Agroecological research cannot make the same mistake. In addition to paying greater attention to the ecological foundation upon which agriculture ultimately depends, agroecological research must understand agriculture within its social context. Understanding agroecosystems as social–ecological systems will permit the evaluation of such qualities of agroecosystems as the long-term effects of different input/output strategies, the importance of the human element to production, and the relationship between economic and ecological components of sustainable agroecosystem management. Ultimately, sustainability is a whole-system, interdisciplinary concept, the highest-order emergent quality of an agroecosystem.

## FOOD FOR THOUGHT

1. In the context of sustainability, what are the differences between the concepts of ecosystem persistence (or resistance) and ecosystem resilience?
2. Describe a characteristic or component of a traditional farming system that would find widespread application in conventional farming systems if sustainability were a primary goal.
3. How might cultural preferences for different kinds of foods affect the choice of appropriate indicators of sustainability?
4. Describe how, as an agroecosystem moves toward sustainability, some components might stay the same while others might change.
5. What is the role of the consumer as an indicator of sustainability?
6. Why are ecological indicators generally easier to measure than social indicators?

## INTERNET RESOURCES

International Institute for Sustainable Development, Measurement and Assessment Initiative
www.iisd.org/measure
A very useful website on sustainability indicators in the social, economic and ecological realms. The objective of the Measurement and Assessment initiative is to facilitate the development of robust sets of indicators for public- and private-sector decision makers to measure progress toward sustainable development and to build an international consensus to promote their systematic use in assessment, reporting, and planning.

International Sustainability Indicators Network
www.sustainabilityindicators.org
A member-driven organization that provides people working on sustainability indicators with a method of communicating with and learning from each other.

Scientific Committee on Problems of the Environment
www.icsu-scope.org
An interdisciplinary group of social and natural scientists addressing current environmental programs, including the development and use of social and ecological indicators.

Sustainable Development Gateway
sdgateway.net/default.htm
A true gateway to a wealth of information relating to sustainable development and sustainability science.

Sustainable Measures
www.sustainablemeasures.com
A web site on indicators of sustainable community: ways to measure how well a community is meeting the needs and expectations of its present and future members.

Sustainability Science
sust.harvard.edu
This is a site created by an interdisciplinary, multi-institution group engaged in the development and application of the emerging field of sustainability science.

## RECOMMENDED READING


Edwards, C.A., R. Lal, P. Madden, R.H. Miller, and G. House, 1990. Sustainable Agricultural Systems. Soil and Water Conservation Society: Ankeny, Ioma. A diverse review of research from around the world on agroecosystems in the context of sustainability.

Filson, G.C. (ed.) 2004. *Intensive Agriculture and Sustainability: A Farming Systems Analysis.* University of British Columbia Press: Vancouver, Canada. A farming systems analysis for the issues associated with sustainable agriculture, including interactions between social, economic, and ecological indicators of sustainability.

Flora, C., (ed.) 2001. Interactions Between Agroecosystems and Rural Communities. CRC Press: Boca Raton, FL, U.S.A. A book addressing the relationship between sustainable agriculture and the human communities that depend on it, with temperate and tropical examples.

Francis, C.A., C. Butler-Flora, and L.D. King. (eds.) 1990. *Sustainable Agriculture in Temperate Zones.* Wiley & Sons: New York. An in-depth examination of approaches to sustainability in temperate agricultural systems.

Giampietro, M. 2004. *Multi-Scale Integrated Analysis of Agroecosystems.* CRC Press: Boca Raton, FL. A challenging look at the need for a holistic approach to the study of agroecosystem sustainability, employing multicriteria analysis and systems theory.

Gliessman, S.R. (ed.) 2001. *Agroecosystem Sustainability: Towards Practical Strategies.* Advances in Agroecology series. CRC Press: Boca Raton, FL. An exploration of the ecological foundation of agroecosystem sustainability, with case studies that provide practical ways to increase, improve, and assess the integration of the social and ecological parameters needed in sustainability analysis.

Magdoff, F. and J. van Es. 2000. Building Soils for Better Crops. Sustainable Agriculture Networks: Beltsville, MD. A practical guide to building and maintaining healthy soils that combines the scientific understanding of soil quality with the practitioner's goal of keeping a soil healthy.

Marten, G.G. 1986. *Traditional Agriculture in Southeast Asia: A Human Ecology Perspective.* Westview Press: Boulder, Colorado. A cultural ecology approach is used to examine the value of traditional agriculture and the needs of such systems in the future.

Mason, J. 2003. *Sustainable Agriculture.* Second Edition. Landlinks Press: Collington, Australia. A volume addressing some of the critical issues facing sustainable agriculture today, from an Australian perspective.

Munasinghe, M. and W. Shearer (eds.) 1995. *Defining and Measuring Sustainability: The Biogeophysical Foundations.* The World Bank: Washington, D.C. A very useful compilation of conference proceedings that explores the biogeophysical foundations for defining and measuring sustainability; provides policy considerations for the international development community.

Pound, B., S. Snapp, S. McDougall, and Braun, A. (eds.) 2003. *Managing Natural Resources for Sustainable Livelihoods: Uniting Science and Participation.* Earthscan: London. A volume demonstrating the need participatory research approaches for natural resources management, which are based on the needs and knowledge of local people.

Schjonning, P., S. Elmholt, and B.T. Christensen, (eds.). 2004. *Managing Soil Quality: Challenges in Modern Agriculture.* CABI Publishing: Cambridge MA. A book that discusses key issues in soil quality and management in our search for more sustainable modern agricultural systems.

Uphoff, N. (ed.) 2001. *Agroecological Innovations: Increasing Food Production With Participatory Development.* Earthscan: London. A presentation of twelve case studies that demonstrate agroecology's potential to produce food in a socially and environmentally viable way.

# 22 Landscape Diversity and Agroecosystem Management

Since the beginning of agriculture, agroecosystems have been altering and displacing naturally occurring terrestrial ecosystems across the face of the earth. The ongoing process of converting land to agricultural production has had a dramatic and usually negative impact on the diversity of organisms and the integrity of ecological processes. Although other forms of human exploitation of the environment, such as urbanization and mining, have also contributed to large-scale habitat modification and the loss of biodiversity and ecosystem function, agricultural production — including grazing and timber production — bears much of the responsibility for causing environmental changes at the biosphere scale that threaten the world's life-support systems.

One of the major goals of developing a sustainable agriculture is to reverse this legacy of destruction and neglect, to conserve biotic resources and protect environmental quality. Indeed, this goal is built into our definition of agricultural sustainability. More-sustainable agroecosystems — more diverse, relying less on external inputs and intensive modification of the environment — will, by their very nature, be more environmentally friendly.

However, a variety of important management principles come to light when we focus on the relationship between agroecosystems and natural ecosystems. In particular, we find that crops and farms can benefit as much as natural ecosystems when we design and manage agroecosystems with natural habitats, native species, and regional ecological processes in mind. Carrying out agricultural production so that it works with, rather than against, natural ecosystem processes is necessary not just for the sake of the natural environmental itself, but for the long-term welfare of human society. We depend on healthy, functioning ecosystems to moderate weather extremes, cycle nutrients, protect riverbanks from erosion, filter our drinking water, detoxify our waste water, generate new soil, pollinate crops, reduce the impacts of droughts and floods, and provide us with a variety of other *ecosystem services*. By replacing most of the earth's natural environments with systems managed for agricultural and timber production, we have seriously threatened the foundations of these ecosystem services. From a sustainability perspective, therefore, we must work toward two goals: (1) protect remaining natural environments, ecosystems, and biodiversity from the effects of our intensive management, and (2) design and manage agroecosystems so that they can function as providers of ecosystem services themselves.

## THE AGRICULTURAL LANDSCAPE

Developing agroecosystems that protect and enhance biotic diversity and ecological processes — and in turn derive benefits from the natural environment — requires a shift of perspective to the regional or landscape level. So first we will examine the basic aspects of the agricultural landscape.

Agricultural development within a formerly natural environment tends to result in a heterogeneous mosaic of varying types of habitat patches spread across the landscape. The bulk of the land may be intensely managed and frequently disturbed for the purposes of agricultural production, but certain parts (wetlands, riparian corridors, and hillocks) may be left in a relatively natural condition, and other parts (borders between fields, areas around buildings, roadsides, and strips between fields and adjacent natural areas) may occasionally be disturbed but not intensely managed. In addition, natural ecosystems may surround or border areas in which agricultural production dominates.

Although the level of human influence on the land varies on a continuum from intense disturbance and management to relatively pristine wildness, we can divide this continuum into three sections to derive three basic kinds of components of the agricultural landscape:

1. Areas of agricultural production. Intensely managed and regularly disturbed, these areas are made up mainly of nonnative, domesticated plant species.
2. Areas of moderate or reduced human influence. This intermediate category includes pastureland, forests managed for timber production, hedgerows and other border areas, and agroforestry systems. These areas are typically made up of some mixture of native and nonnative plant species and are able to serve as habitat for many native animal species.
3. Natural areas. These areas retain some resemblance of the original ecosystem structure and species composition naturally present in the location, although they may be small in size, contain some nonnative species, and be subject to some human disturbance.

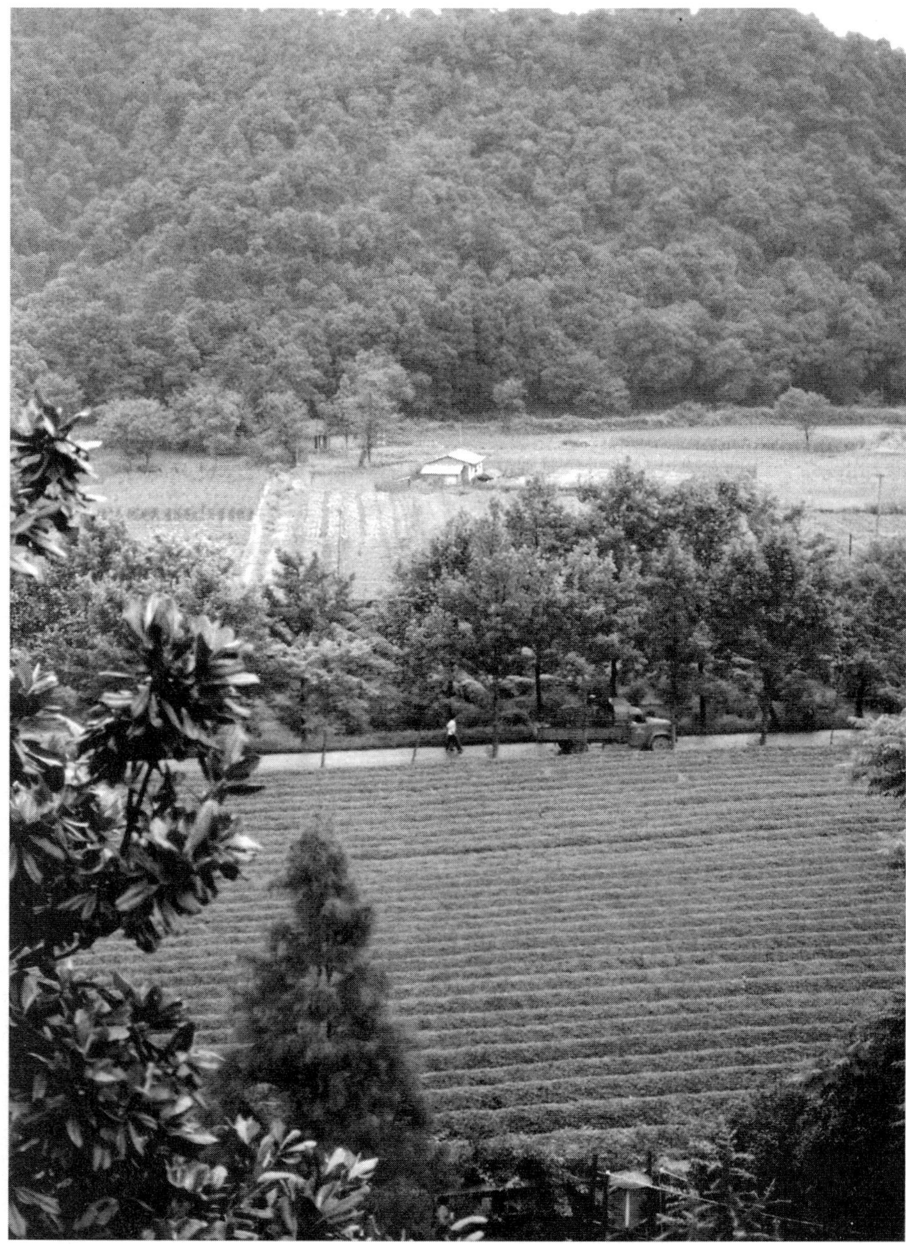

**FIGURE 22.1 A diverse agricultural landscape near Nanjing, China.** Natural ecosystems interface with a variety of human land use activities in an agricultural setting.

These three landscape components, in various combinations and arrangements, form the mosaic pattern of the typical agricultural landscape.

## Landscape Patterns

Within the landscape mosaic, there are three common, recognizable patterns in how the three components are arranged in relation to each other: (1) a natural area and an area managed for agricultural production are separated by an area of moderate or reduced human influence; (2) natural areas form strips, corridors, or patches within an area of agricultural production; and (3) areas of moderate or reduced human influence are dispersed within an area of agricultural production. These three patterns, illustrated in Figure 22.2, can be combined and arranged in many different ways.

An important variable in the mosaic patterning of the agricultural landscape is its degree of heterogeneity or diversity. Landscapes are relatively homogenous when areas of agricultural production predominate, unbroken by patches or strips of the other two kinds of landscape components. Heterogeneous landscapes, in contrast, have an abundance of noncrop and natural patches.

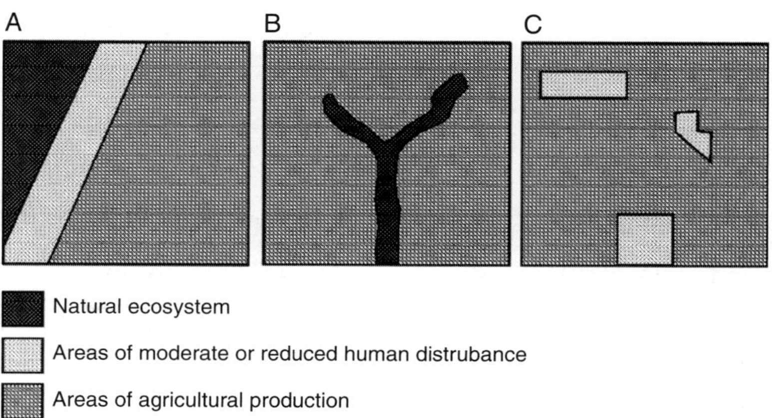


**FIGURE 22.2 Examples of three common patterns in the arrangement of the components of the agricultural landscape.** A natural ecosystem and an agroecosystem can be separated by an area of intermediate human influence (a); a natural ecosystem can form a corridor, strip, or patch within an agroecosystem (b); and areas of less-intense human management can be dispersed within a larger area of agricultural production (c).

The heterogeneity of the agricultural landscape varies greatly by region. In some parts of the world (e.g., the Midwestern U.S.), the heavy use of agricultural chemicals, mechanical technology, narrow genetic lines, and irrigation over large areas have made the landscape relatively homogenous. In such areas, the agricultural landscape is made up mostly of large areas of single-crop agricultural production. In other areas (e.g., the Jiangsu Province of the Yangtze in China), the use of traditional farming practices with minimal industrial inputs has resulted in a varied, highly heterogeneous landscape — possibly even more heterogeneous than would exist naturally.

The typical agricultural landscape, because of its mosaic makeup, is ecologically a fragmented environment. Each patch is a fragment, isolated from other similar patches by some other type of ecologically dissimilar community. On one hand, this fragmentation can have negative effects on populations restricted to a particular type of habitat. On the other hand, a fragmented, heterogeneous landscape has high gamma diversity. As we will explore in the next section, effective management at the level of the landscape involves enhancing gamma diversity and taking advantage of its benefits, while at the same time mitigating the possible negative consequences of habitat fragmentation.

### Analyzing the Landscape

At the landscape level, the movement of organisms and substances between habitat patches becomes a critical factor in the maintenance of overall ecological processes. Also important is the interaction of organisms and physical processes located in different habitat patches. What happens in one area of the landscape can have an impact on other areas. The study of these factors, and how they are shaped by the spatial patterning of the landscape, is known as *landscape ecology.* Because it helps us understand how the different parts of the landscape mosaic are formed and how they interact, landscape ecology provides a good basis for management of the agricultural landscape (Turner et al., 2001; Odum and Barrett, 2004).

Three important tools of landscape ecology are aerial photography, satellite imagery, and geographic information system (GIS) analysis. Using these tools, present landscape patterns can be contrasted with those that were observed in the past. The changes that have occurred can then be correlated with farming systems data to understand the role of agroecosystems in maintaining the stability and sustainability of landscape systems, which provides a basis for designing management schemes that take into account all landscape elements (Ellis, 2004).

Any form of historical data on landscape patterns can be useful in analyzing the agricultural landscape. Census data, such as that from the U.S. Census of Agriculture, can be particularly important in determining the types of crops that have been grown in a region and where they were grown. These data can be given quantifiable values when combined with aerial photographs, allowing the analyst to determine the number of landscape elements present at different times (e.g., crop fields, pastures, riparian corridors, and forest patches). When these data are subjected to GIS analysis, they can become a dynamic way of visualizing the patterns and relationships of landscape structure through time.

For example, the GIS images in Figure 22.3 show changes that have occurred over several decades in an agricultural region of Guangdong Province in China. As this region underwent a shift from a primarily agricultural economy to a more industrialized economy, agricultural land underwent significant change. Through a combination of forest recovery, planted forestry, and the development of orchard crops, woody vegetation recovered, and

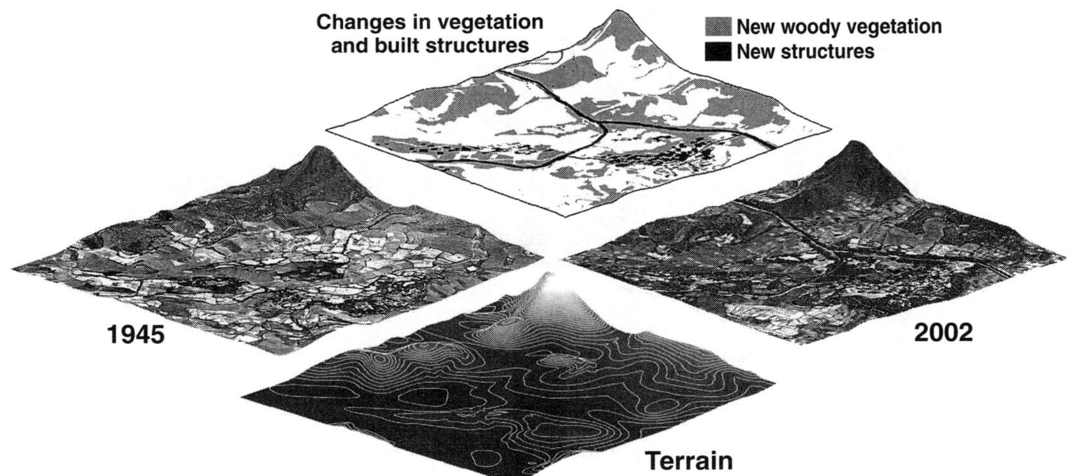


**FIGURE 22.3 GIS analysis of a 1 km² area in western Guangdong Province, China (Dianbai County) showing changes in the agricultural landscape over time.** In the transition from a traditional to a more industrialized economy, built structures increased, much agricultural land was abandoned, and woody vegetation — previously been burned and harvested for fuel — recovered in formerly agricultural areas and in the hills. Images and data courtesy of Erle C. Ellis (www.ecotope.org for more information).

in many formerly agricultural lands, built structures increased. As the images in Figure 22.3 indicate, multiple layers of data that vary in content and time can be integrated to understand the drivers and consequences of changes such as these.

Knowledge of the farming practices that have been used in the past in any particular landscape, combined with knowledge of how different components of the landscape interact, makes it possible to understand how farming practices impact the nonfarm elements of a landscape, and vice versa. Soil erosion rates, fertilizer inputs, pesticide applications, irrigation, crop types and diversity, and other practices and processes can be understood in terms of landscape patterns. Based on this knowledge, recommendations for change in either cropping patterns or farming practices can be made, and decisions on agroecosystem design can move beyond the farm and into the larger landscape context.

## MANAGEMENT AT THE LEVEL OF THE LANDSCAPE

When agroecosystem management is carried out at the level of the larger agricultural landscape, the antagonism that so often exists between the interests of natural ecosystems and those of managed production systems can be replaced by a relationship of mutual benefit. Natural and seminatural ecosystem patches included in the landscape can become a resource for agroecosystems, and agroecosystems can begin to assume a positive rather than negative role in preserving the integrity of natural ecosystems.

The concept of landscape-level management does not necessarily mean coordinated management among the many different stakeholders in an agricultural area (different farmers, governmental agencies, conservation interests, etc.). Its essence is the inclusion of natural ecosystems and local biodiversity in management decisions and planning. Thus, landscape-level management can be implemented by an individual farmer who has direct control over only a small part of the agricultural landscape of a region.

The implementation of landscape-level management has two guiding principles:

1. Diversify the agricultural landscape by increasing the density, size, abundance, variety of noncrop habitat patches, and by creating more connections between them. These patches can vary in their level of disturbance and "naturalness;" what they share in common is the ability to be sites where natural ecological processes can occur and where native or beneficial plant and animal species can find suitable habitat.
2. Manage cropping areas to reduce their negative impacts on the natural environment and maximize their value as habitat for native species. This means eliminating or reducing the use of pesticides, inorganic fertilizer, and irrigation, and finding alternatives to farming practices that interfere with ecosystem processes, such as frequent tilling, leaving fields without soil cover for long periods, planting large-scale monocultures, and mowing or spraying roadsides and ditches.

The latter principle goes hand-in-hand with everything discussed in this text up to this point. Reducing nonfarm inputs, relying on biological controls, diversifying cropping systems, allowing successional processes to proceed further — all these practices contribute to creating more environmentally friendly agroecosystems. Assuming this agroecologically based management, we will focus first on the first principle — diversifying the agricultural landscape — and then later in the section, explore the ways that the alternative management described in the second principle can enhance the ability of the landscape to provide environmental services.

The noncrop habitat patches in a diverse agricultural landscape can interact with areas of agricultural production in a variety of ways. An area of noncrop habitat adjacent to a crop field, for example, can harbor populations of a native parasitic wasp species that can move into the field and parasitize a pest. A riparian corridor vegetated by native plant species provides an example of a more complex relationship: the corridor can filter out dissolved fertilizer nutrients leaching from crop fields, promote the presence of beneficial species, and allow the movement of native animal species into and through the agricultural components of the landscape.

As can be seen in these examples, landscape-level diversification offers benefits to both native species and agroecosystems. When diversification is carefully planned and managed, these benefits can be maximized, and the possible negative effects minimized. Effective landscape-level management is thus an important part of achieving sustainability.

### On-Farm Diversification

The farmer can actively encourage and maintain the presence of native species on the intensively managed areas of the farm by establishing and protecting appropriate habitats (Jackson and Jackson, 2002). These habitats can be within the farm fields, between fields, along roadways, in ditches, along property lines, or at the boundary separating farm fields from housing areas. The habitats can be permanent strips or blocks planted to diverse noncrop perennials, or temporary patches within the farm fields. Methods of creating such habitats include the following:

- Plant a cover crop during the winter months. The crop may provide critical food or cover for a range of animal species, especially ground nesting birds.
- Leave strips of unharvested crops such as corn or wheat; these can provide resources for native animal species.
- Where erosion control is necessary on a farm, plant grassed waterways to enhance diversity and achieve important environmental protection goals.
- On terraced hillsides, plant perennial grasses or shrubs on the walls separating the terraces.
- Plant perennials on land that is marginal or susceptible to erosion, or restore this land to a more natural state by allowing natural succession of native species.
- Restore poorly drained or semipermanent wetland sites on the farm to natural wetlands.
- Retain native trees in and around fields as nesting, perch, and hunting sites for native birds.
- Provide artificial perches for native raptors, and bird boxes for other potentially beneficial bird species.

### LANDSCAPE DIVERSITY IN TLAXCALA, MEXICO

In Tlaxcala, Mexico, rain comes in periodic heavy bursts capable of causing severe erosion. In addition, many local farmers must grow their food on steep, erosion-prone slopes. To deal with this situation, they cultivate hillside terrace systems that not only prevent soil erosion, but also effectively conserve rainfall runoff and provide the basis for exceptional landscape diversity (Figure 22.4). These systems, which make use of water- and sediment-trapping catchment basins called *cajetes*, have enabled traditional farmers in this region to maintain the integrity and fertility of the soil for centuries without relying on imported, commercially produced inputs such as fertilizers (Mountjoy and Gliessman, 1988).

The high degree of landscape diversity in the Tlaxcala terrace systems comes from having a large amount of permanent border space between cultivated terraces covered in natural vegetation. The border areas occupy the edges of the terraces, above and below the cajetes. They are vegetated with a highly diverse mixture of perennials, trees, and weeds, achieved by allowing natural succession to occur. The plants in the borders help cycle nutrients, prevent erosion, and provide habitat for beneficial organisms. Wild relatives of the crop plants often flourish in the border areas also, providing a potential source of gene flow that may help the crops maintain their hardiness and resistance.

Because the terraces are long and narrow, no crop plant is ever more than 6.5 m from a field border. Approximately 30% of the farming landscape is made up of border vegetation, while at any one time about 60% or less of the land is being farmed and 10% or more left fallow. By all measures, these hillside systems are very diverse, and designed to take full advantage of all that landscape-level diversity has to offer.

**FIGURE 22.4 Borders of native perennials and trees alongside cultivated terraces, Tlaxcala, Mexico.** Strips of mostly natural vegetation are prominent and ecologically important components of the agricultural landscape in this hilly farming region. Note the animals grazing the border edge, and corn stalks stacked for future use as feed.

In a highly modified agricultural landscape where very little if any of the natural habitat is left, all of these kinds of measures can be important for restoring the landscape's biodiversity and its ability to provide ecosystem services.

## Farm Borders and Edges

Where relatively extensive nonfarmed natural ecosystems exist around and within the agricultural landscape, the shared boundary, or interface, between these areas and those managed for agricultural production takes on an important ecological significance. This is especially true in regions where considerable topographic, geologic, and microclimatic variability existed before agricultural conversion. Depending on management history, these borders and edges can be abrupt and sharply defined or broad and ill defined. When there is a gradual transition between a crop area and natural vegetation (as occurs, for example, between a shade-tree–covered cacao plantation and the surrounding natural forest), an ecotone is created. Such transitional zones are often recognized as distinct habitats of their own, able to support unique mixtures of species. In many situations, they are made up of successional species from both the natural ecosystem and the manipulated agroecosystem.

### Creating Benefits for the Agroecosystem

Edges that are ecotonal in nature, even if they are relatively narrow, can play important roles in an agricultural landscape. Because the environmental conditions existing within the edge are transitional between the farm habitat and the natural habitat, species from both can occur there together, along with other species that actually prefer the intermediate conditions. Very often the variety and density of life is greatest in the habitat of the edge or ecotone, a phenomenon that has been called the *edge effect*. Edge effect is influenced by the amount of edge available, with length, width, and degree of contrast between adjoining habitats all being determining factors.

Benefits of the edge habitat for cropping systems are becoming more well known. In a thorough review of the topic of the influence of adjacent habitats on insect populations in crop fields, Altieri and Nicholls (2004b) suggest that edges are important habitats for the propagation and protection of a wide range of natural biological control agents of agricultural pest organisms. Some beneficial organisms are not attracted to or able to survive long in the disturbed environment of the crop field, especially those where pesticides are applied; they choose instead to move back and forth from the edge to the farm fields, using the fields mainly for feeding or egg laying. Other beneficials depend on alternate hosts in the edge system to survive times when the agricultural fields do not have populations of their primary host such as during a dry season or when the crop is not present. As we learn more about the conditions needed in edge areas to ensure diverse and effective populations of beneficial organisms, actual management of these transitional areas can become part of the landscape management practices (Figure 22.5)

The management of edges will depend in part on determining their appropriate spatial relationship with farmed areas. What is the ideal proportion of edge habitat area to crop area? How close to the edge habitat does a crop plant need to be for it to benefit from edge-dependent beneficials? Can intermediate habitats such as flowering plant corridors effectively extend edges into a crop area? Such issues will need to be addressed to optimize benefits for the agroecosystem and to enhance regional biodiversity.

### *Protecting Adjacent Natural Ecosystems*

If we shift our perspective to the health of the natural ecosystems on the other side of the edge from the farm fields, the edge can be seen to function as a *buffer zone* that protects the natural system from the potential negative impacts of farming, forestry, or grazing. As a buffer, the edge modifies the wind flow, moisture levels, temperature, and solar radiation characteristic of the farm field so that these environmental conditions do not have as great an impact on the adjacent natural ecosystem (Laurance et al., 2002). This modification is especially important for species that live in the understory of forest vegetation; an abrupt edge might allow wind, heat, and stronger light to penetrate into the forest and disrupt species composition.

Buffer zones can serve other important roles as well. For example, they can prevent fire from moving from the open habitat of the cropping system into the natural ecosystem. Such protection is especially important in areas where fire is used to burn slash left from shifting cultivation practices.

Studies on the central coast of California have demonstrated how buffer zones can effectively mitigatethe impacts of agriculture on the adjacent natural environment (Brown, 1992; Los Huertos, 1999). At and around the study site, hills with highly erosion- and leaching-prone soils slope down to fingers of a wetland estuary. Strawberries are typically planted right down to the edge of the wetland. Erosion rates in excess of 150 t/ha of soil occur in wet years. In addition, nitrates are leached into the estuary by rainfall and irrigation water, and phosphates and pesticide residues that are adhered to eroded soil particles move into the estuary as well, contributing to the degradation of the wetland ecosystem (Soil Conservation Service, 1984). In an attempt to prevent these negative impacts, a buffer zone was planted between the intensively farmed strawberry fields and the estuary (Figure 22.6). Because coastal grass and scrubland occupied the farmed sites originally, native perennial grasses were planted in dense strips varying from 20 to 50 m wide. Once established, the grass cover effectively trapped sediments and took up soluble nutrients, limiting both erosion and the flow of nitrates, phosphates, and pesticides into the estuary. The buffer zone also served as a potential reservoir of beneficial insects for the farm fields.

Buffer zones have become very important parts of ecologically based development (ecodevelopment) projects in many parts of the rural world (Gregg, 1991; Koziell and Saunders, 2001). In regions where forests are being encroached upon by farming and grazing systems that replace the natural ecosystems with agricultural activities, buffer zones can protect the forest from further incursions yet provide an area where human activities can occur. Traditional land use activities, including nonextractive forestry, understory cropping, agroforestry, and collection of native plant or animal material, are permitted in the buffer zone as long as the structure of the forest in the buffer is retained the adjacent forest is protected. In an ideal situation, the forest ecosystem is preserved, limited economic activity goes on in the buffer, and intensive agricultural activities take place in adjacent cleared areas. The success of such programs has been limited due to a range of social, economic, and political reasons (Oldfield and Alcorn, 1991; Naughton-Treves & Salafsky, 2004), but the concept holds promise as an important way of integrating the goals of sustainable agriculture and biodiversity conservation.

## THE ECOLOGY OF PATCHINESS

The patchiness of the agricultural landscape has a profound influence on the ecological processes occurring throughout the landscape. Similar habitat patches are isolated from each other, yet gamma diversity is potentially high. In such a context, the size and shape of patches, and the distance between them, are important factors determining biodiversity at the landscape level.

When highly modified agricultural lands separate natural ecosystem patches, the patches are ecologically analogous to islands. Following the theory of island biogeography presented in Chapter 16, agricultural "oceans" can block or selectively block — that is, *filter* — the movement of different plant and animal species between the natural islands. Thus, a population of a particular species

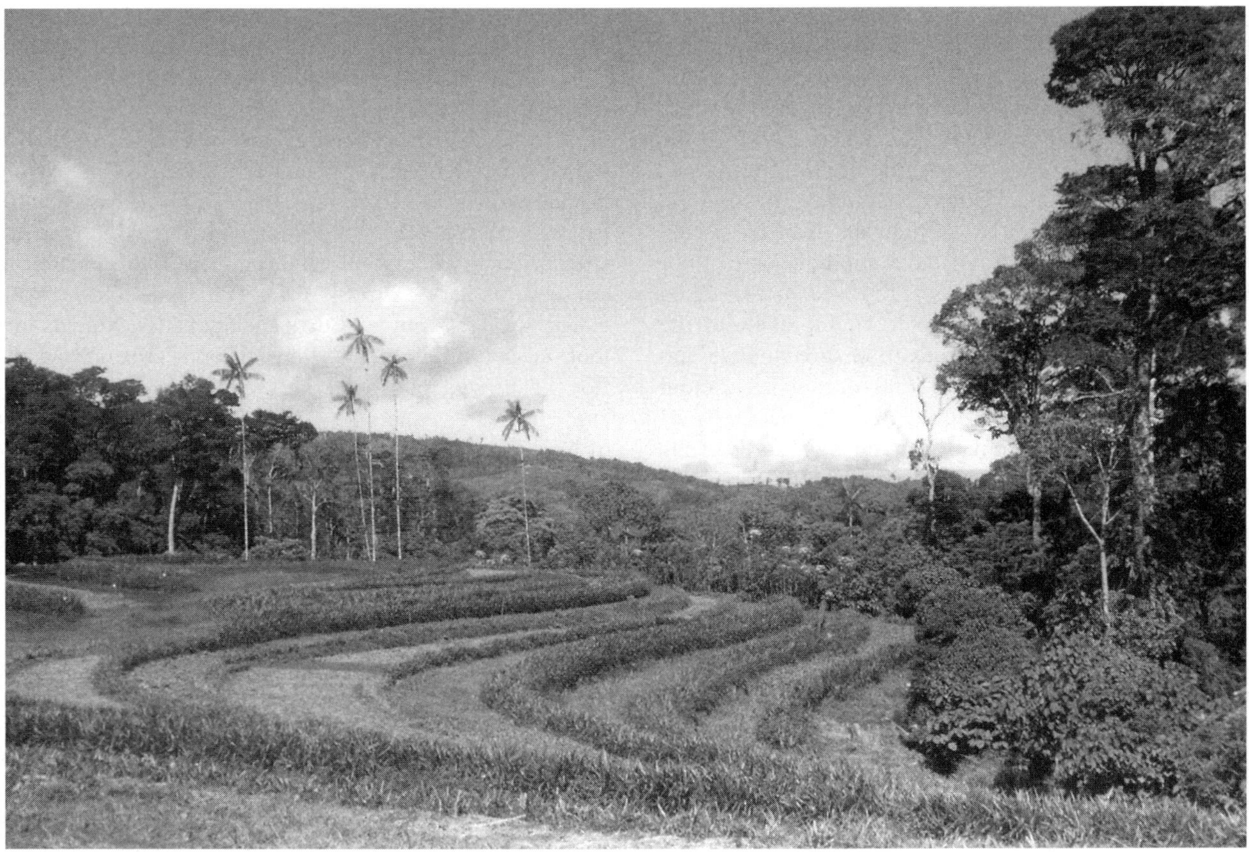

**FIGURE 22.5 A second-growth edge habitat at Finca Loma Linda, Coto Brus, Costa Rica.** Low, diverse vegetation at the forest edge can serve as a habitat for beneficial organisms that, once established there, can move out into the crops.

existing in one patch may be isolated from other populations; unless frequent interchange of individuals can occur between patches, each subpopulation can become subject to either genetic isolation or extirpation.

Because natural ecosystem patches provide refugia for agriculturally beneficial organisms and can provide various other environmental services, there is considerable advantage in determining the optimum density, abundance, and configuration of natural ecosystem patches in relation to areas of agricultural production. Corridors linking habitat patches may be necessary for facilitating movement of beneficial organisms across the landscape. A certain width of edge may provide the optimal edge effect without creating pest problems for both natural and agricultural systems. Promoters of integrated pest management often claim that successful pest management without the use of pesticides will require regional or landscape-level management programs that strive to take advantage of both the isolating mechanisms and facilitating mechanisms of a patchy environment (Settle et al., 1996). Ecologists are being called upon to apply their knowledge of ecological processes in natural ecosystems to solving such problems (Kareiva, 1996).

## THE AGRICULTURAL LANDSCAPE AS A PROVIDER OF ECOLOGICAL SERVICES

When the agricultural landscape is viewed as an integrated whole, combining all of the nonfarmed and farmed areas in a region, it can be managed so that it functions as an integrated ecosystem and provides environmental services in much the same way that natural ecosystems would provide alone. The agroecological knowledge and practices described in Sections II and III of this book provide much of the theoretical and practical basis of this management.

Environmental services are the many "goods" and services provided by natural ecosystems that are essential for human survival and welfare and the global biosphere (Costanza et al., 1997; Matson et al., 1997; Millennium Ecosystem Assessment, 2003). Until recently, we have tended to take them for granted because they are perceived as free and abundant. Ecosystem services that are particularly important for sustainable agroecosystem function include nutrient cycling, biological control of pests and diseases, erosion control and sediment retention, water regulation, and maintenance of the genetic diversity essential for successful crops and animal breeding. Outside of the

**FIGURE 22.6 A native perennial grass buffer strip between strawberry fields and a wetland estuary, Elkhorn Slough, California.** When strawberries are planted to the edge of the estuary (top), the estuary is impacted by erosion and leaching. The perennial grass buffer (bottom) mitigates these impacts while restoring native species diversity to the region.

direct agroecosystem context, ecosystem services are important at a global scale. They regulate the gaseous composition of the atmosphere (especially through sequestration of CO2), create and maintain biodiversity, affect climate and weather, and maintain watershed function. Table 22.1 provides a list of ecosystem services important in an agroecosystem context, each paired with the ecological processes responsible for it.

A natural ecosystem provides ecosystem services when its biochemical, biophysical, and biological processes are functioning in a healthy manner, allowing it to be biologically productive (Swift et al., 2004). The same principle holds for agroecosystems. If an agroecosystem is to be a provider of ecosystem services, it must be designed and managed so that its diversity, stability, and complexity approach that of a natural ecosystem. In other words, increasing agroecosystem diversity (Chapter 16) and allowing greater successional development (Chapter 17) are the bases for creating an agricultural landscape that can attain its potential for full ecosystem function.

**TABLE 22.1**
**Ecosystem Services and the Ecosystem Processes that Provide them in an Agroecosystem Setting**

| Ecosystem Services | Responsible Ecosystem Processes |
|---|---|
| Production of food | Primary production, herbivore consumption, pollination |
| Production of fiber and latex | Primary production, secondary metabolism |
| Production of pharmaceuticals | Secondary metabolism |
| Production of agrochemicals | Secondary metabolism |
| Nutrient cycling | Herbivore consumption, predation, decomposition, mineralization, other elemental transformations |
| Regulation of water flow and storage, flood control | Soil organic matter synthesis, physical and biological soil processes, plant growth above and below ground |
| Regulation of soil and sediment movement, erosion control | Soil organic matter synthesis, physical and biological soil processes, plant growth above and below ground |
| Regulation of biological populations | Plant secondary metabolism, pollination, herbivory, parasitism, microsymbiosis, predation |
| Water and soil purification | Metabolism, decomposition, elemental transformations |
| Regulation of atmospheric composition and climate | Photosynthesis, metabolism, and primary production |

*Source:* Modified from Swift, M. J., A. M. N. Izac and M. van Noordwijk. 2004. *Agri Ecosyst Environ* 104: 113–134.

Diversification of agroecosystems, as we know, comes about through multiple cropping, rotations, fallows, mulching, minimum tillage, and livestock integration, and successional development can be achieved through agroforestry, more extensive use of perennials, and the creation of successional mosaics. And when diverse, successionally developed agroecosystems are managed in concert with the noncrop components of the landscape through the practices discussed earlier in this chapter, the ecological processes of nutrient cycling, population regulation, and energy exchange are integrated across the whole landscape, ensuring the robust functioning from which ecosystem services arise.

When we use ecologically based management practices to enhance the ability of agroecosystems to provide ecosystem services, we are clearly working toward the goal of agricultural sustainability at the same time. But it is only when we expand our thinking to the landscape level that sustainability and ecosystem services converge with the conservation of biodiversity (Swift et al., 2004).

## ROLE OF AGRICULTURE IN PROTECTING BIODIVERSITY AND ECOSYSTEM FUNCTION

Agricultural development has fundamentally changed the relationship between human culture and the natural environment. Not long ago in human history, when all agriculture was traditional and in a small scale, agroecosystems were interspersed as small patches across the larger natural landscape. Managed habitats maintained the integrity of natural ecosystems while diversifying the landscape. Today, in contrast, agricultural land uses predominate, making natural habitats the dispersed patches over much of the earth's land surface.

As a consequence, much of the terrestrial earth is now covered by a cultural landscape rather than a natural one. According to some estimates, 95% of the world's terrestrial environment is urbanized, managed, or used in some way for agriculture, animal husbandry, or forestry (Daily et al., 2003; Pimentel et al., 1992; Vitousek et al., 1997). More than half of the land devoted to agricultural production depends on large-scale, intensive, monocultural, or irrigated management, maintaining very little if any of the ecological processes characteristic of the natural landscape that preceded it. Less than 5% of the world's terrestrial surface area is in protected parks or preserves (Figure 22.7).

On an earth with a cultural landscape, efforts to preserve our remaining ecosystem diversity can no longer be focused primarily on the small areas of land that are still wild. Managed lands, particularly those that are agricultural, have an enormous untapped potential for supporting a diversity of native species and providing

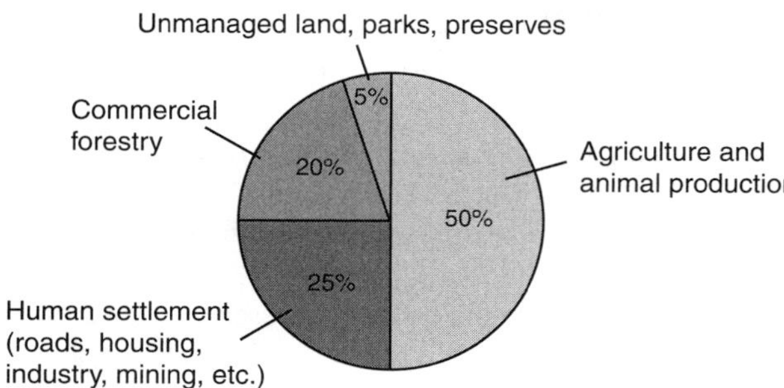


**FIGURE 22.7 Use of the world's land area.** Data from Pimentel, D., U. Stachow, D. A. Takacs, J. W. Brubaker, A. R. Dumas, J. J. Meaney, J. A. S. O'Neil, D. E. Onsi, and D. B. Corzilius. 1992. Conserving biological diversity in agricultural and forestry systems. *BioScience* 42: 354–362, and Vitousek, P. M., H. A. Mooney, J. Lubchenco, and J. Melillo. 1997. Human domination of Earth's ecosystems. *Science* 277: 494–499.

ecosystem services, thus contributing to conservation of global biodiversity.

Potentially, more species of plants and animals might be located on land under some degree of human management. Numbers per unit area might be quite small, but total numbers will eventually be high because we are dealing with such a large land surface. If agroecosystems in particular are managed and designed in ways that make them more friendly to native species — diversifying cropping systems in time and space, using agroforestry designs, reducing pesticide use, and so on — the landscapes of which they are a major part can support a greater diversity of organisms. Vertebrates can be provided with larger habitats, better food sources, and corridors for movement. Native plants can have more suitable habitats and find fewer barriers to dispersal. Smaller organisms, such as belowground microbes and insects, can flourish in the less adverse conditions and thus benefit other species because they are such important elements in ecosystem structure and function.

In short, by managing agricultural landscapes from the point of view of biodiversity conservation as well as production, all organisms can benefit in the long term, including humans. Learning how to manage in this way will require collaboration between conservation biologists, agricultural researchers, farmers, rural sociologists, and others, as well as new directions in research. Examples of necessary research include the following:

- Determining how to design and manage agroecosystems so that they provide habitats for species other than agricultural species.
- Understanding the complex relationships between ecosystem diversity and the ecological processes necessary for providing essential ecosystem services.
- Carrying out conservation studies on agricultural lands so that larger-scale projects can begin to take place, linking more resources and covering larger areas.
- Making agroecology the bridge between conservation and land use in order to sustainably manage the natural resource base upon which all plants, animals, and humans depend.
- Finding ways to support farmers and their organizations, so that they benefit from the use of practices that are conducive to providing and conserving ecosystem services in the landscape.
- Searching for institutional and social arrangements that support integrated landscape management.
- Developing more interdisciplinary approaches to research and problem solving.

The full potential for linking agroecosystems and natural ecosystems, however, can be realized only with fundamental changes in the nature of agriculture itself. The bottom line is that agriculture must adopt ecologically sound management practices, including diversification and the use of biological controls and integrated pest management as replacements for synthetic pesticides, fertilizers, and other chemicals. Only with such a foundation can we attain the goal of a sustainable biosphere.

## THE ECOLOGICAL VISIONS PROJECT

"Within the discipline of ecology, our thinking has … evolved from a focus on humans as intruders on the natural world to humans as part of the natural world."

**–Margaret Palmer et al. (2005:4)**

In 1988, the Ecological Society of America initiated an effort to define research priorities in ecology for the close of the 20th century. This effort became The Sustainable Biosphere Initiative (SBI), published as a special report in the journal *Ecology* in 1991 (Lubchenco et al., 1991). The SBI described "the necessary role of ecological science in the wise management of the Earth's resources and the maintenance of the Earth's life support system," and defined a "critical research agenda" to fill that role.

Significantly, the authors of the SBI moved well beyond a concern with the natural and least-modified ecosystems that have historically been the focus of ecology. They saw managed ecosystems — including agroecosystems — as integral arts of the biosphere, with an important role to play in preserving global biodiversity. In addition, recognizing the complex nature of environmental problems, they called on ecologists to forge alliances with researchers in other disciplines, both in the natural and in the social sciences.

The SBI, therefore, represented a breaking down of the barriers that have long separated the science of ecology from agricultural research, agroecology, and other areas of applied research. The SBI was a clear indication that ecologists were beginning to recognize the importance of research that addresses the ecology of highly modified systems in which humans are major ecological actors.

For more than a decade, the SBI served as an important framework for refocusing ecological research, and it was an impetus for bringing ecological knowledge into the policy-making arena. Then in 2002, the Ecological Society of America began taking a fresh look at the role the science of ecology should take in moving society toward sustainability. In addition to establishing a section for Agroecology within the society, it launched the Ecological Visions Project (EVP), outlined in *Ecological Science and Sustainability for a Crowded Planet: 21st Century Vision and Action Plan for the Ecological Society of America* (Palmer et al. 2004a), published in April 2004 by the ESA. The action-oriented EVP takes the SBI a step further by encouraging ecologists to focus even more strongly on how humans interface with nature.

The EVP recommends three areas of action: "building an informed public; advancing innovative, anticipatory research; and stimulating cultural changes [within ecology] that foster a forward looking and international ecology" (Palmer et al. 2004a:5).

The authors of the EVP document note that previous work on the ecology of natural systems has "uniquely positioned researchers to create an ecology of the future: a science that reshapes the view of key ecological questions and a science that explicitly recognizes and incorporates the dominant influence of humans" (Palmer 2004a:8). They also observe that "many ecologists who once concentrated their research on systems with "minimal" human influence are now focusing on how humans interface with nature … and are now considering the relevance of their work for policy" (Palmer et al 2005:8).

These same authors see a role for ecology that closely aligns it with agroecology:

Our future environment will largely consist of human-influenced ecosystems, managed to varying degrees, in which the natural services that humans depend on will be harder and harder to maintain. The role of science in a more sustainable future must involve an improved understanding of how to design ecological solutions, not only through conservation and restoration, but also by purposeful invention of ecological systems to provide vital services. Shifting from a focus primarily on historical, undisturbed ecosystems to a perspective that acknowledges humans as components of ecosystems, together with new research on ecosystem services and ecological design, will lay the groundwork for sustaining the quality and diversity of life on Earth (Palmer et al. 2004b:1252).

At the same time that ecologists are recognizing the importance of "human-influenced systems," agroecological researchers and others who study these systems are putting ever greater emphasis on ecologists' traditional

concerns by studying the role highly managed systems play in preserving the integrity of natural ecosystems and protecting global biodiversity. The ultimate goal of research in both disciplines is fundamentally the same — sustained functioning of the biosphere as a whole.

## FOOD FOR THOUGHT

1. What are some of the possible ways that organisms typical of natural ecosystems can contribute to the sustainability of agroecosystems?
2. What principal changes must occur in the way present-day conventional agroecosystems are managed in order for them to contribute to the conservation of biodiversity as well as to satisfy human needs for food production?
3. Why is the biodiversity of smaller, less obvious organisms in ecosystems, such as fungi and insects, of potentially greater importance to sustainability than that of the larger, more obvious mammals and birds?
4. Why are the small-scale, integrated farming systems of traditional farmers in a better position to provide important ecosystem services than large-scale conventional systems?
5. What kind of criteria should be used to determine which species in the agricultural landscape are the most important to preserve and enhance?
6. How is the landscape perspective important in sustainable agriculture management?

## INTERNET RESOURCES

The Ecotope Mapping Working Group
www.ecotope.org
The site of the landscape agroecologist Erle

Ellis, demonstrating the exciting integration of landscape ecology, biogeochemistry, global change, and sustainable ecosystem management.

Millennium Ecosystem Assessment
www.millenniumassessment.org/en/index.aspx
An international work program focused on studying the human consequences of ecosystem change and options for responding to those changes. Uses ecosystem services as a foundation to analyze environmental change and sustainability.

Communicating Ecosystem Services
www.esa.org/ecoservices/
A joint project of the Ecological Society of America and the Union of Concerned Scientists. Provides scientists with tools for more effectively communicating the concept of ecosystem services.

International Association of Landscape Ecology
www.landscape-ecology.org
Valuable information on research, conferences, publications, and links related to landscape ecology.

Agroecology Research Group at UCSC
www.agroecology.org
An excellent source of information in agroecology, with an emphasis on linking biodiversity conservation, rural livelihoods, and agroecology.

EPA Landscape Ecology Program
www.epa.gov/esd/land-sci/default.htm
Provides, along with other landscape ecology resources, a comprehensive list of landscape ecology project across the US, with project descriptions.

## RECOMMENDED READING


Bernhardsen, T. 2002. *Geographic Information Systems: an Introduction.* John Wiley & Sons: New York. A comprehensive overview of geographic information systems (GIS), covering theory, applications, and basic techniques.

Büchs, W. (ed.) 2003. *Biotic Indicators for Biodiversity and Sustainable Agriculture.* Elsevier: Amsterdam, the Netherlands. A comprehensive compilation of research papers from different regions of the world, focusing on the interactions between agriculture and biodiversity.

Buck, L.E., J.P. Lassoie, and E.C.M. Fernandes. 1999. *Agroforestry in Sustainable Agricultural Systems. Advances in Agroecology Series.* CRC/Lewis Publishers: Boca Raton, Florida. A broad introduction to the environmental and social conditions that affect the roles and performance of trees in field- and forest-based agricultural production systems.

Gaston, K.J. and J.I. Spicer. 2004. *Biodiversity: An Introduction.* 2nd ed. Blackwell Science: Malden, MA. An overview of what biodiversity is, its relevance to humanity, and issues related to its conservation.

Groom, M.J., G.K. Meffe, and C.R. Carroll. 2005. *Principles of Conservation Biology.* 3rd ed. Sinauer: Sunderland, MA. A very balanced treatment of the field of conservation biology, combining theory and practice.

Hilty, J.A., W.Z. Lidicker, Jr., and A.M. Merenlender. 2006. *Corridor Ecology: The Science and Practice of Linking Landscapes for Biodiversity Conservation.* Island Press: Washington, D.C. Draws on conservation science and practical experience to develop, maintain and improve the connectivity of high biodiversity areas in landscapes.

Leopold, A. 1933. *Game Management.* Scribner. A classic text on the important role of edge effects in maintaining the abundance of certain species of wildlife in an heterogeneous landscape.

Loreau, M., S. Naeem and P. Inchausti. 2002. *Biodiversity and Ecosystem Functioning: Synthesis and Perspectives.* Oxford University Press: New York. A comprehensive and critical overview of recent empirical and theoretical research on the relationship between biodiversity and ecosystem function.

Millennium Ecosystem Assessment. 2003. *Ecosystems and Human Well-Being: A Framework for Assessment.* Island Press: Washington, D.C. A comprehensive and interdisciplinary analysis of the function, value and importance of global ecosystem services, by a distinguished panel of international researchers.

National Research Council. 2005. Valuing Ecosystem Services: Toward Better Environmental Decision-Making. A thorough analysis of the state of ecosystem services in the U.S. and the opportunities and limitations for their valuation and conservation.

Oldfield, M.L. and J.B. Alcorn. 1991. *Biodiversity: Culture, Conservation, and Ecodevelopment.* Westview Press: Boulder, CO. A challenging collection of approaches for combining the conservation of biological resources with rural development using traditional resource management systems as a basis.

Rosa, H., S. Kandel, and L. Dimas. 2003. *Compensation for Environmental Services and Rural Communities: Lessons from the Americas and Critical Themes to Strengthen Community Strategies.* PRISMA: San Salvador, El Salvador. A critical review of the concept of ecosystem services, and its compensation within the context of rural development, as illustrated by case studies from the Americas.

Schroth, G., G.A.B. da Fonseca, C.A. Harvey. C. Gascon, J.L. Vasconcelos, and A-M.N. Izac. 2004. *Agroforestry and Biodiversity Conservation in Tropical Landscapes.* Island Press: Washington, D.C. A very thorough review of the role of agroforestry practices in helping promote biodiversity conservation in human-dominated landscapes of the tropical world.

Thrupp, L.A. 1997. Linking Biodiversity and Agriculture: Challenges and Opportunities for Sustainable Food Security. World Resources Institute: Washington, D.C. A critical analysis of how to integrate biodiversity conservation and agricultural production, taking into account social, economic and ecological parameters.

Turner, M., R.H. Gardner, and R.V. O'Neill. 2003. *Landscape Ecology in Theory and Practice: Pattern and Process.* Springer-Verlag: New York. An integrated perspective of the field of landscape ecology, including theory, empirical applications, and future directions.

# 23 Culture, Community, and Sustainability

In his book *Radical Agriculture*, published in 1976, Rich Merrill wrote about the need to "get culture back into agriculture" (Merrill, 1976). His was an early voice-calling attention to the negative effects of a process that had already been underway for decades: the transformation of agriculture into agribusiness.

Merrill was playing with the dual meaning of *culture*, substituting the meaning having to do with the tilling of the soil with the meaning we have in mind when we use the phrase *human culture*. In this latter sense, culture is an integrated system of human knowledge, belief, and behavior. So Merrill was essentially warning us that agriculture was being drained of its humanity — that the values, behaviors, and social relationships that once supported a stewardship orientation to farmland were falling away.

Now, three decades later, Merrill's plea is as relevant as ever. The agribusiness model, with its drive toward industrialization of food production, has been remarkably successful by many measures, but it has completely changed the social and economic relationships surrounding the production and consumption of food. In reducing farmers to sources of farm products, farm workers to labor costs, and the purchasers and eaters of food to consumers, it has ensured that the real people who populate our food systems will interact only through the medium of money, in a system organized to meet the demands of capital and little else.

Agriculture has not lost its grounding in human culture, as one reading of Merrill's statement might suggest; the problem is that the knowledge, beliefs, behavior, and relationships that have grown up around the production and consumption of food have become major obstacles to sustainability. Consumers have no idea where the food they eat comes from, how their choices affect agroecosystems, the environment, and farmers and farm workers. "Eating is an agricultural act," according to Wendell Berry, but consumers eat as if they are only satisfying their hunger. On the production side, farmers are increasingly at the mercy of a system that separates them from consumers and leaves them with little choice but to play by agribusiness rules, often at the expense of their values.

In order to be sustainable, agriculture needs a "culture" surrounding it that promotes sustainable practices rather than helping to destroy them. To put this kind of culture back into agriculture, we need to reestablish the connections between farm and table, form human relationships around food that are more than economic, and promote values in relation to food consumption that look beyond narrow self-interest.

## THE DECLINE OF FARMING IN THE AGE OF THE GLOBAL SUPERMARKET

When human cultures depended primarily on hunting and gathering, foods directly reflected the local environment. This connection between diet and the local environment remained even after the development of agriculture (Chapter 14), producing remarkable differences and diversity in diets, consumption patterns, and cuisines around the world. But as trade in food grew over time, and profit could be made in such trades, cultures grew less distinct in their diets, and in the quantities and qualities of the foods they consumed. At the same time, the universe of what was available to eat expanded for many people. In the last several decades, these trends have accelerated (Menzel and D'Aluisio, 2005). The food industry now functions in a global marketplace in which food moves quickly from one part of the world to another, the raw materials from farmers purchased at low prices converted into an incredible array of processed, packaged, and preserved food items that hardly resemble the products they were made from.

Anyone who recalls shopping for groceries just two decades ago knows that our diet has changed dramatically, especially in the U.S., but few people realize that this transformation is occurring all around the world. Globalization of the economy is the main causative factor, as large-scale capitalism reaches new parts of the globe. These changes are also the result of economic growth and increasing affluence: as people gain new means to purchase food items they did not consume much before, they eat more meat and fish, and more of the wide array of processed, convenience, and fast-food items now on the market. The global movement of people also plays a part in the transformation of our diets, as travelers, immigrants, and refugees bring their own foods to new lands and learn new food preferences in return.

## The Agrarian Crisis

One would think that with the development of these diverse and dynamic market structures for food and changes in diets, farmers would be enjoying a time of plenty. But as we saw in Chapter 1, rural farm communities are in decline around the world. Once thriving assemblages of people from all walks of life, livelihoods, and outlooks, today they are increasingly aging and depopulating. In the U.S., less than 1% of the population is made up of full-time farmers, and of those, farmers over 65 years old outnumber those under 35 by nearly six to one (USDA, 2002). In developing countries, farmers and their families are leaving rural areas and their farms in alarming numbers, because the land can no longer viably absorb more population, because local prices for farm products collapse as cheaper products are imported under free trade agreements, or subsidies are removed.

When the economies of rural communities decline, their social fabric begins to unravel as well. This unraveling has been documented in the powerful writings of many authors (e.g., Wendell Berry, Gene Logsdon, Donald Worster, Wes Jackson, and others). When a way of life is restricted to merely making a living, many of the reasons for being and doing are lost. When a person feels like they are nothing more than a link in a commodity chain, and less a member of a vibrant, interactive, and healthy community, the indicators of decline appear. Poverty, crime, high school dropout rates, spousal and child abuse, mental stress, and substance abuse — all signs of social dysfunction — soon approach levels similar to those of crowded urban areas. The consequences are ecological as much as they are social, affecting the farmers, their communities, and the landscapes in which they live (Figure 23.1).

In developing countries, farmers are increasingly affected by the double impacts of cheap, heavily subsidized imports of foods from outside of their traditional local markets, coupled with exclusion from opportunities to sell their products for export to distant markets. Locally livelihoods suffer, agroecosystems degrade, and rural communities begin to unravel.

Of course, the declining numbers of farmers do not imply that there has been a decline in the importance of the farm sector. The world still has to eat, and there are 70 million more mouths to feed each year. Despite the drop in people making their living from farming, half of the world's people still depend on farming for their livelihoods. In some parts of the world, such as much of South Asia, over 70% do, and in these regions, agriculture accounts in many places for half of the total economic activity (FAO, 2004). In both developed and developing countries, the decline in the number of people making their living through farming has been accompanied by farm modernization. This, of course, is what has created large increases in farm production per acre and allowed rural people to move into the more advanced economies of the cities. This modernization is typically referred to as "progress," and the substitution of tractors for people is usually seen as a way to a more abundant and affordable food supply for the consumer. But urban life too often does not meet the hopes and expectations of displaced farmers as they encounter unhealthy city living conditions, poor housing, and few opportunities for full employment, putting them in worse situations than they were in on the farm. It would appear that farmers are less lured from their farms by the promise of the city, than they are driven off from their farms by a variety of changes in the way the global food chain is structured and operates, making it more difficult for farmers to survive every day.

**FIGURE 23.1 Suburban sprawl encroaching upon agricultural land in the south bay region near Sa n Jose, California.** Farmers find it very difficult to resist the spread of urban development onto their farmland.

## Economic Imperatives of the Global Food System

Recall the graph in Chapter 1 entitled "Farmers' declining share of the consumer food dollar" (page xx). It shows that the money spent on food has gone increasingly to the processing, shipping, and marketing side of the food system, leaving farmers, at the end of the 20th century, with less than eight cents of every food dollar spent. This by itself is a major reason for the decline of the farming occupation — as a basic economic reality it leaves farmers with little option but to follow the mantra of "get big or get out." But the 73% share of the consumer food dollar going to the processing, packaging, shipping, and marketing middlemen indicates just how much our food system has changed (this share was well under 50% in 1919) and how thoroughly it is now stacked against the small-scale farmer.

With much of the profit in the "marketing" segment of agriculture, it is no surprise that most of the processing, brokerage, shipping, packaging, and marketing functions are performed by transnational corporations and the firms they own or control. Further, these large international corporations have taken full advantage of vertical integration — each corporation owns firms at every link in the food-system chain, from seeds to shipping to processing to distribution to marketing. Over time, there has been a tendency for the overall number of these firms to decrease. This economic concentration, combined with vertical integration, allows a relative handful of agribusiness corporations to dominate the agricultural sector of most nations' economies (Table 23.1).

The farmer, therefore, faces a virtual agricultural oligopoly. For example, consider a typical corn farmer in U.S. Midwest buying seed for next year's crop. That farmer has little choice of what seed to buy and who to buy it from, because he is confronted with a system where the only buyer of corn in his region is a large transnational corporation that is in partnership with another large corporation that provides seeds for the only variety of corn that the buyer will purchase. The bank that provides the production loan most likely is part of the same transnational's portfolio, and will probably have the same requirements of which seed variety to use, and recommend very highly that the farmer use fertilizers and pesticides from sources the transnational also owns or controls. Once the farmer has grown the corn, and does not want to sell to the transnational at the fixed price, he could choose to feed the corn to hogs for sale at auction. But the transnational will be there bidding on the hogs as well. And finally, if the farmer gives up and decides to plant a crop other than corn, he will find that there are very few, if any, other crops that are not controlled by the system of food "cartels." (Halweil, 2004).

**TABLE 23.1**
**Examples of Concentration in the Agricultural Sector**

| Product or Activity | Proportion of all Firms | What These Firms Control |
|---|---|---|
| All seeds | Top 10 firms | One-third of global market |
| Vegetable seeds | 5 firms 3 | 75% of global market |
| Cereal grains | 2 firms (Archer Daniels Midland & Cargill) | 75% of world trade |
| Coffee | 4 largest firms | 40% of world trade |
| Cocoa and pineapples | Handful of multinationals | 90% of world trade |
| Bananas | Handful of multinationals | 80% of world trade |
| Sugar | Handful of multinationals | 60% of world trade |
| Chickens | 1 firm | 60% of purchases in Central America |
| | Fewer than 40 firms | 97% of the US market |
| Beef | 4 firms | 80% of packing in the US |
| Milk | Top 5 firms | 41% of global processing |
| Animal feed | 3 firms | Majority of global production |
| Food retailing | Top 30 grocery chains | 33% of global sales |
| Pesticides | 10 firms | 80% of world market |

*Source:* Adapted from Halweil, B. 2004. A WorldWatch book. Norton: New York.

There is little room for small-scale or family farmers in a system in which the farmers' product is a commodity in a global market controlled by vertically integrated transnationals. Therefore, such farmers are increasingly forced to sell out. Their land is eagerly bought up by developers, or by the larger-scale farmers who have learned to adapt to the system.

One common way of "adapting" to the system is to grow under contract for the larger and larger corporations formed by the mergers and consolidations that are common in the marketplace. As Brian Halweil (2004) points out, over the past two decades, the share of American agricultural output produced under contract has more than tripled, from 10 to 35%, and this does not include contracts that farmers must sign to plant genetically engineered seeds (Chapter 14). When the control of the food system becomes so centralized, the farmer is essentially reduced to a hired hand in a commodity chain. We end up with large-scale farms managed by distant corporations interested in extracting the maximum output at the minimum cost (Figure 23.2).

According to conventional wisdom, agricultural modernization and larger-scale farming improve the efficiency of the food system. Bigger farms can produce more at lower economic costs. Production and equipment costs can be spread over greater area, inputs purchased at bulk rates, and loans negotiated at lower interest. Such advantages are

**FIGURE 23.2 A monoculture of pineapples growing near Buenos Aires, Costa Rica, in an area once covered by tropical rain forest.** Fruit will be exported through a vertically integrated commodity chain, where a transnational owns or controls most of the steps from the field to the table.

indeed increasingly important as agriculture becomes more capital intensive. However, as we have seen throughout this book, most of the ecological elements of sustainability on farms are lost or compromised as scale becomes too large. Rarely are the ecological costs of this scale of farming taken into account in normal cost accounting; but when this is done, it turns out that for many crops, actual production costs are lower when the crops are grown on relatively smaller farms. But larger farms, because of their volume, can afford to sell at a lower margin if forced to do so — as indeed they are — by the food buyers in the food chain. As a result, the bulk of the financial benefit gained by the giant farm over the small one goes to the buyers and processors, not the farmer or the farm community.

## The Isolated Consumer

From the standpoint of food choice and availability, consumers have never had it better. But the same global food system that forces out small-scale farmers and exploits Third-World peasants has brought a variety of negative changes to food consumers. Many of these changes have happened so slowly that we are not conscious of them.

Because much of the food we eat must travel a long distance to get to us, it is not particularly fresh. Even produce, shipped rapidly by air or truck, often under refrigeration, is often picked before it is ripe. And when surviving transport and storage is the major consideration, the breeding (or genetic engineering) process that produces the seeds is likely to have sacrificed taste and nutritive content. Other types of food must be packaged and processed to survive long-distance transport and storage. They have added preservatives and a variety of other added ingredients — such as salt, sugar, and fats — that are linked to obesity, cancer, and other health problems. As a result, we have more food choices than ever before, but much of that food is less fresh, less tasty, and less healthy.

Other losses are less concrete but no less important. Regional and cultural differences in cuisine and diet are slowly disappearing with the homogenization of the food supply. Related to this is the loss of place-based identity. The regional foods that define the places we live in are either being lost or overly hyped as marketing tools.

And finally, there is the unquantifiable but real loss we experience when food consumption is completely detached from the processes that got it to our tables. When we lose all connection with the people who grow our food, we lose touch with all the biological and social facts of the food's existence, and eating is stripped of much of the context and meaning it has had since the long-ago origins of the human species.

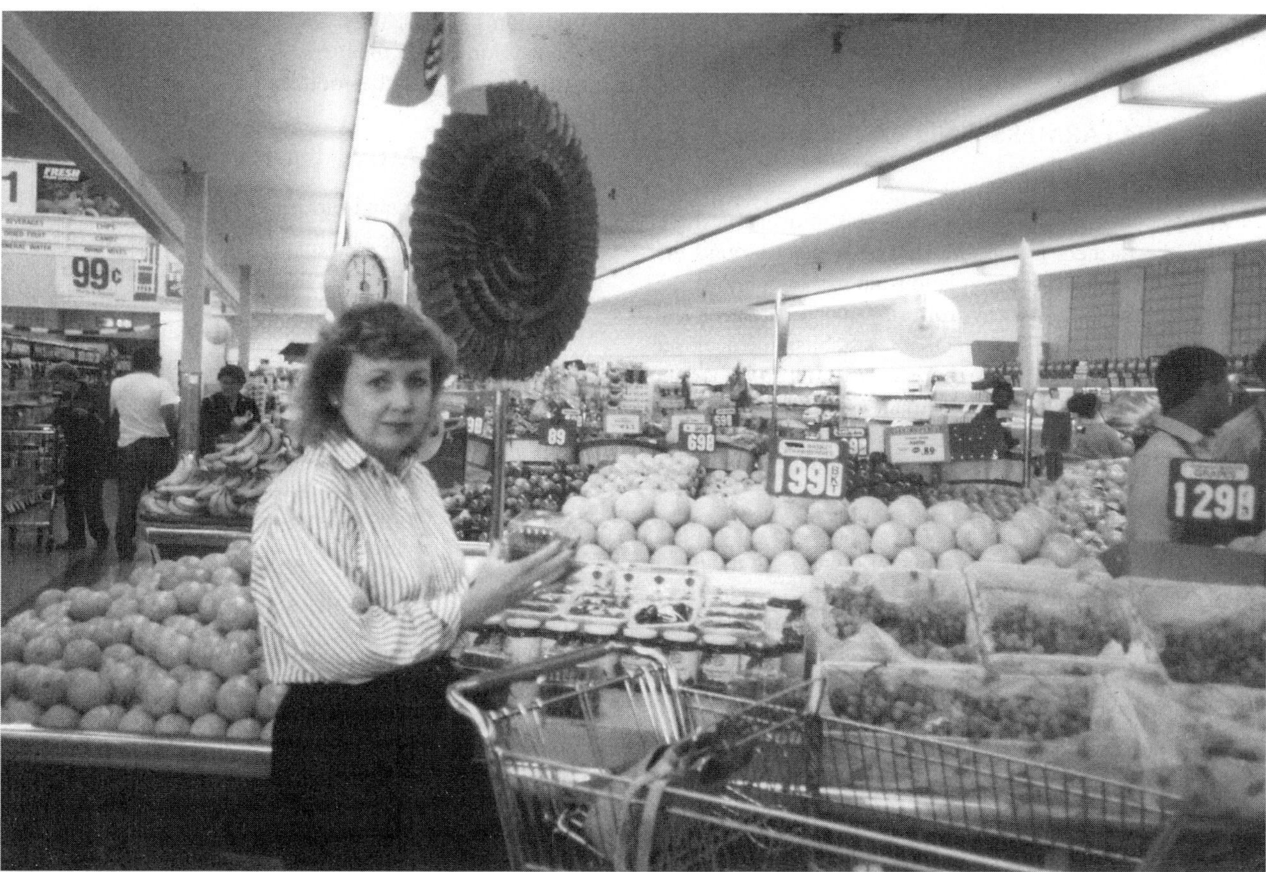


**FIGURE 23.3 Shopper in a typical supermarket.** The consumer has many food choices, but the only information conveyed by the labels is price. Origin, conditions of production, date of harvest, the farmer's share of the profit, and other facts remain unknown.

## Consequences of the Commodification of Food

The food system just described — in which food is grown in large-scale agroecosystems as a commodity in a global market for consumers completely isolated from the production process — has an enormous bearing on sustainability (Figure 23.3).

All the unsustainable on-the-ground practices of present-day industrial agriculture described in Chapter 1 — monoculture, intensive tillage, reliance on external inputs, planting of genetically engineered seeds, and so on — exist in part because of how well they serve this food system. When food is a mere commodity and the only goal of its production is extraction of profit, unsustainable practices flourish. Farms grow larger, industrial methods of production dominate, and more sustainable smaller-scale, traditional, and agroecologically based practices are marginalized.

As a result, what were once self-regulating systems for transforming solar energy, moving nutrients, balancing member populations, and maintaining a dynamic equilibrium through time have become management-intensive systems dependent on nonrenewable fossil energy, synthetic chemical fertilizer inputs, and external population-regulating practices.

How exactly has this happened? The decline of agrarian communities, family farms, and the farming occupation provides one immediate answer. Small-scale farmers are the only good stewards of the natural resource base upon which their farms function; they are the only ones with extensive knowledge of local soils, weather, land races, noncrop plants, pollinators, local sources of soil amendments, ecosystem characteristics, and community needs. When they leave their farms, their knowledge and stewardship values go with them.

Another part of the answer comes from the consumer side of the food system. Agribusiness has manipulated consumer tastes and behaviors, creating more demand for the products with the highest profit potential, such as beef, pork, fast food, high processed snacks, exotic fruits, and out-of-season vegetables — exactly the products that have the highest environmental costs and rely the most on unsustainable practices. Isolated from the production and distribution process, consumers are also isolated from the information and knowledge that might allow them to become conscious of the negative impacts of their behaviors, diets, and food choices.

## BRINGING FARMERS AND CONSUMERS BACK TOGETHER

As we have seen, strong interests have taken over the space between the farmers in the field and the eaters around the table. The dissolution of this relationship has been one of the root causes of the trend away from sustainable practices and relationships. It follows, then, that reestablishing a closer relationship between farmers and consumers is an important part of building a path back toward sustainability.

Considered from a broad perspective, creating a more sustainable food system has two fundamental requirements:

- Diets must change to reduce demand for meat, other animal products, and all food products that require excessive transport, processing, and packaging.
- More land must be cultivated under a stewardship ethic, by independent, relatively small-scale farmers with a stake in their communities, able to make a decent living, unconstrained by the demands of the agribusiness oligopoly.

Both of these goals can be accomplished at the same time by reestablishing the connection between farmers and consumers. If farmers have alternatives to the agribusiness model and the food-system oligopoly, they can remain on the land and farm profitably using the best, most sustainable practices. If consumers are in touch with the food production process, they are aware of how their choices and behaviors affect the growing of food, the environment, and the working of the food system.

### ELEMENTS OF AN ALTERNATIVE FOOD SYSTEM

Bringing consumers and farmers back together is really the same thing as creating an alternative food system. In such a system, (1) food production and consumption has a bioregional basis; (2) the food supply chain has a minimum number of links; (3) farmers, consumers, retailers, distributors and other actors exist in the context of an interdependent community and have the opportunity for establishing real relationships; and (4) opportunities exist for the exchange of knowledge and information among all those who participate in the food system. These aspects of an alternative food system are closely interrelated. Although they are likely to exist together, they are conceptually distinct.

#### Agricultural Bioregionalism

It can be said that as the physical distance between the people who grow food and the people who eat it grows, the chance for the exploitation of both grows as well. An important way to ensure that this exploitation does not happen is to bring "localness" back into agriculture.

Localness depends on physical proximity. When the people who consume food are not far from the people who produce it, that food system is local. Local food systems are identified with a place and contribute to the environmental, social, economic, and cultural development of the communities in that place (Figure 23.4).

When the people living in a particular area or region eat mostly food that is grown or raised locally, they shift

**FIGURE 23.4 An area in rural Germany, near Witzenhausen, that has retained a bioregional agriculture.** The residents of the town can eat food grown nearby.

the focus of their diets. Food that cannot be grown locally is not eliminated from what they eat, but its role is reduced in favor of more local food. In temperate climates, this also implies eating what is available in season and relying more on traditional food-caching techniques such as root cellaring, as well as food preservation and storage techniques such as drying and canning. Although this means "giving up" some of the choice and convenience we have come to expect in the global supermarket, it brings many benefits, including renewed connection to place.

The concept of the *watershed* — an area drained by a single interconnected network of streams — plays a role in discussions of bioregionalism generally. In the context of agricultural bioregionalism, it makes sense to use the parallel concept of the *foodshed*, which can be defined as a geographically limited sphere of land, people, and businesses tied together by food relationships.

Many benefits can be derived from a food system in which foodsheds are the primary functional units. From an ecological perspective, growing and consuming food locally reduces the amount of fossil fuel energy needed to transport food to the consumer. Less energy need be expended to process or store food once it is harvested because food can be consumed sooner following harvest. Food waste can be more easily returned to the farm, promoting nutrient cycling and reducing the dependence on outside nutrient inputs. Diversity at the level of the farm (Chapter 16) and the level of the landscape (Chapter 22) is more easily supported, creating a healthy integration of urbanized areas, working landscapes, and natural ecosystems.

Economically, local economies thrive on local food systems. Money spent on locally grown food can generate nearly twice as much income for the local economy as money spent on food from afar (Halweil, 2004). Money recirculates within the community rather than being siphoned off by distant companies. All sectors of the community benefit from this local flow: local farmers, local businesses, local service agencies, and even local schools and hospitals. Bioregionally based agriculture, therefore, is the key element in any effort to rebuild and restore economically and socially distressed rural communities and regions.

## Shorter Food Supply Chains

One of the problematic aspects of the present global food system is the large number of "links" in the chain between the farmer and consumer. These often include brokers, processors, distributors, transporters, packagers, wholesalers, and retailers. The greater the number of links, the more disconnected the farmer and consumer, the greater the amount of the consumer food dollar siphoned away from the farmer, and the greater the demand for food production to be large scale and driven solely by production criteria. A more sustainable alternative food system requires food supply chains with fewer links. The importance of *short food supply chains*, or SFSCs, has been recognized in the area of rural development (Renting et al., 2003), and the concept is gaining attention as a component of food-system sustainability.

The shortest food supply chain is not even a chain because it has no links at all: consumption of food by the same person, family, or group who grew it. Although growing one's own food is often rejected as impractical, it is practiced to a surprising extent all over the world, even in urban settings. From cities in China to towns all over Europe, the backyard or rooftop kitchen garden is an important source of food. Community gardens — providing gardening plots for those without access to land — are common in cities around the world, and are becoming increasingly popular in U.S. and Western Europe.

The next shortest food-supply chain, of course, is provided by a direct relationship between farmer and consumer. These face-to-face chains occur with farmers' markets, box schemes, roadside sales, farm stores, pick-your-own farms, and the like (Figure 23.5).

Traditional food-retailing arrangements can incorporate shorter food supply chains, too, particularly when restricted to a local foodshed. Supermarkets, food stores, restaurants, and institutions can purchase a large portion of their food direct from local growers. This adds only one link between farmer and consumer. Even if a distributor or other wholesaler is involved, the links are still fewer than those that exist in the global food system, and the distance the food travels is greatly reduced.

Finally, direct or nearly direct farmer–consumer commerce can occur over greater distances, facilitated by present-day communication technology and the transportation infrastructure. Through direct-purchase cooperatives, e-commerce, and subscription plans, consumers can buy high-value products, such as coffee, directly from the farmers who grow them. Even though the products may travel long distances, the long food supply chain of the global food system is effectively short circuited.

**FIGURE 23.5 A rural farmer's association selling organic produce at a market in Porto Alegre, Brazil.** More than a dozen farmers own a truck together and take turns going to the market with their pooled products.

## Food-based Community

The impersonal global food system has inexorably diminished the role of food as a cohesive force in the creation and maintenance of communities. Because food is the most fundamental human need, humans have always come together to ensure food supplies. Throughout our biological and cultural evolution, the need to cooperate in the procurement, production, storage, distribution, and protection of food has caused humans to form hunting bands, villages, towns, cities, and societies. The religious ideas, ways of life, values, and mores that have held these social formations together have always — until recently in human history — been grounded to a great extent in food.

Restoring the fundamental role of food as a bonding force for community is beneficial not just for communities, but for the food system as well. When the production, distribution, and consumption of food occur in a community context, in which people have interdependent relationships, factors that cause imbalance in the system are more readily apparent and more easily adjusted or repaired. It becomes a community concern — something that has a potential effect on everyone — if farmland is being lost to development, if soil erosion is causing productivity declines, if too much food-related money is leaving the community, and if farmers are getting economically squeezed.

### Democratic Information Exchange

In separating farmers and consumers, the global food system has also fundamentally changed the nature of information exchange and communication among the actors in the system. The information that flows through the present system is mostly controlled and mediated by the corporate interests that receive up to 92% of the consumer dollar. These interests want consumers to know as little as possible about the origins, nutritive content, processing of, and economic circumstances of the food they eat, and to be concerned as much as possible with the fetishized aspects of food consumption — how it fits into diet fads, how it is more convenient, and how it helps form one's image and identity. What consumers "want" is thus manipulated to a great extent by the food-supply oligopoly, and this information filters down to farmers as impersonal economic imperatives.

In political terms, democracy is dependent on the free flow of information and open communication. For a democracy to function effectively as the "will of the people," the people must have full access to knowledge about alternatives, possible consequences, the lessons of the past, and so on. In contrast, coercive political systems always rely in part on restricting the flow of information and shaping what gets to count as truth and knowledge. Food systems work the same way. An alternative food system that empowers the eating public and the people who actually grow food — a *food democracy* — requires a free flow of undistorted, unfiltered information and channels of communication among the people in different parts of the system. Democratic information exchange becomes the basis for active, engaged consumers who understand the significance of their choices.

## Building Alternative Food Networks

Farmers, consumer cooperatives, neighborhood associations, groups advocating sustainable development, green entrepreneurs, and others have been quietly building the foundations of a more sustainable food system for decades. Making use of different combinations of the four elements discussed above, they have set up farmers' markets, farm stores, direct-marketing schemes, and many other types of businesses, programs, and institutions that give farmers and consumers alternatives to the global food system.

These alternative food networks (AFNs) are diverse, varying in size, scope, and intent. What they share is a desire to bring many of the missing elements of sustainability back to our food system. They provide real-world, working models of a different, decentralized approach to the ecology and economy of food, thereby helping to create a new culture of sustainability.

Like life forms, AFNs have "evolved" along different paths to exploit different niches. There are abundant niches in the local or regional context. These have been filled by farmers' markets, community-supported agriculture (CSA) schemes, other types of direct-marketing arrangements, local food-focused restaurants, and so on. These AFNs are generally able to incorporate all four elements of alternative food systems at once: they operate in a strictly local context, create short food-supply chains, build food-based community, and allow for democratic information exchange. Many of them are based on face-to-face contact between consumers and producers.

But localness has its limitations. Not all farm products can be grown or produced in every farm community around the world. Climate, soils, geography, and local culture can all restrict what can be grown or raised in a certain area. Coffee, cocoa, vanilla, and mangos, for example, can only be produced in the tropics, and then only in specific parts of the tropics. Cranberries and olive oil can only be produced in temperate regions, and then only in specific parts of the temperate zone. Even if they are committed to "eating locally," consumers will always want to have available some food products that are out of season or impossible to grow locally. Creating a way for consumers to purchase such products, outside of the current global food system, has been the goal of various other types of AFNs. These "extended networks" typically connect consumers and producers more directly, often through the Internet, greatly shortening the supply chain that would otherwise be involved, and at the same time promoting the democratic flow of information.

Table 23.2 lists a variety of AFNs and indicates for each type how it makes use of the four elements of alternative food systems. Some of the more important of these AFN types are discussed in more detail below.

**TABLE 23.2**
**Types of AFNs and Their Relative Contributions to the Four Elements of Sustainable Food Systems**

| | Encompassed With in a Locality | Shortens Food-Supply Chain | Builds Food-Based Community | Promotes Democratic Flow of Information |
|---|---|---|---|---|
| Farmers' markets. Farmers sell their products directly to consumers | • | • | • | • |
| Pick your own. Consumers do their own harvest on the farm | • | • | • | • |
| Farm stores. On-farm store for direct sale, open all year | • | • | • | • |
| Community-supported agriculture. Subscription sales to consumer groups | • | • | • | • |
| Box schemes. Farmer prepares a box on order for consumer | • | • | 0 | 0 |
| Consumer cooperatives. Centralized food buying by consumers | 0 | • | • | • |
| Local-food restaurants. Promotion of local food by restaurants | • | • | • | 0 |
| "Dedicated" retailers. Shops that sell local or regional products | • | • | • | 0 |
| Catering for institutions. Using local and regional products in food service | • | • | 0 | 0 |
| Mail order sales. Long-distance purchase from farmer | | • | 0 | • |
| e-Commerce. Direct purchase through on-line mechanisms | | • | 0 | • |

*Note:* • denotes primary importance. 0 denotes secondary importance or potential.

## Farmers' Markets

At a farmer's market, farmers, growers, or producers from a specific local area are present in person to sell their own produce directly to the public. All products sold are certified to be grown, reared, caught, brewed, pickled, baked, smoked, gathered, or processed by the seller. In the direct sale of their produce to consumers, farmers can take back some of the profits captured by the agribusiness supply chain. Perhaps even more importantly, long-term personal relationships between the farmer and consumer can develop that ultimately keep bringing the consumer back to that farmer. The public can be confident in the origins of the food, ask questions, and stay close to the source of production. The producers get valuable feedback from customers. The absence of middlemen can also mean lower prices to the consumer. Case studies from places as diverse as Costa Rica, the United Kingdom, and U.S. show that a basket of produce purchased at a farmers' market often costs less than the same products purchased commercially (Halweil, 2004).

Over the past two decades, interest in farmers' markets has soared. The number of registered farmers' markets in U.S. has grown by an order of magnitude in 30 years, from about 300 in the mid-1970s to more than 3100 today (Henry Wallace Center, 2001). The city of Santa Cruz, CA, with a current population of about 60,000 people, started its first certified farmers' market in 1976. Today there is at least one market open every day in some part of the city, and on some days more than one, with many of them operating on a year-round basis. Most towns in surrounding communities outside the city limits now have their own markets as well. In the United Kingdom, a national organization provides support, representation, education, and certification for more than 300 markets (National Association of Farmers Markets, 2006). In a food system defined by standardization, mass distribution, and economies of scale, farmers' markets seem to be ideally suited for smaller-scale and beginning farmers. These farmers have the opportunity to begin by marketing relatively small amounts of produce, experiment with new crops and products, and getting into farming even when they have limited access to economic resources.

**FIGURE 23.6 Customers pick up their weekly CSA box.** Subscribers receive a box of fresh produce directly from the farmer during the growing season. Photo courtesy of Martha Brown.

### Community-Supported Agriculture

Compared to the farmers' market model, which is actually an ancient form of direct farmer distribution, CSA is a brand-new innovation. As the name implies, the social and economic bonds associated with the CSA model differ greatly from those in the global food system.

In basic terms, a CSA consists of a community of individuals who pledge support to a farm operation so that the farmland becomes, either directly or indirectly, the community's farm, with the growers and consumers providing mutual support and sharing the risks and benefits of food production. Typically, members or "shareholders" of the farm pledge either to pay a regular subscription cost through the season, or pay in advance to cover the anticipated costs of the farm operation and farmer's salary. In return, members receive a weekly box or basket share in the farm's bounty throughout the growing season (Figure 23.6).

Everyone benefits: the grower receives better prices for his or her crops, gains some financial security, and is relieved of much of the burden of marketing. Consumers receive produce that is fresher, tastier, harvested at the peak of ripeness, and also not fumigated, refrigerated, or packaged.

Beyond the obvious economic benefits of dealing directly with the customer, the CSA arrangement allows the farmer to receive working capital when it is most needed, reducing the need for bank loans and improving cash flow. The farmer also has a secure market for in-season produce and extra yields that might occur. In addition, it is not the farmer alone who takes on the risks of farming, including poor harvests due to unfavorable weather or pests.

While many CSA arrangements do not build in face-to-face contact between the farmer and the consumer, all CSAs create abundant opportunity for the democratic flow of information. The farmer, for example, can include educational information sheets and recipes along with the produce, and members can provide feedback about produce quality and preferences. Some CSAs provide the option of actually working on the farm. Even when members do not participate directly in production, however, their connection to the land and the production process is concrete and meaningful.

Many CSAs donate shares to needy families, soup kitchens, and food banks, or offer sliding-scale memberships so that their clientele are not just those with more resources. Each CSA is structured to meet the needs of the participants, so many types exist, with variation in the level of financial commitment and active participation by the shareholders, financing, land ownership, payment plans, and food distribution systems (Imhoff, 2001).

Most CSAs offer a diversity of vegetables, fruits, and herbs in season; some provide a full array of farm produce, including eggs, meat, milk, baked goods, and even firewood. Some farms team up with others so that members receive goods on a more nearly year-round basis. There is excellent opportunity for the design and management of the farm to reflect this diversity, providing opportunity and impetus for the application of the agroecological concepts and principles presented throughout this book.

The number of CSA operations in U.S. has grown rapidly. The first recognized CSA began in 1985, and by the turn of the century there were more than 1000 (Henry Wallace Center, 2001). In the United Kingdom, CSA-type arrangements have mushroomed in the past decade.

### Extended Networks

Alternatives to the global food system need not be restricted to local networks. An AFN that extends beyond an agricultural bioregion can still create shorter food supply chains, allow for democratic information exchange, and even — in a virtual sense — promote food-based community. Such extended networks take advantage of the communication and distribution infrastructures to allow consumer and producer (or the producer's representative) to transact their exchange directly despite their physical separation.

In extended AFNs, the product matters. It would make no sense, practically or environmentally, for an extended network to deal in a product such as lettuce. The best products for extended networks have no locally produced alternatives, are not rapidly perishable, have a high value, and can be shipped easily. Examples include chocolate, spices, and coffee.

Coffee is the prime example of such a product. It is the second most valuable commodity traded globally after oil. It is grown in one part of the world, and primarily consumed in another, distant from the site of production. This distance has allowed the coffee trade to develop into one of the most exploitative food chains known, with several transnational corporations controlling the roasting, sale, and distribution of the coffee produced by more than 25 million, mostly small-scale growers. The past decade has seen coffee prices paid to the farmer reach their lowest levels in history, while prices paid by consumers climbed higher. Exploitation is occurring on both ends of the food chain (Bacon et al., 2007).

Two types of extended networks have developed around the goal of providing coffee to consumers in developed countries without contributing to the exploitation of coffee growers in developing countries. In one type, the consumer purchases coffee directly from a cooperative of growers, with the transaction facilitated by a nonprofit organization. An example is provided by the coffee subscription program employed by the Community Agroecology Network (CAN).

In the second type of extended network, traditional retail channels are used, but links are eliminated from the distribution chain, and growers are guaranteed a much higher rate of return than they would get selling their coffee in the mainstream commodity market. An example is the Fair Trade Certified coffee for sale in many U.S. food stores and online (Figure 23.8).

Both types of networks can provide consumers with knowledge about the circumstances of the production, distribution of the product, and how it contrasts with that of the global food system. CAN, for example, sends subscribers a regular newsletter with information about the growers and their cooperative, and news about the global coffee economy. In this way, consumers are educated about the importance of their choices, and are connected with the growers. In addition to providing growers with a decent wage, AFNs focused on the coffee trade empower growers to use sustainable, low-external-input practices, such as growing coffee plants under the shade of the modified rainforest canopy.

### Promoting Local Food

Farmers' markets and community-CSA form the basis of an alternative local food system, but they are not likely to replace the traditional distribution and retail system. For this reason, it is important to change this system from within and have it concentrate as much as possible on local food. In any particular agricultural bioregion, many food retailers, restaurant owners, and managers of institutions serving food may be open to purchasing more of their food from local farmers, dairies, breweries, and other producers. In doing so, they may be able to reduce costs, increase their customer base, and stimulate the local economy. A small but growing number of restaurants and retailers in the U.S. and Europe have demonstrated the economic viability of serving or selling food that is almost entirely local in origin.

A coordinated campaign promoting local food can gain the support of chambers of commerce, business organizations, merchant's associations, farm bureaus, and the like. It can consist of any of the following elements:

- Farmers form cooperative arrangements for creating a regional identity in stores or markets.
- Local stores or restaurants offer products that reflect farmer practice or regional production, communicating the uniqueness or special focus of local production systems.
- Local producers and the regional food identity are promoted at special events such as fairs and farmers' markets.
- A common local-identity label is developed for local food products, to help inform consumer choice, and to promote the local food identity at the same time.
- The produce at food stores and supermarkets is labeled with its origin, whether it is local or not.
- Thematic tours of local farms and producers are arranged for both local residents and tourists.

#### Facilitating Informed Consumer Choice

The face-to-face contact between consumers and farmers at farm stands and farmers' market is an ideal occasion for sharing of understanding, farming practices, consumer desires, mutual needs and beliefs, and so on. In AFNs without opportunities for one-on-one communication, the major issue — in terms of democratic flow of information — is consumer education. The consumer needs to have available information that will allow him or her to make informed choices. This is equally important outside of AFNs, where it helps to challenge the abuses of the global food system and the alienation of the consumer.

Various means of facilitating informed consumer choice have been developed by consumer groups, organizations of farmers, extended alternative networks, and governments. In a bioregional context, labels of origin can help consumers distinguish local from nonlocal food, and become more aware of the difference. In the global food market, certification labels have become an important means of educating consumers. The U.S. government's certified organic label, and the Fair Trade certification mentioned above are two examples. The simple existence of such labels raises consciousness of the fact that consumer choices matter.

## FROM CONSUMERS TO FOOD CITIZENS

In our discussion of natural ecosystems in Chapter 2, a consumer was defined as an organism that ingests other organisms (or their parts or products) to obtain nutrients and food energy. Economics texts define the consumer as one who acquires goods or services, or simply a buyer. Neither of these definitions is adequate for describing the role that a buyer and eater of food must play in a sustainable food system. We need a different concept, one that points to the "consumer" as informed, responsible, and engaged. The term *food citizen* does the job well. According to Jennifer Wilkins, food citizenship is "the practice of engaging in food-related behaviors that support, rather than threaten, the development of a democratic, socially and economically just, and environmentally sustainable food system" (Wilkins, 2005).

People can practice food citizenship in many ways. We must first think about the food-system implications of

## COMMUNITY AGROECOLOGY NETWORK

The Community Agroecology Network (CAN) directly links farm communities in Central America with consumers in North America. By reducing the links in the coffee supply chain to the minimum, CAN is able to provide the farmers who grow the coffee with a much higher rate of return than they could get in the conventional coffee market. This fairer economic return supports farmers' efforts to grow their coffee using more ecologically benign methods, and it promotes sustainable livelihoods and economic development in the producer communities.

CAN originated in discussions among six researchers with more than 65 collective years of experience working with communities and farmer groups in Latin America. Concerned about the environmental and social impacts of the deepening coffee crisis, they explored ways of supporting the coffee-growing communities where they had developed long-term relationships. The direct-trade system that is now the core of CAN's efforts seemed the best foundation for accomplishing all their goals.

The name *Community Agroecology Network* was chosen because each word describes an important feature of the organization and its mission:

**Community:** The organization strives to improve the social and economic health of the producer communities, working with farmers and their families, farmer cooperatives, women's organizations, churches, and schools to help them implement their vision of integrating sustainable livelihoods and conservation practices. In North American communities, CAN works with universities and fair and alternative trade organizations and builds a membership network linking people interested in more conscientious consumption.

**Agroecology:** Through research and education, CAN promotes an agroecologically based approach to growing coffee in the mountainous tropical rainforest ecosystems of Latin America. With CAN's direct trade system freeing them from the need to emphasize short-term production over sustainability just to survive, farmers in the producer communities can apply agroecological principles that protect watersheds, soils, biodiversity, and the health of their communities. A direct link is established between an improved economic return and protection of environmental resources (Figure 23.7).

**Network:** CAN works to form networks and alliances among consumers, within producer communities, among different producer communities, and between consumers and producers. Local networks in the producer communities are based on face-to-face interaction. The broader network established between consumers and producers relies on the Internet and other forms of information technology. Through this latter network, the coffee drinker gains an understanding of the individuals and the ecosystems that produce his or her coffee, and farmers learn about the people drinking their coffee. Together they can forge relationships of mutual concern and commitment to sustainability.

CAN is part of a growing movement that uses the communication and technology tools of globalization to build extended alternative food networks. With these tools, the exchange between farmers and consumers that takes place at farmers' markets can now be scaled up to an international level — creating the beginnings of a kind of global farmer's market.

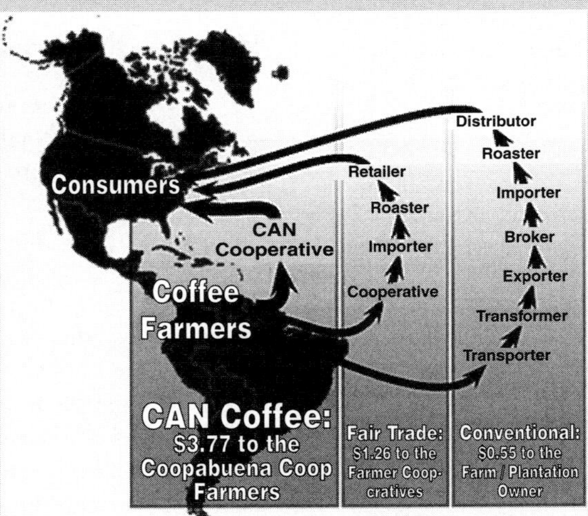


**FIGURE 23.7 Comparison of three coffee supply chains.** Each pound of CAN coffee sold directly to consumers returns up to \$3.77 to farmers. In comparison, farmers received \$1.26 for Fair Trade coffee and only \$0.55 for conventional coffee during the depths of the coffee crisis in 2002–3. Graphic courtesy of CAN.

**FIGURE 23.8 Fair Traded chocolate on a grocery shelf.** The Fair Trade certification tells consumers that the farmers who grew the cacao received fair compensation for their labor. Photo by Eric Engles.

how we eat, and change anything that negatively impacts sustainability. We can choose foods grown or raised in a manner that maintains and regenerates the natural resource base of agriculture, foods processed and marketed in ways that distribute rather than concentrate profits, and foods that are as much as possible transported locally. At the current time, this can best be done by buying outside of the mainstream food system, as we have discussed above.

Beyond being very intentional with one's daily food buying, food citizenship can also involve other actions that send signals for the need for change. One such action is requesting local or sustainably grown produce at mainstream markets and restaurants. Sometimes simply asking questions about where and how items were produced can have an effect. Other important actions include engaging in public policy development from the local to the global level, working to create a culture of sustainability, and educating others about how the present food system works to encourage unsustainable practices, consumer alienation, and agrarian decline.

There are many challenges we face in being truly good food citizens. First, the current food system offers few food options that meet the criteria of local and sustainable. Secondly, current federal policy promotes a narrow range of commodities and this has resulted in an abundance of cheap food components, rather than health or sustainability of the land or the people connected to that land. Third, institutional food buying policies at all levels, from local to federal, makes the purchase of local or sustainable food products extremely difficult. Fourth, we still lack a critical analysis of how health and nutrition systems have been impacted by current food market consolidation and policy. These barriers only emphasize the need for change at all levels.

In previous chapters, we have been calling for a new relationship between the farmer and the land; here we extend the change agenda to the relationship between the eater and the land. Just as we must maintain the land's ability to produce, we must also maintain our relationships as humans with that land and each other. We all have a responsibility for the land, the people it supports, and the culture of our food system (Freyfogle, 2001).

The scale of changes needed in our food system is daunting. Some change has occurred since Rich Merrill's *Radical Agriculture* appeared in 1976. Many of the ecological concepts and approaches to farming that he proposed in his chapter "Toward a Self-Sustaining Agriculture" appear in this textbook, and are being implemented broadly by farmers and researchers alike. But most of this change has been — following the distinctions made in Chapter 20 — at the second and third levels of the transition process, and too restricted to the farm and the farmer. Meanwhile, the separation between the eater and the farmer continues to grow, agriculture becomes more capital intensive, and consolidation puts it under the control of fewer people. To reach the fourth level of the transition process, more radical change — involving the entire food system — is required.

## FOOD FOR THOUGHT

1. When you go to your local supermarket, how much information is available on who grew the food, how it was grown, and how far away it came from?
2. What are the cultural differences in food preferences around the world, and how are these being changed by advertising and the Internet?
3. Food quality is a complex subject. What are some of the components of food quality that extend beyond nutritional aspects and incorporate more of the components of food system sustainability?
4. What part of the food that you eat every day could you grow yourself?
5. How many farmers' markets are there in your community?
6. How many farmers do you know?
7. If you were to change your diet in order to reflect the "culture of sustainability," what would you add or remove from what you eat now?

## INTERNET RESOURCES

Agricultural Marketing Service, Farmers Market Site
www.ams.usda.gov/farmersmarkets
A valuable source of information about the

growing network of farmers' markets in the United States.

Community Agroecology Network
www.communityagroecology.net
A source of information about the opportunities for developing sustainable relationships and alternative food networks that link consumers in the North with producers in the South.

National Association of Farmers Markets
www.farmersmarkets.net
A guide to the expanding network of certified farmers markets in the United Kingdom, and the work of The National Farmers' Retail and Markets Association in fostering the link between farmers and consumers.

Local Food Works
www.localfoodworks.org
A project of the Soil Association of the United Kingdom intended to foster sustainable local food systems through the development of local food networks.

Local Harvest
www.localharvest.org
A remarkable site that links the conscious consumer to a nationwide network of alternative food and farm products, including farmers markets, CSAs, farms, grocery stores, restaurants, and even an online store.

Alternative Farming Systems Information Center, CSA section
www.nal.usda.gov/afsic/csa
A CSA information resource that helps the consumer find a nearby CSA, learn what CSAs are and how they work. Provides links to other alternative farming systems information.

National Agricultural Statistics Service
www.nass.usde.gov
Access to an extensive database about agriculture.

Old Dog Documentaries
www.olddogdocumentaries.com
An organization that uses its documentary film skills to provoke grassroots solutions to some of society's most pressing problems, including food and environmental issues.

## RECOMMENDED READING

Bacon, C., V.E., Méndez, S.R., Gliessman, D. Goodman, and J. Fox, (eds.) *Confronting the Coffee Crisis: Sustaining Livelihoods and Ecosystems in Mexico and Central America*. MIT Press: Boston, Massachusetts, 2007. A probing look at the impact of commodity chains on rural communities in the global South, and alternative steps that can be taken by consumers and consumer organizations in the global North.

Freyfogle, E.T. (ed.) 2001. *The New Agrarianism: Land, Culture, and the Community of Life*. Island Press: Washington, DC. A gathering of powerful writings by well-known authors in the field of food and the environment that shows how there is a groundswell of change in the direction of strengthening our roots in the land, while bringing greater health to families, neighborhoods, and communities in rural as well as urban places.

Halweil, B. 2004. *Eat Here: Reclaiming Homegrown Pleasures in a Global Supermarket.* A World Watch book. Norton: New York. A highly engaging account of where our food comes from, why food system change is needed, and what the alternatives are.

Henderson, E. and R. Van En. 1999. *Sharing the Harvest: A Guide to Community Supported Agriculture*. Chelsea Green Publishing: White River Junction, VT. An informative guide to the history, development, implementation, and benefits of community-supported agriculture.

Magdorf, F, J.B. Foster, and F.H. Buttel (eds.) 2000. *Hungry for Profit*. Monthly Review Press: New York. A complete analysis of the issues and debates surrounding the global commodification of agriculture and the extent to which our environmental, social, and economic problems are intertwined with the structure of global agriculture as it now exists. The book demystifies the reasons why hunger proliferates in the midst of plenty and points the way we can all work towards sustainable solutions.

Menzel, P. and F. D'Aluisio 2005. *Hungry Planet: What the World Eats*. Material World Books: Napa, CA and Ten Speed Press: Berkeley, CA. A beautiful photographic essay of what families eat from around the world, placed in an important context of cultural diversity and the impacts of the global market place on food and diets.

Merrill, R. (ed.) 1976. *Radical Agriculture*. Harper Colophon Books, Harper & Row Publishers: New York. A thought-provoking analysis of the problems as well as a presentation of visionary solutions for moving towards a self-sustaining agriculture, written before most of us were promoting sustainability.

# 24 From Sustainable Agroecosystems to Sustainable Food Systems

Throughout most of this book, we have focused on the ecological processes in agriculture, with a view towards making agriculture sustainable in ecological terms. We have examined the development of practices and technologies that improve crop yields, reduce dependence on external inputs, and protect the on-farm environment. Implementation of these practices, and the ecological concepts and principles on which they are based, is critical to achieving sustainability. But it is also not enough.

As suggested in the previous chapter, all aspects of food production, distribution, and consumption must be included in the picture if agriculture as a whole is to become truly sustainable. This means transforming global food systems, which reach into nearly every aspect of human society and the built environment. Food systems are much bigger than farming, which makes sustainability about more than just farms (Buttel, 1993; Faeth, 1993). It is the complex interaction of *all* the ecological, technical, social, and economic parts of our food systems that will determine if these systems can be sustained over the long term. Our exploration of the consumer–farmer connection in Chapter 23 offered one important window into the issue of food system sustainability; in this chapter we will take a step back to take a broader look at how the principles and concepts presented in this text play into meeting what may be humankind's greatest challenge.

## A BROADER AGENDA

Many conventional agricultural research and extension institutions have begun to make the concept of sustainability a part of their programs, but they continue to suffer from a narrowness of approach. They usually focus on ways to improve yields and increase profits while using less energy and fewer inputs, giving little emphasis to protecting the environment beyond the farm, and failing to take into account the many and complex social and economic conditions that affect farms and farming communities. It is time for them to expand their focus to include entire food systems, and agroecology provides the foundation for doing so.

### Beyond the Individual Farm

The current discussions about sustainable agriculture go much beyond what happens within the fences of any individual farm (Gliessman, 2001; Boody, et al., 2005; Pretty, 2005). A farmer who has converted to sustainable practices knows that agriculture is more than a production activity in which the only goal is to achieve a high yield of a crop in a single season — it must also maintain the conditions on the farm that allow those yields to be produced from one season to the next. But a farmer can no longer only pay attention to the needs of his or her farm and expect to adequately deal with the concerns of long-term sustainability.

In many ways, agriculture is like a stream, with individual farms being different pools along that stream. Many things flow into a farm from upstream, and many things flow out of it as well. Farmers work hard to keep their own farms productive, being careful with the soil and what they add to the farm environment and take out as harvest; thus each pool in the agricultural stream has its own caretaker. In times past, each farmer could keep his or her pool in the stream functioning fairly well, and did not have to worry very much about what was going on either upstream or downstream.

But this "take care of your own" approach has its limits today. One reason is that each individual farmer has less and less control over what flows into his or her pool from upstream. Many unwanted things come from upstream, including pesticides, weed seeds, diseases, and polluted water from other farms. In addition, many things from upstream that the farmer needs he or she has little control over. These needs include labor, a market for the farm products, irrigation water, and farmland. As a result of these upstream influences — further complicated by legislated farm policies and the vagaries of the weather and the market — the stream becomes quite muddied and the job of keeping one's own pool clean very difficult.

Increasingly, each farmer must also consider a second problem: the way he or she takes care of the farm can have many effects downstream. Soil erosion and groundwater depletion can negatively affect surrounding farms. Poor or inefficient use of pesticides and fertilizers can contaminate the water and air, as well as leave potentially harmful residues on the food that others will consume. How well each farmer does on his or her own farm also has an influence on the viability of rural farm economies and cultures broadly. Both upstream and downstream factors are linked in complex ways that in different ways impinge upon the sustainability of each farm.

**FIGURE 24.1 A diverse organic farm in Davenport, CA.** On-farm diversity meshes with the off-farm environment as a farmer extends his view "downstream" from the farm.

The necessity of looking at the entire "stream" means adopting a whole-systems approach to achieving sustainability. We cannot be content with focusing primarily on the development of practices and technologies that are designed for the individual farm. When new technologies are evaluated primarily on their ability to increase yields and reduce costs, and only secondarily on how they reduce negative environmental impacts, they have little hope of contributing to long-term sustainability. The more complex impacts on the entire agricultural system must be included in the evaluation (Figure 24.1).

## Beyond the Bottom Line

Agriculture is very much an economic activity. A farming operation that is not economically viable will not exist for very long. Nevertheless, if economic factors — narrowly defined — remain the most import criteria for determining what is produced and how it gets produced, agriculture can never be sustainable over the long term.

The forces at work in a market-based economy, along with the various political structures put in place to regulate them, are very often at odds with the goals of sustainability (Ikerd, 2005). Market-determined variations in the costs of agricultural inputs and in the prices farmers receive for their production constantly introduce uncertainty and fluctuations into agricultural production. In response, farmers are forced to make decisions based on present economic realities rather than on ecological principles. Many governments, reluctant to let market forces alone set food prices, and influenced by various interest groups, employ price regulations and commodity subsidies that create various incentives and disincentives not necessarily in line with sound agroecological practice. Irrigation and reclamation projects, import/export policies, and agricultural research programs — all of which affect agriculture directly or indirectly — are generally undertaken on the basis of the short-term economic gain they can realize. In the developing world, governmental concerns about food security, the balance of trade, the development of export markets, and the attraction of foreign investment can result in policies with direct impact on farmers and their ability to continue producing food sustainably (Figure 24.2).

A basic problem with market economics is that it creates a context in which the short-term view completely overwhelms the long term. Even when there is agreement that long-term needs are important, economic realities dictate that short-term goals — this year's profit, next

**FIGURE 24.2 Monoculture sunflowers grown for vegetable oil production, Andalucia, Spain.** Prioritizing specific cash crops over diverse, local crops has altered greatly the agricultural landscape in many parts of the world.

year's production quotas — be given priority. Sustainability, in contrast, requires that planning and decision making take place in a time frame much longer than most economic programs ever consider. The environmental impacts of current practices and policies will be fully manifested only over a time span of many decades; likewise, the restoration of damaged ecosystems and the recovery of unproductive, degraded agricultural land will require decades if not centuries.

Another problematic aspect of market-economy forces is that the negative effects of economic activity on the environment, on people's health, and on people's livelihoods are bracketed off as "externalities." They are not counted as costs in the agricultural economic calculus, and are thus disregarded.

If agriculture is to continue as an economic activity into the long-term future, the economic context in which it is practiced must undergo a fundamental shift (Ikerd, 2005). We must recognize, first of all, that a healthy economy ultimately depends on a healthy environment — that agricultural production has an ecological foundation that can be destroyed. Then we must fashion an economics of sustainability, one that rewards ecologically sound practices in the marketplace, values the natural ecosystem processes that contribute to agricultural production, and takes into account now-externalized costs. The alternative food networks described in Chapter 23 are an important — and already-existing — part of this new economics.

Under the criteria of sustainability, long-term consequences become as important — or more so — than short-term economic gain, and nothing is considered an externality. Natural resources, usually exploited by agriculture, are treated as finite social goods. Inputs carry a purchase price based not only on the costs of their production, distribution, and application, but also on their environmental and social costs. Government food policies are based as much on their contribution to sustainability as on their ability to lower food prices (Figure 24.3).

## Beyond Technological Fixes

Part of the reason it is so easy for people to ignore the long-term time frame and the future consequences of our actions is that we have an abiding faith in technology. We trust that progress in technology will always solve our problems. In agriculture, the best example of our naive faith in technology is the Green Revolution. Through the development of higher-yielding plant varieties, Green

**FIGURE 24.3 Consumer choices at the marketplace.** Food prices and labels tend to hide rather than to reveal the actual consequences of food consumption decisions. (Photo by Eric Engles.)

Revolution scientists "solved" the problem of producing food for a rapidly growing world population. In the process, however, they created and exacerbated a host of other problems, including dependence on energy-intensive fertilizers and polluting pesticides, and more-rapid degradation of the soil resource around the world. Furthermore, the underlying problems — rapid population growth and its social causes, unequal distribution of food and agricultural resources — were hidden and not addressed.

This example demonstrates that technology may be an aid in solving a problem, but it can never be the entire solution. Social problems such as the unsustainability of our food systems always have deeper causes than can be addressed by technological innovation alone.

Today, biotechnology — especially the practice of DNA transfer from one unrelated organism to another — is being held out as the technological savior for problems in agriculture. We cannot let the promises of transgenic manipulation distract us from efforts to transform agriculture in ways that address the underlying causes of unsustainability or blind us to the problem of developing technologies that are out of reach of most of the small and resource-limited farmers of the world. Nor can we ignore the environmental and health risks of biotechnology discussed in Chapter 14. With so many unknowns, it would seem that the precautionary rule should apply, and that all transgenic applications be subject to a multifaceted sustainability analysis before use.

## Beyond Ecological Sustainability

Although we must define sustainability primarily in ecological terms, it is also true that ecological sustainability cannot come about in a social and economic context that does not support it. Even though agroecosystems function ecologically, they are manipulated to a high degree by humans. Because of this human impact, the ecological characteristics of an agroecosystem are intimately connected to human social and economic systems. Some of these connections, such as the influence of economic forces on agriculture, have just been discussed.

Recognizing the influence of social, economic, cultural, and political factors on agriculture, we must eventually shift our focus from the sustainability of agroecosystems to the sustainability of our *food systems*. Food systems have a global breadth and comprise all aspects of food production, distribution, and consumption. They include the economic relationships among landowners and farmworkers, farmers and food consumers, citizens of developed countries and citizens of developing countries; the political systems that control these relationships; the social structures that influence how people relate to food production and consumption; and the cultural systems that influence what people believe and value. For food systems to be sustainable, all of their human aspects must support the sustainability of their ecological aspects.

## THE PERMACULTURE IDEA

In the early 1970s, Australians Bill Molison and David Holmgren recognized that agriculture was heading in a direction that could not be sustained forever. Specialization, reliance on fossil fuel-based inputs, soil-eroding practices — all pointed to problems in the present and crises in the future. In response, they developed the idea of "permanent agriculture," or *permaculture*, a way of growing food that worked with nature instead of against it and could thus be sustained into the indefinite future.

Since the 1970s, the concept and practice of permaculture has matured and expanded, becoming a whole-system approach that integrates both the natural world and human society and moves beyond agriculture to include all the relationships between people and the natural environment. According to the Permaculture Research Institute of Australia (www.permaculture.org.au), permaculture is "the harmonious integration of landscape and people providing their food, energy, shelter, and other material and nonmaterial needs in a sustainable way."

In its early conception, permaculture was mainly applied toward household and community self-reliance. But its practitioners came to realize that self-reliance was meaningless to the majority of people, who for financial, practical, and geographic reasons lacked access to land. So permaculture expanded to embrace practical strategies for financing land acquisitions, building economically viable farming enterprises, and empowering communities and regions to create settlement patterns with less impact on the environment. In this way, permaculture became a holistic framework for creating a more sustainable human society. Today's permaculture can be applied at any level from the household to the bioregion, and used to work toward goals ranging from restoring damaged environments to building a more equitable and just society.

A core principle of permaculture is the beneficial design of systems. In essence, this involves carefully observing the elements of a system, noting their needs and potential contributions and the ways that energy can flow among them, and then positioning and interconnecting the elements so that they work in mutual benefit, becoming more diverse and stable over time. In many respects, this is what agroecologists do, too — but there is a crucial difference.

Agroecology springs from ecological principles, derived from observing the workings of natural systems, and through agroecological analysis of farm fields, communities, and landscapes. Agroecology therefore develops an in-depth understanding of the mechanisms and processes operating in sustainable agroecosystems. Permaculture's foundation, in contrast, is more philosophical. It is based on a conviction that observant and thoughtful humans can design ways of living on the earth based on harmony and mutual benefit.

Although permaculture endorses scientific analysis and observation, an agroecological approach requires greater depth in the actual analysis of systems. Agroecologists, therefore, are often better able to determine and monitor the elements of ecological sustainability in agriculture. But the philosophical basis of permaculture has its strengths as well. It inspires many people and spurs much-needed creative thinking and innovative problem solving. In addition, its holistic approach emphasizes to good effect the commonalities among systems big and small, natural, and social.

Ultimately, permaculture and agroecology are highly complementary. They share the goal of sustainability and the principle that well-designed human and cultivated systems are based on the patterns and processes operating in healthy natural systems. Their fusion has resulted in much good work and will undoubtedly produce more in the future.

## TOWARD FOOD SYSTEM SUSTAINABILITY

Much recent discussion in the agroecological and development communities has centered on formulating a definition of sustainability broad enough to include all the forces at work in global food systems (Gliessman, 2001; Pretty, 2002; Giampietro, 2004; Ikerd, 2005). Based on the accumulated experience presented in this text, we offer the following definition: *A sustainable food system is one that recognizes the whole-systems nature of food, feed, and fiber production in balancing the multifaceted concerns of environmental soundness, social equity, and economic viability among all sectors of society, across all nations and generations.* Inherent in this definition is the idea that agricultural sustainability has no limits in space or time — it involves all nations of the world, all living organisms, and all the globe's ecosystems, and extends into the future indefinitely.

### Conceptualizing the Social and Ecological Interconnections of the Food System

Working toward creating the sustainable food system described by this definition requires that we understand how society shapes, and is in turn shaped by, agricultural production, and how agriculture's ecological foundation is harnessed and affected by the process. A number of researchers have developed theoretical models that help us conceptualize these complex relationships. These models offer alternative conceptions of how agroecosystems exist in the intersection between nature and society — how they come about and change as a result of a complex interaction between ecological and social/economic factors. And each model suggests particular requirements for sustainability. Three of these models are presented below.

In the 1970s, when agriculture was dominated by Green Revolution thinking and most farmers and researchers were ignoring both the ecological basis of agriculture and its social ramifications, the Mexican agronomist and ethnobotanist Efraím Hernandéz Xolocotzi introduced a model that pictured agroecosystems as the outcome of constant coevolution between ecological, technological, and socioeconomic factors (Figure 24.4). Each factor influences the design and management of the agroecosystem while simultaneously interacting with the others. A complex set of feedbacks between factors and influences shapes the agroecosystem over time. Implicit in this model is the notion of balance, and imbalance becomes a matter of concern and a barrier to sustainability. For example, if market demands (a socioeconomic factor) cause farmers to put irrigation (a technological factor) into place so that limited rainfall (an ecological factor) can be ignored, the system can be thrown out of balance. The factor interaction may be confined to a back and forth interplay between socioeconomic pressures and technological fixes, and ecological sustainability is sacrificed.

Cornelia Flora offers a model focused on the rural communities in which agricultural production is based. In her model (Figure 24.5), agroecosystems come about as human communities use the potential of the local landscape to extract value and maintain themselves. In this process, they use four basic types of resources. Since these basic resource types are used to create new resources, she defines them as forms of capital. As with the Hernandez X. model, balance is required for sustainability. Specifically, all four forms of capital must be constantly replenished, not depleted. For example, a strategy that emphasizes the short-term accumulation of financial capital through intensive irrigated agriculture can decrease natural capital by causing salinization or loss of natural wetlands.

From the perspective of environmental sociology, Graham Woodgate proposes a model in which nature (the ecological foundation of agroecosystems) is wholly enclosed within society, and the relationship between the two recalls the coevolution of Herndandez X.'s model. In this model, agroecosystems — which occupy the interface

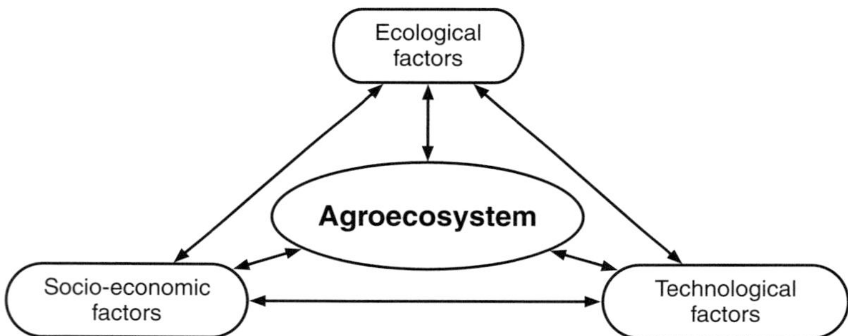


**FIGURE 24.4 Efraím Hernandéz Xolocotzi's model of the factors influencing the design and management of agroecosytems.** [Modified from Hernandez Xolocotzi, E. (ed.) 1977. *Agroecosistemas de México: Contribucciones a la Enseñanza, Investigación, y Divulgación Agrícola*. Colegio de Postgraduados: Chapingo, México.]

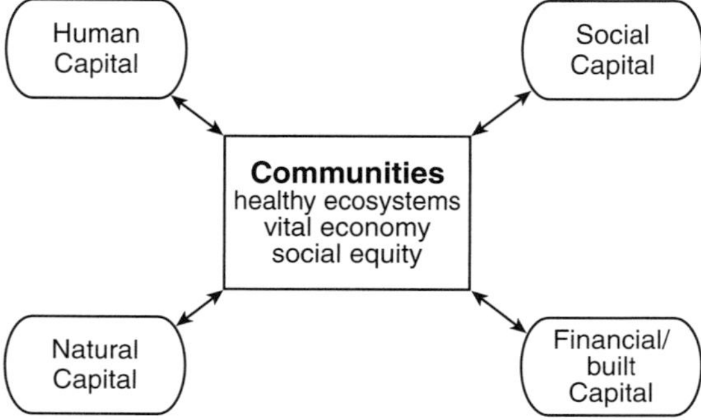


**FIGURE 24.5 Cornelia Flora's model of agroecosystems as products of human communities mediated by factors conceived as varying forms of capital.** [Modified from Flora, C. ed., 2001. *Interactions between Agroecosystems and Rural Communities*. Advances in Agroecology. CRC Press: Boca Raton, FL.]

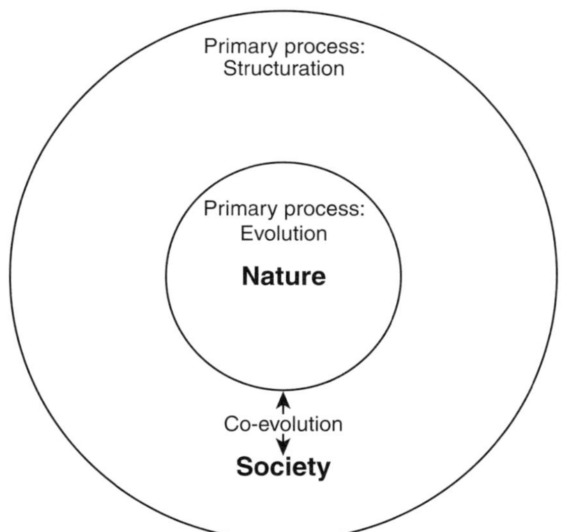


**FIGURE 24.6 Graham Woodgate's model of agroecosystems as part of the interface between society and nature.** [Modified from Gaulthier, R. and G. Woodgate. 2001. In S.R. Gliessman ed., *Agroecosystem Sustainability: Developing Practical Strategies*. Advances in Agroecology. Lewis Publishers: Boca Raton, FL; Woodgate, G., B. Ambrose-Oji, R. Fernandez Durán, G. Guzmán, and E. Sevilla Guzmán. 2005. *New Developments in Environmental Sociology.* Edward Elgar Publishing: Chetanham, U.K.]

between society and nature — can be sustainable only when their social "structuration" processes work in concert with the evolutionary processes of nature. When applied to specific localities, this model emphasizes the importance of place-based coevolution, in which the genetic diversity and ecological dynamics of local agroecosystems are linked with the political and ethical goals and struggles of rural peoples as they strive to maintain their land-based cultural identity (Figure 24.6) (Woodgate et al., 2005).

As the differences among these models suggest, there is no one right way to conceptualize the interactions among the social and ecological factors of agroecosystems, or of the food system as a whole. In an attempt to integrate some of the ideas in the models above, and to focus explicitly on sustainability and change, we present the diagram in Figure 24.7. As shown in the diagram, every agroecosystem develops in the context of, and draws upon, a broad social and ecological foundation. This foundation has a natural ecosystem context — what can be called the ecological background — and a social context — or social background. Any specific agroecosystem is shaped by local, regional, and global factors from both the social and ecological parts of the foundation. Humans can manipulate and manage many features of the ecological parts of the foundation, but the agroecosystem that is developed has to operate within the context of the social foundation upon which each culture is based. As change occurs in either the social or ecological foundation, the stage is set for changes in the agroecosystems that emerge from that foundation.

A sustainable agroecosystem develops when components from both the social and ecological foundation (Table 24.1) are combined into a system with a structure and function that reflects the interaction of human knowledge and preferences with the ecological components of the agroecosystem. The constant interaction between the social and ecological components occurs

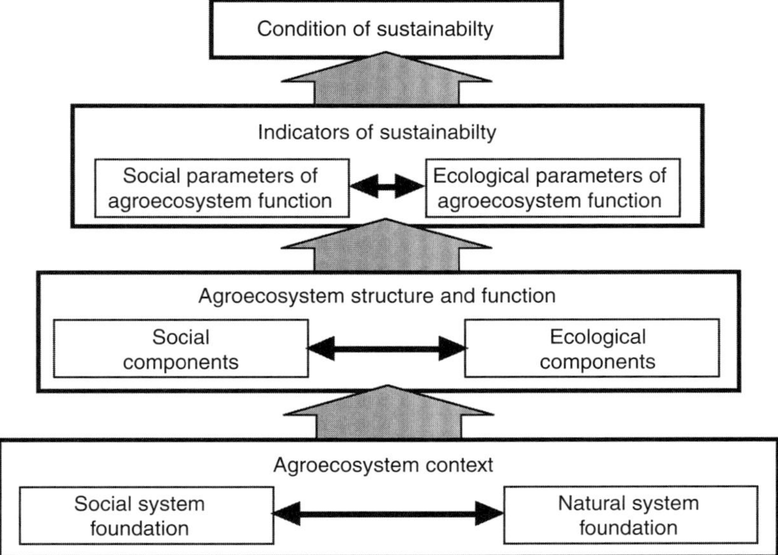


**FIGURE 24.7 The interaction among the social and ecological components of sustainable agroecosystems.** Applied to sets of interconnected agroecosystems, this model may represent the integrated structure of a sustainable food system.

as management techniques, practices, and strategies shift and change. The dynamic nature of agroecosystems sets the stage for a constant interplay between the organization and functioning of farms and the organization and interaction of the social, economic, and cultural components of the society within which the farms are embedded.

Over time, specific parameters (elements or properties) can be measured as indicators of sustainability. The measurable ecological parameters have been described throughout this book; the social parameters of sustainable agroecosystem function remain more difficult to identify and measure. The most useful and most easily measured parameters will undoubtedly vary with time, especially as knowledge and preferences shift and change, the environmental elements develop and mature, and the interactive processes of resistance and resilience combine to guide the rate and direction of such change. One of our greatest challenges is learning how to monitor the impacts of one indicator on another as social and ecological parameters interact, as well as finding ways to link indicators together in some kind of functional or causal relationship.

Ultimately the interaction between the social and ecological components of sustainable agroecosystems lead to the condition of sustainability itself. Sustainability becomes a complex set of conditions that are less dependent on the individual ecological or social components themselves than they are on the emergent qualities that come from their interaction.

Such a framework for defining sustainable agriculture incorporates a whole-systems way of looking at interactions between subsystems. The impact of a new input or practice in an agricultural system can be followed beyond its ecological effects to the societal level. By looking at each farm as an agroecosystem unto itself, and then as part of regional, national, and transnational food systems, we look beyond bottom-line economics for new ways to promote sustainability. Food systems become ecological-based systems that also maintain the societal needs of food security, social equity, and the quality of life that sustainability both engenders and requires.

**TABLE 24.1**
**Some of the Important Aspects of Social and Ecological Systems that Interact at Each Level in Sustainable Food Systems**

| Social System | Ecological System |
|---|---|
| **Social Conditions of Custainability** | **Ecological Conditions of Sustainability** |
| Equitability | Stability |
| Quality of life | Resilience |
| Satisfaction | Efficiency |
| Efficiency | Health |
| Cultural stability | Permanence |
| **Social Parameters of Agroecosystem Function** | **Ecological Parameters of Agroecosystem Function** |
| Dependence on external forces | Biotic diversity |
| Land tenure relationships | Soil fertility and structure |
| Role in food product economy | Moisture availability |
| Food quality | Rates of erosion |
| Share of return to workers | Rates of nutrient recycling |
| **Social Components of Agroecosystem Structure and Function** | **Ecological Components of Agroecosystem Structure and Function** |
| Farmers and farmworkers | Crop plants, livestock, and their genomes |
| Landowners | Biotic interactions |
| Consumers of food products | Noncrop organisms |
| Technical and practical knowledge | Soil quality |
| Ecocultural knowledge | Nutrient cycling |
| **Social System Foundation** | **Natural System Foundation** |
| *Shapes and constrains how human actors design and manage agroecosystems* | *Provides the raw materials for and physical context of agroecosystems* |
| Cultural components: values, ways of life, language | Local components: soil, soil microorganisms, native flora and fauna, ecological relationships, weather and climate, topography |
| Social components: class structure, societal institutions | Global components: biogeochemical cycles, solar radiation, climatic patterns |
| Economic components: market forces, position in global economy | |
| Political components: legislated policies, structure of governance | |

## Key Social Factors in Food System Sustainability

Changes in the social structures and relationships involved in the production, distribution, and consumption of food can be argued for on moral grounds alone. Inequities are present at all levels of the food system, and these have a dramatic effect on people's lives. But it is also possible to argue that certain social conditions and relationships are incompatible with agricultural sustainability. The present social components of global food systems work in concert with unsustainable and resource-degrading practices; a differently-organized set of social components and relations is required to support sustainable, resource-conserving practices. The full development and support of such an argument requires its own book, but here we will touch on some of the important issues involved.

### Equitability

Food system sustainability may very well require greater equity among people in terms of economic power, ownership and control of land, and access to and control of agricultural knowledge and resources (Figure 24.8). Today, inequity exists between citizens of developing nations and those of developed nations, between those who own agricultural land and those who don't, between those who make their living through labor and those who hold vast amounts of agricultural capital. It is important to recognize how this inequity may affect how agroecosystems are designed and managed. To what extent do those with relatively greater power feel compelled to make sure that agricultural production works in ways that maintain their power and control? What role does the structure of inequality have in making them more concerned with making a profit than with taking care of the land? Is the economic insecurity of farmers in developing countries causing them to be more concerned with short-term survival and improvement of their economic position than with conservation and sound agroecological practice?

### Sustainable Diet Patterns

Modernization and development throughout the world are causing an ongoing transformation of people's diets and food consumption patterns. People all over the world are eating more animal products, more food high in oil and fat, and more vegetables and fruit. Because high-protein and high-fat agricultural products are far more costly to produce — in terms of energy use, environmental impact, and land needs — than staple grains, we must carefully

**FIGURE 24.8 Lettuce harvesting for distant markets, Watsonville, CA.** Sustainability requires that all people share equitably in the fruits of the food system.

examine how these current trends in diet across the globe may be exacerbating food supply and production problems. Very difficult choices may confront human society if we fail to critically examine the impacts of human diet on the earth's ecological foundations and our ability to feed our growing numbers.

## Control of Population Growth

Underlying the problem of many people eating a richer and richer diet is the problem of a growing number of people. Experts from various disciplines disagree about what the human "carrying capacity" of the earth is, but few deny that a rapidly growing human population makes it increasingly difficult to simultaneously feed people, protect agricultural resources, and protect the integrity of the natural environment. Any coordinated effort to develop sustainable food systems, therefore, must take on the problem of how to better control human population growth.

## Self-sufficiency and Bioregionalism

Food systems around the world are fast becoming linked into one massive global food system. Although this trend

**FIGURE 24.9 Making corn tortillas in Tlaxcala, Mexico.** Local diets, local culture, and local agriculture are closely linked in sustainable food systems.

has many benefits, it also has many negative consequences for agricultural sustainability. A major problem is that global production and distribution of food requires large amounts of energy for transportation. Perhaps more significant, however, is that a global food system may help create the ideal conditions for exacerbation of the problem of inequity and may help erode sustainable traditional agroecosystems around the world.

In a globalizing food system, producers of inputs such as seeds, fertilizers, pesticides, and machinery are able to expand their influence on agroecosystems, and farmers become increasingly dependent on them and their products and knowledge. Agricultural land becomes more highly valued for its ability to produce for export than for local food needs. Human labor is increasingly replaced with mechanization. The overall consequences are greater integration of agroecosystems into technology- and input-based conventional agriculture, less autonomy, a diminished ability to grow food for local needs, and destruction of traditional and agricultural communities.

But globalization may also have the potential to counter these effects if it can be used instead to promote and support local control of land, the use of local knowledge, direct human involvement in agricultural production, and economic independence. These important aspects of ecologically based farm management may be key parts of designing a sustainable future (Figure 24.9).

### Links Between Agroecology and Changes in the Social Context of Agriculture

Even though the science of agroecology is focused on the ecological aspects of agroecosystems, the principles it seeks to apply can encourage positive changes in the social aspects and contexts of agroecosystems as well. At one level, agroecology is based on a variety of principles that have application in the social realm. Important ecological concepts such as interaction, networks, integration, mutualisms, feedbacks, cycles, relationships, and stability can all have parallels in social interaction and the structures of social institutions. Ecology is also fundamentally antireductionist, encouraging whole-systems thinking and long-time-span planning. At a more concrete level, when agroecological principles are applied in actual farming practice, opportunities arise for creating more-sustainable social relations and economic structures. Below are some examples of this connection.

- The drastic reduction in external inputs involved in agroecological management decreases a system's dependence on external economic forces and makes it less vulnerable to increases in the price of inputs. The farmer can benefit economically, and simultaneously enhance the ecological health of the agroecosystem and economic health of the local community.
- Agroecological principles require that management be based as much on practical knowledge of what works in the field as on theoretical knowledge. This requirement valorizes the practical knowledge of farmers and farmworkers, giving them greater power in demanding equitable treatment.
- The agroecological focus on knowledge of local conditions, local ecosystems, and locally adapted crop plants encourages a bioregional approach to agriculture and gives those who own and work the land more of a personal stake in the long-term ecological integrity of the agroecosystem.
- Agroecological management demands the farmer take a long-term view, which balances the need to prioritize annual yields and profit.
- Agroecological principles are best applied at a relatively small scale. This encourages production for regional consumption instead of for export; it is also more consistent with equitable land ownership and economic benefits than with concentration of farmland in the hands of a few.
- Agroecology recognizes the value of traditional systems that have proven to be stable in both ecological and social terms, and thus supports the social and economic structures and communities that make them possible.
- Agroecological management is best accomplished through intensification of human labor rather than intensification of the use of machinery. Because this labor requires a high degree of knowledge, judgment, and technical skill, agroecologically managed farming can provide many people with dignified and satisfying livelihoods.

The linkages described above demonstrate that changes in farming practices and techniques go hand-in-hand with changes in the overall social context of agriculture. Neither can happen entirely independently of the other, and agroecology has a role to play in both.

### Making Change Happen

Problems in agriculture create the pressures for changes that will support a sustainable agriculture. But it is one thing to express the need for sustainability, and another to bring about the changes that are required. Change must happen at many different levels, and in many different institutions, but ultimately all these aspects of the transition to sustainability are linked.

## CASE STUDY

### The Roots of Change

*"Leaders must shift from a perspective of incremental change to a perspective of transformation. The key is to paint a Vivid Picture. The point of the picture is not to accommodate the present; it's to accelerate the arrival of a particular version of the future. We must stand in the future and look back at the way we came."*

**–Paul Dolan, *True to Our Roots* (2003)**

All indicators tell us that the vast food system of the state of California, which plays a significant role both globally and nationally, is not sustainable. Although parts of the system are changing, moving toward more sustainable practices, the rate of change overall is slow. The impetus for change comes from organic growers, grass-fed beef ranchers, and other alternative farmers, who — together with CSAs, farmers' markets, "natural food" stores, and a segment of the state's consumers — form a kind of alternative food system. But this system is only a niche industry, existing at the margins of the conventional food system.

Over the last 20 years or more, a variety of foundations and grant makers have been putting considerable resources into helping spur the growth of more sustainable farming practices in California, and the alternative food system generally. In 1999, representatives of about a dozen of these philanthropic organizations got together to discuss the future. While they acknowledged progress toward the goal of sustainability, many were not satisfied with the pace of change. To pursue a collaborative strategy for food system change in California, they founded the Funders' Agriculture Working Group, which later became the Funders for Sustainable Food Systems (FSFS).

In 2003, after the release of a report identifying the barriers to systemic change in agriculture, FSFS created the Roots of Change (ROC) Fund as a source of funding for projects that could address and overcome the barriers to change. Top-level food and farming experts were brought together to form the ROC Council, which was given the role of providing advice and strategic counsel to the fund managers.

The ROC Council advised developing a bold, comprehensive, actionable vision for a new food system that could challenge and truly transform the conventional system. This idea became the Vivid Picture project, which was launched in March 2004. Focusing on answering the question "how do we move sustainability from niche to mainstream?" the Vivid Picture project team conducted 159 interviews (including 65 with food system experts and 27 with sustainable producers) and reviewed more than 700 datasets. In November 2005, the team presented its results in a report entitled *The New Mainstream.*

*The New Mainstream* serves as a systemic blueprint for a sustainable food system. The scope of change it advocates and envisions is broader and more ambitious than almost anything yet attempted. The report consists of four parts.

1. A *theory* of transformative change that can be embraced by the public and a broad coalition of stakeholders, leaders, and policy makers. This theory comes from an "opportunities-based" perspective, which posits that change strategies must, among other things, create more winners than losers, be based on incentives rather than penalties or fear, and address "mutual vested interests."
2. A comprehensive *vision* for a sustainable food system that can be in place by the year 2030. Under this system (1) all people in California will have access to quality food; (2) regional producers, manufacturers, distributors and purveyors of food will enjoy economic vitality; (3) food will help all people attain better health, experience well-being, and build their communities; (4) natural resources will be used in a way that maintains them in perpetuity; and (5) food-based regional and cultural identities will be enhanced throughout California.
3. A set of *implementation strategies* for realizing the vision. This agenda for change identifies policies, economic plans, and education and communication programs that can shift the entire system or entire components of the system. These change strategies are designed not simply to create demonstration programs, but to change the game itself. As they are put into place, these strategies will be tested and evaluated for viability and impact so it is possible to focus on those that provide maximum leverage.
4. A set of *success indicators* by which progress towards a sustainable food and farming system can be assessed. The indicators are based on existing, credible data sets and allow for the generation of additional data sets as needed. In addition to helping measure movement towards a sustainable system, they will help motivate it.

The ROC Council has embraced many of the ideas and strategies contained in *The New Mainstream* and is now considering how to implement them. Readers wanting an update can check the ROC website at www.rocfund.org.

### Agricultural Research

Agricultural research institutions and other sites involved in expanding agricultural knowledge can thwart change or support it. In order to move agriculture towards sustainability, research needs to be more anticipatory, allowing us to analyze both the immediate and the future ecological and social impacts of agriculture so that we can identify the key points in the systems on which to focus the search for alternatives or solutions to problems. In addition, research must promote long-term, systems-level research at multiple scales; it must better integrate the natural and social sciences; and it must use sustainability analysis as its guiding principle (Robertson et al., 2004). By following these principles — and understanding the ecological processes in sustainable agriculture — we can enter a new era in agricultural research.

### Agricultural Education

Introducing agroecology into agricultural curricula can be a challenge for many conventional, discipline-oriented, reductionist educators, who will have to learn how to deal with whole systems, rather than the component parts, and ask questions that go beyond normal experimental design for answers (Francis et al., 2003). Agroecological curricula are interdisciplinary, integrate the numerous and complex elements of food systems, focus on understanding the structures and functions of systems, and include the goals of food security and food equity. Agricultural education must shift its emphasis from teaching how to maximize single crop production in an environment of unlimited fossil fuels to fostering an understanding of food systems in all their complexity. Experiential learning in farming communities is another essential component of agroecological education. Students educated with such curricula will become agents of change.

### Farmer Practice

Change must also occur on the ground. Farmers making the transition to more sustainable practices, and farmers in traditional farming communities in developing countries fighting to preserve their ways of life, are leading the way in forging changes in agriculture. The more examples we have of sustainable, economically viable farming, the more likely the remaining parts of food production systems will follow their lead.

### The Food System

Each farm in the food system can be a focal point for changing how we do agriculture, but change must also occur in the global system within which agriculture is now practiced. As we have seen, it is all too easy for individual, piecemeal changes in practice to be captured and co-opted by those who control the processing, distribution, and marketing channels. Changes in the food system, as discussed in Chapter 23, begin with the development of food networks that link farmers and consumers, connect food production and environmental services, and create landscapes that promote healthy environments and healthy communities.

### Consumer Behavior

The decisions that consumers make about what to eat and where to buy it are a crucial factor influencing the nature of the food system. The more aware that consumers become of what's at stake, the more likely they are to support the kinds of alternative food systems outlined in Chapter 23. As consumers become knowledgeable about who grows their food, how, and where, they will make different choices.

### Food Policy

Incentive-based policies that encourage desirable practices are much more positive than penalties or taxes for nonsustainable farming practices, although a combination of both may be needed in order to hasten the change process. Policy should be guided by a vivid picture of what sustainable food systems will look like in the future, rather than trying to hold on to what has not worked in the past.

Overall, we must remember that agricultural systems are a result of the coevolution that occurs between culture and environment, and that humans have the ability to direct that coevolution. A sustainable agriculture values the human as well as the ecological components of food production, and recognizes their linkages and interdependencies.

## FOOD FOR THOUGHT

1. What are some examples of incentive-based policies that would encourage farmers to transition to sustainable farming practices?
2. How do the ecological and social components of agroecosystems change as we expand the scale of analysis from the farm, to the local community, to the region, to the nation, and ultimately, to the entire earth?
3. How are many economic support programs and pricing policies — originally designed to help agriculture — now hindering the development of sustainable agriculture?
4. What is your vision of an agroecosystem that can best integrate all of the components of sustainability we have been discussing?

5. What are the most significant changes that must occur in human attitudes towards agriculture and the food system in order to move towards sustainability?

## INTERNET RESOURCES

Center for Rural Affairs
cfra.org
A nonprofit working to strengthen small business, family farms and ranches, and rural communities.

Community Food Systems and Sustainable Agriculture Program at the University of Missouri
agebb.missouri.edu/sustain

Roots of Change Fund
www.rocfund.org
A transformative collaboration supporting work to catalyze the transition to a healthier food system and a healthier environment in California.

The Vivid Picture Project
www.vividpicture.net
An example of an innovative project that is designed to generate a blueprint for a sustainable food system in California and create a change agenda to achieve the new vision.

USDA Sustainable Agriculture Research and Education Program
www.sare.org
The USDA's resource database for sustainable agriculture.

## RECOMMENDED READING

Allen, P. 2004. *Together at the Table: Sustainability and Sustenance in the American Agrifood System.* Pennsylvania State University Press: Pennsylvania. A critical analysis of contemporary alternatives to the conventional food system, with a focus on the sustainable agriculture and community food security movements in the U.S.

Bell, M.M. 2004. *Farming for Us All: Practical Agriculture and the Cultivation of Sustainability.* Pennsylvania State University Press: University Park, PA. A discussion on the challenges and potential for sustainable farming in Iowa, based on years of research with family farmers.

Berry, W. 1977. *The Unsettling of America: Culture and Agriculture.* Sierra Club Books: San Francisco. A compelling story of the loss of rural farms and rural communities, and the cultural impacts of this loss. Perhaps even more relevant now than when it was published.

Clay, J. 2003. *World Agriculture and the Environment: A Commodity-by-Commodity Guide to Impacts and Practices.* Island Press: Washington, D.C. An analysis of global agricultural commodity production, its environment impacts, and a discussion of potential solutions.

Dolan, P. 2003. *True to Our Roots: Fermenting a Business Revolution.* Bloomberg Press: New York. A visionary book that calls for linking business and sustainability through the use of the 3 E's — economics, environment, and equity — and employing a triple bottom line of profits, planet, and people.

Dresner, S. 2002. *The Principles of Sustainability. Earthscan: London.* A book addressing some of the most challenging issues surrounding the quest for sustainability, including historical development of the concept, contemporary debates, and obstacles.

Flora, C. 2001. (ed.) *Interactions between Agroecosystems and Rural Communities. Advances in Agroecology.* CRC Press: Boca Raton, FL. An innovative social science perspective on agroecosystem design and management, with a focus on alternative ways of working with human communities to increase food system sustainability.

Ikerd, J. 2005. *Sustainable Capitalism: A Matter of Common Sense. Kumarian Press, Inc.: Bloomfield, CT.* A probing critique of capitalist economics, as well as a proposal for a new economics of sustainability based on social and ethical values whereby communities the world over can benefit and thrive.

Norton, B.G. 2005. *Sustainability: A Philosophy of Adaptive Ecosystem Management.* University of Chicago Press: Chicago. A discussion on the need for interdisciplinary approaches, drawing from philosophical and linguistic analyses, and focusing on cooperation and adoption through social learning

Pretty, J. 2002. *Agri-Culture: Reconnecting People, Land and Nature.* Earthscan: London. An urgent call for a new agricultural revolution, in which people matter, diversity of all types is valued, and sustainability is the goal.

Pretty, J. (ed.) 2005. *The Earthscan Reader in Sustainable Agriculture.* Earthscan: London. An essential reference collection of the most influential writers in sustainable agriculture.

Redclift, M. (ed.) 2000. *Sustainability: Life Chances and Livelihoods.* Routledge: London. A comprehensive volume calling for an expansion of the approaches to sustainability, with a focus on public policy in both developed and developing countries.

Reijntjes, C., B. Haverkort, and A. Waters-Bayer. 1992. *Farming for the Future: An Introduction to Low-External-Input and Sustainable Agriculture.* McMillan Press Limited: London. An interdisciplinary approach to sustainable agriculture for developing countries that combines theory and practice in the search for site-specific, farmer-friendly strategies for raising and maintaining productivity using low-cost local resources.

Röling, N.G. and M.A.E. Wagemakers. 2000. *Facilitating Sustainable Agriculture: Participatory Learning and Adaptive Management in Times of Environmental Uncertainty.* Cambridge University Press: Cambridge, UK. A book analyzing the implications of adopting sustainable agricultural practices, at both the farm and the landscape scale, with an emphasis on social aspects.

Strange, M. 1989. *Family Farming: A New Economic Vision.* University of Nebraska Press: Lincoln, NE. A vision for achieving healthy and sustainable family farms.

Wordwatch Institute. 2006. *State of the World.* W.W. Norton: New York. An annual report from the Worldwatch Institute about threats to the global environment and its ability to sustain life, as well as recommendations and strategies for change.

van der Ploeg, J.D., A. Long and J. Banks (eds.) 2002. *Living Countrysides.* Elsevier: Doetinchem, Netherlands. The state of art of the rural development process in Europe, with sustainability anchored in the multifunctionality of agriculture, landscape and culture.

# References


Alcorn, J. B. 1991. Ethics, economics, and conservation. In M. L. Oldfield and J. B. Alcorn (eds.), *Biodiversity: Culture, Conservation, and Ecodevelopment*. pp. 317–349. Westview Press: Boulder, CO.

Allison, J. 1983. An ecological analysis of home gardens (*huertos familiares*) in two Mexican villages. M.A. Thesis, University of California, Santa Cruz.

Altieri, M. A. 1983. *Agroecology*. University of California Press: Berkeley.

Altieri, M. A. 1994. *Biodiversity and Pest Management in Agroecosystems*. Food Products Press: New York.

Altieri, M. A. 1995a. Cover cropping and mulching. In M. A. Altieri (eds.), *Agroecology: The Science of Sustainable Agriculture*. 2nd ed. pp. 219–232. Westview Press: Boulder, CO.

Altieri, M. A. 1995b. *Agroecology: The Science of Sustainable Agriculture*. 2nd ed. Westview Press: Boulder, CO.

Altieri, M. A. 1999. Applying agroecology to enhance the productivity of peasant farming systems in Latin America. *Environment, Development, and Sustainability* 1: 197–217.

Altieri, M. A. 2002. Agroecology: the science of natural resource management for poor farmers in marginal environments. *Agriculture, Ecosystems, and Environment* 93: 1–24.

Altieri, M. A. and L. C. Merrick. 1987. *In situ* conservation of crop genetic resources through maintenance of traditional farming systems. *Economic Botany* 41: 86–96.

Altieri, M. A. and C. I. Nicholls. 2004a. An agroecological basis for designing diversified cropping systems in the tropics. In D. Clements and A. Shrestha (eds.), *New Dimensions in Agroecology*. Food Products Press/The Haworth Press: New York.

Altieri, M. A. and C. I. Nicholls. 2004b. *Biodiversity and Pest Management in Agroecosystems*. 2nd ed. Howarth Press: Binghamton, NY.

Amador, M. F. 1980. *Comportamiento de tres especies (Maiz, Frijol, Calabaza) en policultivos en la Chontalpa, Tabasco, Mexico*. Tesis Profesional, CSAT, Cardenas, Tabasco, Mexico.

Amador, M. F. and S. R. Gliessman. 1990. An ecological approach to reducing external inputs through the use of intercropping. In S. R. Gliessman (ed.), *Agroecology: Researching the Ecological Basis for Sustainable Agriculture*. pp. 146–159. Springer-Verlag: New York.

Anaya, A. L., L. Ramos, R. Cruz, J. G. Hernandez, and V. Nava. 1987. Perspectives on allelopathy in Mexican traditional agroecosystems: A case study in Tlaxcala. *Journal of Chemical Ecology* 13: 2083–2102.

Andersen, M. K., H. Hauggaard-Nielsen, P. Ambus and E. S. Jensen. 2004. Biomass production, symbiotic nitrogen fixation and inorganic N use in dual and tri component annual intercrops. *Plant and Soil* 266(1–2): 273–287.

Andow, D. A. 1991. Vegetational diversity and arthropod population responses. *Annual Review of Entomology* 36: 561–586.

Aoki, M., A. Gilbert, A. Hull, M. Karlberg, and M. MacDonald. 1989. *Multi-variable ecological analysis in a broccoli–lettuce intercrop*. Unpublished data compiled for Environmental Studies 130. University of California, Santa Cruz.

Azzi, G. 1956. *Agricultural Ecology*. Constable Press: London.

Bacon, C. 2005. Confronting the coffee crisis: Can fair trade, organic and specialty coffees reduce small-scale farmer vulnerability in northern Nicaragua? *World Development* 33(3): 497–511.

Bacon, C. M., V. E. Méndez, S. R. Gliessman, D. Goodman and J. A. Fox (eds.), 2007. *Confronting the Coffee Crisis: Sustaining Livelihoods and Ecosystems in Mexico and Central America*. MIT Press: Boston, MA.

Bais, H. P., R. Vepachedu, S. Gilroy, R. M. Callaway, and J. M. Vivanco. 2003. Allelopathy and exotic plant invasion: From molecules and genes to species interactions. *Science* 301: 1377–1380.

Baldwin, C. S. 1988. The influence of field windbreaks on vegetable and specialty crops. *Agriculture, Ecosystems and Environment* 22/23: 191–203.

Baldwin, C. S. and Johnston. 1984. *Windbreaks on the Farm*. Report #527. Publications of the Ontario Ministry of Agriculture and Food Provision: Ontario, Canada.

Barnes, J. P., A. R. Putnam, and B. A. Burke. 1986. Allelopathic activity of rye (*Secale cereale* L.). In A. R. Putnam and C. S. Tang (eds.), *The Science of Allelopathy*. pp. 271–286. John Wiley and Sons: New York.

Batish, D. R., H. P. Singh, R. K. Kohli, and S. Kaur. 2001. Crop allelopathy and its role in ecological agriculture. In R. K. Kohli, H. P. Singh, and D. R. Batish (eds.), *Allelopathy in Agroecosystems*. pp. 121–161. Food Products Press (imprint of The Howarth Press): New York.

Bazzaz, F. 1975. Plant species diversity in old field successional ecosystems in southern Illinois. *Ecology* 56: 485–488.

Bazzaz, F. A. 1996. *Plants in Changing Environments: Linking Physiological, Population, and Community Ecology*. Cambridge University Press: Cambridge, U.K.

Bender, Martin. 2002. Energy in Agriculture: Lessons from the Sunshine Farm Project. The Land Institute www.landinstitute.org/vnews/display.v/ART/2002/09/24/3dbeba6338ac3?in_archive=1. Accessed 29 March 2006.

Bellow, J. G. and P. K. R. Nair. 2003. Comparing common methods for assessing understory light availability in shaded–perennial agroforestry systems. *Agricultural and Forest Meteorology* 114(3–4): 197–211.

Berkes F., J. Colding, and C. Folke. 2000. Rediscovery of traditional ecological knowledge as adaptive management. *Ecological Applications* 10(5): 1251–1262.

Bethlenfalvay, G. J., M. G. Reyes-Solis, S. B. Camel, and R. Ferrera-Cerrato. 1991. Nutrient transfer between the root zones of soybean and maize plants connected by a common mycorrhizal inoculum. *Physiologia Plantarum* 82: 423–432.

Bilbro, J. D. and D. W. Fryrear. 1988. Annual herbaceous windbarriers for protecting crops and soils and managing snowfall. *Agriculture, Ecosystems and Environment* 22/23: 149–161.

Billings, W. D. 1952. The environmental complex in relation to plant growth and distribution. *Quarterly Review of Biology* 27: 251–265.

Blank, S. C., K. Jetter, C. M. Wick, and J. F. Williams. 1993. Incorporating rice straw into soil may become disposal option for farmers. *California Agriculture* 47: 8–12.

Boody, G., B. Vondracek, D. Andow, M. Krinke, J. Westra, J. Zimmerman, and P. Welle. 2005. Multifunctional agriculture in the United States. *Science* 55: 27–38.

Bottrell, D. G. 1996. The research challenge for integrated pest management in developing countries: A perspective for rice in southeast Asia. *Journal of Agricultural Entomology* 13(3): 185–193.

Boucher, D. H. 1985. The idea of mutualism, past and future. In D. H. Boucher (ed.), *The Biology of Mutualism*. pp. 1–28. Oxford University Press: New York.

Boucher, D. and J. Espinosa. 1982. Cropping systems and growth and nodulation responses of beans to nitrogen in Tabasco, Mexico. *Tropical Agriculture* 59: 279–282.

Brady, N. C., and R. R. Weil. 1996. *The Nature and Properties of Soils*. 11th ed. Prentice Hall: Upper Saddle River, New Jersey.

Brandle, J. R. and D. L. Hintz. 1988. Special issue: windbreak technology. *Agriculture, Ecosystems and Environment* 22/23: 1–598.

Brandle, J. R., L. Hodges, and X. H. Zhou. 2004. Windbreaks in North American agricultural systems. *Agroforestry Systems* 61(1): 65–78.

Briske, D. D. 1996. Strategies of plant survival in grazed systems: a functional interpretation. In J. Hodgson and A. W. Illius (eds.), *The Ecology and Management of Grazing Systems*. pp. 37–67. CAB International: Wallingford, U.K.

Brookfield, H. 2001. *Exploring Agrobiodiversity*. Columbia University Press: New York.

Brown, M. 1992. Agriculture and wetlands study initiated. *Cultivar* 10: 1–3.

Brown, L. 2001. *Eco-economy: Building an Economy for the Earth*. Norton and Co.: New York.

Brush, S. B. 1995. *In-situ* conservation of landraces in centers of crop diversity. *Crop Science* 35(2): 346–354.

Brush, S. B. 2004. *Farmer's Bounty: Locating Diversity in the Contemporary World*. Yale University Press: New Haven, CT.

Bruun, H. H. 2000. Patterns of species richness in dry grassland patches in an agricultural landscape. *Ecography* 23(6): 641–650.

Buck, L. E., J. P Lassoie, and E. C. M. Fernandes (eds.) 1999. *Agroforestry in Sustainable Agricultural Systems*. Advances in Agroecology Series. CRC Lewis Publishers: Boca Raton, FL.

Buttel, F. H. 1993. The sociology of agricultural sustainability: some observations on the future of sustainable agriculture. *Agriculture, Ecosystems, and Environment* 46: 175–186.

California Department of Food and Agriculture. 2004. County organic crop value and acreage reports. www.cdfa.ca.gov/is/fveqc/organic.htm.

Carlquist, S. 1965. *Island Life*. The Natural History Press: Garden City, NY.

Chacón, J. C. and S. R. Gliessman. 1982. Use of the "non-weed" concept in traditional tropical agroecosystems of southeastern Mexico. *Agro-Ecosystems* 8: 1–11.

Chang, T. T. 1984. Conservation of rice genetic diversity: luxury or necessity? *Science* 224: 251–256.

Chen, Y., M. De Nobili, and T. Aviad. 2004. Stimulatory effects of humic substances on plant growth. In F. Magdoff and R.R. Weil. *Soil Organic Matter in Sustainable Agriculture*. Advances in Agroecology Series. pp: 103–129. CRC Press: Boca Raton, FL.

Chou, C. H. 1990. The role of allelopathy in agroecosystems: studies from tropical Taiwan. In S. R. Gliessman (ed.), *Agroecology: Researching the Ecological Basis for Sustainable Agriculture*. pp. 104–121. Springer Verlag: New York.

Clark, E. A. 2004. Benefits of re-integrating livestock and forages in crop production systems. In D. Clements and A. Shrestha (eds.), *New Dimenions in Agroecology*. pp. 405–436. Food Products Press: Binghamton, NY.

Clements, D. and A. Shrestha (eds.), 2004. *New Dimensions in Agroecology*. Food Products Press: New York.

Colvin, W. I. and S. R. Gliessman. 2000. Fennel *(Foeniculum vulgare)* management and native species enhancement on Santa Cruz Island, California. pp. 184–189 in *Proceedings of the Fifth California Islands Symposium,* March 29–April 1, 1999. U.S. Department of the Interior, Minerals Management Service, Pacific OCS Region, OCS Study, MMS 99-0038.

Colwell, R., and D. Futuyma. 1971. On the measurement of niche breadth and overlap. *Ecology* 52: 567–576.

CGER (Commission on Geosciences, Environment and Resources). 2001. *Climate Change Science: An Analysis of Some Key Questions*. National Academies Press.

Connell, J. H. 1978. Diversity in tropical rain forests and coral reefs. *Science* 199: 1302–1310.

Connell, J. H. and R. O. Slayter. 1977. Mechanisms of succession in natural communities and their role in community stability and organization. *American Naturalist* 111: 1119–1144.

Conway, M. and L. Liston (ed.), 1990. *The Weather Handbook*. Conway Data: Atlanta.

Coombes D. S. and N. W. Sotherton. 1986. The dispersal and distribution of polyphagous predatory Coleoptera in cereals. *Annals of Applied Biology* 108: 461–474.

Corbett A. and R. E. Plant. 1993. Role of movement in the response of natural enemies to agroecosystem diversification: a theoretical evaluation. *Environmental Entomology* 22: 519–531.

Costanza, R., R. d'Arge, R. de Groot, S. Farber, M. Grasso, B. Hannon, K. Limburg, S. Naeem, R. V. O'Neil, J. Paruelo, R. G. Raskins, P. Sutton, and M. van den Belt.

1997. The value of the world's ecosystem services and natural capital. *Nature* 387: 253–260.

Cox, T. S., M. Bender, C. Picone, D. L. Van Tassel, J. B. Holland, E. C. Brummer, B. E. Zoeller, A. H. Paterson, and W. Jackson. 2002. Breeding perennial grain crops. *Critical Reviews in Plant Sciences* 21(2): 59–91.

Cox, G. W. and M. D. Atkins. 1979. *Agricultural Ecology.* Freeman: San Francisco.

Crews, T. E. and S. R. Gliessman. 1991. Raised field agriculture in Tlaxcala, Mexico: an ecosystem perspective on maintenance of soil fertility. *American Journal of Alternative Agriculture* 6: 9–16.

Cromack, K., R. E. Miller, O. T. Helgerson, R. B. Smith, and H. W. Anderson. 1999. Soil carbon and nutrients in a coastal Oregon Douglas-fir plantation with red alder. *Soil Science Society of America Journal* 63(1): 232–239.

Cunningham, A. B. 2001. *Applied Ethnobotany: People, Wild Plant Use and Conservation*. Earthscan: London.

Daily, G. C. 1995. Restoring value to the world's degraded lands. *Science* 269: 350–354.

Daily, G. C., G. Ceballos, and J. Pachecho. 2003. Country biogeography of neotropical mammals: conservation opportunities in agricultural landscapes of Costa Rica. *Conservation Biology* 17: 1814–1826.

Dalgaard, R., N. Halberg, and J. R. Porter. 2001. A model for fossil energy use in Danish agriculture used to compare organic and conventional farming. *Agriculture, Ecosystems, and Environment* 87: 51–65.

Dalgaard, T., T. Heidmann, and L. Mogensen. 2002. Potential N-losses in three scenarios for organic farming in a local area of Denmark. *European Journal of Agronomy* 16: 207–217.

Daubenmire, R. F. 1974. *Plants and Environment.* 3rd ed. John Wiley and Sons: New York.

Davies, B. 2003. Improving plant water use efficiency by physiological measures. Presented at the Biological Mechanisms of Water-Saving Agriculture session of the International Conference on *Water-Saving Agriculture and Sustainable Use of Water and Land Resources*, held in Yangling, Shaanxi, P.R. China, October 2003. *Journal of Experimental Botany* 54: i3.

Davis, J. H. C., J. N. Woolley, and R. A. Moreno. 1986. Multiple cropping with legumes and starchy roots. In C. A. Francis (ed.), *Multiple Cropping Systems*. pp. 133–160. MacMillan: New York.

DeHaan, L. R., D. L. Van Tassel, and T. S. Cox. 2005. Perennial grain crops: a synthesis of ecology and plant breeding. *Renewable Agriculture and Food Systems* 20: 5–14.

Del Angel-Perez, A. L. and M. A. Mendoza. 2004. Totonac homegardens and natural resources in Veracruz, Mexico. *Agriculture and Human Values* 21(4): 329–346.

Dell, J. D. and F. R. Ward. 1971. Logging residues on Douglas-fir region clearcuts: weights and volumes. Report 115: 1–10. Pacific Northwest Forest and Range Experiment Station, USDA Forest Service.

Denevan, W. M., and C. Padoch. 1987. *Swidden-fallow agroforestry in the Peruvian Amazon*. New York Botanical Gardens: Bronx, NY.

Denich, M., P. L. G. Vlek, T. D. D. Sa, K. Vielhauer, and W. G. Lucke. 2005. A concept for the development of fire-free fallow management in the Eastern Amazon, Brazil. *Agriculture Ecosystems and Environment* 110(1–2): 43–58.

Denys, C. and T. Tscharntke. 2002. Plant-insect communities and predator-prey ratios in field margin strips, adjacent crop fields, and fallows. *Oecologia* 130(2): 315–324.

Dewey, K. G. 1979. Agricultural development: impact on diet and nutrition. *Ecology of Food and Nutrition* 8: 247–253.

Dharmasena, K. H. and M. Wijeratne. 1996. Analysis of nutritional contribution of homegardening. *Tropenlandwirt* 97(2): 149–158.

Douglass, G. (ed.), 1984. *Agricultural Sustainability in a Changing World Order.* Westview Press: Boulder, CO.

Ellingson, L. J., J. B. Kauffman, D. L. Cummings, R. L. Sanford, and V. J. Jaramillo. 2000. Soil N dynamics associated with deforestation, biomass burning, and pasture conversion in a Mexican tropical dry forest. *Forest Ecology and Management* 137(1–3): 41–51.

Ellis, E. C. and S. M. Wang. 1997. Sustainable traditional agriculture in the Tai Lake region of China. *Agriculture, Ecosystems, and Environment* 61: 177–193.

Ellis, E. C. 2004. Long-term ecological changes in the densely populated rural landscapes of China. In R. S. DeFries, G. P. Asner, and R. A. Houghton (eds.), *Ecosystems and Land Use Change*. Geophysical Monograph Series. Vol. 153. pp. 303–320. American Geophysical Union: Washington, D.C.

Elton, C. 1927. *Animal Ecology.* Sidgwick and Jackson: London.

Entz, M. H., W. J Bullied, and F. Katepa-Mupondwa. 1995. Rotational benefits of forage crops in Canadian prairie cropping systems. *Journal of Production Agriculture* 8: 521–529.

Espinosa, J. 1984. The allelopathic effects of red root pigweed (*Amaranthus retroflexus*) and lambsquarters (*Chenopodium album*) on growth and nodulation of beans (*Phaseoleus vulgaris*). M.A. Thesis, Biology, University of California, Santa Cruz.

Etherington, J. R. 1995. *Environment and Plant Ecology.* 3rd ed. John Wiley and Sons: New York.

Evenari, M., D. Koller, L. Shanan, N. Tadmor, and Y. Ahoroni. 1961. Ancient agriculture in the Negev. *Science* 133: 979–996.

Ewel, J. 1999. Natural systems as models for the design of sustainable systems of land use. *Agroforestry Systems* 45: 1–21.

Ewel, J., C. Berish, B. Brown, N. Price, and J. Raich. 1981. Slash and burn impacts on a Costa Rican wet forest site. *Ecology* 62: 816–829.

Ewel, J., F. Benedict, C. Berish, B. Brown, S. R. Gliessman, M. Amador, R. Bermudez, A. Martinez, R. Miranda, and N. Price. 1982. Leaf area, light transmission, roots and leaf damage in nine tropical plant communities. *Agro-Ecosystems* 7: 305–326.

Ewert, D. and S. Gliessman. 1972. Regeneration under a tree in the tropical wet forest, Osa Peninsula. *Field Problem Report, Tropical Biology Course Book*, pp. 306–310. Organization for Tropical Studies.

Faeth, P. 1993. An economic framework for evaluating agricultural policy and the sustainability of production systems. *Agriculture, Ecosystems, and Environment* 46: 161–174.

FAO (Food and Agriculture Organization of the United Nations). 1998. *Special: Biodiversity for Food and Agriculture: Farm Animal Genetic Resources*. FAO: Rome.

FAO (Food and Agriculture Organization of the United Nations). 1999. Agricultural Biodiversity. Multifunctional Character of Agriculture and Land Conference. Background Paper 1. Maastricht, Netherlands. September 1999.

FAO (Food and Agriculture Organization of the United Nations). 2002. *Crops and Drops: Making the Best Use of Water for Agriculture*. FAO: Rome.

FAO (Food and Agriculture Organization of the United Nations). 2003. *Food self-reliance of developing countries and trade-distorting subsidies*. FAO Fact Sheet No. 1. Input for the WTO Ministerial Meeting in Cancun. FAO: Rome.

FAO (Food and Agriculture Organization of the United Nations). 2004. *The State of Food Insecurity in the World 2004*. FAO: Rome.

FAOSTAT (Food and Agriculture Organization of the United Nations, Statistics Database). www.apps.fao.org. Accessed January 2004.

FAOSTAT (Food and Agriculture Organization of the United Nations, Statistics Database). www.apps.fao.org. Accessed January to May 2005.

Farrell, J. 1990. The influence of trees in selected agroecosystems in Mexico. In S. R. Gliessman (ed.), *Agroecology: Researching the Ecological Basis for Sustainable Agriculture*. pp. 169–183. Springer-Verlag: New York.

Fearnside, P. M. 1986. *Human Carrying Capacity of the Brazilian Rainforest*. Columbia University Press: New York.

Finegan, B. 1996. Pattern and process in neotropical secondary rain forests: the first 100 years of succession. *Trends in Ecology and Evolution* 11(3): 119–124.

Fitter, A. 2003. Making allelopathy respectable. *Science* 301: 1337–1338.

Flietner, D. 1985. Don Ignacio's home garden. In R. Jaffe and S. R. Gliessman (eds.), *Proceedings of OTS course 85-4*. pp. 57–67. San Jose, Costa Rica.

Flora, C. (ed.), 2001. Introduction, and shifting agroecosystems and communities. In Flora, C. *Interactions between Agroecosystems and Rural Communities*. Advances in Agroecology. Chapters 1 and 2. pp. 1–13. CRC Press: Boca Raton, FL.

Fluck, R. C. (ed.), 1992. *Energy in Farm Production*. Vol. 6. Elsevier: Amsterdam.

Fowler, C. and P. Mooney. 1990. *Shattering: Food, Politics, and the Loss of Genetic Diversity*. Univ. Arizona Press: Tucson, AZ.

Francis, C., G. Lieblein, S. Gliessman, T. A. Breland, N. Creamer, R. Harwood, L. Salomonsson, J. Helenius, D. Rickerl, R. Salvador, M. Wiendehoeft, S. Simmons, P. Allen, M. Altieri, J. Porter, C. Flora, and R. Poincelot. 2003. Agroecology: the ecology of food systems. *Journal of Sustainable Agriculture* 22: 99–118.

Francis, C. A. (ed.), 1986. *Multiple Cropping Systems*. MacMillan: New York.

Franzluebbers, A. J. 2004. Tillage and residue management effects on soil organic matter. In Magdoff, F. and R. R. Weil. *Soil Organic Matter in Sustainable Agriculture*. Advances in Agroecology Series. pp. 227–268. CRC Press: Boca Raton, FL.

Freyfogle, E. T. 2001. *The New Agrarianism: Land, Culture, and the Community of Life*. Island Press: Washington, D.C.

Fujiyoshi, P. 1997. *Ecological aspects of inteference by squash in a corn/squash intercropping agroecosystem*. Unpublished data from Ph.D. thesis in Biology, in progress. University of California, Santa Cruz.

Fujiyoshi, P., S. R. Gliessman, and J. H. Langenheim. 2002. Inhibitory potential of compounds released from squash (*Cucurbita* spp.) under natural conditions. *Allelopathy Journal* 9(1): 1–8.

Furuno, T. 2001. *The Power of Duck*. Tagari Publications: Tasmania, Australia.

García-Espinosa, R., R. A. Robinson, J. A. Aguilar-P, S. Sandoval-I, and R. Guzmán-P. 2003. Recurrent selection for quantitative resistance to soil-borne diseases in beans in the Mixteca region, Mexico. *Euphytica* 130: 241–247.

Gardner, G. and B. Halweil. 2000. *Overfed and Underfed: the Global Epidemic of Malnutrition*. Worldwatch paper #150.

Gaulthier, R. and G. Woodgate. 2001. Coevolutionary Agroecology: a policy-oriented analysis of socio-environmental dynamics, with special reference to forest margins in north Lampung, Indonesia. In S.R. Gliessman (ed.), *Agroecosystem Sustainability: Developing Practical Strategies*. Advances in Agroecology. Lewis Publishers: Boca Raton, FL.

Gause, G. F. 1934. *The Struggle for Existence*. Williams and Wilkins: Baltimore.

Giampietro, M. 2004. *Multi-Scale Integrated Analysis of Agroecosystems*. CRC Press: Boca Raton, FL.

Gliessman, S. R. 1978a. Sustained yield agriculture in the humid lowland tropics. *INTECOL Newsletter* 7: 1.

Gliessman, S. R. 1978b. Memorias del Seminario Regional sobre la Agricultura Agricola Tradicional. CSAT, Cardenas, Tabasco, Mexico.

Gliessman, S. R. 1978c. Unpublished research report. Colegio Superior de Agricultura Tropical.

Gliessman, S. R. 1978d. The establishment of bracken following fire in tropical habitats. *American Fern Journal* 68: 41–44.

Gliessman, S. R. 1979. Allelopathy in crop/weed interactions in the humid tropics. In A. Amador (ed.), *Memoirs of Seminar Series of Ecology*. pp. 1–8. Colegio Superior de Agricultura Tropical: Cardenas, Tabasco, Mexico.

Gliessman, S. R. 1982. Nitrogen cycling in several traditional agroecosystems in the humid tropical lowlands of southeastern Mexico. *Plant and Soil* 67: 105–117.

Gliessman, S. R. 1983. Allelopathic interactions in crop–weed mixtures: applications for weed management. *Journal of Chemical Ecology* 9: 991–999.

Gliessman, S. R. 1987. Species interactions and community ecology in low external-input agriculture. *American Journal of Alternative Agriculture* 11: 160–165.

Gliessman, S. R. 1988. *Allelopathic effects of crops on weeds*. Unpublished manuscript. University of California, Santa Cruz.

Gliessman, S. R. 1989. Allelopathy and agricultural sustainability. In C. H. Chou and G. R. Waller (eds.), *Phytochemical Ecology: Allelochemicals, Mycotoxins and Insect*

*Pheromones and Allomones*. pp. 69–80. Institute of Botany: Taipei, Taiwan.

Gliessman, S. R. (ed.), 1990a. *Agroecology: Researching the Ecological Basis for Sustainable Agriculture*. Springer-Verlag: New York.

Gliessman, S. R. 1990b. Integrating trees into agriculture: the home garden agroecosystem as an example of agroforestry in the tropics. In S. R. Gliessman (ed.), *Agroecology: Researching the Ecological Basis for Sustainable Agriculture*. pp. 160–168. Springer-Verlag: New York.

Gliessman, S. R. 1992a. Agroecology in the tropics: achieving a balance between land use and preservation. *Environmental Management* 16: 681–689.

Gliessman, S. R. 1992b. Unpublished data. Agroecology Program, University of California.

Gliessman, S. R. 1995. Sustainable agriculture: an agroecological perspective. In J. S. Andrews and I. C. Tommerup (ed.), *Advances in Plant Pathology*. Vol. 11. pp. 45–56.

Gliessman, S. R. (ed.) 2001. *Agroecosystem Sustainability: Developing Practical Strategies*. Advances in Agroecology. CRC Press: Boca Raton, FL.

Gliessman, S. R. 2002a. Allelopathy and agroecology. In Inderjit and A.U. Mallik (eds.), *Chemical Ecology of Plants: Allelopathy in Aquatic and Terrestrial Ecosystems*. pp. 173–185. Birkhauser Verlag: Switzerland.

Gliessman, S. R. 2002b. Making the conversion to sustainable agroecosystems: getting from here to there with agroecology. *California Certified Organic Farmers Newsletter* 19(3): 6–8.

Gliessman, S.R. 2004. Integrating agroecological processes into cropping systems research. *Journal of Crop Improvement* 11: 61–80.

Gliessman, S. R. and M. F. Amador. 1980. Ecological aspects of production in traditional agroecosystems in the humid lowland tropics of Mexico. In J. I. Furtado (ed.), *Tropical Ecology and Development*. International Society for Tropical Ecology: Kuala Lumpur, Malaysia.

Gliessman, S. R. and R. Garcia-Espinosa. 1982. A green manure crop for the lowland tropics. In National Research Council (ed.), *Tropical Legumes: Resources for the Future*. National Academy of Science: Washington, D.C.

Gliessman, S. R., R. Garcia-Espinosa, and M.F. Amador. 1981. The ecological basis for the application of traditional agricultural technology in the management of tropical agroecosystems. *Agro-Ecosystems* 7: 173–185.

Gliessman, S. R., M. R. Werner, S. Sweezy, E. Caswell, J. Cochran, and F. Rosado-May. 1996. Conversion to organic strawberry management changes ecological processes. *California Agriculture* 50: 24–31.

Glover, J. D. 2005. The necessity and possibility of perennial grain production systems. *Renewable Agriculture and Food Systems* 20: 1–4.

Gomez-Pompa, A., M. F. Allen, S. L. Fedick, and J. J. Jimenez-Osornio (eds.), 2003. *The Maya Area: Three Millennia at the Human-wildland Interface*. The Howarth Press, Inc.: New York.

Gonzalez Jácome, A. 1985. Home gardens in Central Mexico. In I. S. Farrington (ed.), *Prehistoric Intensive Agriculture in the Tropics*. BAR: Oxford, England.

Gonzalez Jácome, A. 1986. *Agroecologia del Suroeste de Tlaxcala, Historia y Sociedad en Tlaxcala*. pp. 201–220. Gobierno del Estado de Tlaxcala: Tlaxcala, Mexico.

Gonzalez Jácome, A. and S. Del Amo Rodriguez (eds.), 1999. *Agricultura y Sociedad en México: Diversidad, Enfoques, Estudios de Caso*. Plaza y Valdes Editores: Mexico, D.F., Mexico.

Gordon, A. M. and S. M. Newman. 1997. *Temperate Agroforestry Systems*. CABI Publishing: London.

Gregg, W. P. Jr. 1991. MAB Biosphere reserves and conservation of traditional land use systems. In M. L. Oldfield and J. B. Alcorn (eds.), *Biodiversity: Culture, Conservation, and Ecodevelopment*. pp. 274–294. Westview Press: Boulder, CO.

Gregson, M. E. 1996. Long-term trends in agricultural specialization in the United States: some preliminary results. *Agricultural History* 70: 90–101.

Grime, J. P. 1977. Evidence for the existence of three primary strategies in plants and its relevance to ecological and evolutionary theory. *American Naturalist* 111: 1169–1194.

Grinnell, J. 1924. Geography and evolution. *Ecology* 5: 225–229.

Grinnell, J. 1928. Presence and absence of animals. *University of California Chronicles* 30: 429–450.

Guo, J. Y. and A. D. Bradshaw. 1991. The flow of nutrients and energy through a Chinese farming system. *Journal of Applied Ecology* 30: 86–94.

Guyot, G. 1989. Les effets aérodynamiques et microclimatiques des brise-vent et des amenagements régionaux. In W. S. Reifsnyder and T. O. Darnhofer (eds.), *Meteorology and Agroforestry*. pp. 485–520. ICRAF: Nairobi.

Hall, S. J. G. and J. Ruane. 1993. Livestock breeds and their conservation: a global overview. *Conservation Biology* 7: 815–825.

Halweil, B. 2004. *Eat Here: Reclaiming Homegrown Pleasures in a Global Supermarket*. A WorldWatch book. Norton: New York.

Hamilton, M. B. 1994. Ex-situ conservation of wild plant species: time to reassess the genetic assumptions and implications of seed banks. *Conservation Biology* 8(1): 39–49.

Hansen, J. W. and J. W. Jones. 1996. A systems framework for characterizing farm sustainability. *Agricultural Systems* 51: 185–201.

Hanson, H. C. 1939. Ecology in agriculture. *Ecology* 20: 111–117.

Hardison, J. R. 1976. Fire and flame for plant disease control. *Annual Review of Phytopathology* 14: 355–379.

Harper, J. L. 1974. The need for a focus on agro-ecosystems. *Agroecosystems* 1: 1–12.

Harper, J. L. 1977. *Population Biology of Plants*. Academic Press: London.

Hart, R. D. 1979. *Agroecosistemas: Conceptos Basicos*. CIAT: Turrialba, Costa Rica.

Hart, R. D. 1984. Agroecosystem determinants. In R. Lowrance, B. R. Stinner, and G. J. House (eds.), *Agricultural Ecosystems: Unifying Concepts*. pp. 105–119. John Wiley and Sons: New York.

Hart, R. D. 1986. Ecological framework for multiple cropping research. In C. A. Francis (ed.), *Multiple Cropping Systems*. pp. 40–56. MacMillan: New York.

Hartwig, N. L. and H. U. Ammon. 2002. Cover crops and living mulches. *Weed Science* 50: 688–699.

Hauggaard-Nielsen H. and E. S. Jensen. 2005. Facilitative root interactions in intercrops. *Plant and Soil* 274(1–2): 237–250.

Hendrix, P. F., R. W. Parmilee, D. A. Crossley, D. C. Coleman, E. P. Odum, and P. M. Groffman. 1986. Detritus food webs in conventional and no-tillage agroecosystems. *BioScience* 36: 374–380.

Henry A. Wallace Center. 2001. Making changes: turning local vision into national solutions. From Henry A. Wallace Center for Agricultural and Environmental Policy. Winrock International: Arlington, VA.

Hernandez Xolocotzi, E. (ed.) 1977. *Agroecosistemas de México: Contribucciones a la Enseñanza, Investigación, y Divulgación Agrícola*. Colegio de Postgraduados: Chapingo, México.

Hill, S. B. 1998. Redesigning agroecosystems for environmental sustainability: a deep systems approach. *Systems Research and Behavioral Science* 15: 391–402.

Hill, S. B. 2004. Redesigning pest management: a social ecology approach. *Journal of Crop Improvement* 12: 491–510.

Hisyam, A., M. Anwar, and Suharto. 1979. Social and Cultural Aspects of Homegardens. Presented at the Fifth International Symposium of Tropical Ecology, Kuala Lumpur.

Hodgson, J. and A. W. Illius (eds.), 1996. *The Ecology and Management of Grazing Systems*. pp. 247–272. CAB International: Wallingford, U.K.

House, G. J. and B. R. Stinner. 1983. Arthropods in no-tillage soybean agroecosystems: community composition and ecosystem interactions. *Environmental Management* 7: 23–28.

Hubbell, S. P., R. B. Foster, S. T. O'Brien, K. E. Harms, R. Condit, S. J. Wechsler, S. J. Wright, and S. Loo de Lao. 1999. Light-gap disturbances, recruitment limitation, and tree diversity in a neotropical forest. *Science* 283: 554–557.

Hunt, S. C., J. L. Sawin, and P. Stair. 2006. Cultivating renewable alternatives to oil. In L. Starke (ed.), *State of the World 2006*. Worldwatch Institute. W.N. Norton & Co.: New York.

Hutchinson, G. E. 1957. Concluding remarks. Population Studies: Animal Ecology and Demography. Presented at the Cold Spring Harbor Symposium on *Quantitative Biology*.

Ikerd, J. 2005. *Sustainable Capitalism: A Matter of Common Sense*. Kumarian Press, Inc.: Bloomfield, CT.

Illic, P. 1989. Plastic tunnels for early vegetable production: are they for you? Family Farm Series. pp. 1–12. Small Farm Center, University of California Cooperative Extension: Davis, CA.

Imhoff, D. 2001. Linking tables to farms. In E. R. Freyfogle (ed.), *The New Agrarianism: Land, Culture, and the Community of Life*. Island Press: Washington, DC.

Innis, Q. 1997. *Intercropping and the Scientific Basis of Traditional Agriculture*. Intermediate Technology Development Group: London, U.K.

INTECOL. 1976. *Report on an International Programme for Analysis of Agro-Ecosystems*. International Association for Ecology.

Jackson, W. 2002. Natural systems agriculture: a truly radical alternative. *Agriculture, Ecosystems and Environment* 88: 111–117.

Jackson, W. and L. L. Jackson. 1999. Developing high seed yielding perennial polycultures as a mimic of mid-grass prairie. In R. LeFroy (ed.), E. C. Lefroy, R. J. Hobbs, M. H. O'Connor, J. S. Pate. *Agriculture as a Mimic of Natural Ecosystems*. pp. 1–37. Kluwer Academic Publishers: Netherlands.

Jackson, D. L. and L. L. Jackson. 2002. *The Farm as Natural Habitat: Reconnecting Food Systems with Ecosystems*. Island Press: Washington, D.C.

James, C. 2003. *Global Status of Commercialized Transgenic Crops: 2003*. International Service for the Acquisition of Agri-Biotech Applications.

Jansen, D. M., J. J. Stoorvogel, and R. A. Schipper. 1995. Using sustainability indicators in agricultural land use analysis: an example from Costa Rica. *Netherlands Journal of Agricultural Science* 43: 61–82.

Janzen, D. H. 1973. Tropical agroecosystems. *Science* 182: 1212–1219.

Jimenez-Osornio, J. J. and V. M. Rorive (eds.), 1999. *Los Camellones y Chinampas Tropicales*. Ediciones de la Universidad Autonoma de Yucatan: Merida, Yucatan, Mexico.

Joern, A. 2005. Disturbance by fire frequency and bison grazing modulate grasshopper assemblages in tallgrass prairie. *Ecology* 86(4): 861–873.

Jordan, C. 1985. *Nutrient Cycling in Tropical Forests*. John Wiley and Sons: New York.

Joshi, L., P. K. Shrestha, C. Moss, and F. L. Sinclair. 2004. Locally derived knowledge of soil fertility and its emerging role in integrated natural resource management. In M. van Noordwijk, G. Cadish, and C. K. Ong (eds.), *Below-ground Interactions in Tropical Agroecosystems: Concepts and Models with Multiple Plant Components*. pp. 17–40. CABI Publishing: Oxfordshire and Cambridge, MA.

Kareiva, P. 1996. Contributions of ecology to biological control. *Ecology* 77: 1963–1964.

Keever, C. 1950. Causes of succession in old fields of the Piedmont, North Carolina. *Ecological Monographs* 20: 229–250.

Kehlenbeck, K. and B. L. Maass. 2005. Crop diversity and classification of homegardens in Central Sulawesi, Indonesia. *Agroforestry Systems* 63(1): 53–62.

Kimbrell, A. 2002. *The Fatal Harvest Reader*. Island Press: Washington.

Klages, K. H. W. 1928. Crop ecology and ecological crop geography in the agronomic curriculum. *Journal of the American Society of Agronomy* 20: 336–353.

Klages, K. H. W. 1942. *Ecological Crop Geography*. MacMillan: New York.

Klee, G. 1980. *World Systems of Traditional Resource Management*. Halstead: New York.

Kohli, R. K., H. P. Singh, and D. R. Batish. 2001. *Allelopathy in Agroecosystems*. Food Products Press (imprint of The Howarth Press): New York.

Koike, S. and K. V. Subbarao. 2000. Broccoli residues can control Verticillium wilt of cauliflower. *California Agriculture* 54: 30–33.

Koizumi, H., Y. Usami, and M. Satoh. 1992. Energy flow, carbon dynamics, and fertility in three double-cropping agroecosystems in Japan. In M. Shiyomi, E. Yano, H. Koizumi, D. A. Andow, and N. Hokyo (eds.), *Ecological Processes in Agro-Ecosystems*. pp. 157–188. National Institute of Agro-Environmental Sciences: Tukaba, Ibaraki, Japan.

Koocheki, A, M. Nassiri, S. R. Gliessman, and A. Zarea. 2006. Agrobiodiversity of field crops in Iran. Unpublished manuscript.

Koocheki, A. and S. R. Gliessman. 2005. Pastoral nomadism: a sustainable system for grazing land management in arid areas. *Journal of Sustainable Agriculture* 25: 113–131.

Kort, J. 1988. Benefits of windbreaks to field and forage crops. *Agriculture, Ecosystems and Environment* 22/23: 165–190.

Koziell, I. and J. Saunders. 2001. *Living Off Biodiversity: Exploring Livelihoods and Biodiversity Issues in Natural Resources Management*. IIED: London.

Kumar, B.M. and P.K.R. Nair. 2004. The enigma of tropical homegardens. *Agroforestry Systems* 61(1): 135–152.

Kuminoff, N., V. Alvin, D. Sokolow, and D. A. Sumner. 2001. *Farmland Conversion: Perceptions and Realities*. AIC Issues Brief #16. University of California Agricultural Issues Center.

Kwabiah, A. B. 2004. Biological efficiency and economic benefits of pea–barley and pea–oat intercrops. *Journal of Sustainable Agriculture* 25(1): 117–128.

Lagemann, J. and J. Heuveldop. 1982. Characterization and evaluation of agroforestry systems: the case of Acosta-Puriscal, Costa Rica. *Agroforestry Systems* 1: 101–115.

Laing, D. R., P. G. Jones, and J. H. C. Davis. 1984. Common bean (*Phaseolus vulgaris* L.). In P. R. Goldsworthy and N. M. Fisher (eds.), *The Physiology of Tropical Field Crops*. pp. 305–351. Wiley and Sons: Chichester, U.K.

Lal, R., E. Regnier, D. J. Exkert, W. M. Edwards, and R. Hammond. 1991. Expectations of cover crops for sustainable agriculture. In W. L. Hargrove (ed.), *Cover Crops for Clean Water*. pp. 1–14. Soil and Water Conservation Society: Iowa.

Landis, D. A., F. D. Menalled, A. C. Costamagna, and T. K. Wilkinson. 2005. Manipulating plant resources to enhance beneficial arthropods in agricultural landscapes. *Weed Science* 53(6): 902–908.

Larcher, W. 1980. *Physiological Plant Ecology*. Springer-Verlag: New York.

Laurance, W. F., T. E. Lovejoy, H. L. Vasconcelos, E. M. Bruna, R. K. Didham, P. C. Stouffer, C. Gascon, R. O. Bierregaard, S. G. Laurance, and E. Sampaio. 2002. Ecosystem decay of Amazonian forest fragments: a 22-year investigation. *Conservation Biology* 16: 605–618.

Leather, G. R. and F. A. Einhellig. 1986. Bioassays in the study of allelopathy. In A. R. Putnam and C. S. Tang (eds.), *The Science of Allelopathy*. pp. 133–146. John Wiley and Sons: New York.

LEISA (*Low External Input and Sustainable Agriculture*). 2005. Special Issue: Small Animals in Focus. Vol. 21(3).

Letourneau, D. K. 1986. Associational resistance in squash monoculture and polycultures in tropical Mexico. *Environmental Entomology* 15: 285–292.

Levins, R. 1968. *Evolution in Changing Environments*. Princeton University Press: Princeton.

Liebig, M. A., M. E. Miller, G. E. Varvel, J. W. Doran, and J. D. Hanson. 2004. AEPAT: a computer program to assess agronomic and environmental performance of management practices in long-term agroecosystem experiments. *Agronomy Journal* 96(1): 109–115.

Linn, L. 1984. The effects of *Spergula arvensis* borders on aphids and their natural enemies in Brussels sprouts. M.A. Thesis, University of California, Santa Cruz.

Liverman, D. M., M. E. Hanson, B. J. Brown, and J. R. W. Merideth. 1988. Global sustainability: towards measurement. *Environmental Management* 12: 133–143.

Loftas, T. and J. Ross (eds.), 1995. *Dimensions of Need: An Atlas of Food and Agriculture*. Food and Agriculture Organization of the United Nations: Rome.

Loomis, R. S. and D. J. Connor. 1992. *Crop Ecology: Productivity and Management in Agricultural Systems*. Cambridge University Press: Cambridge.

Los Huertos, M. 1999. *Nitrogen Dynamics in Vegetated Buffer Strips between Estaurine Wetlands and Rowcrops on the Central Coast of California*. Ph.D. dissertation, Environmental Studies, University of California, Santa Cruz.

Lotter, D. 2004. Beyond GMO — the REAL answer to healthy, disease resistant crops. *The New Farm* (USA) (electronic form). www.newfarm.org/international/panam_don/nov04/chapingo.shtml. Accessed 3 February 2006.

Loucks, O. L. 1977. Emergence of research on agro-ecosystems. *Annual Review of Ecology and Systematics* 8: 173–192.

Lowrance, R., B. R. Stinner, and G. J. House. 1984. *Agricultural Ecosystems: Unifying Concepts*. Wiley: New York.

Lu, B. R. and A. A. Snow. 2005. Gene flow from genetically modified rice and its environmental consequences. *Bioscience* 55(8): 669–678.

Lubchenco, J., A. Olson, L. Brubaker, S. Carpenter, M. Holland, S. Hubbell, S. Levin, J. MacMahon, P. Matson, J. Melillo, H. Mooney, C. Peterson, H. Pulliam, L. Real, P. Regal, and P. Risser. 1991. The sustainable biosphere initiative: an ecological research agenda. *Ecology* 72: 371–412.

Ludlow, M. M. 1985. Photosynthesis and dry matter production in C3 legumes and C4 grasses. *Australian Journal of Plant Physiology* 12: 557–572.

Lyon, T. L., H. O. Buckman, and N. C. Brady. 1952. *The Nature and Properties of Soils*. 5th ed. Macmillan: New York.

Ma, M. H., S. Tarmi, and J. Helenius. 2002. Revisiting the species-area relationship in a semi-natural habitat: floral richness in agricultural buffer zones in Finland. *Agriculture, Ecosystems and Environment* 89(1–2): 137–148.

MacArthur, R. H. 1962. Generalized theorems of natural selection. *Proceedings of the National Academy of Sciences* 43: 1893–1897.

MacArthur, R. H. and E. O. Wilson. 1967. *The Theory of Island Biogeography*. Princeton Univ. Press: Princeton.

MacRae, R. J., S. B. Hill, G. R. Mehuys, and J. Henning. 1990. Farm-scale agronomic and economic conversion from conventional to sustainable agriculture. *Advances in Agronomy* 43: 155–198.

Magdoff, F. and H. van Es. 2000. *Building Soils for Better Crops*. 2nd ed. Sustainable Agriculture Network: Washington D.C.

Magdoff, F. and R. R. Weil. 2004. *Soil Organic Matter in Sustainable Agriculture*. Advances in Agroecology Series. CRC Press: Boca Raton, FL.

Maingi, J. M., C. A. Shisanya, N. M. Gitonga and B. Hornetz. 2001. Nitrogen fixation by common bean (*Phaseolus vulgaris* L.) in pure and mixed stands in semi-arid south-east Kenya. *European Journal of Agronomy* 14(1): 1–12.

Major, J., C. R. Clement, and A. DiTommaso. 2005. Influence of market orientation on food plant diversity of farms located on Amazonian dark earth in the region of Manaus, Amazonas, Brazil. *Economic Botany* 59(1): 77–86.

Mamolos, A. P. and K. L. Kalburtji. 2001. Significance of allelopathy in crop rotation. In R. K. Kohli, H. P. Singh, and D. R. Batish (eds.), *Allelopathy in Agroecosystems*. pp. 197–218. Food Products Press (imprint of The Howarth Press): New York.

Marks, R. and R. Knuffke. 1998. *America's Animal Factories: How States Fail to Prevent Pollution from Livestock Waste*. Natural Resources Defense Council.

Marshall, E. J. P., V. K. Brown, N. D. Boatman, P. J. W. Lutman, G. R. Squire, and L. K. Ward. 2003. The role of weeds in supporting biological diversity within crop fields. *Weed Research* 43: 77–89.

Mascaro, J., I. Perfecto, O. Barros, D. H. Boucher, I. G. de la Cerda, J. Ruiz and J. Vandermeer. 2005. Aboveground biomass accumulation in a tropical wet forest in Nicaragua following a catastrophic hurricane disturbance. *Biotropica* 37(4): 600–608.

Masera, O. and S. López-Ridaura. 2000. *Sustentabilidad y Sistemas Campesinos*. Mundi-Prensa México: D.F., México.

Matson, P.A, W. J. Parton, A. G. Power, and M. J. Swift. 1997. Agricultural intensification and ecosystem properties. *Science* 277: 504–509.

McIntyre, S. and R. Hobbs. 1999. A framework for conceptualizing human effects on landscapes and its relevance to management and research models. *Conservation Biology* 13(6): 1282–1292.

McNaughton, K. G. 1988. Effects of windbreaks on turbulent transport and microclimate. *Agriculture, Ecosystems and Environment* 22/23: 17–39.

McNaughton, S. J. 1985. Ecology of a grazing ecosystem: the Serengeti. *Ecological Monographs* 55: 259–294.

McNaughton, S. J. 1990. Mineral nutrition and seasonal movements of African migratory ungulates. *Nature* 345: 613–615.

Menzel, P. and F. D'Aluisio. 2005. *Hungry Planet: What the World Eats*. Material World Books: Napa, CA and Ten Speed Press: Berkeley, CA.

Méndez, V. E. 2000. An assessment of tropical homegardens as examples of local sustainable agroforestry systems. In S. R. Gliessman (ed.), *Agroecosystem Sustainability: Developing Practical Strategies*. Advances in Agroecology series. pp. 51–66. CRC Press: Boca Raton, FL.

Méndez, V. E., R. Lok, and E. Somarriba. 2001. Interdisciplinary analysis of homegardens in Nicaragua: micro-zonation, plant use and socioeconomic importance. *Agroforestry Systems* 51(2): 85–96.

Merrill, R. (ed.) 1976. *Radical Agriculture*. Harper and Row, Publishers: New York.

Millennium Ecosystem Assessment. 2003. *Ecosystems and Human Well-Being: A Framework for Assessment*. Island Press: Washington, D.C.

Minchin, F. R., R. J. Summerfield, A. R. J. Eaglesham, and K. A. Stewart. 1978. Effects of short-term waterlogging on growth and yield of cowpea (*Vigna unguiculata*). *Journal of Agricultural Science* 90: 355–366.

Monteith, J. L. 1973. *Principles of Environmental Physics*. Edward Arnold, Ltd.: London.

Moritz, M. A. and D. C. Odion. 2005. Examining the strength and possible causes of the relationship between fire history and Sudden Oak Death. *Oecologia* 144(1): 106–114.

Mountjoy, D. C. and S. R. Gliessman. 1988. Traditional management of a hillside agroecosystem in Tlaxcala, Mexico: an ecologically-based maintenance system. *American Journal of Alternative Agriculture* 3: 3–10.

Muller, C. H. 1974. Allelopathy in the environmental complex. In B. R. Strain and W. D. Billings (eds.), *Vegetation and Environment*, Part VI. pp. 73–85. W. Junk B. V. Publisher: The Hague.

Muramoto, J., S. R. Gliessman, S. T. Koike, C. Shennan, D. Schmida, R. Stephens, and S. Swezey. 2005. Maintaining agroecosystem health in an organic strawberry/vegetable rotation system. www.agroecology.org/people/joji/research/elkhorn_4.htm.

Murray, M. G. and A. W. Illius. 1996. Multispecies grazing in the Serengeti. In J. Hodgson and A. W. Illius (eds.), *The Ecology and Management of Grazing Systems*. pp. 247–272. CAB International: Wallingford, U.K.

Nair, P. K. R. 1983. Tree integration on farmlands for sustained productivity of small holdings. In W. Lockeretz (ed.), *Environmentally Sound Agriculture*. pp. 333–350. Praeger: New York.

Nair, P. K. R. 1984. *Soil Productivity Aspects of Agroforestry: Science and Practice in Agroforestry*. International Council for Research in Agroforestry (ICRAF): Nairobi, Kenya.

Nair, P. K. R. 2001. Do tropical homegardens elude science, or is it the other way around? *Agroforestry Systems* 53(2): 239–245.

NAPAP (National Acid Precipitation Assessment Program). www.oar.noaa.gov/organization/napap.html. Accessed January 2005.

National Association of Farmers Markets. www.farmersmarkets.net. Accessed 20 January 2006.

National Research Council. 1989. *Alternative Agriculture*. National Academy Press: Washington, D.C.

Natural Resources Conservation Service (NRCS). 2005. Soil organic matter: managing soil organic matter — the key to air and water quality. http://soils.usda.gov/sqi/concepts/som.html. Accessed 8 November 2005.

Naughton-Treves, L. and N. Salafsky. 2004. Wildlife conservation in agroforestry buffer zones: opportunities and conflict. In G. Schroth, G. A. B. da Fonseca, C. A. Harvey, C. Gascon, H. L. Vasconcelos, and A. M. N. Izac (eds.), *Agroforestry and Biodiversity Conservation in Tropical Landscapes*. pp. 319–345. Island Press: Washington, D.C.

Naylor, R. E. L. 1984. Seed ecology. *Advances in Research and Technology of Seeds* 9: 61–93.

Nevo, E. 1998. Genetic diversity in wild cereals: regional and local studies and their bearing on conservation *ex situ* and *in situ*. *Genetic Resources and Crop Evolution* 45(4): 355–370.

Nicholls, C., M. Parrella, and M. A. Altieri. 2000. Establishing a plant corridor to enhance beneficial insect biodiversity in an organic vineyard. *Organic Farming Research Foundation Information Bulletin*. Winter 2000, number 7: 7–9.

Nierenberg, D. 2005. *Happier Meals: Rethinking the Global Meat Industry.* WorldWatch Paper 171. WorldWatch Institute: Washington, D.C.

Nierenberg, D. and B. Halweil. 2004. Cultivating Food Security. Chapter 4 in *State of the World 2004*. Worldwatch Institute/W.W. Norton and Co.: Washington, DC and New York.

Nordstrom, K. F. and S. Hotta. 2004. Wind erosion from cropland in the USA: a review of problems, solutions, and prospects. *Geoderma* 121(3–4): 157–167.

Norton, R. L. 1988. Windbreaks: benefits to orchard and vineyard crops. *Agriculture, Ecosystems and Environment* 22/23: 205–213.

Nye, P. H. and D. J. Greenland. 1960. *The Soil Under Shifting Cultivation.* Commonwealth Bureau of Soils: Harpenden.

Odum, E. P. 1969. The strategy of ecosystem development. *Science* 164: 262–270.

Odum, E. P. 1971. *Fundamentals of Ecology.* W. B. Saunders: Philadelphia.

Odum, E. P. 1993. *Ecology and Our Endangered Life-Support Systems.* Sinauer Associates Incorporated: Sunderland, MA.

Odum, H. T. 1996. *Environmental Accounting: EMERGY and Environmental Decision Making*. Wiley: New York.

Odum, E. P. and G. W. Barrett. 2004. *Fundamentals of Ecology.* 5th ed. Thomson Brooks/Cole: Belmont, CA.

Ong, C. K., R. M. Kho, and K. Onnesstraat. 2004. Ecological interactions in multispecies agroecosystems: concepts and rules. In M. van Noordwijk, G. Cadish, C. K. Ong (eds.), *Below-ground Interactions in Tropical Agroecosystems: Concepts and Models with Multiple Plant Components*. pp. 1–16. CABI Publishing: Cambridge, MA.

Overland, L. 1966. The role of allelopathic substances in the "smother crop" barley. *American Journal of Botany* 53: 423–432.

Palmer, M. A., E. S. Bernhardt, E. A. Chornesky, S. L. Collins, A. P. Dobson, C. S. Duke, B. D. Gold, R. Jacobson, S. Kingsland, R. Kranz, M. J Mappin, M. Luisa Martinez, F. Micheli, J. L. Morse, M. L Pace, M. Pascual, S. Palumbi, O. J. Reichman, A. Townsend, and M. G. Turner. 2004a. *Ecological Science and Sustainability for a Crowded Planet: 21st Century Vision and Action Plan for the Ecological Society of America.* Report from the Ecological Visions Committee to the Governing Board of the Ecological Society of America, April 2004.

Palmer, M. A., E. S. Bernhardt, E. A. Chornesky, S. L. Collins, A. P. Dobson, C. S. Duke, B. D. Gold, R. Jacobson, S. Kingsland, R. Kranz, M. J Mappin, M. Luisa Martinez, F. Micheli, J. L. Morse, M. L Pace, M. Pascual, S. Palumbi, O. J. Reichman, A. Simmons, A. Townsend, and M. Turner. 2004b. Ecology for a crowded planet. *Science* 304: 1251–1252.

Palmer, M. A., E. S. Bernhardt, E. A. Chornesky, S. L. Collins, A. P. Dobson, C. S. Duke, B. D. Gold, R. Jacobson, S. Kingsland, R. Kranz, M. J Mappin, M. Luisa Martinez, F. Micheli, J. L. Morse, M. L Pace, M. Pascual, S. Palumbi, O. J. Reichman, A. Townsend, and M. Turner. 2005. Ecological science and sustainability for the 21st century. *Frontiers in Ecology and the Environment* 3(1): 4–11.

Papadakis, J. 1938. *Compendium of Crop Ecology.* Buenos Aires, Argentina.

Paulus, J. 1994. *Ecological aspects of orchard floor management in apple agroecosystems of the central California coast.* Ph.D. dissertation, University of California, Santa Cruz.

Pianka, E. R. 1970. On *r* - and *K*-selection. *American Naturalist* 104: 592–597.

Pianka, E. R. 1978. *Evolutionary Ecology.* 2nd ed. Harper and Row: New York.

Pianka, E. R. 2000. *Evolutionary Biology.* 6th ed. Benjamin Cummings: San Francisco, CA.

Pickett, S. T. A. and P. White (ed.) 1985. *The Ecology of Natural Disturbances and Patch Dynamics*. Academic Press: Orlando, FL.

Pimentel, D. 2005. Environmental and economic costs of the application of pesticides, primarily in the United States. *Environment, Development and Sustainability* 7: 229–252.

Pimentel, D. and W. Dazhong. 1990. Technological changes in energy use in U.S. agricultural production. In C. R. Carrol, J. H. Vandermeer, and P. M. Rosset (eds.), *Agroecology.* pp. 147–164. McGraw Hill: New York.

Pimentel, D. and M. Pimentel. 2003. Sustainability of meat-based and plant-based diets and the environment. *American Journal of Clinical Nutrition* 78: 3.

Pimentel, D. and M. Pimentel (eds.), 1997. *Food, Energy, and Society.* 2nd ed. University Press of Colorado: Niwot, Colorado.

Pimentel, D., D. Nafus, W. Vergara, D. Papaj, L. Jaconetta, M. Wulfe, L. Olsvig, K. French, M. Loye, and E. Mendoza. 1978. Biological solar energy conversion and U.S. energy policy. *Bioscience* 28: 376–382.

Pimentel, D., W. Dahzhong, and M. Giampietro. 1990. Technological changes in energy use in U.S. agricultural production. In S. R. Gliessman (ed.), *Agroecology: Researching the Ecological Basis for Sustainable Agriculture*. pp. 305–321. Springer-Verlag: New York.

Pimentel, D., L. McLaughlin, A. Zepp, B. Latikan, T. Kraus, P. Kleinman, F. Vancini, W. Roach, E. Graap, W. Keeton, and G. Selig. 1991. Environmental and economic effects of reducing pesticide use. *BioScience* 41: 402–409.

Pimentel, D., U. Stachow, D. A. Takacs, J. W. Brubaker, A. R. Dumas, J. J. Meaney, J. A. S. O'Neil, D. E. Onsi, and D. B. Corzilius. 1992. Conserving biological diversity in agricultural and forestry systems. *BioScience* 42: 354–362.

Pimentel, D., M. Pimentel, and M. Karpenstein-Machan. 1998. Energy Use in Agriculture: An Overview. International Commission of Agricultural Engineering Ejournal (cigr-ejournal.tamu.edu)

Pimentel, D., P. Hepperly, J. Hanson, D, Douds, and R. Seidel. 2005. Environmental, energetic, and economic comparisons of organic and conventional farming systems. *BioScience* 55: 573–582.

Pinchinat, A. M., J. Soria, and R. Bazan. 1976. Multiple cropping in tropical America. In R. I. Papendick, P. A. Sanchez, and G. B. Triplett (eds.), *Multiple Cropping*. pp. 51–62. American Society of Agronomy: Madison, Wisconsin.

Piper, J. K. 1994. Perennial polycultures: grain agriculture fashioned in the prairie's image. *The Land Report* (51): 7–13.

Piper, J. K. 1999. Natural systems agriculture. In W. W. Collins and C. O. Qualset (eds.), *Biodiversity in Agroecosystems*. pp. 167–195. CRC Press: Boca Raton, FL.

Postel, S. and B. Richter. 2003. *Rivers for Life: Managing Water for People and Nature*. Island Press: Washington, D.C.

Postel, S. and A. Vickers. 2004. Boosting water productivity. Chapter 3 in *State of the World 2004*. Worldwatch Institute/W.W. Norton and Co.: Washington, DC and New York.

Practical Farmers of Iowa. 2002. *Alternatives in Agriculture: Thompson On-farm Research*. PFI: Boone, Iowa.

Pretty, J. 2002. *Agri-Culture: Reconnecting People, Land and Nature*. Earthscan: London.

Pretty, J. (ed.) 2005. *The Earthscan Reader in Sustainable Agriculture*. Earthscan: London.

Pretty, J. and R. Hine. 2000. Feeding the world with sustainable agriculture: a summary of new evidence. *Final Report from SAFE-World Research Project*. University of Essex: Colchester, U.K.

Price, P. W. 1976. Colonization of crops by arthropods: non-equilibrium communities in soybean fields. *Environmental Entomology* 5: 605–611.

Putnam, A. R. and L. A. Weston. 1986. Adverse impacts of allelopathy in agricultural systems. In A. R. Putnam and C. S. Tang (eds.), *The Science of Allelopathy*. pp. 43–56. John Wiley and Sons: New York.

Putnam, A. R. and W. B. Duke. 1974. Biological suppression of weeds: evidence for allelopathy in accessions of cucumber. *Science* 185: 370–372.

Qasem, J. R. and C. I. Foy. 2001. Weed allelopathy, its ecological impacts and future prospects: a review. In R. K. Kohli, H. P. Singh, and D. R. Batish, *Allelopathy in Agroecosystems*. pp. 44–119. Food Products Press (imprint of The Howarth Press): New York.

Qualset, C. and H. Shands. 2005. *Safeguarding the Future of U.S. Agriculture: The Need to Conserve Threatened Collections of Crop Diversity Worldwide*. University of California, Division of Agriculture and Natural Resources, Genetic Resources Conservation Program. Davis, CA, USA.

Radosevich, S. R., J. S. Holt, and C. Ghersa. 1997. *Weed Ecology: Implications for Vegetation Management*. 2nd ed. John Wiley and Sons: New York.

Raeburn, P. 1995. *The Last Harvest: The Genetic Gamble That Threatens to Destroy American Agriculture*. Simon and Shuster: New York.

Raison, R. J. 1979. Modification of the soil environment by vegetation fires, with particular reference to nitrogen transformations: a review. *Plant and Soil* 51: 73–108.

Ramrao, W. Y., S. P. Tiwari, and P. Singh. 2005. Crop-livestock integrated farming system for augmenting socio-economic status of smallholder tribal farmers of Chhattisgarh in Central India. *Livestock Research for Rural Development*. Vol. 17. Art. #90. Retrieved November 4, 2005, from http://www.cipav.org.co/lrrd/lrrd17/8/ramr17090.htm.

Rappaport, R. A. 1971. The flow of energy in an agricultural society. *Scientific American* 224: 117–132.

Rasul, G. and G. B. Thapa. 2003. Sustainability analysis of ecological and conventional agricultural systems in Bangladesh. *World Development* 31(10): 1721–1741.

Reganold, J. P., L. F. Elliot, and Y. L. Unger. 1987. Long-term effects of organic and conventional farming on soil erosion. *Nature* 330: 370–372.

Reifsnyder, W. S., and T. O. Darnhofer. 1989. *Meteorology and Agroforestry*. International Centre for Research in Agroforestry: Nairobi, Kenya.

Renting, H., R. K. Mardsen, and J. Banks. 2003. Understanding alternative food networks: exploring the role of short food supply chains in rural development. *Environment and Planning* 35: 393–411.

Rice, E. L. 1984. *Allelopathy*. 2nd ed. Academic Press: Orlando, FL.

Rickerl, D. and C. Francis (eds.), 2004. Agroecosystems analysis. *Agronomy Monograph*. No. 43. America Society of Agronomy: Madison, WI.

Risch, S. 1980. The population dynamics of several herbivorous beetles in a tropical agroecosystem: the effect of intercropping corn, beans, and squash in Costa Rica. *Journal of Applied Ecology* 17: 593–612.

Risser, P. G. 1995. Indicators of grassland sustainability: a first approximation. In M. Munasinghe and W. Shearer (eds.), *Defining and Measuring Sustainability: The Biophysical Foundations*. pp. 310–319. World Bank: Washington, D.C.

Robertson, G.P., J.C. Broome, E. A. Chornesky, J. R. Frankenberger, P. Johnson, M. Lipson, J. A. Miranowski, E. D. Owens, D. Pimentel, and L. A. Thrupp. 2004. Rethinking the vision for environmental research in US agriculture. *Bioscience* 54: 61–65.

Robinson, R. A. 1996. *Return to Resistance: Breeding Crops to Reduce Pesticide Dependence*. AgAccess: Davis, CA.

Ruiz-Rosado, O. 1984. Effects of weed borders on the dynamics of insect communities in a cauliflower agroecosystem. M.A. Thesis, University of California, Santa Cruz.

Russell, W. M. S. 1968. The slash-and-burn technique. *Natural History* 78: 58–65.

Sahagian, D. 2000. Global physical effects of anthropogenic hydrological alterations: sea level and water redistribution. *Global and Planetary Change* 25(1–2): 39–48.

Sahagian, D. L., F. W. Schwartz, and D. K. Jacobs. 1994. Direct anthropogenic contributions to sea level rise in the twentieth century. *Nature* 367: 54–56.

Saito, T. and T. Miyata. 2005. Situation and problems on transgenic technology for insect pest control. *Japanese Journal of Applied Entomology and Zoology* 49(4): 171–185.

Salick, J. and L. C. Merrick. 1990. Use and maintenance of genetic resources: crops and their wild relatives. In C. R. Carroll, J. H. Vandermeer, and P. M. Rosset (eds.), *Agroecology*. pp. 517–548. McGraw-Hill: New York.

Sances, F. 1982. Spider mites can reduce strawberry yields. *California Agriculture* 36: 6–9.

Sanchez, P. A. 1976. *Properties and Management of Soils in the Tropics*. John Wiley and Sons: New York.

Santalla M., J. M. Amurrio, A. P. Rodino, and A. M. de Ron. 2001. Variation in traits affecting nodulation of common bean under intercropping with maize and sole cropping. *Euphytica* 122(2): 243–255.

Schafer, K., M. Reeves, S. Spitzer, and S. Kegley. 2004. *Chemical Trespass: Chemicals in Our Bodies and Corporate Accountability*. Pesticide Action Network North America.

Schierea, J. B., M. N. M. Ibrahim, and H. van Keulenc. 2002. The role of livestock for sustainability in mixed farming: criteria and scenario studies under varying resource allocations. *Agriculture, Ecosystems and Environment* 90(2): 139–153.

Scriber, J. M. 1984. Nitrogen nutrition of plants and insect invasion. In R. D. Hauck (ed.), *Nitrogen in Crop Production*. pp. 441–460. American Association of Agronomy: Madison, WI.

Seiter, S. and W.P. Horwath. 2004. Strategies for managing soil organic matter to supply plant nutrients. In F. Magdoff and R.R. Weil. *Soil Organic Matter in Sustainable Agriculture*. Advances in Agroecology Series. pp. 269–293. CRC Press: Boca Raton, FL.

Serageldin, I. 1995. *Toward Sustainable Management of Water Resources*. World Bank: Washington, D.C.

Settle, W. H., H. Ariawan, E. T. Astuti, W. Cahyana, A. L. Hakim, D. Hindayana, A. S. Lestari, Pajarningsih, and Sartanto. 1996. Managing tropical rice pests through conservation of generalist natural enemies and alternative prey. *Ecology* 77: 1975–1988.

Seubert, C. E., P. A. Sanchez, and C. Valverde. 1977. Effects of land clearing methods on soil properties of an ulltisol and crop performance in the Amazon jungle of Peru. *Tropical Agriculture* (Trinidad) 54: 434–437.

Sevilla Guzman, E. 1999. Introducción. In Instituto de Sociología y Estudios Campesinos ed., *Dicen los Ganaderos: Taller para el cuidado del la dehesa*. Sociedad Cooperativa Andaluza Corpedroches: Seville, Spain.

Shan, L. and X. Deng. 2003. Biologic water-saving approaches and future perspectives. Presented at the Biological Mechanisms of Water-Saving Agriculture session of the International Conference on *Water-Saving Agriculture and Sustainable Use of Water and Land Resources*, held in Yangling, Shaanxi, P.R. China, October 2003. *Journal of Experimental Botany* 54: i3.

Simberloff, D. S. and E. O. Wilson. 1969. Experimental zoogeography of islands: the colonization of empty islands. *Ecology* 50: 278–296.

Smith, M. E. and C. A. Francis. 1986. Breeding for multiple cropping systems. In C. A. Francis (ed.), *Multiple Cropping Systems*. pp. 219–249. Macmillan Publishing Company: New York.

Soil Conservation Service. 1984. Strawberry Hills target area: watershed area study report, Monterey County, CA. USDA, River Basin Planning Staff, Soil Conservation Service.

Soule, J. D. and J. K. Piper. 1992. *Farming in Nature's Image*. Island Press: Washington, D.C.

Stern, W. R. and C. M. Donald. 1961. Light relationships in grass/clover swards. *Australian Journal of Agricultural Research* 13: 599–614.

Stinner, B. R., D. A. Crossley, E. P. Odum, and R. L. Todd. 1984. Nutrient budgets and internal cycling of N, P, K, Ca, and Mg in conventional, no-tillage, and old-field ecosystems on the Georgia Piedmont. *Ecology* 65: 354–369.

Sullivan, P. 2003. *Overview of Cover Crops and Green Manures*. ATTRA — National Sustainable Agriculture Information Service: Fayetteville, AK.

Swezey, S. L. 2004. Trap-cropping the western tarnished plant bug, *Lygus hesperus* Knight, in California organic strawberries. Proceedings. California Organic Production and Farming in the New Millenium: A Research Symposium. UC Berkeley, CA. July 15, 2004.

Swezey, S. L., J. Rider, M. W. Werner, M. Buchanan, J. Allison, and S. R. Gliessman. 1994. Granny Smith conversions to organic show early success. *California Agriculture* 48: 36–44.

Swezey, S. L., P. Vossen, J. Caprile, and W. Bentley. 2002. *Organic Apple Production Manual*. Publication 3403. University of California, Agriculture and Natural Resources: Oakland, CA.

Swift, M. J., A. M. N. Izac and M. van Noordwijk. 2004. Biodiversity and ecosystem services in agricultural landscapes — are we asking the right questions? *Agriculture, Ecosystems and Environment* 104: 113–134.

Tansley, A. G. 1935. The use and abuse of vegetational concepts and terms. *Ecology* 16: 284–307.

Tellarini, V. and F. Caporali. 2000. An input/output methodology to evaluate farms as sustainable agroecosystems: an application of indicators to farms in Central Italy. *Agriculture, Ecosystems, and Environment* 77: 111–123.

Theunissen, J. and H. van Duden. 1980. Effects of intercropping with *Spergula arvensis* on pests of Brussels sprouts. *Entomologia Experimentalis et Applicata* 27: 260–268.

Thomas M. B., S. D. Wratter, and N. W. Sotherton. 1991. Creation of "island" habitats in farmland to manipulate populations of biological arthropods, predator densities, and emigration. *Journal Applied Ecology* 28: 906–917.

Thrupp, L. A. 2004. The importance of biodiversity in agroecosystems. *Journal of Crop Improvement* 12: 315–337.

Tibke, G. 1988. Basic principles of wind erosion control. *Agriculture, Ecosystems and Environment* 22/23: 103–122.

Tischler, W. 1965. *Agrarökologie*. Fischer Verlag: Jena.

Treschow, M. 1970. *Environment and Plant Response*. McGraw-Hill: New York.

Tsegaye, B. 1997. The significance of biodiversity for sustaining agriculture and the role of women in the traditional sector: the Ethiopian experience. *Agriculture, Ecosystems, and Environment* 62: 215–227.

Turner, B. L., R. E. Kasperson, P. A. Matson, J. J. McCarthy, R. W. Corell, L. Christensen, N. Eckley, J. X. Kasperson, A. Luers, M. L. Martello, C. Polsky, A. Pulsipher, and A. Schiller. 2003. A framework for vulnerability analysis in sustainability science. *Proceedings of the National Academy of Sciences of the United States of America* 100(14): 8074–8079.

Turner, M. G., R. H Gardner, and R. V. O'Neill. 2001. *Landscape Ecology in Theory and Practice: Pattern and Process.* Springer Verlag: New York.

Tuxill, J. and G. P. Nabhan. 2001. *People, Plants and Protected Areas: A Guide to In Situ Management.* Earthscan: London.

Ullstrup, A. J. 1972. The impact of the southern corn leaf blight epidemics of 1970-1971. *Annual Review of Phytopathology* 10: 37–50.

U.S. Census Bureau database. www.census.gov. Accessed January 2005.

USDA (United States Department of Agriculture). 2002. Census of Agriculture, Washington, D.C.

USDA (United States Department of Agriculture). 2003. "Organic Agriculture: Gaining Ground." February 2003 issue of *Amber Waves.* USDA Economic Research Service. www.ers.usda.gov/AmberWaves/Feb03/Findings/OrganicAgriculture.htm.

Van der Pijl, L. 1972. *Principles of Dispersal in Higher Plants.* Springer-Verlag: Berlin.

Van Tuijl, W. 1993. *Improving Water Use in Agriculture: Experiences in the Middle East and North Africa.* Report # 201. World Bank.

Vandermeer, J. 1989. *The Ecology of Intercropping.* Cambridge University Press: New York.

Vandermeer, J., I. G. de la Cerda, D. Boucher, I. Perfecto, and J. Ruiz. 2000. Hurricane disturbance and tropical tree species diversity. *Science* 290(5492): 788–791.

van Noordwijk, M., G. Cadish, C.K. Ong (eds.), 2004. *Below-ground Interactions in Tropical Agroecosystems: Concepts and Models with Multiple Plant Components.* CABI Publishing: Cambridge, MA.

Verkerk R. H. J., S. R. Leather, D. J. Wright. 1998. The potential for manipulating crop-pest–natural enemy interactions for improved insect pest management. *Bulletin of Entomological Research* 88(5): 493–501.

Vitousek, P. M., H. A. Mooney, J. Lubchenco, and J. Melillo. 1997. Human domination of Earth's ecosystems. *Science* 277: 494–499.

Waldon, H. 1994. Resilience, equilibrium, and sustainability in three ecosystems. Ph.D. dissertation, University of California, Santa Cruz.

Weiner, J. 1990. Plant population ecology in agriculture. In C. R. Carroll, J. H. Vandermeer, and P. M. Rossett (eds.), *Agroecology.* pp. 235–262. McGraw Hill: New York.

Went, F. 1944. Thermoperiodicity in growth and fruiting of the tomato. *American Journal of Botany* 31: 135–150.

Whittaker, R. H. 1975. *Communities and Ecosystems.* 2nd ed. MacMillan: New York.

Wiersum, K. F. 1981. Introduction to the agroforestry concept. In Wiersum (ed.), *Viewpoints in Agroforestry.* Agricultural University of Wageningen, The Netherlands.

Wilken, G. C. 1969. Drained-field agriculture: an intensive farming system in Tlaxcala, Mexico. *The Geographical Review* 59:215–241.

Wilken, G. C. 1988. *Good Farmers: Traditional Agricultural Resource Management in Mexico and Central America.* University of California Press: Berkeley.

Wilkins, J. L. 2005. Eating right here: moving from consumer to food citizen. *Agriculture and Human Values* 22: 269–273.

Willey, R. W. 1981. A scientific approach to intercropping research. *Proceedings, International Workshop on Intercropping.* pp. 4–14. ICRISAT: India.

Willis, R. J. 1985. The historical bases of the concept of allelopathy. *Journal of the History of Biology* 18(1): 71–102.

Wilsie, C. P. 1962. *Crop Adaptation and Distribution.* Freeman: San Francisco.

Wilson, E. O. 1992. *The Diversity of Life.* W. W. Norton and Co.: New York and London.

Woodgate, G., B. Ambrose-Oji, R. Fernandez Durán, G. Guzmán, and E. Sevilla Guzmán. 2005. Alternative food and agriculture networks: an agroecological perspective on responses to economic globalizations and the 'new' agrarian question. In M.R. Redclift and G. Woodgate. *New Developments in Environmental Sociology.* pp. 586–612. Edward Elgar Publishing: Chetanham, U.K.

World Commission on Environment and Development. 1987. *Our Common Future.* Oxford University Press: Oxford.

World Congress on Conservation Agriculture. 2005. www.ecaf.org/English/Congress.htm. Accessed 7 Jan 2005.

WMO (World Meteorological Organization). 2003. *Scientific Assessment of Ozone Depletion: 2002,* Global Ozone Research and Monitoring Project — Report No. 47. WMO: Geneva.

Young, A. 1989. The environmental basis of agroforestry. In W. S. Reifsnyder and T. O. Darnhofer (eds.), *Meteorology and Agroforestry.* pp. 29–48. International Council for Research in Agroforestry: Nairobi, Kenya.

Zelitch, I. 1971. *Photosynthesis, Photorespiration, and Plant Productivity.* Academic Press: New York.

Zhen, L. and J. K. Routray. 2003. Operational indicators for measuring agricultural sustainability in developing countries. *Environmental Management* 32: 34–46.

Zhengfang, L. 1994. Energetic and ecological analysis of farming systems in Jiangsu Province, China. Presented at the 10th International Conference of the International Federation of Organic Agriculture Movements (IFOAM), Lincoln University, Lincoln, New Zealand, December 9–16.

# Glossary

**abiotic factor** A nonliving component of the environment, such as soil, nutrients, light, fire, or moisture.

**adaptation** (1) Any aspect of an organism or its parts that is of value in allowing the organism to withstand the conditions of the environment. (2) The evolutionary process by which a species' genome and phenotypic characteristics change over time in response to changes in the environment.

**agrobiodiversity** The component of biodiversity related to food and agriculture production. The term encompasses diversity within species, among species, within agroecosystems, within regions, and in the world food system as a whole.

**agroecology** The science of applying ecological concepts and principles to the design and management of sustainable food systems.

**agroecosystem** An agricultural system understood as an ecosystem.

**agroforestry** The practice of including trees in crop- or animal-production agroecosystems.

**agrosilvopastoral system** An agroecosystem combining trees, livestock grazing, and crops.

**allelopathy** An interference interaction in which a plant releases into the environment a compound that inhibits or stimulates the growth or development of other plants.

**alluvium** Soil that has been transported to its present location by water flow.

**alpha diversity** The variety of species in a particular location in one community or agroecosystem.

**alternative food network** A business, program, or institution that promotes a more sustainable relationship between the growing of food and its consumption.

**amensalism** An interorganism interaction in which one organism negatively impacts another organism without receiving any direct benefit itself.

**animal husbandry** The practice of breeding and caring for livestock animals such as goats, cattle, sheep, camels, etc.

**autotroph** An organism that satisfies its need for organic food molecules by using the energy of the sun, or of the oxidation of inorganic substances, to convert inorganic molecules into organic molecules.

**beta diversity** The difference in the assemblage of species from one location or habitat to another nearby location or habitat, or from one part of an agroecosystem to another.

**biogeochemical cycle** The manner in which the atoms of an element critical to life (such as carbon, nitrogen, or phosphorus) move from the bodies of living organisms to the physical environment and back again.

**biological control** The use of natural enemies for the control of pests.

**biomass** The mass of all the organic matter in a given system at a given point in time.

**bioregionalism** Integration of human activities within the ecological limits of a landscape.

**biotic factor** An aspect of the environment related to organisms or their interactions.

**boundary layer** A layer of air saturated with water vapor (from transpiration) that forms next to a leaf surface when there is no air movement.

**buffer zone** A less-intensively managed and less-disturbed area at the margins of an agroecosystem that protects the adjacent natural system from the potential negative impacts of agricultural activities and management.

**bulk density** The mass of soil per unit of volume.

**capillary water** The water that fills the micropores of the soil and is held to soil particles with a force between 0.3 and 31 bars of suction. Much of this water (that portion held to particles with less than 15 bars of suction) is readily available to plant roots.

**carbon dioxide compensation point** The concentration of carbon dioxide in a plant's chloroplasts below which the amount of photosynthate produced fails to compensate for the amount of amount of photosynthate used in respiration.

**carbon fixation** The part of the photosynthetic process in which carbon atoms are extracted from atmospheric carbon dioxide and used to make simple organic compounds that eventually become glucose.

**carbon partitioning** The manner in which a plant allocates to different plant parts the photosynthate it produces.

**carbon sequestration** Capturing or locking up of carbon dioxide from the atmosphere in terrestrial or marine sinks (eg. soil, trees, animals, microorganisms).

**catabatic warming** The process that occurs when a large air mass expands after having been forced over a mountain range and becomes warmer and dryer as a result of the expansion.

**cation exchange capacity** A measurement of a soil's ability to bind positively charged ions (cations), which include many important nutrients.

**climax** In classical ecological theory, the end point of the successional process; today, we refer instead to the stage of maturity reached when successional development shifts to dynamic change around an equilibrium point.

**clone** An individual produced asexually from the tissues, cells, or genome of another individual. A clone is genetically identical to the individual from which it was derived.

**cold air drainage** The flow of cold air down a slope at night, when reradiation of heat (and therefore cooling of air) occurs more rapidly at higher elevations.

**colluvium** Soil that has been transported to its present location by the actions of gravity.

**commensalism** An interorganism interaction in which one organism is aided by the interaction and the other is neither benefited nor harmed.

**community** All the organisms living together in a particular location.

**compensating factor** A factor of the environment that overcomes, eliminates, or modifies the impact of another factor.

**competition** An interaction in which two organisms remove from the environment a limited resource that both require, and both organisms are harmed in the process. Competition can occur between members of the same species and between members of different species.

**consumer** Ecologically, an organism that ingests other organisms (or their parts or products) to obtain its food energy. Agroecologically, a person who obtains food or food products from a farmer for his or her sustenance.

**continental influence** The climatic effect of being distant from the moderating effects of a large body of water.

**Coriolis effect** The deflection of air currents in atmospheric circulation cells due to the rotation of the earth.

**cross-pollination** The fertilization of a flower by pollen from the flower of another individual of the same species.

**CSA** **Community-supported agriculture**. A subscription arrangement in which a farm regularly delivers its products to a central pickup point, or directly to the consumer.

**cultural energy inputs** Forms of energy used in agricultural production that come from sources controlled or provided by humans.

**cytosterility** A genetically controlled condition of male sterility in the breeding line of a self-pollinating crop variety. A breeding line with cytosterility is used as the seed-producing parental line in the production of hybrid seed.

**dark reactions** The processes of photosynthesis that do not require light; specifically, the carbon-fixing and sugar-synthesizing processes of the Calvin cycle.

**decomposer** A fungal or bacterial organism that obtains its nutrients and food energy by breaking down dead organic and fecal matter and absorbing some of its nutrient content.

**density-dependent** Directly linked to population density. This term is usually used to describe growth-limiting feedback mechanisms in a population of organisms.

**density-independent** Not directly linked to population density. This term is usually used to describe growth-limiting feedback mechanisms in a population of organisms.

**detritivore** An organism that feeds on dead organic and fecal matter.

**dew point** The temperature at which relative humidity reaches 100% and water vapor is able to condense into water droplets. The dew point varies depending on the absolute water vapor content of the air.

**directed selection** The process of controlling genetic change in domesticated plants through manipulation of the plants' environment and their breeding process.

**disturbance** An event or short-term process that alters a community or ecosystem by changing the relative population levels of at least some of the component species.

**diversity** (1) The number or variety of species in a location, community, ecosystem, or agroecosystem. (2) The degree of heterogeneity of the biotic components of an ecosystem or agroecosystem (see *ecological diversity*).

**domestication** The process of altering, through directed selection, the genetic makeup of a species so as to increase the species' usefulness to humans.

**dominant species** The species with the greatest impact on both the biotic and abiotic components of its community.

**dry farming** The practice of conserving natural rainfall so as to facilitate farming without irrigation in a normally dry environment or season.

**dynamic equilibrium** A condition characterized by an overall balance in the processes of change in an ecosystem, made possible by the system's resiliency, and resulting in relative stability of structure and function despite constant change and small-scale disturbance.

**easily available water** That portion of water held in the soil that can be readily absorbed by plant roots—usually capillary water between 0.3 and 15 bars of suction.

**ecological diversity** The degree of heterogeneity of an ecosystem's or agroecosystem's species makeup, genetic potential, vertical spatial structure, horizontal spatial structure, trophic structure, ecological functioning, and change over time.

**ecological energy inputs** Forms of energy used in agricultural production that come directly from the sun.

**ecological niche** An organism's place and function in the environment, defined by its utilization of resources.

**ecosystem** A functional system of complementary relations between living organisms and their environment within a certain physical area.

**ecosystem services** The processes by which the environment produces essential resources, such as clean water and air, that we often take for granted.

**ecotone** A zone of gradual transition between two distinct ecosystems, communities, or habitats.

**ecotype** A population of a species that differs genetically from other populations of the same species because local conditions have selected for certain unique physiological or morphological characteristics.

**edge effect** The phenomenon of an edge community, or ecotone, having greater ecological diversity than the neighboring communities.

**emergent property** A characteristic of a system that derives from the interaction of its parts and is not observable or inherent in the parts considered separately.

**environmental complex** The composite of all the individual factors of the environment acting and interacting in concert.

**environmental resistance** The genetically based ability of an organism to withstand stresses, threats, or limiting factors in the environment.

**eolian soil** Soil that has been transported to its current location by the actions of wind (*aeolian* is an acceptable alternative spelling).

**epiphyll** A plant that uses the leaf of another plant for support, but that draws no nutrients from the host plant.

**epiphyte** A plant that uses the trunk or stem of another plant for support, but that draws no nutrients from the host plant.

**eutrophication** Nutrient enrichment of water that leads to algal blooms, disruption of food webs, and in the worst cases, complete eradication of life through deoxygenation.

**evapotranspiration** All forms of evaporation of liquid water from the earth's surface, including the evaporation of bodies of water and soil moisture and the evaporation from leaf surfaces that occurs as part of transpiration.

**externalized cost** In economic terms, a negative consequence that is put outside (made external to) the system being considered. Conventional agriculture has many externalized costs, including degradation of ecological resources, hazards to human health, and disintegration of social systems. Every externalized cost involves privatizing a gain and socializing its associated costs.

**field capacity** The amount of water the soil can hold once gravitational water has drained away; this water is mostly capillary water held to soil particles with at least 0.3 bars of suction.

**food citizen** a consumer who makes food decisions that support a democratic, economically just, and environmentally sustainable food system.

**food democracy** a food system in which consumers are empowered to make informed choices and farmers can make a living using sustainable practices.

**food security** Access to sufficient food of appropriate diversity for a healthy diet.

**foodshed** a geographically limited sphere of land, people, and businesses tied together by food relationships.

**food system** The interconnected meta-system of agroecosystems, their economic, social, cultural, and technological support systems, and systems of food distribution and consumption.

**generalist** A species that tolerates a broad range of environmental conditions; a generalist has a broad ecological niche.

**genetic engineering** Transfer, by biotechnological methods, of genetic material from one organism to another. See *transgenic*.

**genetic erosion** The loss of genetic diversity in domesticated organisms that has resulted from human reliance on a few genetically uniform varieties of food crop plants and animals.

**genetic vulnerability** The susceptibility of genetically uniform crops to damage or destruction caused by outbreaks of a disease or pest or unusually poor weather conditions or climatic change.

**genotype** An organism's genetic information, considered as a whole.

**GEO** A genetically engineered organism.

**glacial soil** Soil that has been transported to its current location by the movement of glaciers.

**gravitational water** That portion of water in the soil not held strongly enough by adhesion to soil particles to resist the downward pull of gravity.

**green manure** Organic matter added to the soil when a cover crop (often leguminous) is tilled in.

**gross primary productivity** The rate of conversion of solar energy into biomass in an ecosystem.

**habitat** The particular environment, characterized by a specific set of environmental conditions, in which a given species occurs.

**hardening** Subjecting a seedling or plant to cooler temperatures in order to increase its resistance to more extreme cold.

**herbivore** An animal that feeds exclusively or mainly on plants. Herbivores convert plant biomass into animal biomass.

**heterosis** The production of an exceptionally vigorous and/or productive hybrid progeny from a directed cross between two pure-breeding plant lines.

**heterotroph** An organism that consumes other organisms to meet its energy needs.

**horizons** Visually distinguishable layers in the soil profile.

**horizontal resistance** The ability of a crop variety to resist generally the threats posed by all possible diseases, pests, and environmental changes, based on the variety's possession of a variety of resistant traits accumulated through population-level breeding and ongoing directed selection at all levels. Contrasted to *vertical resistance*, the ability of a variety to resist a specific pathogen or pest.

**humification** The decomposition or metabolization of organic material in the soil.

**humus** The fraction of organic matter in the soil resulting from decomposition and mineralization of organic material.

**hybrid vigor** The production of an exceptionally vigorous and/or productive hybrid progeny from a directed cross between two pure-breeding plant lines. A synonym for *heterosis*.

**hydration** The addition of water molecules to a mineral's chemical structure.

**hydrological cycle** The process encompassing the evaporation of water from the earth's surface, its condensation in the atmosphere, and its return to the surface through precipitation.

**hydrolysis** Replacement of cations in the structure of a silicate mineral with hydrogen ions, resulting in the decomposition of the mineral.

**hydroxide clay** A mineral component of the soil without definite crystalline structure composed of hydrated iron and aluminum oxides.

**hygroscopic water** The moisture that is held the most tightly to soil particles, usually with more than 31 bars of suction; it can remain in soil after oven drying.

**importance value** A measure of a species' presence in an ecosystem or community—such as number of individuals, biomass, or productivity—that can be used to determine the species' contribution to the diversity of the system.

**insolation** Expsosure to sunlight, or, more technically, the rate of solar radiation received per unit area.

**integrated farm** A farm on which livestock animals and crop plants are combined to take advantage of the synergisms that arise from this combination.

**integrated pest management** The use of a variety of methods and approaches to manage pests and diseases, with a goal of eliminating pesticide use.

**intermediate disturbance hypothesis** The theory that diversity and productivity in natural ecosystems are highest when moderate disturbance occurs periodically but not too frequently.

**interspecific competition** Competition for resources among individuals of different species.

**intraspecific competition** Competition for resources among individuals of the same species.

**inversion** The sandwiching of a layer of warm air between two layers of cold air in a valley.

***K*-strategist** A species that lives in conditions where mortality is density-dependent; a typical *K*-strategist has a relatively long lifespan and invests a relatively large amount of energy in each of the few offspring it produces.

**landrace** A locally adapted strain of a species bred through traditional methods of directed selection.

**landscape ecology** The study of environmental factors and interactions at a scale that encompasses more than one ecosystem at a time.

**leaf-area index** A measure of leaf cover above a certain area of ground, given by the ratio of total leaf surface area to ground surface area.

**light compensation point** The level of light intensity needed for a plant to produce an amount of photosynthate equal to the amount it uses for respiration.

**light reactions** The components of photosynthesis in which light energy is converted into chemical energy in the form of ATP and NADPH.

**limiting nutrient** A nutrient not present in the soil in sufficient quantity to support optimal plant growth.

**living mulch** A cover crop that is interplanted with the primary crop(s) during the growing season.

**lodging** The flattening of a crop plant or crop stand by strong wind, usually involving uprooting or stem breakage.

**macronutrient** A nutrient plants need in large quantities; the macronutrients include carbon, nitrogen, oxygen, phosphorus, sulfur, and water.

**maritime influence** The moderating effect of a nearby large body of water, such as an ocean, on the weather and climate of an area.

**mass selection** The traditional method of directed selection, in which seed is collected from those individuals in a population that show one or more desirable traits and then used for planting the next crop.

**microclimate** The environmental conditions in the immediate vicinity of an organism.

**micronutrient** A nutrient necessary for plant survival but needed in relatively small quantities.

**mineralization** The process by which organic residues in the soil are broken down to release mineral nutrients that can be utilized by plants.

**mountain wind** The downslope movement of air at night that occurs as the upper slopes of a mountain cool more rapidly than those below.

**multifunctionality** The ability of agroecosystems to perform a variety of functions in addition to food and fiber production, including land conservation, maintenance of landscape structure, biodiversity conservation, environmental services, economic viability, and social good.

**mutualism** An interaction in which two organisms impact each other positively; neither is as successful in the absence of the interaction.

**mycorrhizae** Symbiotic fungal connections with plant roots through which a fungal organism provides water and nutrients to a plant and the plant provides sugars to the fungi.

**natural selection** The process by which adaptive traits increase in frequency in a population due to the differential reproductive success of the individuals that possess the traits.

**net primary productivity** The difference between the rate of conversion of solar energy into biomass in an ecosystem and the rate at which energy is used to maintain the producers of the system.

**niche amplitude** The size or range of one or more of the dimensions of the multidimensional space encompassed by a particular species' niche. The niche amplitude of a generalist species is larger than that of a specialist species.

**niche breadth** Essentially a synonym for *niche amplitude*.

**niche diversity** Differences in the resource-use patterns of similar species that allow them to coexist successfully in the same environment.

**niche** See ecological niche.

**open pollination** The natural dispersal of pollen among all the members of a cross-pollinating crop population, resulting in the maximum degree of genetic mixing and diversity.

**organism** An individual of a species.

**overyielding** The production of a yield by an intercrop that is larger than the yield produced by planting the component crops in monoculture on an equivalent area of land.

**oxidation** The loss of electrons from an atom that accompanies the change from a reduced to an oxidized state.

**parasite** An organism that uses another organism for food and thus harms the other organism but usually does not kill the host.

**parasitism** An interaction in which one organism feeds on another organism, harming (but generally not killing) it.

**parasitoid** Insect parasites whose larvae live within and consume their host, usually another insect.

**patchiness** A measurement of the diversity of successional stages present in a specific area.

**patchy landscape** A landscape with a diversity of successional stages or habitat types.

**percolation** Water movement through the soil due to the pull of gravity.

**permanent wilting point** The level of soil moisture below which a plant wilts and is unable to recover.

**phenotype** The physical expression of the genotype; an organism's physical characteristics.

**photoperiod** The total number of hours of daylight.

**photorespiration** The energetically wasteful substitution of oxygen for carbon dioxide in the dark reactions of photosynthesis, which occurs when plant stomata close and carbon dioxide concentration declines.

**photosynthate** The simple-sugar end products of photosynthesis.

**polyploid** Having three or more times the haploid number of chromosomes.

**population** A group of individuals of the same species that live in the same geographic region.

**potential niche** The maximum possible distribution of a species in the environment.

**predation** An interaction in which one organism kills and consumes another.

**predator** An animal that consumes other animals to satisfy its nutritive requirements.

**prescribed burn** A fire set and controlled by humans to achieve some management objective, such as improving pasture in grazing systems.

**prevailing winds** The general wind patterns characteristic of broad latitudinal belts on the earth's surface.

**primary production** The amount of light energy converted into plant biomass in a system.

**primary succession** Ecological succession on a site that was not previously occupied by living organisms.

**producer** An organism that converts solar energy into biomass.

**production** Harvest output or yield.

**productivity index** A measure of the amount of biomass invested in the harvested product in relation to the total amount of standing biomass present in the rest of the system.

**productivity** The ecological processes and structures in an agroecosystem that enable production.

**protocooperation** An interaction in which both organisms are benefited if the interaction occurs, but neither are harmed if it does not occur.

***r*-strategist** A species that exists in relatively harsh environmental conditions and whose mortality is generally determined by density-independent factors; an *r*-strategist allocates more energy to reproduction than to growth.

**rainfed agroecosystem** A farming system in which crop water needs are met by natural precipitation.

**realized niche** The actual distribution of an organism in the environment (compare with *potential niche*).

**regolith** The layer or mantle of unconsolidated material (soil and mineral subsoil) between the soil surface and the solid bedrock of the earth below.

**relative humidity** The ratio of the actual water content of the air to the amount of water the air is capable of holding at a particular temperature.

**relative rate of light transmission** The percentage of the total incident light at the canopy of a system that reaches the ground.

**residual soil** Soil formed at its current location.

**response** A physiological change in a plant that is induced by an outside, usually environmental, condition.

**rhizobia** Nitrogen-fixing soil microorganisms that form mutualistic root interactions with plants (primarily legumes).

**safe site** A specific location that provides the environmental conditions necessary for seed germination and initial growth of the seedling.

**salinization** The process of salt build-up in soils, associated with high evaporation following irrigation and salt deposition at the soil surface.

**saltation** The transport of small soil particles just above the soil surface by wind.

**saturation point** The level of light intensity at which photosynthetic pigments are completely stimulated and unable to make use of additional light.

**secondary succession** Succession on a site that was previously occupied by living organisms but that has undergone severe disturbance.
**seed bank** The total seed presence in the soil.
**self-pollination** The fertilization of the egg of a plant by its own pollen.
**Shannon index** A measure of the species diversity of an ecosystem based on information theory.
**short food supply chain** A route from production of a food product to consumption by the consumer that requires a minimum number of steps.
**silicate clay** A soil component made up primarily of microscopic aluminum silicate plates.
**silvopastoral system** An agroecosystem that combines trees and livestock grazing.
**Simpson index** A measure of the species diversity of an ecosystem based on the concept of dominance.
**slope wind** Air movement caused by the different heating and cooling rates of mountain slopes and valleys.
**soil creep** The movement of large soil particles along the soil surface by wind.
**soil health** The overall picture of the soil's ability to support crop growth without degradation.
**soil profile** The set of observable horizontal layers in a vertical cross section of soil.
**soil solution** The liquid phase of the soil, made up of water and its dissolved solutes.
**solution** The process by which soluble minerals in the regolith are dissolved into water.
**specialist** A species with a narrow range of environmental tolerance.
**species evenness** The degree of heterogeneity in the spatial distribution of species in a community or ecosystem.
**species richness** The number of different species in a community or ecosystem.
**standing crop** The total biomass of plants in an ecosystem at a specific point in time.
**stomata** The openings on a leaf surface through which gases enter and leave the internal leaf environment.
**succession** The process by which one community gives way to another.
**successional mosaic** A patchwork of habitats or areas in different stages of succession.
**symbiosis** A relationship between different organisms that live in direct contact.
**synthetic variety** A crop or horticultural variety produced through the cross-pollination of a limited number of parents that cross well and have certain desirable traits.
**tilth** The combination of the characteristics of soil crumb structure, porosity, and ease of tillage.
**transgenic** A descriptive term applied to organisms developed by transferring genes from one organism to another.
**transpiration** The evaporation of water through the stomata of a plant, which causes a flow of water from the soil through the plant and into the atmosphere.
**transported soil** Soil that has been moved to its current location by environmental forces.
**trophic level** A location in the hierarchy of feeding relationships within an ecosystem.
**trophic structure** The organization of feeding and energy-transfer relationships that determine the path of energy flow through a community or ecosystem.
**valley wind** Air movement that occurs when the heating of a valley causes warm air to rise up adjacent mountain slopes.
**vernalization** The process in which a seed is subjected to a period of cold, causing changes that allow germination to occur.
**water of hydration** Water that is chemically bound to soil particles.
**watershed** A portion of the landscape draining to a single point.

# Index

## B



## C

## D



## E

# O



# P

## Q



## R



## S

## T

## U



## V



## W



## X



## Y



## Z